T0332512

Basic Elements of Differential Geometry and Topology

Mathematics and Its Applications (*Soviet Series*)

Volume 60

Basic Elements of Differential Geometry and Topology

by

S. P. NOVIKOV and A. T. FOMENKO
Moscow State University, U.S.S.R.

KLUWER ACADEMIC PUBLISHERS
DORDRECHT / BOSTON / LONDON

Library of Congress Cataloging in Publication Data

Novikov, Sergeĭ Petrovich.
[Èlementy different͡sial'noĭ geometrii i topologii. English]
Basic elements of differential geometry and topology / by S.P. Novikov and A.T. Fomenko ; [translated from the Russian by M. Tsaplina].
p. cm. -- (Mathematics and its applications (Soviet series) ; 60)
Translation of: Èlementy different͡sial'noĭ geometrii i topologii.
Includes bibliographical references and index.
ISBN 0-7923-1009-8 (alk. paper)
1. Geometry, Differential. 2. Topology. I. Fomenko, A. T. II. Title. III. Series: Mathematics and its applications (Kluwer Academic Publishers). Soviet series ; 60.
QA641.N6413 1990
516.3'6--dc20 90-48303

ISBN 0-7923-1009-8

Published by Kluwer Academic Publishers,
P.O. Box 17, 3300 AA Dordrecht, The Netherlands.

Kluwer Academic Publishers incorporates
the publishing programmes of
D. Reidel, Martinus Nijhoff, Dr W. Junk and MTP Press.

Sold and distributed in the U.S.A. and Canada
by Kluwer Academic Publishers,
101 Philip Drive, Norwell, MA 02061, U.S.A.

In all other countries, sold and distributed
by Kluwer Academic Publishers Group,
P.O. Box 322, 3300 AH Dordrecht, The Netherlands.

Printed on acid-free paper

This is the translation of the original book
ЭЛЕМЕНТЫ ДИФФЕРЕНЦИАЛЬНОЙ ГЕОМЕТРИИ И ТОПОЛОГИИ
Published by Nauka © 1987

Translated from the Russian by M. V. Tsaplina

Printed in the Netherlands

SERIES EDITOR'S PREFACE

'Et moi, ..., si j'avait su comment en revenir, je n'y serais point allé.'

Jules Verne

The series is divergent; therefore we may be able to do something with it.

O. Heaviside

One service mathematics has rendered the human race. It has put common sense back where it belongs, on the topmost shelf next to the dusty canister labelled 'discarded non-sense'.

Eric T. Bell

Mathematics is a tool for thought. A highly necessary tool in a world where both feedback and non-linearities abound. Similarly, all kinds of parts of mathematics serve as tools for other parts and for other sciences.

Applying a simple rewriting rule to the quote on the right above one finds such statements as: 'One service topology has rendered mathematical physics ...'; 'One service logic has rendered computer science ...'; 'One service category theory has rendered mathematics ...'. All arguably true. And all statements obtainable this way form part of the raison d'être of this series.

This series, *Mathematics and Its Applications*, started in 1977. Now that over one hundred volumes have appeared it seems opportune to reexamine its scope. At the time I wrote

> "Growing specialization and diversification have brought a host of monographs and textbooks on increasingly specialized topics. However, the 'tree' of knowledge of mathematics and related fields does not grow only by putting forth new branches. It also happens, quite often in fact, that branches which were thought to be completely disparate are suddenly seen to be related. Further, the kind and level of sophistication of mathematics applied in various sciences has changed drastically in recent years: measure theory is used (non-trivially) in regional and theoretical economics; algebraic geometry interacts with physics; the Minkowsky lemma, coding theory and the structure of water meet one another in packing and covering theory; quantum fields, crystal defects and mathematical programming profit from homotopy theory; Lie algebras are relevant to filtering; and prediction and electrical engineering can use Stein spaces. And in addition to this there are such new emerging subdisciplines as 'experimental mathematics', 'CFD', 'completely integrable systems', 'chaos, synergetics and large-scale order', which are almost impossible to fit into the existing classification schemes. They draw upon widely different sections of mathematics."

By and large, all this still applies today. It is still true that at first sight mathematics seems rather fragmented and that to find, see, and exploit the deeper underlying interrelations more effort is needed and so are books that can help mathematicians and scientists do so. Accordingly MIA will continue to try to make such books available.

If anything, the description I gave in 1977 is now an understatement. To the examples of interaction areas one should add string theory where Riemann surfaces, algebraic geometry, modular functions, knots, quantum field theory, Kac-Moody algebras, monstrous moonshine (and more) all come together. And to the examples of things which can be usefully applied let me add the topic 'finite geometry'; a combination of words which sounds like it might not even exist, let alone be applicable. And yet it is being applied: to statistics via designs, to radar/sonar detection arrays (via finite projective planes), and to bus connections of VLSI chips (via difference sets). There seems to be no part of (so-called pure) mathematics that is not in immediate danger of being applied. And, accordingly, the applied mathematician needs to be aware of much more. Besides analysis and numerics, the traditional workhorses, he may need all kinds of combinatorics, algebra, probability, and so on.

In addition, the applied scientist needs to cope increasingly with the nonlinear world and the extra mathematical sophistication that this requires. For that is where the rewards are. Linear models are honest and a bit sad and depressing: proportional efforts and results. It is in the non-linear world that infinitesimal inputs may result in macroscopic outputs (or vice versa). To appreciate what I am hinting at: if electronics were linear we would have no fun with transistors and computers; we would have no TV; in fact you would not be reading these lines.

There is also no safety in ignoring such outlandish things as nonstandard analysis, superspace and anticommuting integration, p-adic and ultrametric space. All three have applications in both electrical engineering and physics. Once, complex numbers were equally outlandish, but they frequently proved the shortest path between 'real' results. Similarly, the first two topics named have already provided a number of 'wormhole' paths. There is no telling where all this is leading - fortunately.

Thus the original scope of the series, which for various (sound) reasons now comprises five subseries: white (Japan), yellow (China), red (USSR), blue (Eastern Europe), and green (everything else), still applies. It has been enlarged a bit to include books treating of the tools from one subdiscipline which are used in others. Thus the series still aims at books dealing with:

- a central concept which plays an important role in several different mathematical and/or scientific specialization areas;
- new applications of the results and ideas from one area of scientific endeavour into another;
- influences which the results, problems and concepts of one field of enquiry have, and have had, on the development of another.

Like algebraic geometry differential geometry is a notoriously hard subject to teach and to 'self-study'. Partly because it is very large and it is not easy to select a coherent basic chunk, partly because it is well developed and advanced. On the other hand the subject is of vast importance in terms of applications especially to modern physics. Indeed it is impossible to do or understand gauge theories for example without a solid differential geometry and topology background. The authors have some 15 years experience in teaching a coherent course on the topic covering all the essentials. This volume is the distilled essence of their course. It is a pleasure and honour to welcome such a nice text by such eminent authors in this series.

The shortest path between two truths in the real domain passes through the complex domain.

J. Hadamard

La physique ne nous donne pas seulement l'occasion de résoudre des problèmes ... elle nous fait pressentir la solution.

H. Poincaré

Never lend books, for no one ever returns them; the only books I have in my library are books that other folk have lent me.

Anatole France

The function of an expert is not to be more right than other people, but to be wrong for more sophisticated reasons.

David Butler

Kyoto, August 1990

Michiel Hazewinkel

Preface

For a number of years, beginning with the early 70's, the authors have been delivering lectures on the fundamentals of geometry and topology in the Faculty of Mechanics and Mathematics of Moscow State University. This text-book is the result of this work. We shall recall that for a long period of time the basic elements of modern geometry and topology were not included, even by departments and faculties of mathematics, as compulsory subjects in a university-level mathematical education. The standard courses in classical differential geometry have gradually become outdated, and there has been, hitherto, no unanimous standpoint as to which parts of modern geometry should be viewed as abolutely essential to a modern mathematical education. In view of the necessity of using a large number of geometric concepts and methods, a modernized course in geometry was begun in 1971 in the Mechanics division of the Faculty of Mechanics and Mathematics of Moscow State University. In addition to the traditional geometry of curves and surfaces, the course included the fundamental priniciples of tensor analysis, Riemannian geometry and topology. Some time later this course was also introduced in the division of mathematics. On the basis of these lecture courses, the following text-books appeared:

S.P. Novikov: Differential Geometry, Parts I and II, Research Institute of Mechanics of Moscow State University, 1972.

S.P. Novikov and A.T. Fomenko: Differential Geometry, Part III, Research Institute of Mechanics of Moscow State University, 1974.

The present book is the outcome of a revision and updating of the above-mentioned lecture notes. The book is intended for the mathematical, physical and mechanical education of second and third year university students. The minimum abstractedness of the language and style of presentation of the material, consistency with the language of mechanics and physics, and the preference for the material important for natural sciences were the basic principles of the presentation.

At the end of the book are several Appendices which may serve to diversify the material presented in the main text. So, for the purposes of mechanical and physical education the information on elementary groups of transformations and geometric elements of variational calculus can be extended using these Appendices. For mathematicians, the Appendices may serve to enrich their knowledge of Lobachevsky geometry and homology theory. We believe that Appendices 2 and 3 are very instructive for those who wish to become acquainted with the simplest geometric ideas fundamental to physics. Appendix 7 includes selected problems and exercises for the course.

The list of references may assist in further independent study. A more detailed text-book which provides deeper insight into geometry and its applications is Modern Geometry [1].

CONTENTS

PART I

BASIC CONCEPTS OF DIFFERENTIAL GEOMETRY

1.1 General Concepts of Geometry

Let us turn to the subject matter of geometry. Our first acquaintance with geometry goes back to school years. School geometry (the geometry of the ancient Greeks) studies the various metrical properties of the simplest geometric figures, that is, basically finds relationships between lengths and angles in triangles and other polygons. Such relationships provide the basis for the calculations of the surface areas and volumes of solids. We would like to pay attention to the fact that the central concepts underlying school geometry are the following: the length of a straight line (or a curve) segment and the angle between two intersecting straight lines (or curves). The angle was always measured at the point of intersection of these lines.

In the university we are given a course in analytic geometry, whose chief aim is to describe geometric figures by means of algebraic formulae referred to a *Cartesian system of coordinates* of a plane or of a three-dimensional space (e.g. an ellipse in a plane is described by the equation $x^2/a^2 + y^2/b^2 = 1$). The words "analytic geometry" are obviously indicative of the method, whereas the objects under study are the same as in elementary Euclidean geometry. Differential geometry is also the same old subject except that here the subtler techniques of analytic geometry, differential calculus and linear algebra are widely used.

We shall systematize our basic concepts of geometry as follows.

First, our geometry develops in a certain space consisting of points $P, Q, \ldots$.

Second, as in analytic geometry, we introduce a system of *Cartesian* coordinates $x^1, \ldots, x^n$ for the space, that is, associate with each point of the space, a set of numbers $(x^1, \ldots, x^n)$ which are the *coordinates* of the point. The number of coordinates n is called the *dimension of the space.*

It is required that distinct points be assigned distinct n-sets. Two points P and Q with coordinates $(x^1, \ldots, x^n) = P$ and $(y^1, \ldots, y^n) = Q$ coincide if and only if $x^i = y^i$ for all i.

Conversely, each set of numbers $(x^1, \ldots, x^n)$ must be assigned to some point P of the space. Then, such a space is called a *Cartesian* space — points of the space are identified with *all* sets of numbers $(x^1, \ldots, x^n)$, where $-\infty < x^i < +\infty$ and the integer n is the *dimension* of the space.

Third, geometry requires that we can define the concept of the *length* of a segment in space and the concept of the *angle* between two intersecting curves at a *point where they intersect.*

To a certain approximation we may say that we live in an Euclidean three-dimensional space in which we have introduced Cartesian coordinates with special properties:

a) each point P is assigned three coordinates (x^1, x^2, x^3);

b) if the coordinates of a point P are (x^1, x^2, x^3) and the coordinates of a point Q are (y^1, y^2, y^3) then the square of the length of the rectilinear segment joining the points P and Q is equal to $l^2 = (x^1 - y^1)^2 + (x^2 - y^2)^2 + (x^3 - y^3)^2$.

In the case where conditions a) and b) are fulfilled, the space is called *Euclidean*, and the Cartesian coordinates with such properties are called *Euclidean coordinates.*

From the course in linear algebra we know that it is convenient that points of a Euclidean space can be associated with vectors. We have a point O as the origin. The vector going from the point O to a given point P will be called the *radius vector* of the point P. The Cartesian coordinates $(x^1, \dots, x^n)$ of the point P will be called the *coordinates of the vector.* We can make a coordinate-wise summation of two vectors $\xi = (x^1, \dots, x^n)$, $\eta = (y^1, \dots, y^n)$, which join the point O, respectively, with the points P and Q, to obtain the vector $\xi + \eta$ with the coordinates $(x^1 + y^1, \dots, x^n + y^n)$. We can also multiply a vector by a number. Vectors e_1, e_2, e_3 coordinatized, respectively, by $e_1 = (1, 0, 0)$, $e_2 = (0, 1, 0)$ and $e_3 = (0, 0, 1)$ clearly have length 1. It is shown below that they are mutually perpendicular and that any vector ξ with coordinates (x^1, x^2, x^3) can be expressed as $\xi = x^1 e_1 + x^2 e_2 + x^3 e_3$. The space is here three-dimensional and $n = 3$. The definition is, of course, similar for any n. Thus, a Euclidean space may be regarded as a linear space (or a vector space), for which the square of the distance between any two points (end-points of radius vectors) $\xi = (x^1, \dots, x^n)$ and $\eta = (y^1, \dots, y^n)$ is measured as $l^2 = \sum_{i=1}^{n} (x^i - y^i)^2$. In the Euclidean 3-space we have $n = 3$, for the Euclidean plane $n = 2$, and the case $n > 3$ is simply an extension to higher dimensions.

In the Euclidean space there exists an operation called the *scalar product* of vectors, which is of fundamental importance.

DEFINITION 1. If we take a vector $\xi = (x^1, \dots, x^n)$ and a vector $\eta = (y^1, \dots, y^n)$, then their *Euclidean scalar product* is the number $\xi\eta = \eta\xi = \sum_{i=1}^{n} x^i y^i = x^1 y^1 + x^2 y^2 + \dots + x^n y^n$; in the literature, the scalar product $\xi\eta$ is often denoted by (ξ, η) or $\langle \xi, \eta \rangle$.

Making use of this concept we can say that the square of the length of the straight line segment going from a point P with the radius vector $\xi = (x^1, \dots, x^n)$ to a point Q with the radius vector $\eta = (y^1, \dots, y^n)$ is the scalar product of the vector $\xi - \eta$ by itself, and the length of any vector $\gamma = (z^1, \dots, z^n)$ is equal to $(\gamma\gamma)^{1/2}$, where $\gamma\gamma$ is a scalar square of the vector γ.

The length of the vector γ is often denoted by $|\gamma| = (\gamma\gamma)^{1/2}$.

From analytic geometry we know that the angle between two vectors $\xi = (x^1, \dots, x^n)$ and $\eta = (y^1, \dots, y^n)$ is also expressed in terms of the scalar product of these vectors, namely:

$$\cos\phi = \frac{\xi\eta}{(\xi\xi)^{1/2}(\eta\eta)^{1/2}} = \frac{\xi\eta}{|\xi|\,|\eta|}.$$

Thus, the concepts of length and angle are closely related with the concept of the scalar product of vectors. Subsequently, it is just the concept of a scalar product that we take as the basic concept of geometry.

Now let there exist a segment of a curve in a Euclidean space given in the parametric form:

$$x^1 = f^1(t), \dots, x^n = f^n(t),$$

where $f^i(t)$ are differentiable functions of the parameter t, and the parameter t runs a segment from a to b. The *tangent* or *velocity vector* of the curve at the instant of time t is the vector:

$$v(t) = \left(\frac{df^1}{dt}, \frac{df^2}{dt}, \dots, \frac{df^n}{dt}\right).$$

A curve is called *regular* if its velocity vector is nonzero at each point of the curve.

DEFINITION 2. The *length of the curve segment* is the number:

$$l = \int_a^b (v(t)\, v(t))^{1/2}\, dt = \int_a^b |v(t)|\, dt.$$

In other words, the *length of a curved segment* is defined as the integral of the length of its velocity vector.

If a curve $x^i, = f^i(t)$, $i = 1, \dots, n$, intersects another curve $x^i = g^i(t)$, $i = 1, \dots, n$, at $t = t_0$, then we can speak of the *angle* between these two curves at the point where they intersect. Denote the tangent vectors to the curves at $t = t_0$, respectively, by:

$$v = \left(\frac{df^1}{dt}, \dots, \frac{df^n}{dt}\right)_{t=t_0},$$

$$w = \left(\frac{dg^1}{dt}, \dots, \frac{dg^n}{dt}\right)_{t=t_0}.$$

DEFINITION 3. The *angle between two curves* at the point of their intersection $t = t_0$, is the angle between two vectors v, w, that is an angle ϕ such that there holds the equality:

$$\cos \phi = \frac{vw}{|v|\,|w|}.$$

The two latter definitions can be regarded as important facts to be included in the course of mathematical analysis. However, they may also be regarded as *basic definitions*. Then we should check the consistency of these definitions with the visual concepts of curve lengths and of angles between any two curves in Euclidean space. By this verification, we wish to demonstrate once again that, from the modern point of view, the whole geometry is based on the concept of the scalar product of tangent vectors.

Why have we preferred here to give definitions rather than to formulate theorems on the length of a curve segment and the angle between two curve segments?

The point is that mathematical theorems can be proved only if some definitions of the basic quantities are given. What was the definition of length that we dealt with earlier? Let us analyze carefully our old concept of length. That was the length of a straight line in Euclidean space. We could, therefore, define the length of a polygonal arc (i.e. a broken straight line segment) as the sum of the lengths of the straight line segments composing it. Next, following the definition of a circumference, familiar to the reader from secondary school, we may represent a curve segment as the limit of a sequence of broken lines and define its length as the limit of the lengths of the broken line segments aproximating our curve. From school mathematics we know that the circumference of a circle of radius R is $2\pi R$. Next, analytic geometry teaches us that the length of a straight line segment — a vector ξ

with coordinates $(y^1, \dots, y^n)$ — is equal to $\left((y^1)^2 + \dots + (y^n)^2\right)^{1/2}$ (by Pythagoras' theorem).

An approximate calculation shows that *our definition of length yields the same result.*

1. The straight line segment. For simplicity we suppose that a segment comes from the origin. Then it is given by the formula $x^i = y^i t$, where $0 \le t \le 1$. For $t = 0$ the coordinates x^i are all zero, while for $t = 1$ all the coordinates $x^i = y^i$; the corresponding point is the end-point of the vector ξ. The length of a straight line segment is conventionally given by the formula:

$$l = \int_0^1 dt \sqrt{\left(\frac{dx^1}{dt}\right)^2 + \dots + \left(\frac{dx^n}{dt}\right)^2} = \left((y^1)^2 + (y^2)^2 + \dots + (y^n)^2\right)^{1/2}.$$

Using our definition of a straight line segment, we have arrived at exactly the same formula.

2. The circle. The circle (in a plane) is given by the equations:

$$x^1 = R \cos t, \; x^2 = R \sin t,$$

where $0 \le t \le 2\pi$.

The circumference is equal to:

$$l = \int_0^{2x} \left(R^2 \sin^2 t + R^2 \cos^2 t\right)^{1/2} dt = 2\pi R.$$

Thus, for the circle also, our definition of length gives the answer it should.

3. Our definition of length clearly satisfies the requirement that the length of an arc made up of two pieces be the sum of the lengths of those two pieces.

It is already apparent that our definition of length satisfies all the necessary requirements to serve our intuitive ideas concerning this quantity.

However, we still have an obstacle in our way; let us examine carefully our definition of the length of a curve segment.

The length of the curve $\{x^i = f^i(t)\}$ is calculated by the formula:

$$l = \int_a^b |v(t)|\, dt\,, \quad v(t) = \left(\frac{df^1}{dt}, \dots, \frac{df^n}{dt}\right),$$

where $|v| = (vv)^{1/2} = \sqrt{\sum_{i=1}^{n} \left(\frac{df^i}{dt}\right)^2}$.

It should be emphasized that our formula for the length of a curve segment refers to parametrized curves $x^i = f^i(t)$, $i = 1, 2, \dots, n$, $a \le t \le b$. Simply speaking, we "run" along the curve with a parameter t, which varies between a and b, at a speed $v(t) = (df^1/dt, \dots, df^n/dt)$, and this speed v of our motion along the curve enters explicitly into our formula.

What will happen if we trace out the same curve segment with a different speed? We are moving from the point $P = (f^1(a), \dots, f^n(a))$ to the point $Q = (f^1(b), \dots, f^n(b))$. Shall we obtain the same number if we move along the same curve from P to Q, but at a different speed?

The precise formulation of this question is as follows. Suppose that we have a new parameter τ varing from a' to b' ($a' \le \tau \le b'$) and that the parameter t is represented as a function of τ: $t = t(\tau)$, where $t(a') = a$, $t(b') = b$ and $dt/d\tau > 0$.

The inequality $dt/d\tau > 0$ implies that we move along the curve with parameter τ *in the same direction* as along the curve with parameter t. For what follows we should remember that $dt/d\tau = |dt/d\tau| > 0$.

Then our curve can be represented in the following form:

$$x^i = f^i(t) = f^i(t(\tau)) = g^i(\tau), \quad i = 1, \dots, n.$$

With the new parametrization, the speed at which we move along the curve is given by:

$$v'(\tau) = \left(\frac{dg^i}{d\tau}, \dots, \frac{dg^n}{d\tau}\right), \text{ where } a' \le \tau \le b'.$$

(The prime here does not indicate differentiation.)

The length of the curve has the form:

$$l = \int_{a'}^{b'} |v'(\tau)|\, d\tau\,.$$

We should prove the following equality:

$$\int_{a'}^{b'} |v'(\tau)|\, d\tau \overset{?}{=} \int_a^b |v(t)|\, dt.$$

Let us verify this equality. Since:

$$|v'(\tau)| = \sqrt{\sum_{i=1}^{n} \left(\frac{dg^i}{d\tau}\right)^2} = \sqrt{\sum_{i=1}^{n} \left(\frac{df^i}{dt}\frac{dt}{d\tau}\right)^2} =$$

$$= \left|\frac{dt}{d\tau}\right| \cdot \sqrt{\sum_{i=1}^{n} \left(\frac{df^i}{dt}\right)^2} = \frac{dt}{d\tau} \sqrt{\sum_{i=1}^{n} \left(\frac{df^i}{dt}\right)^2},$$

we have:

$$\int_{a'}^{b'} |v'(\tau)|\, d\tau = \int_{a'}^{b'} |v(t(\tau))| \frac{dt}{d\tau} d\tau = \int_a^b |v(t)|\, dt.$$

Thus, we have arrived at the result:

$$\int_{a'}^{b'} |v'(\tau)|\, d\tau = \int_a^b |v(t)|\, dt. = l \qquad \text{(the curve length).}$$

Conclusion. The length of an arc of a curve is independent of the speed at which the arc is traced out.

It is even simpler to show that the length of a curve segment does not depend on the direction in which the segment is traced out, and that the angle between two curves does not depend on the way in which the parameter on the curves is chosen (but it does depend on the direction).

If a curve in a plane is given by the equation $x^1 = f(x^2)$, then we express x^2 in terms of t to obtain:

$$\frac{dx^1}{dt} = \frac{df}{dx^2}, \quad \frac{dx^2}{dt} = 1.$$

Therefore $v = \left(\frac{df}{dx^2}, 1\right)$, and the length of the curve is calculated by the formula:

$$l = \int_a^b |v(t)|\, dt = \int_a^b \sqrt{1 + (\frac{df}{dx^2})^2}\, dx^2$$

(the coordinates are customarily denoted as $x^1 = x, x^2, = y$).

In a space with coordinates $(x^1, x^2, x^3) = (x, y, z)$, for a curve $z = f(x)$, $y = g(x)$ we obtain the following expression for the length:

$$l = \int_a^b \sqrt{1 + (\frac{df}{dx})^2 + (\frac{dg}{dx})^2}\, dx,$$

if $a \le x \le b$.

REMARK. We may dispose of the choice of the parameter t along the curve in a different way, namely, we may choose the parameter t such that $|v| = c$, where c is a constant; then the length is given by:

$$l = \int_a^b |v|\, dt = c(b - a).$$

A parameter t, such that $|v(t)| = 1$, is called a *natural parameter* — it is equal to the segment length which we trace out.

We have discussed the basic, simplest concepts of classical and analytic geometry such as *lengths, angles, Cartesian coordinates, Euclidean space*; it has also been shown that the most convenient basic concept which determines Euclidean geometry is the concept of the *scalar product of vectors* in terms of which we can express the length of a curve segment and the angle between two curves. We have given the *formula for calculating the length of a curve arc* in terms of the integral of the length of the velocity vector and established the correspondence of this formula with the usual intuitive idea of length.

In Euclidean (and general Riemannian) geometry we encounter only a positive scalar product. We shall adduce an example of a non-positive scalar product of vectors of a four-dimensional space $(x^1, x^2, x^3, x^0 = ct)$ which plays a fundamental role in the theory of relativity:

$$\xi\eta = x^1y^1 + x^2y^2 + x^3y^3 - x^0y^0,$$

where $\xi = (x^1, x^2, x^3, x^0)$, $\eta = (y^1, y^2, y^3, y^0)$. Such a space is called a *pseudo-Euclidean* (*Minkowski*) *space*.

We have here three types of vectors:

$\xi\xi > 0$ (space-like)
$\xi\xi < 0$ (time-like)
$\xi\xi = 0$ (light-like).

We can readily see that the lengths of the vectors determined here by the usual formulae may appear to be imaginary or zero, and the angles may appear to be complex. For this reason it is more convenient to use a scalar product. Strictly speaking, this example was given just as an illustration of the general assertion that the most important basic concept of modern geometry is a scalar product.

These concepts are not yet enough for the development of modern geometry. We shall now discuss such useful, and later on, necessary concepts as function in Cartesian space $(x^1, \dots, x^n)$, its *gradient* and *directional derivative*, the concept of a *region* in space and its *boundaries*, and finally go over to general coordinates in a region of space. All these concepts are not, of course, new for us; they are familiar to the reader in this or that measure from the course in mathematical analysis where they are likely to have been introduced formally — axiomatically.

Our goal is to treat these concepts from the point of view of geometry.

The concept of function is clear enough: the majority of physical functions can be measured by numbers in a certain system of units, and the value of this quantity is a function of the position of the object (system) in space. The position of a mechanical system of n material points in a Euclidean three-dimensional space is described by a set of coordinates of points $(x^{11}, x^{12}, x^{13}; x^{21}, x^{22}, x^{23}; \dots ; x^{n1}, x^{n2}, x^{n3})$ and velocities of points $(\dot{x}^{11}, \dot{x}^{12}, \dot{x}^{13}; \dot{x}^{21}, \dot{x}^{22}, \dot{x}^{23}; \dots ; \dot{x}^{n1}, \dot{x}^{n2}, \dot{x}^{n3})$, where $\dot{x}^{ij} = dx^{ij}/dt$ (one of the indices, namely, the first one indicates the number of the point. Let us put $\dot{x}^{ij} = v^{ij}$; then we see that the state (position) of the system is described by the point of a $6n$-dimensional Cartesian space:

$$(x^{ij}, v^{ij}) \quad j = 1, 2, 3, \quad i = 1, 2, \dots, n.$$

Besides, we often consider *constraints* on the position of points — especially *holonomic constraints* of the form:

$$f_q(x^{11}, x^{12}, x^{13}, \dots, x^{n1}, x^{n2}, x^{n3}) = 0, \quad q = 1, 2, \dots, s,$$

involving no velocities v^{ij}.

(For velocities we shall derive the relation $\sum_{ij} \left(\partial f_q/\partial x^{ij}\right) v^{ij} = 0$.)

To describe these constraints imposed on a system, we shall need the concept of functions f_q.

Recall that the *holonomic* constraints in mechanics are the equations $f_q = 0$ relating the *coordinates* of a system.

As an example, we shall say that in classical mechanics an *ideal rigid body* is understood as a system of n points (n is large) with the following constraints: the distance between any pair of points is constant.

Sometimes it is possible to impose the following constraints:

$$f_q(x^{11}, \dots, x^{n3}) < 0, \quad q = 1, 2, \dots, s,$$

or

$$f_q(x^{11}, \dots, x^{n3}) \leq 0, \quad q = 1, 2, \dots, s,$$

which define *regions* (with or without boundary) in a Cartesian $3n$-dimensional space of positions in the system. We encounter many such examples in mechanics. Now we must introduce the general concept of a region.

Suppose we are given an m-dimensional Cartesian space with coordinates $x^1, \dots, x^m$.

DEFINITION 4. A *region without boundary* is a set of points, in an m-dimensional space, such that together with each point of this set it also contains all points of the space sufficiently close to it.

In terms of "$\varepsilon - \delta$" we have: for any point P of a region there exists a small $\delta > 0$, generally depending on this point, such that all points of the space are contained in the region provided that their distance from the point P is smaller than δ.

A *region with boundary* is obtained from a region without boundary by simply adjoining all *boundary points*, that is, points that can be reached from within the region, by sequences of interior points converging to them.

The whole space is, of course, a region. Another simple example of a region without boundary is a region, in a plane, consisting of points of this plane (x^1, x^2), such that $(x^1)^2 + (x^2)^2 < 1$ (an open disc).

The corresponding region with boundary consists of all points of the plane, such that $(x^1)^2 + (x^2)^2 \leq 1$ (this can be verified).

This example is in a sense typical.

The following simple theorem holds.

THEOREM 1. *Let in an n-dimensioanl Cartesian space* $(x^1, \ldots, x^n)$ *there exist a family of continuous functions* $f_1(x), \ldots, f_q(x)$, *where* $x = (x^1, \ldots, x^n)$. *Consider the set of points satisfying the inequalities:*

$$f_1(x) < 0, \ldots, f_q(x) < 0.$$

Then this set of points is a region without boundary.

Proof. Suppose in a space there exists a point $P = (x_0^1, \ldots, x_0^n)$ with coordinates satisfying the inequalities:

$$f_1(x_0^1, \ldots x_0^n) < -\varepsilon_1 < 0, \quad f_2(x_0^1, \ldots x_0^n) < -\varepsilon_2 < 0, \ldots$$

$$\ldots, f_q(x_0^1, \ldots x_0^n) < -\varepsilon_q < 0.$$

Since all the functions f_i are continuous, by the definition of continuous functions, there exists a small number $\delta > 0$ such that the values of all functions $f_1, \ldots, f_q$ are still negative at all points Q whose distance to the point P is less than δ. (Recall the "$\varepsilon - \delta$"-definition of continuous function).

Thus, we choose the number δ in such a way that at points Q with the distance to the point P smaller than δ the inequality $|f_j(P) - f_j(Q)| < \min(\varepsilon_1, \ldots, \varepsilon_q)$, $j = 1, \ldots, q$ holds. At all such points Q we have $f_j(Q) < 0, j = 1, \ldots, q$. Therefore in the space, all sufficiently close points surrounding the point P belong to our set of points, and the result follows.

Note that when moving along curves from within the region we can, by virtue of continuity of the functions f_j, reach only those points at which $f_j \leq 0$ (perhaps not all of them).

EXERCISE. Solutions of the set of inequalities $f_j \leq 0$ may also include, besides a region with boundary, some *extra points*. For the case of one inequaltiy $f \leq 0$ show that these extra points are those of the *local minimum* of the function f.

Usually in applications, if a region has a *smooth* boundary, it is given by one inequality $f(x) < 0$ ($f(x) \leq 0$). If the boundary has angles, edges, faces, etc. then it (the boundary) is given by *several equations* and the region by *several inequalities.*

EXAMPLE. A plane (x^1, x^2) and a region

$$ax^1 + bx^2 < A, \; cx^1 + dx^2 < B$$

$$(ad - bc \neq 0).$$

Here a pair of half lines is the boundary of the region (Figure 1).

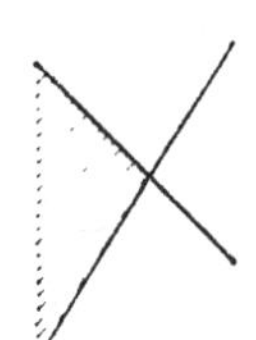

Figure 1.

Very important, and frequently encountered, is the concept of a *bounded region* in space, i.e. a region such that all points sufficiently far from the origin do not belong to it. The simplest example is a ball $\{\sum_{i=1}^{n} (x^i)^2 < R^2\}$ with a sphere as boundary.

Having discussed this general, and intuitively obvious, concept of a region in space, we now proceed to the *gradient of a function* and the *directional derivative.*

Suppose we are given a smooth (i.e. continuously differentiable) function $f(x^1, \dots, x^n)$ in Euclidean space with Cartesian coordinates. The function may be given only in a region of space — we now have the right to use the mathematically rigorous concept of a region.

DEFINITION 5. The *gradient of the function* $f(x^1, \dots, x^n)$ at a point $P = (x_0^1, \dots, \dots, x_0^n)$ in a given Cartesian system of coordinates of Euclidean space (or of its region) is the vector grad f with coordinates:

$$\operatorname{grad} f\big|_P = \left(\frac{\partial f}{\partial x^1}, \dots, \frac{\partial f}{\partial x^n}\right)_{x^i = x_0^i} = \sum_{i=1}^{n} \frac{\partial f}{\partial x^i} e_i,$$

where e_i are the basis unit vectors, and all the derivatives $\partial f/\partial x_i$ are calculated at the point $P = (x_0^1, \dots, x_0^n)$.

If we regard the gradient as a function of the point P, we shall obtain the so-called *vector field*, that is, the situation frequently encountered in mechanics and physics when at each point of a space or of its region a vector is given which is fixed to this point — in our case this is the vector grad f at the point P.

The reader is acquainted with the geometrical meaning of the gradient: if the motion originates at the point P, then the function f increases in the same direction in which grad f goes from the point P. In mechanics, for example, for *conservative systems*, the forces have the form of the gradient of a certain function (called *potential*), only if taken with a minus sign since the force hampers the motion of the system up the levels of the constant height of this potential.

Let us ask a question. Suppose we are given the function $f(x^1, \dots, x^n)$ and a certain curve $x^i = g^i(t)$, $i = 1, \dots, n$ in space (the parameter t varies between the a and b values).

If we consider the function $f(x)$ only at the points of this curve, then it will naturally become the function $\phi(t)$ of the time t. *What is the rate with which this function $f(g^1(t), \dots, g^n(t)) = \phi(t)$ varies with varying parameter t?* This question can be easily answered. By the differentiation rule of composite function we have:

$$\frac{d\phi}{dt} = \frac{\partial f}{\partial x^1}\frac{dg^1}{dt} + \dots + \frac{\partial f}{\partial x^n}\frac{dg^n}{dt}.$$

We can readily see that this is a scalar product of the gradient f by the velocity vector of the curve:

$$\frac{d\phi}{dt} = (\operatorname{grad} f) \cdot v ,$$

where:

$$\operatorname{grad} f = \sum_{i=1}^{n} \frac{\partial f}{\partial x^i} e_i , \quad v = \sum \frac{dg^1}{dt} e_i ,$$

e_i being the basis unit vectors.

On the basis of this result, we deduce the following definition.

DEFINITION 6. A *derivative of the function $f(x^1, \dots, x^n)$ with respect to the direction of the vector* $\xi = (y^i, \dots, y^n)$ calculated at the point $P = (x_0^1, \dots, x_0^n)$ is the scalar product of the gradient of the function f (calculated at the point P) by the vector ξ. Formally, we have:

$$\left.\frac{df}{d\xi}\right|_P = \sum_{i=1}^{n} \frac{\partial f}{\partial x^i} y^i(x^i = x_0^i).$$

The *derivative with respect to the direction of the vector* will be denoted by $df/d\xi$.

EXAMPLE. The directional derivative of the i-th coordinate unit vector is simply a partial derivative with respect to the i-the coordinate:

$$\frac{df}{de_i}\Big|_P = \frac{\partial f}{\partial x^i}\,(x^i = x_0^i).$$

One of the main properties of differentiation with respect to direction is as follows.

PROPOSITION 1. *If we are given a curve $x^i = g^i(t)$, $i = 1, \ldots, n$, such that at points of this curve the scalar product of the gradient f by the velocity vector is equal to zero, the function f is constant along the curve.*

Proof. If $x^i = g^i(t)$, $i = 1, \ldots, n$ is our curve and $\phi(t) = f(g^1(b), \ldots, g^n(t))$, then

$$\frac{df}{dv} = \frac{d\phi}{dt} = (\text{grad}\, f) \cdot v\ ,$$

(at points of the curve), where

$$v = \left(\frac{dg^1}{dt}, \ldots, \frac{dg^n}{dt}\right)$$

is the velocity vector of the curve. Since by the condition $(\text{grad}\, f) \cdot v \equiv 0$, it follows that $d\phi/dt \equiv 0$ and $\phi(t) = \text{const.}$, and the assertion follows.

When we begin to study differential equations and we meet the equations of mechanics and physics, we come across the concept of the "integral of equation" which is a function constant along the trajectories — the solutions of the equation.

1.2 Coordinates in Euclidean Space

In the preceding section we have defined the concept of a region in an n-dimensional Euclidean space — a region without boundary and a region with boundary — and proved the theorem stating that a family of continuous functions $f_1(x^1, \dots, x^n)$, $f_2(x^1, \dots, x^n)$, ... , $f_q(x^1, \dots, x^n)$ specifies, with the help of the rigorous inequalities $f_i(x) < 0$, a region without boundary. By means of the inequalitites $f_i(x) \leq 0$ the family of functions often determines a region with boundary.

As mentioned above, a region with a smooth boundary (without angles) is usually given by one inequality $f(x) < 0$, whereas a region with angles, edges, etc. is given by several functions:

$$f_1(x) < 0, \dots, f_q(x) < 0, \quad q > 1.$$

As an example, we shall consider the regions of the type of polyhedra given by a set of linear inequalities:

$$f_1(x) = a_{11}x^1 + \dots + a_{1n}x^n < A_i,$$

$$\dots\dots\dots\dots\dots\dots$$

$$f_1(x) = a_{q1}x^1 + \dots + a_{qn}x^n < A_q,$$

where A_q are numbers, x^i are coordinates and a_{ij} are numbers. We have also defined an important concept of a bounded region.

We have introduced the concept of the gradient of a function and the derivative of a function with respect to the direction of a vector as a scalar product of the gradient of this function by this vector. The following property of the directional derivative has been proved.

If the gradient of a function has a zero scalar product with the velocity vector of a certain curve (i.e. they are orthogonal), then the function is constant along this curve. More generally, if a function $f(x)$ is considered only at points of the curve $x^i = g^i(t)$, it becomes a function of the parameter t: $\phi(t) = f(g^i(t), \dots, g^n(t))$, $d\phi/dt$ being equal to (grad f) $\cdot$ v and to df/dv (the derivative with respect to the direction of v), where $v = (dg^1/dt, \dots, dg^n/dt)$ is the velocity vector.

Note that in connection with the concept of directional derivative we shall, additionally, consider some properties of vector fields, introduce the concept of a dynamical system and its integrals, and then describe the important special classes of dynamical systems.

In this section we shall be concerned with a very important concept of *general regular coordinates in Cartesian space* or in a region of this space.

We shall recall the well-known types of coordinates with which the reader is already acquainted:

1) Cartesian coordinates $x^1, \dots, x^n$;

2) in the plane — polar coordinates r, ϕ, where $x^1 = r\cos\phi$, $x^2 = r\sin\phi$; with a special choice of the polar axis we always have $r \geq 0$.

Next, the pairs (r, ϕ) and $(r, \phi + 2\pi k)$ for an integer k describe one and the same point $P = (x^1, x^2)$. All the pairs $(0, \phi)$ describe one and the same point (the origin of coordinates). We can see that the angular coordinate ϕ is *multi-valued* ($\phi \approx \phi + 2\pi k$), and at the origin there arises a singularity. This point will be called a *singular point* of the system of coordinates. If we expresse r in terms of x^1, x^2, then $r = \left((x^1)^2 + (x^2)^2\right)^{1/2}$. This function is non-differentiable when $x^1 = 0, x^2 = 0$ (which is obvious).

Considering the derivative of the function r with respect to the direction of the vector $\xi = (y^1, y^2)$ we obtain (at the point x^1, x^2):

$$\frac{dr}{d\xi} = \frac{\partial r}{\partial x^1}y^1 + \frac{\partial r}{\partial x^2}y^2 = \frac{x^1y^1 + x^2y^2}{r},$$

$$r = \left((x^1)^2 + (x^2)^2\right)^{1/2}.$$

The limit of this expression for $x^1 \to 0, x^2 \to 0$ does not exist: it depends on the choice of a line (with a direction preserved) along which we move towards the point $(0, 0)$. If we move along a straight line $x^1 = 0$ (varying $x^2 > 0$), then:

$$\frac{dr}{d\xi} = \frac{x^2y^2}{r} = y^2 \quad (x^1 = 0,\ x^2 > 0).$$

If $x^2 = 0$, then $dr/d\xi = x^1y^1/r = y^1$ $(x^2 = 0, y^1 > 0)$. Thus, moving towards the point $(0, 0)$ along these two curves, we obtain two distinct limits y^2 or y^1, respectively.

We may regard the function $\rho = r^2 = (x^1)^2 + (x^2)^2$, rather than r, to be a coordinate. This function is differentiable when $x^1 = 0, x^2 = 0$, and we have $x^1 = (\rho)^{1/2}\cos\phi$, $x^2 = (\rho)^{1/2}\sin\phi$.

However, $\operatorname{grad}\rho = 0$ at the point $(0, 0)$.

We have the choice of two versions:

a) either a radial coordinate r non-differentiable at the point (0, 0),

b) or a radial coordinate $\rho = r^2$ which is everywhere differentiable, but $\operatorname{grad} \rho|_{(0,\,0)} = 0$.

In doing so we of course assume either r or $\rho = r^2$ to be a function of Cartesian coordinates (x^1, x^2).

Let us now consider *cylindrical* and *spherical* systems of coordinates in a three-dimensional space $(x^1, x^2, x^3) = (x, y, z)$.

The cylindrical coordinates r, ϕ, z, where $z = z, x = r \cos \phi, y = r \sin \phi$, are polar coordinates in the (x, y)-plane.

Here $r = 0$ gives a straight line — the z-axis along which the coordinate system "spoils".

For the spherical system of coordinates (Figure 2) we have r, ϕ, θ, for which:

$$z = r \cos \theta, \quad x = r \sin \theta \cos \phi, \quad y = r \sin \theta \sin \phi,$$

$$0 \leq \theta \leq \pi, \quad \cos \theta = \frac{z}{\left(x^2 + y^2 + z^2\right)^{1/2}}$$

$$0 \leq \phi \leq 2\pi, \quad \operatorname{tg} \phi = y/x$$

$$r = \left(x^2 + y^2 + z^2\right)^{1/2}.$$

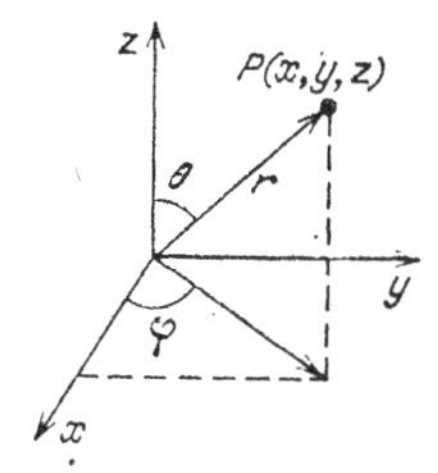

Figure 2.

We can see again that the function $r = r(x, y, z)$ is non-differentiable at the point (0, 0, 0).

Furthermore, the function $\sin \theta = \dfrac{\left(x^2 + y^2\right)^{1/2}}{\left(x^2 + y^2 + z^2\right)^{1/2}}$ is non-differentiable when $x = 0, y = 0$ (and for any z), i.e. along the z-axis.

We see that all these coordinate systems, as distinguished from the Cartesian one, have points which may be thought of as *singular* in the sense that at these points one of the coordinates is either non-differentiable (as a function of a Cartesian coordinate) or differentiable, but its gradient is equal to zero.

So, in a region of space let there be given initial (Cartesian) coordinates $x^1, \dots, x^n$.

Let there also be given some other coordinates $(z^1, \dots, z^n)$ in the same region.

By definition, we can write the equality:

$$x^i = x^i(z^1, \dots, z^n), \; i = 1, \dots, n,$$

or

$$z^j = z^j(x^1, \dots, x^n), \; j = 1, \dots, n.$$

These equalities imply that each point of the region can be assigned either a set of Cartesian coordinates $^j(x^1, \dots, x^n)$ or a set of new coordinates $(z^1, \dots, z^n)$, and therefore the Cartesian coordinates can be expressed in terms of the new ones and vice versa.

Let us analyze the linear coordinates in space:

$$x^i = \sum_{i=1}^{n} a^i_j z^j, \; i = 1, \dots, n \, .$$

For z to be expressible in terms of x, it is necessary and sufficient, as the reader knows from linear algebra, that the matrix $A = (a^i_j)$ has the inverse $A^{-1} = B = (b^i_j)$ or else the determinant of the matrix A is different from zero.

The inverse matrix B is defined as follows: $B = (b^i_j)$, where:

$$\sum_{j=1}^{n} b^i_j a^j_k = \delta^i_k \, ;$$

$$\delta^i_k = \begin{cases} 1, & i = k, \\ 0, & i \neq k; \end{cases} \qquad E = (\delta^i_k) \text{ is a unit matrix.}$$

Thus, the Cartesian coordinates (x) of the point P are expressed through the new set of numbers (z) by means of the matrix $A = (a^i_j)$.

Briefly, we can write:

$$X = Az \; \left(x^j = \sum_{j=1}^{n} a^i_j z^j\right).$$

An important agreement: to avoid repetition of the Σ sign, we shall henceforth imply it in any formula where one and the *same index is twice repeated*: once as a *lower* and then as an *upper* one. For example, $\sum_{j=1}^{n} a^i_j z^j$ is written as $a^i_j z^j$.

If to the point P there corresponded the set of coordinates $(x^1, \dots, x^n)$, in the new coordinates to this point there corresponds the set $(z^1, \dots, z^n)$, such that $x^i = a^i_j z^j$, $i = 1, \dots, n$.

It should be noted that $a^i_j = \partial x^i/\partial z^j$, and these numbers are constant. The determinant of the matrix A is not equal to zero (the matrix is said to be *non-degenerate*).

Let us now examine arbitrary new coordinates:

$$x^i = x^i(z^1, \dots, z^n), \ i = 1, \dots, n,$$

and the point $P = (x_0^1, \dots, x_0^n)$.

We assume that the new coordinates determine each point P, which means that to any set of numbers $(x_0^1, \dots, x_0^n)$ in the space region we are studying there corresponds at least one set $(z_0^1, \dots, z_0^n)$, such that $x_0^i = x^i(z_0^1, \dots, z_0^n)$, $i = 1, \dots, n$.

DEFINITION 1. The point $P = (x_0^1, \dots, x_0^n)$ is called *a non-singular point* of the coordinate system $(z^1, \dots, z^n)$ for $z^1 = z_0^1, \dots, z^n = z_0^n$ (where $x_0^i = x^i(z_0^1, \dots, z_0^n)$) if and only if the matrix:

$$A = \left(\frac{\partial x^i}{\partial z^j}\Big|_{z^q = z_0^q}\right) = (a^i_j)$$

has a non-zero determinant (or if there exists an inverse matrix).

This matrix is called the *Jacobian matrix*, and its determinant is called the *Jacobian* (the Jacobian matrix is denoted as $(\partial x/\partial z)$, and the Jacobian as $|\partial x/\partial z| = J$).

The *inverse transformation theorem* (a particular case of the general *implicit function theorem*) is proved in mathematical analysis.

Given the new coordinates $x^i = x^i(z^1, \dots, z^n)$, $i = 1, \dots, n$, $x_0^i = x^i(z_0^1, \dots, z_0^n)$ and the Jacobian $J = |\partial x/\partial z| \neq 0$ for $z^i = z_0^i$, $i = 1, \dots, n$, we can express the coordinates $z^1, \dots, z^n$ in terms of $x^1, \dots, x^n$ within a sufficiently small neighbourhood of the point $P = (x_0^1, \dots, x_0^n)$, so that $z^i = z^i(x^1, \dots, x^n)$, $z_0^i = z^i(x_0^1, \dots, x_0^n)$, $i = 1, \dots, n$. Given this, the matrix $b^i_j = \partial z^i/\partial x^j$ is inverse to the matrix $a^k_q = \partial x^k/\partial z_q$, so that:

$$\frac{\partial z^i}{\partial x^j}\frac{\partial x^i}{\partial z^k} = \delta^i_k = \begin{cases} 1, & i = k, \\ 0, & i \neq k. \end{cases}$$

This assertion for $n = 1$ looks like this: if $x = x(z)$ and $dx/dz \neq 0$ ($z = z_0$), then we can express $z = z(x)$ in such a manner that $\frac{dz}{dx}\frac{dx}{dz} = 1$ in a sufficiently small neighbourhood of the point x_0, where $x_0 = x(z_0)$.

This assertion is already familiar to the reader in the case of linear changes of coordinates $x = Az$, $x = (x^1, \ldots, x^n)$, $z = (z^1, \ldots, z^n)$, where $x^i = a^i_j z^j$; then the numbers $a^i_k = \partial x^i/\partial z^k$ are *constant*. In this case, $z = Bx$, where B is the inverse matrix to A.

Let us now look at polar, cylindrical and spherical coordinates.

1. *Polar coordinates*: $x^1 = x$, $x^2 = y$, $n = 2$, where:

$$x = r\cos\phi,\ y = r\sin\phi,\ z^1 = r \geq 0,\ z^2 = \phi.$$

Let us construct a Jacobian matrix $A = (\partial x^i/\partial z^j)$:

$$A = \begin{pmatrix} \frac{\partial x}{\partial r} & \frac{\partial x}{\partial \phi} \\ \frac{\partial y}{\partial r} & \frac{\partial y}{\partial \phi} \end{pmatrix} = \begin{pmatrix} \cos\phi & -r\sin\phi \\ \sin\phi & r\cos\phi \end{pmatrix}$$

For the Jacobian we have:

$$J = \left|\frac{\partial x}{\partial z}\right| = r \geq 0.$$

Hence, the Jacobian is equal to zero at the point $r = 0$ only. In the region $r > 0$ (ϕ is arbitrary) the new coordinates do not have singular points.

2. *Cylindrical coordinates*: for cylindrical coordinates $r, \phi\ z$ in a space $x^1 = x$, $x^2 = y$, $x^3 = z$ we shall have $z = z$, $x = r\cos\phi$, $y = r\sin\phi$. By analogy with the

polar coordinates, we find the Jacobian matrix $A = (\partial x/\partial z)$:

$$A = \begin{pmatrix} \cos\phi & -r\sin\phi & 0 \\ \sin\phi & r\cos\phi & 0 \\ 0 & 0 & 1 \end{pmatrix}$$

The Jacobian is equal to zero for $r = 0$ only. In the region $r > 0$ the coordinate system does not have singular points.

3. *Spherical coordinates*: $x^1, = x, x^2 = y, x^3 = z, z^1 = r, z^2, = 0, z^3 = \phi$, where:

$$x = r\sin\theta\cos\phi, \quad 0 \le \theta \le \pi,$$

$$y = r\sin\theta\sin\phi, \quad 0 \le \theta \le 2\pi,$$

$$z = r\cos\theta, \qquad r \ge 0.$$

The Jacobian matrix is:

$$A = \begin{pmatrix} \sin\theta\cos\phi & r\cos\theta\cos\phi & -r\sin\theta\sin\phi \\ \sin\theta\sin\phi & r\cos\theta\sin\phi & r\sin\theta\cos\phi \\ \cos\theta & -r\sin\theta & 0 \end{pmatrix}$$

The Jacobian $J = |\partial x/\partial z| = r^2\sin\theta$ (check!).

We can see that in the region $r > 0, \theta \neq 0, \pi$ this Jacobian is not equal to zero, and therefore the spherical system of coordinates does not have singular points here. Points $r = 0$ (for any θ, ϕ) or $\theta = 0, \pi$ (for any r, ϕ) are *singular*. Here, on the contrary, we cannot express z^i in terms of $x^1, \dots, x^n$, at least so as to obtain differentiable functions $z = z(x)$ (at singular points) since the Jacobian $|\partial x/\partial z|$ is equal to zero at these points.

Let us now set our initial problem: to calculate the length of a curve in general coordinates z, where $x^i = x^i(z^1, \dots, z^n)$, $(x^1, \dots, x^n)$ are Cartesian coordinates. The curve is given parametrically: $z^i = z^i(t)$, $i = 1, \dots, n$ or $x^j(z^1(t), \dots, z^n(t))$:

$$x^j = g^j(t), \quad g^j(t) = x^j(z^1(t), \dots, z^n(t)).$$

According to Section 1.1, we have for the length of a curve segment:

$$l = \int_a^b |v(t)|\, dt,$$

where $v(t) = (dg^1/dt, \dots, dg^n/dt)$ is the velocity vector,

$$|v| = \sqrt{\sum_{i=1}^{n} \left(\frac{dg^i}{dt}\right)^2}.$$

Since

$$g^i(t) = x^i(z^1(t), \dots, z^n(t)),$$

it follows that:

$$\frac{dg^i}{dt} = \frac{\partial x^i}{\partial z^j}\frac{dz^j}{dt}, \quad i = 1, \dots, n.$$

Let $a^i_j(z^1, \dots, z^n) = \partial x^i/\partial z^j$. Then $dx^i/dt = dg^i/dt = a^i{}_j\, dz^j/dt = a^i{}_j\, v^j{}_z$, where, by definition, $v_z = (v^1_z, \dots, v^n_z)$ is the velocity vector in the coordinates $z^1, \dots, z^n$, i.e.

$$v^j_z = \frac{dz^j}{dt}, \quad j = 1, \dots, n.$$

The velocity vector in the initial Cartesian (Euclidean) coordinates $v = (dg^i/dt)$, $i = 1, \dots, n$, will be denoted by v_x. The length of the vector in Cartesian coordinates has the form:

$$|v|^2 = \sum_{j=1}^{n} \left(\frac{dg^i}{dt}\right)^2.$$

Since $\frac{dg^i}{dt} = a^i_j \frac{dz^j}{dt}$, we have:

$$|v|^2 = \sum_{i=1}^{n} \left(\frac{dg^i}{dt}\right)^2 = \sum_{i=1}^{n} \left(a^i_j \frac{dz^j}{dt}\right)^2 = g_{ik}\frac{dz^i}{dt}\frac{dz^k}{dt},$$

where:

$$g_{ik} = g_{ki} = \sum_{j=1}^{n} a_i^j \cdot a_k^j.$$

Conclusion. In coordinates $z^1, \dots, z^n$, where $x^i = x^i(z^1, \dots, z^n)$ the scalar square of the velocity vector $v_z = (\frac{dz^1}{dt}, \dots, \frac{dz^n}{dt})$ of a curve segment is given by the formula:

$$|v_z|^2 = |v_x|^2 = g_{ij} \frac{dz^i}{dt} \frac{dz^j}{dt},$$

$$g_{ij} = \sum_{k=1}^{n} a_i^k a_j^k = \delta_{kq} a_i^k a_j^q,$$

$$a_i^k(z^1, \dots, z^n) = \frac{\partial x^k}{\partial z^i}.$$

How shall we describe the class of coordinates $z^1, \dots, z^n$ such that the length of the vector is expressed in them by the formula:

$$|v_z|^2 = \sum_{j=1}^{n} (y^1)^2, \text{ where } v_z = (y^1, \dots, y^n) = (\frac{dz^1}{dt}, \dots, \frac{dz^n}{dt})?$$

Such coordinates are called *Euclidean.*

If $x^j, = x^j(z^1, \dots, z^n)$, $(a^i{}_j) = (\partial x^i / \partial z^j) = A$, then it is necessary and sufficient that for Euclidean coordinates there holds the property:

$$g_{ij} = \delta_i = \begin{cases} 1, & i = j, \\ 0, & i \neq j, \end{cases} \qquad \text{(by definition).}$$

Since $g_{ij} = \sum_{k=1}^{n} a_i^k a_j^k$, the property $g_{ij} = \delta_{ij}$ is called, as the reader knows, *orthogonality of the matrix* $(a_i^k) = A$. Under such conditions, $x = x(z)$ is a linear change, that is, the functions a_i^k are constant and this change is an *orthogonal transformation.*

1.3 Riemannian Metric in a Region of Euclidean Space

We shall briefly recall the material of the preceding section. Suppose we are given a space (or a region of space) with Carteisian coordinates $(x^1, \dots, x^n)$ and some new coordinates $(z^1, \dots, z^n)$, $x^i = x^i(z^1, \dots, z^n)$ or $x = x(z)$, the new coordinate system possessing no singular points: $J \neq 0$, $J = |\partial x/\partial z|$ is a Jacobian.

If the length of an arc $x^i = x^i(t)$ is measured by the formula:

$$l = \int_a^b \sqrt{\sum_{i=1}^n \left(\frac{dx^i}{dt}\right)^2}\, dt,$$

we are dealing with Euclidean coordinates. In the new coordinates $z^1, \dots, z^n$ we have $z^i = z^i(t)$, $i = 1, \dots, n$, where $x^i(t) = x^i(z^1(t), \dots, z^n(t))$.

For the length of the same arc, but already in the new coordinates, we have:

$$l = \int_a^b \sqrt{g_{ij}\frac{dz^i}{dt}\frac{dz^j}{dt}}\, dt,$$

where $x^i = x^i(z^1(t), \dots, z^n(t))$, and t varies from a to b and:

$$\sqrt{g_{ij}\frac{dz^i}{dt}\frac{dz^j}{dt}} = \sqrt{\sum_{i=1}^n \left(\frac{dx^i}{dt}\right)^2},$$

$$\frac{dx^i}{dt} = \frac{\partial x^i}{\partial z^j}\frac{dz^j}{dt},$$

whence:

$$g_{ij} = \sum_{k=1}^n \frac{\partial x^k}{\partial z^i}\frac{\partial x^k}{\partial z^j} = \delta_{kq}\frac{\partial x^k}{\partial z^i}\frac{\partial x^q}{\partial z^j}.$$

In matrix notation $G = A \circ A^T$, where $G = (g_{ij})$, $A = (\partial x^k/\partial z^i)$ and A^T is a transposed matrix. Note that $dz/dt = (dz^1/dt, \dots, dz^n/dt)$ is the velocity vector of the same arc referred to the same parameter t, but this vector is measured in the new coordinates $z^1, \dots, z^n$.

By definition, we assume that $dz/dt = \xi$ is the same vector as $dx/dt = \eta$ at the point $P = (z^1(t), \ldots, z^n(t)) = (x^1(t), \ldots, x^n(t))$ written in two coordinate systems, (z) and (x).

If we have two curves $z^i = f^i(t)$ and $z^i = g^i(t)$, $i = 1, \ldots, n$, which intersect when $t = t_0$ and have the angle ϕ between their velocity vectors, then:

$$\cos \phi = \frac{\xi_1 \xi_2}{|\xi_1|\,|\xi_2|},$$

where

$$\xi_1 = \left(\frac{df^i}{dt}\right)_{t=t_0}, \quad \xi_2 = \left(\frac{dg_i}{dt}\right)_{t=t_0}.$$

In the coordinates $z^1, \ldots, z^n$ the formula for the scalar product is:

$$\xi_1 \xi_2 = g_{ij}\, \xi_1^i \xi_2^j,$$

$$g_{ij} = \sum_{k=1}^{n} \frac{\partial x^k}{\partial z^i} \frac{\partial x^k}{\partial z^j} = \delta_{kq} \frac{\partial x^h}{\partial z^i} \frac{\partial x^q}{\partial z^j},$$

$$\xi_1^i = \left(\frac{df^i}{dt}\right)_{t=t_0}, \quad \xi_2^i = \left(\frac{dg^i}{dt}\right)_{t=t_0},$$

where $i = 1, \ldots, n$, $t = t_0$ is the point of intersection of the two curves.

On the basis of this result, we shall introduce the concept of Riemannian metric (see Definition 1).

A *Riemannian metric* in a region of space relative to arbitrary regular coordinates $z^1, \ldots, z^n$ is given by a family of functions $g_{ij}(z^1, \ldots, z^n) = g_{ji}(z^1, \ldots, z^n)$, and if we are given a curve $z^i = z^i(t)$, $i = 1, \ldots, n$, the square of the length of its velocity vector $v_z = \left(dz^i/dt\,\big|_{t=t_0}\right)$ at the point $t = t_0$ is the number:

$$l^2 = g_{ij} \frac{dz^i}{dt} \frac{dz^j}{dt}.$$

DEFINITION 1. A family of functions $g_{ij}(z) = g_{ji}(z)$ is said to define a *Riemannian metric* (relative to coordinates $(z^1, \dots, z^n)$) if for any $z^1, \dots, z^n$ the form $g_{ij}(z)\, \eta^i \eta^j$ is positive. If $\det(g_{ij}) \neq 0$ but the form has no fixed sign, then the family g_{ij} is said to determine a *pseudo-Riemannian metric.*

We define the *arc length* relative to the Riemannian metric or pseudo-Riemannian metric g_{ij} to be:

$$l = \int_a^b \sqrt{g_{ij} \frac{dz^i}{dt} \frac{dz^j}{dt}}\, dt .$$

If we have two curves $z^i = f^i(t)$ and $z^i = g^i(t)$ which intersect when $t = t_0$, then the *angle between the curves* is a number ϕ such that:

$$\cos \phi = \frac{\xi \cdot \eta}{|\xi|\, |\eta|},$$

where $\xi \cdot \eta = g_{ij}\, \xi^i \eta^i$, $|\xi| = (\xi \cdot \xi)^{1/2}$, $|\eta| = (\eta \cdot \eta)^{1/2}$, ξ, η are the velocity vectors at the intersection point $t = t_0$.

If, in the same region, we take new coordinates $y^1, \dots, y^n$, such that $z^i = z^i(y^1, \dots, y^n)$, $i = 1, \dots, n$, and $|\partial_z/\partial_y| \neq 0$ (the Jacobian $J \neq 0$), then relative to the new coordinates $y^1, \dots, y^n$, the Riemannian metric is represented by a family of functions $g_{ij}(y^1, \dots, y^n)$, $g'_{ij} = g'_{ji,}$ where:

$$g'_{ij} = \frac{\partial z^k}{\partial y^i} g_{kl} \frac{\partial z^l}{\partial y^j} = g'_{ij}(y^1, \dots, y^n).$$

In matrix language:

$$g' = A \circ g \circ A^{\mathrm{T}},$$

where

$$A = \left(\frac{\partial z^i}{\partial y^k}\right), \quad g' = (g'_{ij}), \quad g = (g_{ij}).$$

The length of arcs and the angles at which they intersect are calculated in the new coordinates $y^1, \dots, y^n$ by the same formula, but now instead of $g_{ij}(z^1, \dots, z^n)$ we should put $g'_{ij}(y^1, \dots, y^n)$. All the above refers to the definition of Riemannian metric.

EXAMPLES. *Euclidean metric.*

1. Let $n = 2$. In the plane, polar and Cartesian coordinates are related as follows: $x^1 = r\cos\phi$, $x^2 = r\sin\phi$; relative to Cartesian coordinates, the metric has the form:

$$g_{ij} = \delta_{ij} = \begin{cases} 1, & i = j, \\ 0, & i \neq j, \end{cases} \qquad (g_{ij}) = \begin{pmatrix} 1 & 0 \\ 0 & 1 \end{pmatrix}$$

while relative to polar coordinates we have:

$$(g'_{ij}) = \begin{pmatrix} 1 & 0 \\ 0 & r^2 \end{pmatrix}.$$

This means that for the curve $r = r(t)$, $\phi = \phi(t)$

$$l = \int_a^b \sqrt{\left(\frac{dr}{dt}\right)^2 + r^2\left(\frac{d\phi}{dt}\right)^2}\, dt.$$

2. $n = 3$. Relative to Cartesian coordinates, we have $g_{ij} = \delta_{ij}$; relative to cynlindrical coordinates $r = y^1$, $\phi = y^2$, $z = y^3$:

$$(g'_{ij}) = \begin{pmatrix} 1 & 0 & 0 \\ 0 & r^2 & 0 \\ 0 & 0 & 1 \end{pmatrix};$$

for the length of an arc we have:

$$l = \int_a^b \sqrt{\left(\frac{dr}{dt}\right)^2 + r^2\left(\frac{d\phi}{dt}\right)^2 + \left(\frac{dz}{dt}\right)^2}\, dt.$$

Relative to spherical coordinates $y^1 = r$, $y^2 = \theta$, $y^3 = \phi$, we have:

$$(g'_{ij}) = \begin{pmatrix} 1 & 0 & 0 \\ 0 & r^2 & 0 \\ 0 & 0 & r^2 \sin^2\theta \end{pmatrix}$$

and for the length of an arc:

$$l = \int_a^b \sqrt{\left(\frac{dr}{dt}\right)^2 + r^2\left[\left(\frac{d\theta}{dt}\right)^2 + \sin^2\theta\left(\frac{d\phi}{dt}\right)^2\right]}\, dt .$$

The Riemannian metric is often given by the formulae for the differential of length dl or $(dl)^2$ as follows:

in Cartesian coordinates:

$$(dl)^2 = \sum_{i=1}^{n} (dx^i)^2,$$

in polar coordinates when $n = 2$:

$$(dl)^2 = (dr)^2 + r^2(d\phi)^2,$$

in cylindrical coordinates:

$$(dl)^2 = (dr)^2 + r^2(d\phi)^2 + (dz)^2,$$

and in spherical coordinates:

$$(dl)^2 = (dr)^2 + \left((d\theta)^2 + \sin^2\theta(d\phi)^2\right) r^2.$$

We have defined the Riemannian metric g_{ij} in a region of space with coordinates $(z^1, \ldots, z^n)$, $g_{ij} = g_{ij}(z^1, \ldots, z^n)$. A *metric* is said to be *Euclidean* if there exist new coordinates $x^1, \ldots, x^n$, $x^i = x^i(z^1, \ldots, z^n)$, $|\partial x^i/\partial z| \neq 0$, such that:

$$g_{ij} = \sum_k \frac{\partial x^k}{\partial z^i} \frac{\partial x^k}{\partial z^j} = \delta_{kq} \frac{\partial x^k}{\partial z^i} \frac{\partial x^q}{\partial z^j} .$$

Relative to coordinates $x^1, \ldots, x^n$ we have:

$$g_{ij} = \delta_{ij} = \begin{cases} 1, & i=j, \\ 0, & i \neq j, \end{cases}$$

and the coordinates $x^1, \dots, x^n$ are termed *Euclidean coordinates.*

We always require that the determinant $|g_{ij}|$ be non-zero or, in other words, that the metric g_{ij} be non-degenerate.

If the matrix $(g_{ij}(z^1, \dots, z^n))$ determines a *positive quadratic form*— that is, the lengths of all non-zero vectors (and, therefore, of all curve segments) are positive, then we say that g_{ij} represents the *Riemannian metric.* If the determinant $|g_{ij}|$ is non-zero, but the form $g_{ij}\,\xi^i\,\xi^j$ has no fixed sign, then we say that there exists a *pseudo-Riemannian metric.*

Of particular importance is the case where $n = 4$ and the form $g_{ij}\,\xi^i\,\xi^j$ at *each* point $z_0^1, \dots, z_0^n$ can be brought to the form $(\xi^1)^2 + (\xi^2)^2 + (\xi^3)^2 - (\xi^4)^2$. These are the metrics on which the general theory of relativity is constructed.

Now we shall consider Riemannian metrics, i.e. $g_{ij}\,\xi^i\,\xi^j > 0$ (at all points).

What metrics do we know for the case $n = 2$? Above, we have already acquainted ourselves with metrics on the Euclidean plane and in the standard two-dimensional sphere given in spherical coordinates by the equation $r = r_0$. Restricting the space metric to a sphere, i.e. putting $r = r_0$, we come to the following metric:

$$dl^2 = \left((d\theta)^2 + \sin^2\theta\,(d\phi)^2\right) r_0^2.$$

Replacing the usual trigonometric function sin by the hyperbolic sh, we shall write another metric $dl^2 = (d\chi)^2 + \mathrm{sh}^2\chi(d\phi)^2$. This metric turns out to be connected with *Lobachevsky geometry* which is treated in Section 1.4.

So, we compare three metrics in two-dimensional space:

1) Euclidean metric: in Cartesian coordinates, x, y, this is:

$$dl^2 = dx^2 + dy^2,$$

or in polar coordinates r, ϕ:

$$dl^2 = dr^2 + r^2 d\phi^2;$$

2) Metric of the sphere: in spherical coordinates θ, ϕ:

$$dl^2 = d\theta^2 + \sin^2\theta\, d\phi^2;$$

3) Lobachevskian metric:

$$dl^2 = d\chi^2 + \mathrm{sh}^2\chi d\phi^2 .$$

This metric may equivalently be given in Cartesian coordinates x, y in a half plane $y > 0$ by the formula:

$$dl^2 = \frac{dx^2 + dy^2}{y^2} .$$

Why do we distinguish these three metrics?

Why is the Lobachevskian metric added to the Euclidean one and to the metric of a sphere? What common property have all these three metrics?

These three metrics appear to be the most *symmetric*. What is the exact meaning of this?

Let us consider transformation of the change of coordinates $z^i = z^i(y^1, \dots, y^n)$. We have:

$$g'_{ij} = \frac{\partial z^k}{\partial y^i} g_{kl} \frac{\partial z^l}{\partial y^j} .$$

If $g'_{ij} \equiv g_{ij}$, the transformation is called the *motion* (or *symmetry*) of the metric — it exactly preserves the form of the scalar product.

EXAMPLE. Suppose $n = 2$, (x, y) are Cartesian coordinates in the plane, and $g_{ij} = \delta_{ij}$ is a Euclidean metric.

a) Consider the *translation*:

$$\begin{aligned} x &= \tilde{x} + x_0, \\ y &= \tilde{y} + y_0 \end{aligned} \qquad \left(\begin{aligned} \tilde{x} &\to \tilde{x} + x_0 = x, \\ \tilde{y} &\to \tilde{y} + y_0 = y \end{aligned} \right);$$

b) consider the *rotation*:

$$x = \tilde{x} \cos\phi + \tilde{y} \sin\phi,$$

$$y = -\tilde{x} \sin\phi + \tilde{y} \cos\phi, \quad \phi = \text{const.}$$

All these transformations are motions of the Euclidean metric. They are described by three numbers (x_0, y_0, ϕ).

Another example is a *sphere*.

It is positioned in a three-dimensional Euclidean space with spherical coordinates r, θ, ϕ and is given by the equation $r = r_0$.

Obviously, any rotation of a Euclidean space about the origin (about any axis) represents the motion of the sphere, that is, on the sphere $r = r_0$, it does not change the lengths of the curves and the angles between them. How many rotations are there?

Rotation is given (in Cartesian coordinates) by an orthogonal matrix:

$$A = \begin{pmatrix} a_1^1 & a_2^1 & a_3^1 \\ a_1^2 & a_2^2 & a_3^2 \\ a_1^3 & a_2^3 & a_3^3 \end{pmatrix}$$

which determines the coordinate transformation:

$$x^i = a_j^i y^j .$$

Given this, we have:

$$\sum_{i=1}^{3} (a_j^i)^2 = 1, \quad 1 \leq j \leq 3,$$

$$a_i^j a_k^i = \delta_k^j = \begin{cases} 1, & j = k, \\ 0, & j \neq k . \end{cases}$$

If the vectors e_1, e_2, e_3 were ortho-normalized: $e_i e_j = \delta_{ij}$, then the vectors $Ae_i = a_i^j e_j$ are also ortho-normalized.

The matrix A is described by nine numbers a_j^i which satisfy the six equations $a_i^j a_k^i = \delta_k^j$.

So, all the rotations are described by three numbers (e.g. by the *Euler angles* ϕ, ψ, θ which the reader will come to know in mechanics).

Thus, the metric of a sphere also has a *three-dimensionaal* set of motions.

The third example is a *Lobachevskian metric*.

Consider the upper half plane (x, y), $y > 0$ and an element of the arc length $(dl)^2 = \dfrac{(dx)^2 + (dy)^2}{y^2}$. Assume that:

$$z = x + iy \quad (i^2 = -1), \quad z' = x' + iy'.$$

If $z = \dfrac{az' + b}{cz' + d}$ and the numbers a, b, c, d are real, then on condition that the determinant of the matrix $\begin{pmatrix} a & b \\ c & d \end{pmatrix}$ is equal to unity, we obtain the transformation $x = x(x', y'), \ y = y(x', y')$,

$$x + iy = \frac{a(x' + iy') + b}{c(x' + iy') + d}.$$

Verify by a direct calculation that this is the motion of the Lobachevskian plane.

How many types of such motions exist? The motions are given by the matrix $\begin{pmatrix} a & b \\ c & d \end{pmatrix}$ under one condition that $ad - bc = 1$. We see again that this is a *three-dimensional* set.

(The matrices $\begin{pmatrix} \lambda a & \lambda b \\ \lambda c & \lambda d \end{pmatrix}$ and $\begin{pmatrix} a & b \\ c & d \end{pmatrix}$ yield the same transformation, and therefore, we can assume that $ad - bc = 1$.)

1.4 Pseudo-Euclidean Space and Lobachevsky Geometry

As emphasized above, from the contemporary point of view the construction of one or another geometry should be started with the introduction of scalar product which is, thus, the basic concept.

Recall the basic properties of the Euclidean (i.e. positive definite) scalar product. If $(\xi, \eta) = \sum_{i=1}^{n} \xi^i \eta^i$ (we are considering a Euclidean space $\mathbb{R}^n$ of dimension n) then:

1) $(\xi, \eta) = (\eta, \xi)$;
2) $(\lambda\xi, \eta) = (\xi, \lambda\eta) = \lambda(\xi, \eta)$;
3) $(\xi, \eta + \phi) = (\xi, \eta) + (\xi\ \phi)$;
4) $(\xi,\xi) \geq 0 \quad (\xi, \xi) = 0$ if and only if $\xi = 0$;
5) $\left((\xi + \eta, \xi + \eta)\right)^{1/2} \leq \left((\xi, \xi)\right)^{1/2} + \left((\eta, \eta)\right)^{1/2}$ (inequality of triangles).

Properties 4 and 5 characterize positive definiteness of the scalar product; they do not hold for pseudo-Euclidean scalar products.

DEFINITION 1. A linear real space of dimension n is called a *pseudo-Euclidean space of index s* if in this space the following bilinear form is given:

$$(\xi, \eta)_s = -\xi^1\eta^1 - \ldots - \xi^s\eta^s + \xi^{s+1}\eta^{s+1} + \ldots + \xi^n\eta^n.$$

If $s = 0$, we obtain a Euclidean space. A pseudo-Euclidean space of index s will be denoted as $\mathbb{R}^n_s$. The space $\mathbb{R}^4_1$ is the *space of the special theory of relativity* and is called the *Minkowski space.*

REMARK. Investigation of the space $\mathbb{R}^n_{n-s}$ is reduced to investigation of the space $\mathbb{R}^n_s$ since all the lengths in $\mathbb{R}^n_{n-s}$ can be multiplied by i; then, obviously, the form $(\xi, \eta)_{n-s}$ will become the form $(\xi, \eta)_s$. We may always assume, therefore, that $s \leq [n/2]$ (integral part).

As in Euclidean space, the length of the vector ξ in the space $\mathbb{R}^n_s$ is determined by the formula:

$$|\xi|_s = \left((\xi, \xi)_s\right)^{1/2},$$

but the lengths of the vectors in $\mathbb{R}^n_s$, as distinguished from $\mathbb{R}^n$, can be *zero* and *purely imaginary*.

In the space $\mathbb{R}^n$, the set of all points ξ, such that $|\xi| = \rho$, forms an $(n-1)$-dimensional sphere S^{n-1} (hypersphere). In the pseudo-Euclidean space $\mathbb{R}^n_s$, we can also consider a set of ponts ξ whose distance from the origin is ρ (but now ρ can be not only a real number, but also purely imaginary or zero). This set of points will be referred to as a *psuedo-sphere of index s* and will be denoted by S^{n-1}_s. Clearly, $S^{n-1}_0 = S^{n-1}$. Indefiniteness of the form $(\xi, \eta)_s$ gives rise to a more diversified geometry on pseudo-spheres S^{n-1}_s as compared with the case $s = 0$. In the sequel we distinguish psuedo-spheres of *real radius, imaginary radius* and *zero radius*. A pseudo-sphere of zero radius is described by the following second-order equation:

$$-(\xi^1)^2 - \dots - (\xi^s)^2 + (\xi^{s+1})^2 + \dots + (\xi^n)^2 = 0,$$

that is, such a pseudo-sphere is a second-order cone in $\mathbb{R}^n_s$ with the vertex in the origin. Clearly, all the vectors emerging from the origin and lying on this cone have zero length, while the vectors going outside this cone have non-zero length. The pseudo-sphere S^{n-1}_s of zero radius is called an *isotropic cone*.

REMARK. In the Minkowski space $\mathbb{R}^n_s$, the isotropic cone is entirely filled with *light vectors* ξ (i.e. $(\xi, \xi)_s = 0$) and is called the *light cone* since a light beam started from the origin will propogate along the generator of this cone.

Let us cosider examples. Let $n = 1$; then $s = 0$ (since we agreed to assume that $s \leq [n/2]$) and the space $\mathbb{R}^1_0$ is a usual real straight line.

Now let $n = 2$; $s = 1$. The isotropic cone consists of two straight lines: $x^1 = \pm x^2$ (we are considering a two-dimensional plane $\mathbb{R}^2$ relative to Cartesian coordinates x^1 and x^2; just in this usual plane we are modelling pseudo-Euclidean geometry of index one). This cone splits $\mathbb{R}^2$ into two regions: in one of them, $(\xi, \xi)_1 > 0$ (namely, this is the region defined by the inequality $|x^2| > |x^1|$); in the other $(\xi, \xi)_1 < 0$ (namely, the region defined by the inequality $|x^2| < |x^1|$) (Figure 3). The pseudo-spheres of real radius are the hyperbolas:

$$-(x^1)^2 + (x^2)^2 = \alpha^2 \quad (\rho = \alpha),$$

and the pseudo-spheres of imaginary radius are the hyperbolas (Figure 4):

$$-(x^1)^2 + (x^2)^2 = -\alpha^2 \quad (\rho = i\alpha).$$

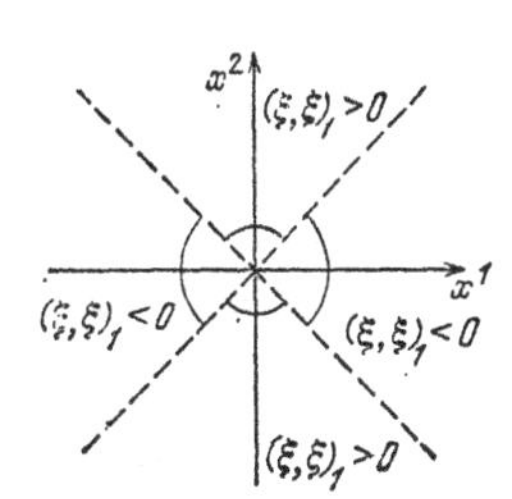

Figure 3.

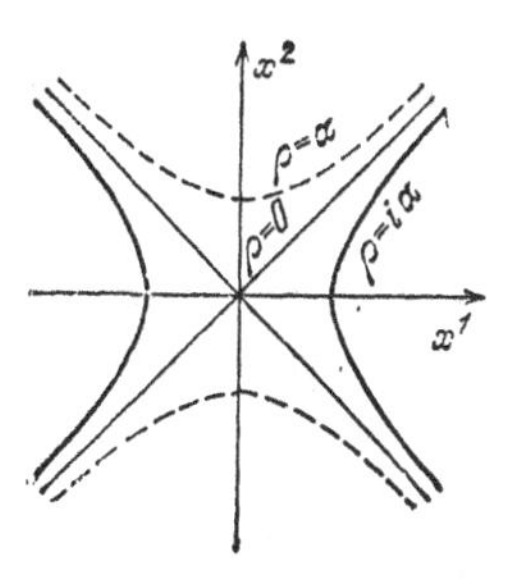

Figure 4.

Now let $n = 3$, $s = 1$ (recall that the study of $\mathbb{R}_2^3$ is reduced to the study of $\mathbb{R}_1^3$). The isotropic cone (a pseudo-sphere of zero radius) is the usual second-order cone, with axis x^1, given by the equation:

$$-(x^1)^2 + (x^2)^2 + (x^3)^2 = 0.$$

It also splits the whole space into two regions (in conventional terms, "into internal and external regions") (Figure 5).

The pseudo-spheres of real radius are one-sheeted hyperboloids,

$$-(x^1)^2 + (x^2)^2 + (x^3)^2 = \alpha^2 \quad (\rho = \alpha),$$

while those of imaginary radius are two-sheeted hyperboloids (Figure 6):

$$-(x^1)^2 + (x^2)^2 + (x^3)^2 = -\alpha^2 \quad (\rho = i\alpha).$$

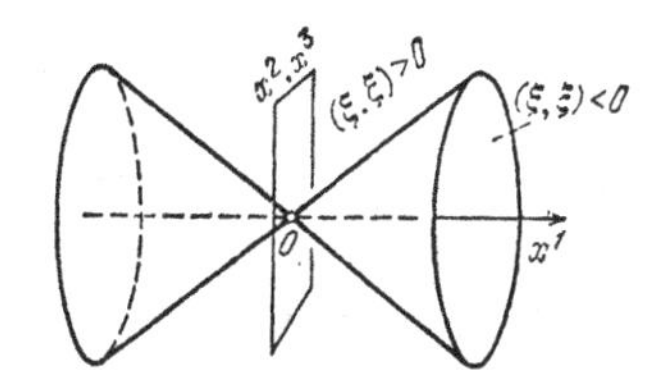

Figure 5.

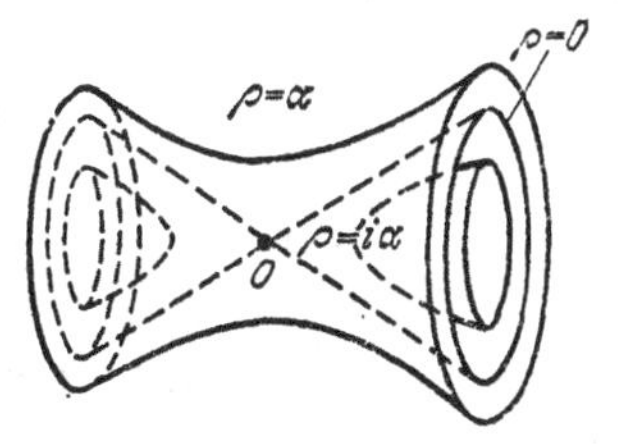

Figure 6.

Let us consider the case $\mathbb{R}_1^2$ in more detail. We shall be concerned with the group of motions of the plane $\mathbb{R}_1^2$, i.e. the set of all linear transformations $C: \mathbb{R}_1^2 \to \mathbb{R}_1^2$ preserving the form $(\xi, \eta)_1$. The transformation C preserves the form if $(C\xi, C\eta)_1 = (\xi, \eta)_1$ for any vectors ξ, η. But before calculating this group, recall a similar calculation for the Euclidean scalar product.

If a linear transformation $C: \mathbb{R}^n \to \mathbb{R}^n$ preserves the form $(\xi, \eta) = \sum_{i=1}^{n} \xi^i \eta^i$, then C is an orthogonal matrix, that it $C^{-1} = C^T$. Then $\det C = \pm 1$. In the case $n = 2$, the set of all orthogonal matrices of order 2 can be written as follows:

$$O(2) = \left\{ \begin{pmatrix} \cos\phi & \sin\phi \\ -\sin\phi & \cos\phi \end{pmatrix}; \quad \begin{pmatrix} \cos\phi & \sin\phi \\ \sin\phi & -\cos\phi \end{pmatrix} \right\}.$$

REMARK. The group of all orthogonal transformations in $\mathbb{R}^n$ is denoted by $O(n)$, and the subgroup containing those orthogonal transformations which have a positive determinant (i.e. preserve orientation of the space $\mathbb{R}^n$) is denoted by $SO(n)$.

Let us consider the $SO(2)$ group (i.e. the set of rotations of a plane preserving the orientation of the plane) and associate with each matrix $\begin{pmatrix} \cos\phi & \sin\phi \\ -\sin\phi & \cos\phi \end{pmatrix}$ a complex number $z = e^{i\phi}$ whose modulus is equal to unity. This correspondence will be denoted by $\mathcal{F}$. Clearly, $\mathcal{F}$ is a one-to-one correspondence and is continuous in either side, $0 \leq \phi < 2\pi$, i.e. $\mathcal{F}$ determines "homeomorphism" between $SO(2)$ group and the circumference $S^1 = \{z = e^{i\phi}\}$. Furthermore, $\mathcal{F}$ also establishes an algebraic isomorphism between the $SO(2)$ group (operations in $SO(2)$ are a multiplication of matrices) and the S^1 group (the operation on S^1 is multiplication of complex numbers:

$$z_1 z_2 = e^{i\phi} \cdot e^{i\phi} = e^{i(\phi+\phi)}).$$

A verification of this fact reduces to calculation of the matrix:

$$\begin{pmatrix} \cos(\phi+\psi) & \sin(\phi+\psi) \\ -\sin(\phi+\psi) & \cos(\phi+\psi) \end{pmatrix} = \begin{pmatrix} \cos\phi & \sin\phi \\ -\sin\phi & \cos\phi \end{pmatrix} \begin{pmatrix} \cos\psi & \sin\psi \\ -\sin\psi & \cos\psi \end{pmatrix}.$$

The whole group $O(2)$ is obviously homeomorphic to a unity of two circumferences (Figure 7).

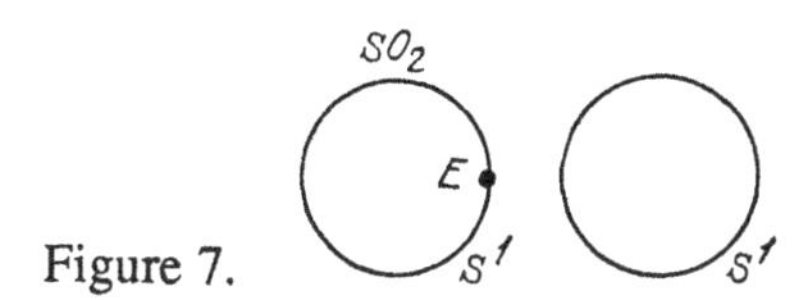

Figure 7.

Now we go over to the plane $\mathbb{R}^2_1$. Let $C = \begin{pmatrix} a & b \\ c & d \end{pmatrix}$ and $(C\xi, C\eta)_1 = (\xi, \eta)_1$. Recall that if $B(\xi, \eta)$ is an arbitrary form of $B(\xi, \eta) = b_i \xi^i \eta^j$; $b_{ij} = b_{ji}$, then this form is assigned the symmetric matrix (b_{ij}) in a one-to-one manner. If C is an arbitrary linear transformation in space, then the matrix B (following the form $B(\xi, \eta)$) transforms to a new matrix B' (corresponding to the transformed form) which is related to the matrix B as:

$$B' = CBC^{\mathrm{T}}.$$

For example, if $B(\xi, \eta) = \sum_{i=1}^{n} \xi^i \eta^i$, then $B = E$ is a unit matrix; and if C preserves this form, then $B' = B = E$, whence we have $E = CEC^{\mathrm{T}}$, i.e. $C^{-1} = C^{\mathrm{T}}$, which implies orthogonality of the matrix C.

In our case, $B(\xi, \eta) = (\xi, \eta)_1 = -\xi^1\eta^1 + \xi^2\eta^2$, i.e. $B = \begin{pmatrix} -1 & 0 \\ 0 & 1 \end{pmatrix}$.

Let $C: \mathbb{R}^2_1 \to \mathbb{R}^2_1$ (that is, C preserves our form); then $B' = B = \begin{pmatrix} -1 & 0 \\ 0 & 1 \end{pmatrix}$, that is:

$$\begin{pmatrix} -1 & 0 \\ 0 & 1 \end{pmatrix} = \begin{pmatrix} a & c \\ b & d \end{pmatrix} \begin{pmatrix} -1 & 0 \\ 0 & 1 \end{pmatrix} \begin{pmatrix} a & b \\ c & d \end{pmatrix} = \begin{pmatrix} -a^2 + c^2 & -ab + cd \\ -ab + cd & -b^2 + d^2 \end{pmatrix},$$

which yields the following relation for the numbers a, b, c, d:

$$c^2 - a^2 = -1, \ ab = cd, \ d^2 - b^2 = 1.$$

In the case of the Euclidean plane $\mathbb{R}^2$, each rotation was determined by the angle of rotation ϕ of the orthogonal frame; an analogous parameter will also be introduced in the case of the pseudo-Euclidean plane $\mathbb{R}^2_1$. Let us consider the frames $e_1 = (1, 0)$, $e_2 = (0, 1)$ and $C(e_1) = ae_1 + ce_2$, $C(e_2) = be_1 + de_2$ and let $\beta = c/a$.

The direct calculation yields:

$$a = \pm \frac{1}{(1-\beta^2)^{1/2}}, \quad c = \pm \frac{\beta}{(1-\beta^2)^{1/2}},$$

$$d = \pm \frac{1}{(1-\beta^2)^{1/2}}, \quad b = \pm \frac{1}{(1-\beta^2)^{1/2}}.$$

Thus, the group of all transformations C preserving the pseudo-Euclidean scalar product $(\xi, \eta)_1$ consists of the following matrices:

$$C = \begin{pmatrix} \dfrac{\pm 1}{(1-\beta^2)^{1/2}} & \dfrac{\pm \beta}{(1-\beta^2)^{1/2}} \\ \dfrac{\pm \beta}{(1-\beta^2)^{1/2}} & \dfrac{\pm 1}{(1-\beta^2)^{1/2}} \end{pmatrix}.$$

Instead of the angle of ordinary rotation ϕ, we introduce the angle of hyperbolic rotation ψ by setting $\beta = \operatorname{th} \psi$: then

$$C = \begin{pmatrix} \pm \operatorname{ch} \psi & \pm \operatorname{sh} \psi \\ \pm \operatorname{sh} \psi & \pm \operatorname{ch} \psi \end{pmatrix},$$

i.e. the group preserving the pseudo-Euclidean metric is the group of hyperbolic rotations (Figure 8). Recall the group of orthogonal rotations of the plane $\mathbb{R}^2$ which consisted of two connected components (two pieces) — two circumferences. The group of hyperbolic rotations has a more complicated organization: it consists of four connected components (four pieces):

$$\left\{ \begin{pmatrix} \operatorname{ch} \psi & \operatorname{sh} \psi \\ \operatorname{sh} \psi & \operatorname{ch} \psi \end{pmatrix}; \begin{pmatrix} -\operatorname{ch} \psi & -\operatorname{sh} \psi \\ -\operatorname{sh} \psi & -\operatorname{ch} \psi \end{pmatrix}; \right.$$

$$\left. \begin{pmatrix} \operatorname{ch} \psi & -\operatorname{sh} \psi \\ \operatorname{sh} \psi & -\operatorname{ch} \psi \end{pmatrix}; \begin{pmatrix} -\operatorname{ch} \psi & \operatorname{sh} \psi \\ -\operatorname{sh} \psi & \operatorname{ch} \psi \end{pmatrix} \right\}.$$

Each of these pieces is homeomorphic to a real straight line $\mathbb{R}^1$.

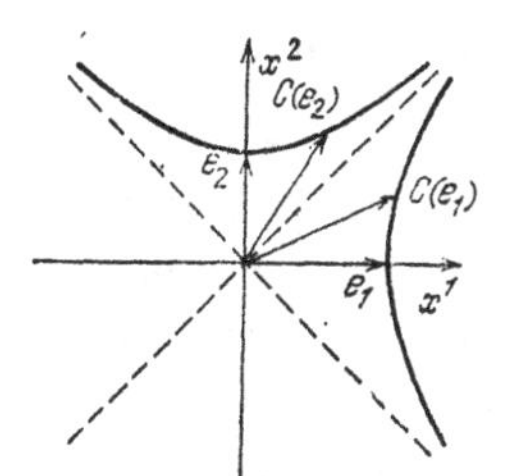

Figure 8.

REMARK. Of those four connected components, only one is a sub-group, namely:

$$\left\{\begin{pmatrix} \text{ch}\,\psi & \text{sh}\,\psi \\ \text{sh}\,\psi & \text{ch}\,\psi \end{pmatrix}\right\};$$

the rest of the components not being sub-groups. We can calculate the quotient group of the complete group of motions from the connected component of unity. To carry out this calculation, we should bear in mind the following general assertion: *the connected component of unity, G_0, in an arbitrary topological group G is always a normal divisor* (prove this!), and therefore the *factor group G/G_0* is defined (its order is equal to the number of connected components in the group G). In our example, the direct calculation (to be carried out by the reader) yields that the factor group G/G_0 is isomorphic to the Abelian group $\mathbb{Z}_2 \otimes \mathbb{Z}_2$. Note that the factor group $O(n)/SO(n)$ is isomorphic to $\mathbb{Z}_2$.

Now we shall turn to the metric properties of the space $\mathbb{R}_1^3$. Consider the form:

$$B(\xi, \eta) = (\xi, \eta)_1 = -\xi^1\eta^1 + \xi^2\eta^2 + \xi^3\eta^3.$$

The space $\mathbb{R}_1^3$ will again be modelled in the space $\mathbb{R}^3$, and therefore the Cartesian coordinates in $\mathbb{R}^3$ will be denoted by x, y, z; then $(\xi, \xi)_1 = -x^2 + y^2 + z^2$. As we have already established, the *hypersphere* (or *pseudo-sphere*) *of imaginary radius $\tau = i\rho$ in a space $\mathbb{R}_1^3$ is a two-sheeted hyperboloid* given by the following equation:

$$-\rho^2 = -x^2 + y^2 + z^2.$$

Since this hyperboloid is imbedded in $\mathbb{R}_1^3$, we can say that "the geometry of the space $\mathbb{R}_1^3$ induces a certain geometry on the hyperboloid".

From the point of view of Riemannian metric, the idea expressed above will be formulated as follows: "the metric of the space $\mathbb{R}_1^3$ induces a certain metric on the hyperboloid". At the present moment, however, even without the general concept of the metric tensor g_{ij}, we can already impart some meaning to the words "a geometry induced on the hyperboloid". Indeed, let us consider a hyperboloid $-\rho^2 = -x^2 + y^2 + z^2$ (for simplicity we shall restrict our consideration to one of its sheets; for example, to the one described by the inequality $x > 0$); quite ordinary points of the hyperboloid will be treated as "points" of the geometry induced on it, and the various lines obtained on the hyperboloid when it is intersected by the planes $ax + by + cz = 0$ passing through the origin of coordinates will be thought of as "straight lines" of the induced geometry (Figure 9). We shall proceed to this geoemery., To do so, we shall make a transformation which will bring into correspondence the geometry of the hyperboloid and the geometry in a ring in the Euclidean plane. This transformation is called a *stereographic projection.* The stereographic projection of the sphere S^2 onto the plane $\mathbb{R}^2$ is described in the theory of functions of one complex variable. Recall the construction of this projection (Figure 10).

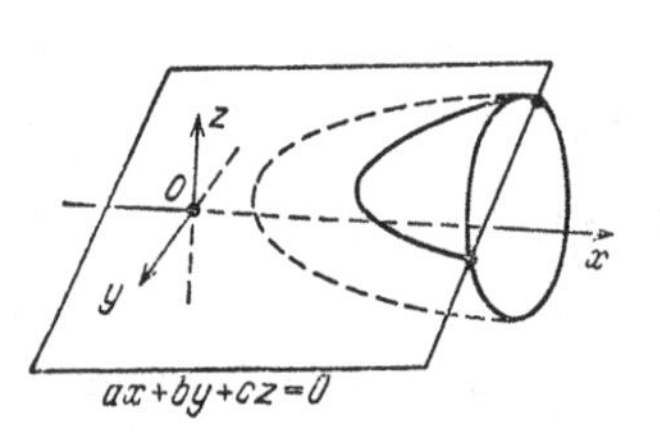

Figure 9.

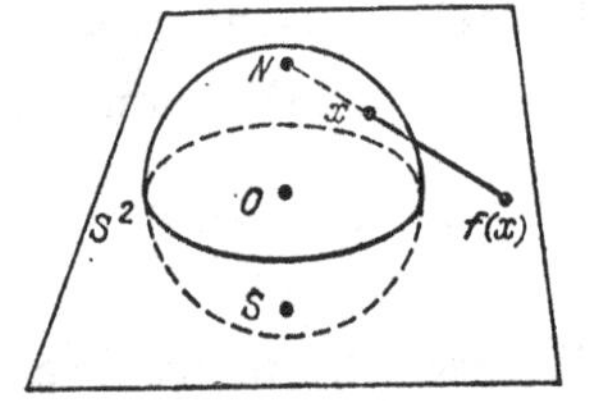

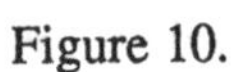

Figure 10.

The plane $\mathbb{R}^2$ (= $\mathbb{C}$) passes through the centre O of the sphere S^2, and the stereographic projection $f: S^2 \to \mathbb{R}^2$ assoociates each point x (which does not coincide with the north pole N) with the point $f(x)$ — the point where the ray Nx meets the plane $\mathbb{C}$. Given this, to the north pole there corresponds an infinitely remote point of the extended complex plane. The south pole goes over to the origin. The analogy between the usual sphere S^2 and the pseudo-sphere S_1^2 is rather widespread. In particular, the stereogrqphic projection of the pseudo-sphere S_1^2 onto the plane $\mathbb{R}^2$ is specified in a quite similar way.

The pseudo-sphere S_1^2: $\{-\rho^2 = -x^2 + y^2 + z^2\}$ is centred at the origin O; the north pole is the point with Cartesian coordinates: $(-\rho, 0, 0)$; the south pole is the point $(\rho, 0, 0)$; the plane onto which we shall make the projection is the YOZ-plane passing through the pseudo-sphere centre (by the way, the restriction of the form $(\xi, \eta)_1$ to the YOZ-plane is the following form: $\xi^2\eta^2 + \xi^3\eta^3$), that is, the pseudo-Euclidean geometry of $\mathbb{R}_1^3$ induces Euclidean geometry in the YOZ-plane. Figure 11 illustrates the cross-section of a hyperboloid by a plane passing through the X-axis. But since we have restricted our consideration to only one hyperboloid sheet $x > 0$, it follows that the image of this sheet under projection f does not cover the whole of the plane $\mathbb{R}^2 = YOZ$, but only an open ring of radius ρ.

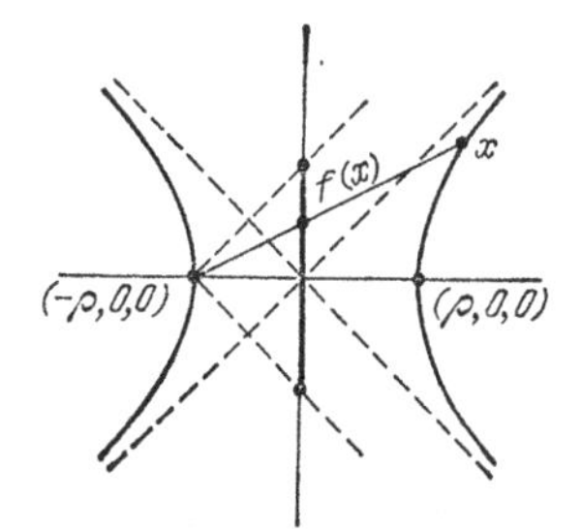

Figure 11.

REMARK. If we consider the whole of the pseudo-sphere S_1^2 (that is, both the sheets of the hyperboloid), then under stereographic projection, the image of S_1^2, as distinct from that of the usual sphere S^2, covers in one-to-one manner only part of the plane YOZ (the north pole passes again into an infinitely remote point, and the circle $y^2 + z^2 = \rho^2$ is not covered).

Let x, y, z be coordinates of the point $x \in S_1^2$ (where $x > 0$) and let (u^1, u^2) be coordinates of the point $f(x) \in YOZ$, where f is a stereographic projection. We calculate the relation between these coordinates in an explicit form.

LEMMA 1. *Let* $x = (x, y, z)$; $u = (u^1, u^2)$. *Then*

$$x = \rho\left(-1 + \frac{2\rho^2}{-(u^1)^2 - (u^2)^2 + \rho^2}\right);$$

$$y = \frac{2\rho u^1}{-(u^1)^2 - (u^2) + \rho^2}; \quad z = \frac{2\rho u^2}{-(u^1)^2 - (u^2) + \rho^2}.$$

Proof. From Figure 11 it is seen that:

$$\frac{y}{u^1} = \frac{x+\rho}{\rho}, \quad \frac{z}{u^2} = \frac{x+\rho}{\rho};$$

that is,

$$y = u^1\left(1 + \frac{x}{\rho}\right); \quad z = u^2\left(1 + \frac{x}{\rho}\right).$$

Since $-\rho^2 = -x^2 + y^2 + z^2$, it follows that substituting y and z into this equation, we obtain:

$$x = -\rho\left(1 + \frac{2\rho^2}{(u^1)^2 + (u^2)^2 - \rho^2}\right).$$

This concludes the proof of the lemma.

Under the stereographic projection $f\colon S_1^2 \to \{y^2 + z^2 < \rho^2\} = D^2$, the points of the hyperboloid are transformed into points of the two-dimensional ring D^2 of radius ρ. Into what curves on the ring D^2 will the "straight lines" of our geometry on the hyperboloid, i.e. lines of intersection of the hyperboloid by the planes $ax + by + cz = 0$ be transformed?

LEMMA 2. *Each line of intersection of* S_1^2 *with the plane* $ax + by + cz = 0$ *transforms under the mapping* f *into an arc of a circumference intersecting the circumference* $y^2 + z^2 = \rho^2$ *at a right angle.*

REMARK. Recall that the angle between smooth curves at the point of their intersection is the angle between their tangents at this point.

Proof of Lemma 2. By virtue of Lemma 1, to clarify the fact into what curve a "straight line" on S_1^2 is carried, it suffices to substitute the expressions of x, y, z in terms of u^1, u^2 into the equation of plane $ax + by + cz = 0$. Let, for example, $a \neq 0$. Then, after simple transformations, the equation:

$$-a - \frac{2a\rho^2}{(u,u)-\rho^2} + \frac{2bu^1}{(u,u)-\rho^2} + \frac{2cu^2}{(u,u)-\rho^2} = 0$$

is reduced to the form:

$$\left(u^1 - \frac{b}{a}\right)^2 + \left(u^2 - \frac{c}{a}\right) = \frac{b^2+c^2}{a^2} - \rho^2,$$

i.e. defines a circumference with centre at the point $\left(\frac{b}{a}, \frac{c}{a}\right)$ and radius $r = \sqrt{\frac{b^2+c^2}{a^2} - \rho^2}$ which intersects the circumference $y^2 + z^2 = \rho^2$ at points A and B at a right angle,

$$\rho^2 + r^2 = \left(\frac{b}{a}\right) + \left(\frac{c}{a}\right)^2.$$

REMARK. The image of the "straight line" from S_1^2 under the mapping f is not the whole of the circumference:

$$\left(u^1 - \frac{b}{a}\right)^2 + \left(u^2 - \frac{c}{a}\right)^2 = \frac{b^2+c^2}{a^2} - \rho^2,$$

but only that part of it which is contained in the ring $y^2 + z^2 < \rho$.

Thus, we have completed the proof of the theorem: *the geometry induced on the pseudo-sphere S_1^2 by the geometry of the pseudo-Euclidean space $\mathbb{R}_1^3$ coincides with the geometry arising in a ring of radius ρ in the Euclidian plane $\mathbb{R}$ provided that, as points of this geometry, we take ordinary points of this ring, and as "straight lines" of this geometry we take arcs of the circumferences intersecting the ring boundary at a right angle*. (In particular, "straight lines" are, of course, all diameters of the ring since they may be treated as arcs of circumferences of infiniely large radius.)

REMARK. A geometry induced on a pseudo-sphere by the geometry of $\mathbb{R}_1^3$ is called *Lobachevsky geometry*, and its model in a ring of radius ρ in a Euclidean plane is called the *Poincaré model of Lobachevsky geometry*.

REMARK. Lobachevsky himself obtained his geometry in quite a different manner, without making use of pseudo-Euclidean spaces.

Using the Poincaré model, we can easily verify the classical Euclidean axioms (postulates) except, of course, the fifth postulate (parallel axiom).

Indeed, from Figure 12 we can see that through a point exterior to a straight line we can, using Lobachevsky geometry, draw *infinitely many straight lines parallel to a given one* (i.e. not intersecting it).

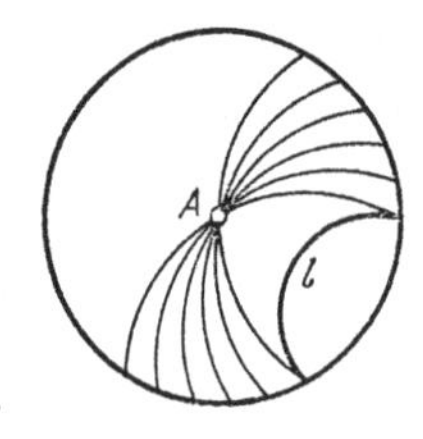

Figure 12.

Note that if $\rho \to \infty$, then Lobachevsky geometry becomes Euclidean geometry (arcs of circumferences become straight lines). Later, when studying Lobachevsky geometry, we assume $\rho = 1$.

Now we ask the question: what geometry, that is, the geometry with what properties, arises if we cleave with planes an ordinary sphere $S^2 \subset \mathbb{R}^3$ rather than a pseudo-sphere $S_1^2 \subset \mathbb{R}_1^3$?

Consider the geometry in which "points" are ordinary points of the sphere S^2: ($|x| = 1$ in $\mathbb{R}^3$) and "straight lines" are the various equators of the sphere S^2 (intersections with the various planes passing through the centre of the sphere). This geometry has, as it stands, the shortcoming that many straight lines (not only one) may pass through two distinct points; this will be the case if we consider two diametrically opposite sides of the sphere. But if, as "points", we consider in our geometry pairs $(x, -x)$ where x spans the whole S^2, then in this geometry there hold Euclidean postulates, except order axioms and the fifth postulate. Namely, through a point exterior to a straight line we can draw not a single straight line parallel to a given one, i.e. any two straight lines intersect (any equators intersect at diametrically opposite points of the sphere). The order axioms do not hold in the absence of the concept "one point lies between two others". The described operation (identification of x and $-x$, where x spans S^2) is equivalent to factorization of the sphere S^2 with respect to the action of the group $\mathbb{Z}_2$, which yields a two-dimensional projective space $\mathbb{R}P^2$; the geometry constructed on $\mathbb{R}P^2$ is called *elliptic geometry*.

Thus, we have distinguished three geometries:

1) Euclidean geometry,
2) Lobachevsky geometry,
3) Geometry on the sphere.

In spite of profound differences among them, all the three geometries can be studied in parallel; they are widely interrelated. We shall return to their study from the point of view of the metric tensor g_{ij}. We have calculated the groups of motions of these three "uniform" geometries; in these geometries, the groups of motions are described by *three* parameters.

The space $\mathbb{R}^4_1$ is called the *space of the special theory of relativity*, and the geometry arising in this space is called *Minkowski geometry*. The coordinates in $\mathbb{R}^4_1$ are conventionally denoted by x, y, z (spatial coordinates) and ct (time coordinate); then $(\xi, \eta)_1 = -c^2tt' + xx' + yy' + zz'$. The isotropic cone $(\xi, \xi)_1 = 0$ is called the *light cone*, vectors ξ such that $(\xi, \xi)_1 > 0$ are called *space-like*, and vectors ξ such that $(\xi, \xi)_1 < 0$ are called *time-like*. Here c is the speed of light.

1.5 Flat Curves

Several of the following sections are devoted to the branch of the classical differential geometry associated with the concepts of *curvature* and *torsion* of curves in the Euclidean plane and in Euclidean 3-space.

Let us consider a Euclidean plane with coordinates (x, y) and basic unit vectors e_1, e_2, where any point P is given by a radius vector $r = xe_1 + ye_2$ with tail at the origin O and tip at a particular point P coordinatized by (x, y). The length of the vector r is given by the Euclidean formula $|r| = (rr)^{1/2} = (x^2 + y^2)^{1/2}$ Suppose we are given a smooth curve:

$$r(t) = (x = x(t), \ y = y(t)),$$

where points of the curve are given as follows: $x(t)e_1 + y(t)e_2$. The length of the curve segment has the form:

$$l = \int_a^b \sqrt{(\dot{x})^2 + (\dot{y})^2}\, dt = \int_a^b dl, \quad \dot{x} = \frac{dx}{dt}, \ \dot{y} = \frac{dy}{dt},$$

where the differential of length:

$$dl = |v|\, dt, \quad |v| = (v \cdot v)^{1/2}, \quad v = \dot{x} e_1 + \dot{y} e_2$$

is the velocity vector.

We shall write $v_t = dr/dt$ indicating explicitly, thereby, the parameter with respect to which the tangent vector is calculated. We shall often find it convenient to consider curves parametrized by the *natural* (length) *parameter*:

$$x = x(l), \ y = y(l).$$

In this case $v = v_l = (dx/dl)e_1 + (dy/dl)e_2$ $|v| = 1$.

If the curve was parametrized by an arbitrary parameter t, $x = x(t)$, $y = y(t)$, we have the relation $dl = (\dot{x}^2 + \dot{y}^2)^{1/2}\, dt$. Two vectors (those of *velocity* and *acceleration*) will play an important role:

$$\frac{dr}{dt} = v_t, \quad \frac{d^2r}{dt^2} = w_t = \ddot{x}e_1 + \ddot{y}e_2.$$

If the parameter is natural ($t = l$), we shall have $|v_l| = 1$. There holds a simple, but frequently encountered, lemma.

LEMMA 1. *If there exists a time-dependent vector* $v = v(t)$, *where* $|v| = 1$, *then the vectors* v *and* $\dot{v} = d/dt$ *are orthogonal.*

Proof. Since $v = v^1 e_1 + v^2 e_2$ and $|v^2| = (v^1)^2 + (v^2)^2 = 1$, we have:

$$d/dt\,(v \cdot v) = v\dot{v} + v\dot{v} = 2v\dot{v} = d/dt\,(|v|^2) = 0,$$

therefore $v \cdot \dot{v} = 0$, which proves the lemma.

REMARK. If there exist any two vectors $v = v(t)$ and $w = w(t)$, then in Euclidean geometry there holds the formula:

$$d/dt\,(vw) = \dot{v}w + v\dot{w}\ .$$

In application to a curve parametrized by the natural parameter $l = t$, $r = r(t) = x(t)e_1 + y(t)e_2$, our lemma suggests:

$$v = dr/dl\,.$$

COROLLARY. *The velocity vector* $v(t)$ *and the acceleration vector* $w(t) = dv/dl$ *are orthogonal if the parameter is natural*: $t = l$ (the arc length).

DEFINITION 1. *The curvature of a flat curve* is a magnitude of the acceleration vector $k = |w(t)|$ provided that $t = l$ (the natural parameter).

DERIVATION. It is immediate that:

$$\frac{dv}{dl} = kn = \frac{d^2 r}{dl^2},$$

where n is the unit vector normal to the curve and

$$n = \frac{w}{|w|} = \frac{1}{\sqrt{\left(\frac{d^2 x}{dl^2}\right)^2 + \left(\frac{d^2 y}{dl^2}\right)^2}}\left(\frac{d^2 x}{dl^2} e_1 + \frac{d^2 y}{dl^2} e_2\right)$$

The radius of curvature R is the number $1/k$.

along the entire curve $r(t)_1$, a smooth field of normals $\hat{n}(l)$ oriented so that the frame $(\hat{n}(l), v(l))$, where $v(l)$ is the unit vector tangent to the curve and directed towards the increase of the natural parameter $t = l$, have orientation coinciding with that fixed in the plane. In that case the curvature $\hat{k}$ is defined as $d^2r/dl^2 = \hat{k}\hat{n}(l)$. If $d^2r/dl^2 \neq 0$ at each point of the curve, then $|\hat{k}| \equiv |k| \neq 0$. But if the acceleration vector vanishes at some points, then the direction of the normal $n = \frac{d^2r}{dl^2} / |\frac{d^2r}{dl^2}|$ may vary as distinct from the direction of the normal $\hat{n}(l)$. Thus, $|\hat{k}| = |k| = k$, but $\hat{k}$ may change sign for the opposite when moving along the curve ($|dr/dl| \neq 0$).

Does this concept of curvature agree with our intuitive ideas?

The curvature has the following properties.

1) The curvature of a straight line is zero.

Proof. Let $x = x_0 + al$, $y = y_0 + bl$ (straight line), the parameter l being natural; this means that:

$$a^2 + b^2 = \left(\frac{dx}{dl}\right)^2 + \left(\frac{dy}{dl}\right)^2 = 1.$$

Then $w = \frac{d^2r}{dl^2} = 0$ and $k = 0$, $R = \infty$.

2) The curvature of a circle of radius R is $k = 1/R$.

Let:

$$x = x_0 + R \cos 1/R, \quad y = y_0 + R \sin 1/R, \quad R = \text{const}.$$

Then $\frac{d^2x}{dl^2} = -\frac{\cos 1/R}{R}$, $\frac{d^2y}{dl^2} = -\frac{\sin 1/R}{R}$. Consequently, we obtain for the curvature $|w| = 1/R = k$. An important theorem holds.

THEOREM 1. *Given the parametric equation* $r = r(l)$ *of a curve, in terms of the natural parameter* l, *the following Frenet formulae hold*:

$$\frac{dv}{dl} = kn = w,$$

$$\frac{dn}{dl} = kv,$$

where $n = \frac{w}{|w|}$ *is the unit normal vector.*

Proof. Since n is a unit vector, $nn = 1$, and the vectors, n and v are orthogonal, according to Lemma 1, we have:

a) $dn/dl \perp n$ (Lemma 1),

b) $dn/dl = \alpha v$ ($n \perp v$ and the dimension = 2).

Given $|v| = 1$, we have $|\alpha| = |dn/dl|$. What is the value of α? Since $vn = 0$, we have:

$$0 = \frac{d}{dl}(vn) = \frac{dv}{dl}n + v\frac{dn}{dl} = k + \alpha(vv) = k + \alpha = 0,$$

$$(nn = 1, \quad vv = 1).$$

Whence $\alpha = -k$, as claimed.

What is the geometric meaning of the Frenet formulae? Since $dv/dl = kn$, $dn/dl = -kv$ and (v, n) is a unit orthonormal frame, it follows that:

$$v + \Delta v = v + (\Delta l)\frac{dv}{dl} = v + (k\Delta l)\, n,$$

$$n + \Delta n = n + (\Delta l)\frac{dn}{dl} = n + (-k\Delta l)\, v$$

with accuracy of the order of the second power of small quantities.

Suppose $k\Delta l = \Delta\phi$ (the increment of the angle). For small angles $\Delta\phi$

$$\cos(\Delta\phi) \cong 1 + O((\Delta\phi)^2),$$

$$\sin(\Delta\phi) \cong \Delta\phi + O((\Delta\phi)^3),$$

and we have:

$$v + (\Delta v) \cong \cos(\Delta\phi)v + \sin(\Delta\phi)\, n,$$

$$n + (\Delta n) \cong -\sin(\Delta\phi)v + \cos(\Delta\phi)\, n,$$

that is, under this transformation the frame is rotated through the small angle $\Delta\phi$.

Hence, *the Frenet formulae determine a rotation of the frame (y, n) in going to a nearby point $l \to l + \Delta l$ with accuracy of the order of the second power of small quantities.*

This fact is sometimes also expressed by the formula:

$$k = \left|\frac{d\phi}{dl}\right|,$$

where ϕ denotes the vector through which the vector v (or n) is rotated in moving along the curve. The sign indicates the direction (clockwise or counter-clockwise) in which the frame (v, n) is rotated when moving along the curve. The parameter t was always taken to be the natural one.

It is now natural to ask how we go about calculating the curvature of a flat curve parametrized as $r(t) = (x(t), y(t))$, where t is not the natural parameter?

In this case $v_t = r' = x' e_1 + y' e_2$ and $|v_t| \neq 1$. The vectors v_t and $v'_t = r''$ (the velocity and acceleration) are not therefore necessarily perpendicular.

Let $\xi = \xi(t) = \xi^1 e_1 + \xi^2 e_2$ be any arbitrary vector. For our curve we had $dl = |r'|\, dt = |v_t|\, dt$. For an arbitrary vector $\xi(t)$ we have:

$$\frac{d\xi}{dl} = \frac{d\xi}{dt}\frac{dt}{dl} = \frac{1}{|v_t|}\frac{d\xi}{dt},$$

where $|v_t| = |r'|$ and the velocity is determined relative to the parameter t given along the curve.

Suppose

$$\xi(t) = \frac{v_t}{|v_t|} = \frac{dr}{dl},$$

where $\xi(t)$ is the unit vector of the tangent (it coincides with the velocity vector v in the case where the parameter is natural).

By the definition of curvature:

$$k = \left|\frac{d^2r}{dl^2}\right| = \left|\frac{d\xi}{dl}\right| = \left|\frac{d}{dl}\left(\frac{v_t}{|v_t|}\right)\right|$$

(the length of the acceleration vector in the natural parameter is equal to the curvature).

By definition,

$$\frac{d}{dl}\left(\frac{v_t}{|v_t|}\right) = \frac{1}{|v_t|}\frac{d}{dt}\left(\frac{v_t}{|v_t|}\right) = \frac{1}{|v_t|^2}\left(\frac{dv_t}{dt} - \frac{v_t}{|v_t|}\cdot\frac{d|v_t|}{dt}\right) =$$

$$= \frac{1}{|v_t|^2}\left(\ddot{r} - \frac{\dot{r}}{2|\dot{r}|^2}\frac{d}{dt}(|\dot{r}|^2)\right);$$

$$v_t = \dot{r}, \ \frac{d}{dt}(|\dot{r}|)^2 = 2\dot{r}\ddot{r}.$$

Thus, we obtain (assuming that $|\dot{r}| \neq 0$):

$$k = \left|\frac{d}{dl}\left(\frac{v_t}{|v_t|}\right)\right| = \left|\frac{d}{dl}\left(\frac{\dot{r}}{|\dot{r}|}\right)\right| = \frac{1}{|\dot{r}|^2}\left|\ddot{r} - \left(\frac{\dot{r}\ddot{r}}{|\dot{r}|^2}\right)\dot{r}\right|.$$

For the curvature, we have:

$$k = \left|\frac{d^2r}{dl^2}\right| = \frac{1}{|\dot{r}|^2}\left|\xi - (\xi\eta)\,\eta\right|,$$

where:

$$\xi = \ddot{r}, \quad \eta = \frac{dr}{dl} = \frac{\dot{r}}{|\dot{r}|} = \frac{v_t}{|v_t|}.$$

The components of the vector $\frac{d^2r}{dl^2} = w = \frac{1}{|\dot{r}|^2}\left(\ddot{r} - \left(\frac{\dot{r}\ddot{r}}{|\dot{r}|^2}\right)\dot{r}\right)$ have the form:

$$|\dot{r}|^2 w = \left(\ddot{x} - \frac{\dot{x}\ddot{x} + \dot{y}\ddot{y}}{\dot{x}^2 + \dot{y}^2}\cdot\dot{x}\right)e_1 + \left(\ddot{y} - \frac{\dot{x}\ddot{x} + \dot{y}\ddot{y}}{\dot{x}^2 + \dot{y}^2}\cdot\dot{y}\right)e_2.$$

Next, we have:

$$|w|^2 = k^2 = \frac{(\ddot{x}\dot{y} - \ddot{y}\dot{x})^2}{(\dot{x}^2 + \dot{y}^2)^3}.$$

For the curvature, we obtain the relation:

$$k^2 = \frac{|\ddot{x}\dot{y} - \ddot{y}\dot{x}|^2}{(\dot{x}^2 + \dot{y}^2)^3} \quad \text{(an important formula).}$$

The numerator is the square of the determinant of the matrix A, where:

$$A = \begin{pmatrix} \dot{x} & \dot{y} \\ \ddot{x} & \ddot{y} \end{pmatrix}.$$

Thus, we have arrived at the following theorem.

THEOREM 2. *If at any point on a curve the velocity vector does not become zero, then for any choice of parameter t for the curve $x = x(t)$, $y = y(t)$ there holds the formula*:

$$k = \left|\frac{d^2 r}{dl^2}\right| = \frac{|\ddot{x}\dot{y} - \ddot{y}\dot{x}|}{(\dot{x}^2 + \dot{y}^2)^{3/2}},$$

where $\dot{x}^2 + \dot{y}^2 \neq 0$ *since* $|\dot{r}| \neq 0$.

Hence the absolute value of the acceleration d^2r/dl^2, i.e. the square root of the sum of squares of the componenets of the acceleration is the number:

$$k = \frac{|\ddot{x}\dot{y} - \ddot{y}\dot{x}|}{(\dot{x}^2 + \dot{y}^2)^{3/2}}.$$

We have obtained the basic theorems of the theory of plane curves in Euclidean geometry. Let us make several remarks. We shall later prove the following property of time-dependent orthogonal transformations.

Given an orthogonal matrix $A = A(t)$, where:

$$A(0) = E \left(a^i_j(0) = \delta^i_j = \begin{cases} 1, & i \neq j \\ 0, & i \neq j \end{cases} \right).$$

The matrix $\frac{dA}{dt}\Big|_{t=0} = \left(\dot{a}^i_j(0)\right)$ is skew-symmetric. i.e. $\dot{a}^i_j(0) = -\dot{a}^j_i(0)$.

This fact is proved below. Its manifestations were the Frenet formulae $dv/dl = kn$, $dn/dl = -kv$, where $k = k(l)$ and $\frac{dA}{dl}\Big|_{l=l_0} = \begin{pmatrix} 0 & k \\ -k & 0 \end{pmatrix}$.

It was shown separately that the matrix $B = dA/dl$ is an infinitesimal rotation through an angle $\Delta\phi = k\Delta l$ of the frame (v, n) in moving along the curve or a rotation of the vector v since the rotation of the vector n orthogonal to it is thereby defined. Thus, $k = d\phi/dl$, where ϕ is the angle of rotation of the vector v.

LEMMA 2. *Let $A(t)$ be a smooth family of orthogonal matrices and let $A(0) = E$. Then the matrix $X = \dot{A}(t)|_{t=0}$, which is the derivative of the family $A(t)$ at the point $t = 0$, is skew-symmetric.*

Proof. The orthogonality condition for the matrices $A(t)$ is $(A(t)a,\ A(t)b) = (a, b)$ for any vectors a and b. Differentiating this identity with respect to t, we obtain the equality $(A'(t)a, A(t)b) + (A(t)a, A'(t)b) = 0$. When $t = 0$, we obtain $(Xa, b) + (a, Xb) = 0$. Setting $a = e_i$, $b = e_j$, we come to $(Xe_i, e_j) = -(e_i, Xe_j)$, i.e. $x^i_j = -x^j_i$, where $X = (x^i_j)$. Here e_i and e_j are orthonormal, basic vectors, as required.

1.6 Space Curves

We now proceed to the theory of space curves.

For any curve $x = x(t)$, $y = y(t)$, $z = z(t)$ or in terms of the vectors $r = r(t)$ there holds the equalities:

$$dl = |\dot{r}|\, dt = |v_t|dt = \left(\dot{x}^2 + \dot{y}^2 + \dot{z}^2\right)^{1/2} dt .$$

As in the planar case, we shall first consider the natural parameter l only, since it is in terms of l that our basic concepts are most conveniently defined. Our curve is thus given by $r = r(l)$, $x = x(l)$, $y = y(l)$, $z = z\,(l)$, where x, y, z are Euclidean coordinates.

By definition, $v = \dot{r} = \dot{x} e_1 + \dot{y} e_2 + \dot{z} e_3$ and $w = \ddot{r} = \dot{v} = \ddot{x} e_1 + \ddot{y} e_2 + \ddot{z} e_3$ (we use the dot to indicate derivatives with respect to l, d/dl, since $t = l$). We define curvature as in the planar case.

DEFINITION 1. The *curvature of a space curve* $r = r(t)$ is the absolute value of the acceleration relative to the parameter l : $k = |w| = |\ddot{r}|$ (where dot stands for d/dl). The *radius of curvature* is $R = 1/k$.

In the three-dimensional case, however, the velocity vector $v = dr/dl$ and the acceleration vector $w = d^2r/dl^2$ are not enough to compile a complete reference frame even if $|w| \neq 0$. We know from Lemma 1, Section 1.5 that $wv = 0$ or $w \perp v$ since $|v| = 1$. Besides, it is obvious that in a three-dimensional space the curvature alone is not enough to characterize the geometrical properties of the curve. Imagine, for example, a curve winding round a cylinder ($x = R \cos t$, $y = R \sin t$, $z = t$) (a circular helix). In addition to curvature, it has a third direction in which it is "contorting" (Figure 13). The third basis vector can be taken orthogonal to v and w.

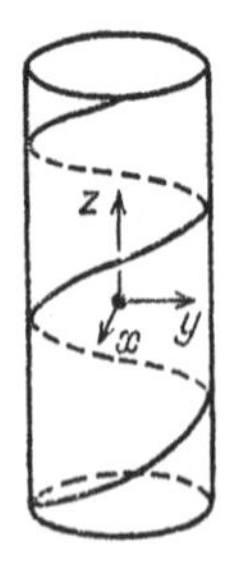

Figure 13.

We remind the reader of the well-known operation, from the linear algebra of Euclidean 3-space, of the *vector product* of vectors.

If ξ, η are vectors in a three-dimensional space $\xi = \xi^i e_i$, $\eta = \eta^i e_i$, where e_i form an orthogonal basis ($e_i \perp e_j$, $|e_i| = 1$), then we can build a vector:

$$\gamma = [\xi, \eta] = -[\eta, \xi], \quad \gamma = \gamma^i e_i,$$

where

$$\gamma^1 = \xi^2\eta^3 - \xi^3\eta^2, \quad \gamma^2 = \xi^3\eta^1 - \xi^1\eta^3, \quad \gamma^3 = \xi^1\eta^2 - \xi^2\eta^1,$$

or $\pm\gamma^i$ is equal to the determinant of the part of the matrix $\begin{pmatrix} \xi^1 & \xi^2 & \xi^3 \\ \eta^1 & \eta^2 & \eta^3 \end{pmatrix}$, which remains after the i-th column is crossed out.

We can readily see that:

$$[\xi, \eta] = -[\eta, \xi], \quad [\xi_1 + \xi_2, \eta] = [\xi_1, \eta] + [\xi_2, \eta], \quad [\lambda\xi, \eta] = \lambda[\xi, \eta],$$

and it can be verified that *Jacobi's identity* holds:

$$\big[[\xi, \eta], \gamma\big] + \big[[\gamma, \xi], \eta\big] + \big[[\eta, \gamma], \xi\big] = 0.$$

The following properties of the vector product are also well-known to the reader: the vector $[\xi, \eta]$ is directed perpendicularly to the plane of the vectors $\lambda\xi + \mu\eta$, the vector length being equal to $|[\xi, \eta]| = |\xi|\,|\eta|\,|\sin\phi|$, where ϕ is the angle between ξ and η,

$$\cos^2\phi = 1 - \sin^2\phi = \left(\frac{\xi\eta}{|\xi|\,|\eta|}\right)^2.$$

REMARK. If the vectors ξ and η lie in the plane (x, y), their vector product is orthogonal to the plane (directed along the z-axis) and $[\xi, \eta] = (\xi^1\eta^2 - \xi^2\eta^1)e_3$ and $|[\xi, \eta]| = |\xi^1\xi^2 - \xi^2\eta^1| = |\xi|\,|\eta|\,|\sin\phi|$.

We can now rewrite the formula for the curvature of a flat curve to obtain:

$$k = \frac{|[\dot{r}\ddot{r}]|}{|\dot{r}|^{3/2}} = \frac{|\dot{x}\ddot{y} - \ddot{x}\dot{y}|^2}{(\dot{x}^2 + \dot{y}^2)^{3/2}},$$

for an arbitrary parameter t.

Thus, the general formula for the curvature of a flat curve is expressed in terms of the length of the vector product $[r', r'']$, and since the curvature is related, by virtue of the Frenet formulae, to the rotation of the reference frame, it is natural to relate the curvature to the angular velocity of the frame (v, n) directed orthogonal to the (x, y)-plane.

Now we are in a position to return to our space curve:

$$r = r(l), \quad r = (x, y, z):$$

$$x = z(l), \quad y = y(l), \quad z = z(l).$$

Then $v(l) = \frac{dr}{dl} = \frac{dx}{dl} e_1 + \frac{dy}{dl} e_2 + \frac{dz}{dl} e_3.$

We assume that $|w| \neq 0$ and $|v| \neq 0$; such points are called *non-degenerate points* of the curve. We assume here $|v| = 1$, $wv = 0$ (or $w \perp v$). Consider the vector $b = [v, n]$, $n = w/|w|$. We shall call b the *vector of binormal* to the curve or the *binormal* to the curve, and n the *vector of the principal normal* to the curve, or the *principal normal*).

We can readily see that:

$$|b| = |v|\,|n|\,|\sin\phi| = 1, \quad b \perp v, \ b \perp n.$$

We thus have an orthonormal frame (v, n, b) at each point of the curve where $|w| \neq 0$ (i.e. at each non-degenerate point).

As in the case of scalar product, we shall find the following lemma useful.

LEMMA 1. *For two arbitrary vectors $\xi(t)$ and $\eta(t)$ in a three-dimensional space there holds a Leibniz type formula:*

$$\frac{d}{dt}[\xi, \eta] = \left[\frac{d\xi}{dt}, \eta\right] + \left[\xi, \frac{d\eta}{dt}\right].$$

The proof is immediate from the general Leibniz formula for differentiation of the product of functions $(fg)' = \dot{f}g + f\dot{g}$.

THEOREM 1. *For any space curve* $r = r(l)$, *where* l *is the natural parameter, the following Frenet formulae hold:*

$$\frac{dv}{dl} = kn,$$

$$\frac{dn}{dl} = -\kappa b - kv,$$

$$\frac{db}{dl} = \kappa n,$$

where $|\kappa| = |db/dl|$ *is a definition, and the number* κ *is called the torsion of the space curve (which need not necessarily be positive). In the planar case, we have* $b =$ const. *and therefore* $\kappa = |db/dl| = 0$.

Relative to the basis (v, n, b) at a given point of the curve where $v = e_1$, $n = e_2$, $b = e_3$ we have, using matrix notation, $de_i/dl = b^j_i e_j$, $i = 1, 2, 3$, the matrix $B = (b^j_i)$, $i, j = 1, 2, 3$ being of the form:

$$B = \begin{pmatrix} 0 & +k & 0 \\ -k & 0 & -\kappa \\ 0 & +\kappa & 0 \end{pmatrix}.$$

We can see again that, as in the two-dimensional case, the matrix is skew-symmetric.

The proof of the theorem is given somewhat later.

In connection with the concept of the vector product $[\xi, \eta]$ of vectors ξ and η we have made several remarks on the algebra which we are now in a position to specify. A skew-symmetric matrix $A = (a_{ij})$, $a_{ij} = -a_{ij}$ in an n-dimensional space $(i = 1, \dots, n; j = 1, \dots, n)$ is determined by $n(n-1)/2$ numbers a_{ij}, where $i < j$.

It can be verified that for two skew-symmetric matrices A, B of order $n \times n$, their commutator $A \circ B - B \circ A = [A, B] = -[B, A]$ is again a skew-symmetric matrix. For $n = 3$, we have $n(n-1)/2 = 3$. Therefore, skew-symmetric matrices in a three-dimensional space also form a three-dimensional linear space

$$A = \begin{pmatrix} 0 & a_{12} & a_{13} \\ -a_{12} & 0 & a_{23} \\ -a_{13} & -a_{23} & 0 \end{pmatrix}$$

with coordinates $a_{12} = X, a_{13} = Y, a_{23} = Z$. Thus, the matrix A is interprested as a vector in the three-dimensional space, (X, Y, Z), where $X = a_{12}, Y = a_{13}, Z = a_{23}$.

The commutator $[AB] = A \circ B - B \circ A$ *appears to be exactly the vector product of the vectors A and B regarded as skew-symmetric matrices for* $n = 3$.

What is the algebraic representation of the scalar product?

It turns out that $2(A, B) = -\mathrm{Sp}\,(A \circ B)$, where $\mathrm{Sp}\, C = \sum_{i=1}^{n} c_{ii}, C = (c_{ij})$, $\mathrm{Sp}\, C$ is the trace (check it!).

Here we have used the notation:

$$\text{"Sp"} \ (= \text{Spur}) = \text{the trace of the matrix.}$$

It is sometimes denoted by "Tr" (= Trace).

Furthermore, the question has arisen why for the natural parameter $l = t$ we have the equality: the absolute value of the acceleration $= |\dot{x}\ddot{y} - \ddot{x}\dot{y}| = |[vw]|$. The derivation of this equaltiy is as follows. The acceleration is equal to $w = (\ddot{x}, \ddot{y})$ and $(\dot{x}\ddot{y} - \ddot{x}\dot{y})e_3$ is the vector product of the velocity vectors $v = \dot{x}e_1 + \dot{y}e_2$ by the acceleration w. We have:

$$|[v, w]| = |v|\,|w|\,|\sin\phi| = |w|\,|\sin\phi| \quad (|v| = 1).$$

Since $w \perp v$, it follows that $|\sin\phi| = 1$, $|[v, w]| = |w| = k$.

We now proceed to the proof of the Frenet formulae for space curves.

Suppose the curve is given in terms of the natural parameter l:

$$r = r(l), \ x = x(l), \ y = y(l), \ z = z(l)$$

$$v = \frac{dr}{dl} = \left(\frac{dx}{dl}e_1 + \frac{dy}{dl}e_2 + \frac{dz}{dl}e_3\right),$$

where v is the velocity of the motion along the curve, $|v| = 1$, and $w = d^2r/dl^2$.

By definition, $k = |w|$, where k is curvature. We should prove the equalities:

$$\frac{dv}{dl} = kn$$

(by definition we have : $n = w/|w|, \quad w \neq 0$),

$$\frac{dn}{dl} = -kv - \kappa b,$$

$$\frac{db}{dl} = \kappa n, \quad b = [v, n].$$

Proof. Introduce the formula $db/dl = \kappa n$. Since $b = [v, n]$, it follows that

$$\frac{d[v, n]}{dl} = [v', n] + [v, n'] \quad \text{(by the Leibniz formula);}$$

then in accordance with the Lemma proved above, we have $v' \perp v$ and $n' \perp n$ ($|v| = 1$, $|n| = 1$). Therefore, $v' = kn$, $n' = \alpha v + \beta b$, where α and β are unknown. Since $[n, n] = 0$, and $[v, v] = 0$ it follows that $[v', n] = 0$,

$$[v, n'] = \beta(v, b] = \pm \beta n.$$

Accordingly, $\dfrac{db}{dl} = \dfrac{d}{dl}[v, n] = \pm\beta n = \kappa n$, where κ is determined from this equality.

Thus $db/dl = \kappa n$. The number κ is called *torsion*. If $v = e_1$, $n = e_2$, $b = e_3$, we have:

$$\frac{de_i}{dl} = b_i^j e_j \qquad (i, j = 1, 2, 3).$$

Next,

$$\frac{dv}{dl} = \frac{de_1}{dl} = kn = ke_2 \qquad (k = b_1^2,\ b_1^1 = b_1^3 = 0),$$

$$\frac{db}{dl} = \frac{de_3}{dl} = \kappa n = \kappa e_2 \qquad (\kappa = b_3^2,\ b_3^3 = b_3^1 = 0).$$

We shall calculate dn/dl directly. Since $n' \perp n$, it follows that $\dfrac{dn}{dl} = \alpha v + \beta b$. Note that $n = -[v, b]$, and, therefore:

$$\frac{d|v, b|}{dl} = [v', b] + [v, b'] = [kn, b] + [v, \kappa n] =$$

$$= k[nb] + \kappa[v, n] = kv + \kappa b,$$

as required.

Now using Lemma 2 of Section 1.5, we shall give another proof of this theorem. For the matrix $B = (b_i^j)$ we have:

$$B = \begin{pmatrix} 0 & k & 0 \\ ? & ? & ? \\ 0 & \kappa & 0 \end{pmatrix}.$$

Recall that if $A(t)$ is an orthogonal matrix, $A(0) = E$, then $\left.\frac{dA}{dt}\right|_{t=0} = B$ is skew-symmetric.

If we make use of this fact, we can fill in all the question marks in the matrix B:

$$B = \begin{pmatrix} 0 & k & 0 \\ ? & ? & ? \\ 0 & \kappa & 0 \end{pmatrix} = \begin{pmatrix} 0 & k & 0 \\ -k & 0 & -\kappa \\ 0 & \kappa & 0 \end{pmatrix}.$$

This implies that $dn/dl = -kv - \kappa b$ if $dv/dl = kn$ and $db/dl = \kappa n$.

EXERCISE. Suppose we are given a helical line

$$x = x_0 + R \sin t, \quad y = y_0 + R \cos t, \quad z = at.$$

Write this curve in terms of the natural parameter l and calculate the curvature and the torsion. It turns out that the curvature and the torsion are constant along this curve.

We have deduced the main facts from the theory of flat and space curves.

In conclusion, we should explain in what sense the curvature and torsion of a curve in Euclidean space make up a complete set of geometric invariants for the given curve.

In the Euclidean plane we have: given the dependence $\hat{k} = \hat{k}(l)$, the curve is uniquely restored to an accuracy of motion of the entire space. The function (equality) $\hat{k} = \hat{k}(l)$ is sometimes called "the natural equation of a planar curve". Here $\hat{k}$ is the curvature with sign, which has been defined in Section 1.5.

In the three-dimensional case we have: given the functions $k = k(l)$, $\kappa = \kappa(l)$, we can reconstruct the curve to an accuracy of motion of the entire space as a rigid integer. This pair of equalities is called "the natural equation of the space curve".

The next topic to be considered is the theory of surfaces in the three-dimensional space.

1.7 The Theory of Surfaces in Three-dimensional Space. Introduction

In this section we shall be concerned with the ways of setting a surface, the Riemannian metric (the first quadratic form)[1]. In the three-dimensional case, the surface is the simplest object with, so to say, intrinsic geometry. What do we mean by this?

We have investigated curves and their metric invariants in a plane and space. These invariants (*normal, binormal, curvature and torsion, Frenet formulae*) depend, however, on the manner the curve is embedded in space, and in this sense they are invariants only of the way of embedding, the shape of the curve, i.e. are concepts of *extrinsic geometry*. A curve has no *intrinsic metric invariants*: obviously, we can coordinatize a curve with the natural parameter l, such that the lengths (on the curve) between two points along both the curve and the straight line are measured in the same way, that is:

$$l = \int_{t_0}^{t} |v_t|\, dt, \quad v_t = \dot{r} = (\dot{x}, \dot{y}, \dot{z}).$$

For surfaces this is not the case: for instance, it is impossible to coordinatize a sphere (or even a piece of a sphere) in such a way that the formula for the curve lengths in these coordinates be the same as those in Cartesian coordinates x, y in the Euclidean plane.

What is the way to determine a surface? There exist three ways for a surface to be specified:

1) as the *graph of a function* (the simplest case)

$$z = f(x, y),$$

2) as the graph of an *equation* (a more general case)

$$F(x, y, z) = \text{const.},$$

3) by *parametric* equations (similar to those for a curve) $r = r(u, v)$:

$$x = x(u, v), \quad y = y(u, v), \quad z = z(u, v),$$

where u, v are running over a certain region in the plane (u, v).

---

[1] In Western literature the first (second) quadratic form is often termed the first (second) fundamental form.

DEFINITION 1. We say that the equation $F(x, y, z) =$ const. *determines a surface which is non-singular at a point* $P(x_0, y_0, z_0)$, where $F(x_0, y_0, z_0) =$ const. if the gradient of the function F at the point P is non-zero, i.e.

$$\frac{\partial F}{\partial x}e_1 + \frac{\partial F}{\partial y}e_2 + \frac{\partial F}{\partial z}e_3 \neq 0 \quad (x = x_0,\ y = y_0,\ z = z_0).$$

By the implicit function theorem if $\left.\frac{\partial F}{\partial z}\right|_{x_0, y_0, z_0} \neq 0$, then near the point $(x_0, y_0, z_0) = P$ the equation $F(x, y, z) = C$ can be solved for z: $z = f(x, y)$, where $z_0 = f(x_0, y_0)$, and in a certain neigbourhood of the point (x_0, y_0) in the (x, y)-plane, $F(x, y f(x, y)) \equiv C$. We shall obtain $F(x, y, z) = C$, $\frac{\partial F}{\partial x}dx + \frac{\partial F}{\partial y}dy + \frac{\partial F}{\partial z}dz = 0$, whence:

$$\frac{\partial f}{\partial x} = -\frac{\partial F/\partial x}{\partial F/\partial z}, \quad \frac{\partial f}{\partial y} = -\frac{\partial F/\partial y}{\partial F/\partial z}.$$

Consequently, for the surface $F(x, y, z) = C$ in a neighbourhood of the point (x_0, y_0, z_0) the parametric representation $z = f(u, v)$, $x = u$, $y = v$ (near the point $x_0 = u_0$, $y_0 = v_0$) holds. We can see that *locally, near a non-singular point on the surface, the surface can always be given parametrically*. In other words, the surface near a (non-singular) point can be parametrized by *local regular* coordinates u, v.

Inversely, let a surface be given parametrically: $r = r(u, v)$, i.e.

$$x = x(u, v), \quad y = y(u, v), \quad z = z(u, v).$$

DEFINITION 2. A point $P = (x_0, y_0, z_0) = (x(u_0, v_0),\ y(u_0, v_0), z(u_0, v_0))$ on the surface is called *non-singular* if the matrix:

$$A = \begin{pmatrix} \frac{\partial x}{\partial u} & \frac{\partial y}{\partial u} & \frac{\partial z}{\partial u} \\ \frac{\partial x}{\partial v} & \frac{\partial y}{\partial v} & \frac{\partial z}{\partial v} \end{pmatrix}_{u=u_0,\ v=v_0}.$$

has rank 2.

THEOREM 1. *If a surface is given parametrically and if the point $P = (u_0, v_0)$ is non-singular, then near this point the surface can be given by the equation $F(x, y, z) = 0$, where $F(x_0, y_0, z_0) = 0$ and $F\big|_{(x_0, y_0, z_0)} \neq 0$, i.e. both the definitions of non-singular points are equivalent.*

Proof. By the definition of non-singular point, the rank of the matrix A equals 2.

Let, for the sake of definiteness, the determinant have the form:

$$\frac{\partial x}{\partial u}\frac{\partial y}{\partial v} - \frac{\partial x}{\partial v}\frac{\partial y}{\partial u} \neq 0.$$

Recall the inverse function theorem: let $x = x(u, v)$, $y = y(u, v)$, let the Jacobian at the point (u_0, v_0) be non-zero and let $x = x(u_0, v_0)$, $y_0 = y(u_0, v_0)$; then at a certain neighbourhood of the point (x_0, y_0) we can find the inverse function:

$$u = u(x, y), \quad u_0 = u(x_0, y_0),$$

$$v = v(x, y), \quad v_0 = v(x_0, y_0),$$

the matrix $\begin{pmatrix} \frac{\partial u}{\partial x} & \frac{\partial v}{\partial x} \\ \frac{\partial u}{\partial y} & \frac{\partial v}{\partial y} \end{pmatrix}$ being the inverse to the matrix $\begin{pmatrix} \frac{\partial x}{\partial u} & \frac{\partial y}{\partial u} \\ \frac{\partial x}{\partial v} & \frac{\partial y}{\partial v} \end{pmatrix}$.

By virtue of this theorem, we shall find the expression $u = u(x, y)$, $v = v(x, y)$ and substitute them into the expression for $z = z(u, v) = z(u(x, y), v(x, y))$, where $z = z(u_0, v_0) = z(u(x_0, y_0), v(x_0, y_0))$. We obtain the expression for the surface in the form $z = f(x, y)$, $z_0 = f(x_0, y_0)$ near the non-singular point (x_0, y_0, z_0), which completes the proof of the theorem.

Thus, we have arrived at the conclusion that *locally, in the neighbourhood of a non-singular point $P = (x_0, y_0, z_0)$ on the surface, all the three ways of defining surfaces (by smooth functions) are equivalent.*

EXAMPLES

1. The *ellipsoid* $x^2/a^2 + y^2/b^2 + z^2/c^2 = 1$ (Figure 14).
This surface

a) has no singular points,
b) is not globally the graph of a function (whereas locally it is),
c) has no global parametrization (such that all the points are non-singular).

2. The *one-sheeted hyperboloid* $x^2/a^2 + y^2/b^2 - z^2/c^2 = 1$ (Figure 15).

This surface

a) is not globally the graph of a function,
b) can globally be given parametrically.

3. The *two-sheeted hyperboloid* $-x^2/a^2 - y^2/b^2 + z^2/c^2 = 1$ (Figure 16).

This surface is such that one half of it can be given both as the graph $z = f(x, y)$ and parametrically.

4. The *cone* $x^2/a^2 + y^2/b^2 - z^2/c^2 = 0$ (Figure 17).

Here the point (0, 0, 0) is singular.

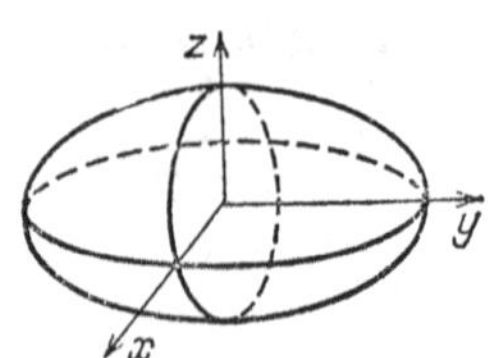

Figure 14.

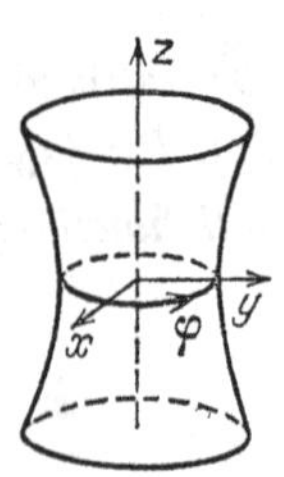

Figure 15.

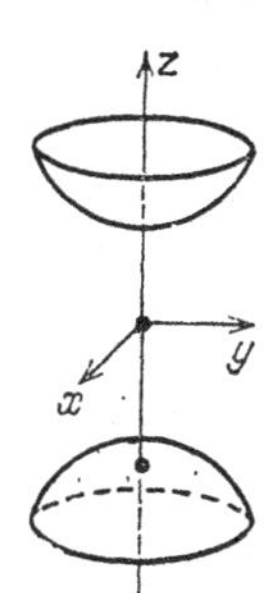

Figure 16.

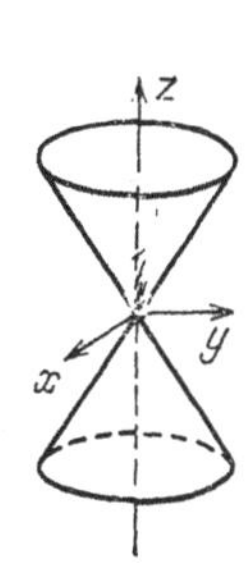

Figure 17.

We now turn to the *intrinsic geometry* on the surface itself.

We have already made some introductory remarks concerning the curvature of the surface. Let us return to this concept. Suppose we are given a surface with (x_0, y_0, z_0) as a non-singular point on it. Choose an orthonrmal frame, where z is normal to the surface while x and y are tangent to it. Then locally near the point (x_0, y_0, z_0) the surface is given in the form $z = f(x, y)$, $z_0 = f(x_0, y_0)$, and

$$\frac{\partial f}{\partial x}\bigg|_{x_0,y_0} = \frac{\partial f}{\partial y}\bigg|_{x_0,y_0} = 0.$$

Consider the second differential of the function $z = f(x, y)$ or

$$2d^2 f = \frac{\partial^2 f}{\partial x^2}(dx)^2 + 2\frac{\partial^2 f}{\partial x\,\partial y}dx\,dy + \frac{\partial^2 f}{\partial y^2}(dy)^2$$

and construct the matrix $a_{ij} = \dfrac{\partial^2 f}{\partial x^i\,\partial x^j}$, where $x^1 = x, x^2 = y$ (the matrix is called *hessian*). We shall view this matrix as quadratic at the point (x_0, y_0, z_0), where $\partial f/\partial x = 0$, $\partial f/\partial y = 0$.

DEFINITION 3. *Principal curvatures of the surface* are eigenvalues of the matrix a_{ij} at the point where the surface is given in the form $z = f(x, y)$ and grad $f = 0$.

The *Gaussian curvature* is the determinant of the matrix (a_{ij}) at this point, and the *mean curvature* is the trace of the matrix at this point. The trace is therefore equal to $a_{11} + a_{22} = k_1 + k_2$, where k_1 and k_2 are eigenvalues; $\kappa = k_1 k_2 = \det(a_{ij}) = a_{11}a_{22} - a_{12}^2$ (*Gaussian curvature*).

The *Gaussian curvature of the surface turns out to depend only on the intrinsic metric properties of the surface* (this assertion is proved below).

We have defined the concept of curvature in special coordinates associated with the point under investigation: the z-axis is normal to the surface, while the x- and y-axes are tangent to the surface at this point. Then locally the surface is written in the form $z = f(x, y)$, grad f being equal to zero at the point (x_0, y_0). Such coordinates can, of course, always be chosen.

We now turn to a consistent presentation of the basic concepts of intrinsic geometry of the surface. To begin with, we introduce the Riemannian metric on the surface; all intrinsic geometric invariants will be expressed in terms of this metric. In this connection it is instructive to recall the concept of Riemannian metric.

Let a surface (or a piece of this surface) be given parametrically: $r = r(u, v)$, $r = (x, y, z)$, where (u, v) are the coordinates of a point on the surface. The point will be thought of as non-singular, i.e. the rank of the matrix $A = \begin{pmatrix} x_u & y_u & z_u \\ x_v & y_v & z_v \end{pmatrix}$ is 2, where $x_u = \frac{\partial x}{\partial u}$, $x_v = \frac{\partial x}{\partial v}$, etc.

How have we defined the Riemannian metric? Suppose we are given the curve $u = u(t)$, $v = v(t)$.

What is the length of the curve $l = \int_a^b |v_t|\, dt$, $v_t = (\dot{u}, \dot{v})$? Here $v_t = (\dot{u}, \dot{v})$ is the velocity vector in the coordinates u, v and

$$|v_t|^2 = g_{ij}\dot{x}^i \dot{x}^j, \quad x^1 = u, \quad x^2 = v, \quad g_{ij} = g_{ij}(u, v).$$

We have called the family of functions $g_{ij}(u, v)$ (in the coordinates (u, v) the Riemannian metric; it determines the arc length, as well as the angles between two curves at the point where they intersect.

How shall we define the arc length? What are the values of $g_{ij}(u, v)$? (Here $u = x^1$, $v = x^2$).

Note that the curve $u = u(t)$, $v = v(t)$ is given in the coordinates (u, v) on the surface, but the surface itself lies in a Euclidean 3-space (x, y, z), where $r = r(u, v)$, $r = (x, y, z)$. Naturally, the *arc (curve segment) length $u = u(t)$, $v = v(t)$ on the surface is understood as the length of this arc in a three-dimensional Euclidean space.* Recall that the surface lies in the space.

We shall write the curve in the form:

$$x = (u(t), v(t)) = x(t),$$

$$y = y(u(t), v(t)) = y(t),$$

$$z = z(u(t), v(t)) = z(t).$$

For the arc length in a three-dimensional Euclidean space we have

$$l = \int_a^b \left(\dot{x}^2 + \dot{y}^2 + \dot{z}^2\right)^{1/2} dt \qquad \text{(by definition).}$$

Since $\dot{x} = \frac{\partial x}{\partial u}\dot{u} + \frac{\partial x}{\partial v}\dot{v}$, etc., then we come to

$$\dot{x}^2 + \dot{y}^2 + \dot{z}^2 = E\dot{u}^2 + 2F\dot{u}\dot{v} + G\dot{v}^2 = g_{ij}\dot{x}^i\dot{x}^j,$$

where

$$E = g_{11}, \quad F = g_{12} = g_{21}, \quad G = g_{22}; \quad u = x^1, \quad v = x^2,$$

and

$$g_{11} = x_u^2 + y_u^2 + z_u^2 = E,$$

$$g_{12} = x_u x_v + y_u y_v + z_u z_v = F,$$

$$g_{22} = x_v^2 + y_v^2 + z_v^2 = G,$$

$$r_u = x_u e_1 + y_u e_2 + z_u e_3 = r_{x^1},$$

$$r_v = x_v e_1 + y_v e_2 + z_v e_3 = r_{x^2},$$

$$g_{ij} = r_{x^i} r_{x^j} \quad (i = 1, 2; \; j = 1, 2).$$

The functions $g_{ij}(u, v) = (E, F, G)$ are defined in the coordinates of the surface.

The expression $g_{ij}\, dx^i\, dx^j = E(du)^2 + 2F\, dudv + G(dv)^2$ is usually called the *first quadratic form* (or the Riemannian metric on the surface). The first quadratic form is a quadratic form defined on tangent vectors to the surface at a given point.

If the surface is given in the form $F(x, y, z) = C$, then the Riemannian metric on the surface (the first quadratic form) is:

$$(dx)^2 + (dy)^2 + (dz)^2$$

under the condition that $F(x, y, z) = C$. This implies that:

$$F_x\,dx + F_y\,dy + F_z\,dz = 0 \quad (F_x = \partial F/\partial x, \dots).$$

If at the point under investigation $\partial F/\partial z \neq 0$, then:

$$dz = -\frac{F_x}{F_z}dx - \frac{F_y}{F_z}dy = A\,dx + B\,dy \quad \text{and } z = f(x, y).$$

Therefore, on the surface $F(x, y, z) = C$, $x = u$, $y = v$

$$(dx)^2 + (dy)^2 + (dz)^2 = (dx)^2 + (dy)^2 + \left(\frac{F_x}{F_z}dx + \frac{F_y}{F_z}dy\right)^2,$$

$$g_{11} = E = 1 + \frac{F_x^2}{F_z^2}, \quad g_{12} = F = \frac{F_xF_y}{F_z^2}, \quad g_{22} = G = 1 + \frac{F_y^2}{F_z^2},$$

where $u = x = x^1$, $v = y = x^2$. If the surface has the form $z = f(x, y)$, then $g_{11} = 1 + f_x^2$, $g_{12} = f_xf_y$, $g_{22} = 1 + f_y^2$.

Thus, the Riemannian metric on the surface appears here as the *way to calculate the lengths of curve segments* in the coordinates u, v which describe points of the surface. Since the surface lies in a three-dimensional (Euclidean) space, we deal here simply with the length of this curve segment in a three-dimensional Euclidean space.

Let Euclidean coordinates x, y, z be given in the form of the function $x = x(u, v)$, $y = y(u, v)$, $z = z(u, v)$ on the surface. By definition:

$$(dx)^2 + (dy)^2 + (dz)^2 = g_{ij}dx^i\,dx^j,$$

$$x^1 = u, \quad x^2 = v; \quad g_{11} = E, \; g_{12} = F, \; g_{22} = G.$$

The metric $g_{ij}(x^1, x^2)$ on the surface is said to be *Euclidean* if there exists a pair of functions on the surface $\bar{u}(x^1, x^2)$, $\bar{v}(x^1, x^2)$, such that $(d\bar{u})^2 + (d\bar{v})^2 = g_{ij}\,dx^i\,dx^j$. For the surface in a *three-dimensional* Euclidean space we have:

$$(dx)^2 + (dy)^2 + (dz)^2 = \sum g_{ij} dx^i dx^j,$$

$$x = x(u, v), \ y = y(u, v), \ z = z(u, v), \ u = x^1, \ v = x^2,$$

i.e. the metric is decomposed into the sum of squared differentials of *three* functions $x(u, v)$, $y(u, v)$, $z(u, v)$ rather than two (generally speaking). If the metric on the surface decmposes into the sum of squared differentials of two fuctions, these two functions are Euclidean coordinates on the surface, and the metric on it is Euclidean.

EXAMPLE. The metric of a cylinder is Euclidean. The equation of the cylinder has the form $f(x, y) = C$ (the coordinate z does not enter). Euclidean coordinates on a cylinder are coordinates z and the natural parameter l of a flat curve $f(x, y) = C$, $\bar{u} = z$, $\bar{v} = l$.

We have:

$$(dx)^2 + (dy)^2 + (dz)^2 = (dz)^2 + (dl)^2.$$

1.8 The Theory of Surfaces. Riemannian Metric and the Concept of Area

In Section 1.7 we introduced three ways in which a surface may be given in a three-dimensional space: by the equation $z = f(x, y)$ or $F(x, y, z) = C$, grad $F \neq 0$, and parametrically by $r = r(u, v)$, $r = (x, y, z)$, where u, v are parameters. Tangent vectors to a surface are the velocity vectors of the curves lying on the surface.

At a non-singular point, where the rank of the matrix $\begin{pmatrix} x_u & y_u & z_u \\ x_v & y_v & z_v \end{pmatrix}$ is equal to 2, these three ways are *locally equivalent.*

We have also defined the Riemannian metric on the surface (given parametrically)

$$(dl)^2 = (dx)^2 + (dy)^2 + (dz)^2 = g_{ij}\, dx^i\, dx^j,$$

where $x^1 = u$, $x^2 = v$, $g_{11} = F$ $g_{12} = g_{21} = F$, $g_{22} = G$. The Riemannian metric has been defined to meet the requirement that for any curve on the surface $u = u(t)$, $v = v(t)$ its length on the surface $l = \int_a^b \left(g_{ij}\dot{x}^i\dot{x}^j\right)^{1/2} dt$ coincides with its length in the three-dimensional Euclidean metric of an envelope space:

$$l = \int_a^b \left(\dot{x}^2 + \dot{y}^2 + \dot{z}^2\right)^{1/2} dt,$$

where

$$x(t) = x(u(t), v(t)), \quad y(t) = y(u(t), v(t)),$$

$$z(t) = z(u(t), v(t)), \quad u = x^2, \quad v = x^2.$$

Thus, by definition:

$$(dx)^2 + (dy)^2 + (dz)^2 = g_{ij}\, dx^i\, dx^j,$$

where

$$dx = x_u\, du + x_v\, dv,$$

$$dy = y_u\, du + y_v\, dv,$$

$$dz = z_u\, du + z_v\, dv,$$

($x_u = \partial x/\partial u, \ldots$). If the surface is given in the form $z = f(x, y)$, then $x = u, y = v$, and therefore

$$(dl)^2 = (1 + f_x^2)(dx)^2 + 2f_xf_y\, dxdy + (1 + f_y^2)(dy)^2$$

or

$$g_{11} = 1 + f_x^2, \quad g_{12} = g_{21} = f_xf_y, \quad g_{22} = 1 + f_y^2 .$$

If the surface is given in the form $F(x, y, z) = C$ and grad $F \neq 0$, then on the surface:

$$dF = F_x\, dx + F_y\, dy + F_z\, dz = 0.$$

In the case $F_z \neq 0$, we can take $u = x, v = y$

$$dz = A\, dx + B\, dy, \; A = -\frac{F_x}{F_z}, \quad B = -\frac{F_y}{F_z},$$

whence $(dl)^2 = (dx)^2 + (dy)^2 + (A\, dx + B\, dy)^2$ (on the surface). The vector ξ is called a *tangent* vector to a surface at a point P if it is the velocity vector of a certain curve lying on the surface which passes through the point P.

If $\xi = \xi^1 e_1 + \xi^2 e_2 + \xi^3 e_3$ is a tangent vector to the surface then we have $F_x\xi^1 + F_y\xi^2 + F_z\xi^3 = 0$ or $\xi \perp$ grad F. From this, we come to the conclusion that the *vector* grad F is *normal to the surface* $F(x, y, z) = C$. When the surface is given parametrically by $r = r(u, v)$, $r\,(x, y, z)$, we have two vectors:

$$\xi = r_u = r_{x1} = (x_u e_2 + y_u e_2 + z_u e_3),$$

$$\eta = r_v = r_{x2} = (x_v e_2 + y_v e_2 + z_v e_3).$$

These are both tangent vectors to the surface. If they are linearly independent, their vector product $[\xi, \eta] = [r_u, r_v]$ is orthogonal to the plane r_u, r_v, that is, gives a normal vector to the surface.

With the Riemannian metric we can measure the length of any curve $u = f(t)$, $v = g(t)$ on the surface and the angle between two curves at the point of their intersection, since the scalar product of the velocity vectors $(\dot{f}_1, \dot{g}_1)$ and $(\dot{f}_2, \dot{g}_2)$ at the intersection point (u_0, v_0) of the curves $(f_1(t), g_1(t))$, $(f_2(t), g_2(t))$, $u = f_i(t)$, $v = g_i(t)$, $i = 1, 2$, is given by the formula (at the point $u_0 = f_i(t_0)$, $v_0 = g_i(t_0)$):

$$\eta_1\eta_2 = g_{ij}\eta_1^i\eta_2^j, \quad g_{ij} = g_{ij}(u_0, v_0),$$

$$\eta_1 = (\dot{f}_1, \dot{g}_1), \quad \eta_2 = (\dot{f}_2, \dot{g}_2).$$

Another question arises: *How shall we measure the area of a region on a surface?*

If we have a Euclidean plane with coordinates x, y and a *region* U in this plane, then the area of U is given, as we know, by the double integral $\sigma(U)$, i.e. the area of the region has the form $\sigma(U) = \iint_U dx\, dy$. If we make a (one-to-one) change of variables:

$$x = x(u, v), \quad y = y(u, v),$$

we come to the formula:

$$\sigma(U) = \iint_V |x_u y_v - y_u x_v|\, du\, dv,$$

where V is a region in the (u, v)-plane corresponding to the region U in the (x, y)-plane (see the theorem in analysis on the change of variables in a double integral).

Thus, we have:

$$\sigma(U) = \iint_V |J|\, du\, dv,$$

where J is the Jacobian of the change of variables $x = x(u, v)$, $y = y(u, v)$

$$J = x_u x_v.$$

A question arises: How shall we calculate the area of a region on the surface $r = r(u, v)$, $r = (x, y, z)$ in a space if we know the Riemannian metric on the surface itself:

$$(dl)^2 = g_{ij}\, dx^i\, dx^j, \quad x^1 = u, \quad x^2 = v.$$

Let us consider the determinant of the matrix:

$$\det(g_{ij}) = g = g_{11}g_{22} - g_{12}^2 = EG - F^2 > 0.$$

DEFINITION 1. The *area of a region U* on the surface $r = r(, v)$, $r = (x, y, z)$ is the expression:

$$\sigma(U) = \iint_V (g)^{1/2}\, du\, dv,$$

where U is the region on the surface given parametrically as the image of the region V in the (u, v)-plane.

So as not to introduce some special notation, we shall, in some cases, write

$$\sigma(U) = \iint_U (g)^{1/2}\, du\, dv$$

implying that U is the image of V.

The expression $(g)^{1/2}\, du\, dv$ is called the *differential of area* with Riemannian metric (g_{ij}).

EXAMPLES.

1. If the metric is Euclidean, $(dl)^2 = (du)^2 + (dv)^2$,

$$g_{ij} = \delta_{ij} = \begin{cases} 1, & i = j, \\ 0, & i \neq j, \end{cases}$$

then $g = 1$ and $(g)^{1/2} = 1$. Therefore $(g)^{1/2}\, du\, dv = du\, dv$.

2. For the metric of the sphere $(dl)^2 = R^2(du)^2 + R^2 \sin^2 u (dv)^2$,

$$g_{ij} = \begin{pmatrix} R^2 & 0 \\ 0 & R^2 \sin^2 u \end{pmatrix}, \quad (g)^{1/2} = R^2 |\sin u|$$

we have

$$\sigma(U) = \iint_U R^2 \sin u\, du\, dv \quad \text{and} \quad 0 \leq u \leq \pi/2.$$

If the sphere is given in the form $x^2 + y^2 + z^2 = R^2$, then $u = \theta$, $v = \phi$,

$$(dl)^2 = R^2\,[(d\theta)^2 + \sin^2\theta (d\phi)^2], \quad z = R\cos\theta,$$

$$y = R\sin\theta\sin\phi, \quad x = R\sin\theta\cos\phi.$$

Therefore, $\sigma(U) = \iint_U R^2 \sin\theta \, d\theta \, d\phi$. We can readily see that the total area of the sphere is equal to $4\pi R^2$.

On what grounds do we define the area of a region in such a manner? Why is the element of area to be taken in the form $(g)^{1/2}\, du\, dv$ if the scalar product of tangent vectors at the point (u, v) is given by the matrix $g_{ij}(u, v)$?

To gain a deeper inight into this problem, we shall consider a pair of vectors of the Euclidean plane ξ, η and a parallelogram $(\lambda\xi + \mu\eta)$, $0 \le \lambda \le 1, 0 \le \mu \le 1$.

The area of the parallelogram is equal to $\sigma = |\xi^1\eta^2 - \xi^2\eta^1| = |\det A|$

$$A = \begin{pmatrix} \xi^1_{\ 1} & \xi^2_{\ 2} \\ \eta^1 & \eta^2 \end{pmatrix},$$

i.e. the matrix A is formed by the components of the vectors:

$$\xi = \xi^1 e_1 + \xi^2 e_2, \quad \eta = \eta^1 e_1 + \eta^2 e_2.$$

Let us now consider another example. Suppose we are given a plane with coordinates (u, v), where any vector has the form $ue_1 + ve_2$ and e_1, e_2 are basis vectors.

Suppose the scalar product of the basis vectors is given by the matrix:

$$e_i e_j = g_{ij}; \ i = 1, 2; \ \ j = 1, 2.$$

Calculate the area of the parallelogram spanned by the vectors e_1 and e_2. The points of the parallelogram are $\lambda e_1 + \mu e_2$, where $0 \le \lambda \le 1, 0 \le \mu, \lambda \le 1$.

If $g_{ij} = \delta_{ij}$, the area of the parallelogram is equal to 1 (unit square).

We assume the matrix (g_{ij}) to be the matrix of positive quadratic form:

$$g_{ij}\, \xi^i \xi^j > 0.$$

PROPOSITION 1. *The area of a parallelogram* $|e_1, e_2|$ *spanned by the vectors* e_1 *and* e_2 *is equal to* $(g)^{1/2}$ *where* $g = \det(g_{ij}) = g_{11}g_{22} - g_{12}g_{21} = g_{11}g_{22} - g^2_{21}$.

Proof. The quadratic form g_{ij} can be reduced to a diagonal form $g'_{ij} = \delta_{ij}$ through a linear transformation A.. More precisely this means the following: there exist vectors e'_1, e'_2

$$e'_1 = a^1_1 e_1 + a^2_1 e_2, \quad e'_2 = a^1_2 e_1 + a^2_2 e_2,$$

such that $e'_i e'_j = g'_{ij} = \delta_{ij}$ (or $|e'_i|^2 = 1$, $e'_1 \perp e'_2$). Then the area of a parallelogram spanned by the vectors e'_1, e'_2 is equal to 1 (unit square).

Since $e'_1 = a_1^1 e_1 + a_1^2 e_2 = a_1^i e_i$, $e'_2 = a_2^1 e_1 + a_2^2 e_2 = a_2^i e_i$, it follows that:

$$g_{11} = e_1 e_2 = (a_1^1)^2 + (a_1^2)^2,$$

$$g_{12} = g_{21} = e_1 e_2 = a_1^1 a_2^1 + a_2^2 a_1^2,$$

$$g_{22} = (a_2^1)^2 + (a_2^2)^2.$$

In the matrix language, we have $(g_{ij}) = A \circ A^{\mathrm{T}}$

$$A = \begin{pmatrix} a_1^1 & a_1^2 \\ a_2^1 & a_2^2 \end{pmatrix}, \qquad A^{\mathrm{T}} = \begin{pmatrix} a_1^1 & a_2^1 \\ a_1^2 & a_2^2 \end{pmatrix} = (a_j^i),$$

$$A \circ A^{\mathrm{T}} = \begin{pmatrix} g_{11} & g_{12} \\ g_{21} & g_{22} \end{pmatrix}.$$

What is the area of the parallelogram spanned by the vectors e_1 and e_2.

Since the basis e_1, e_2 is orthonormal, the area of the parallelogram $|e_1 e_2|$ is equal to det A, i.e. to the determinant of the matrix A.

1) The determinant of A^{T} = the determinant of A.
2) The determinant of $A \circ A^{\mathrm{T}} = (\det A) \cdot (\det A^{\mathrm{T}}) = (\det A)^2 = g$.

We obtain $(g)^{1/2} = \det A$ = the area of $|e_1 e_2|$, and the assertion follows.

We have used the fact that in an n-dimensional Euclidean space with orthonormal basis $e_1, \dots, e_n$, $e_i e_j = \delta_{ij}$, the *volume of a parallelipiped spanned by the vectors* $e_1, \dots, e_n$, $e_i = a^j{}_i e_j$, $i = 1, \dots, n$, *is equal to the determinant of the matrix* $A = (a^j_i)$ (in the absolute value).

For $n = 2$ this fact is already known, but we leave to the reader to verify that it is valid for all n.

If the matrix of scalar products is given in the form $e_i e_j = g_{ij}$, then the volume of the parallelipiped spanned by the vectors $e_1, \dots, e_n$ is equal to the root of the determinant of the matrix (g_{ij}), i.e.

$$\text{the volume} = (g)^{1/2} = \left(\det\,(g_{ij})\right)^{1/2}.$$

These two facts are the basic ones for some theorems from mathematical analysis and geometry:

1) In the Euclidean plane ($n = 2$) after the change of coordinates ($i = 1, 2$) $x^i = x^i(z^1, z^2)$ the area of a region is calculated by the formula:

$$\sigma(U) = \iint_U dx^1\,dx^2 = \iint_V |J|\,dz^1\,dz^2, \quad \hat{J} \circ \hat{J}^{\mathrm{T}} = (g_{ij});$$

2) In the Riemannian metric (g_{ij}) the element of area has the form:

$$\sigma(U) = \iint_U (g)^{1/2} du\,dv.$$

REMARK. The matrix $\hat{J} = \begin{pmatrix} \dfrac{\partial x^1}{\partial z^1} & \dfrac{\partial x^2}{\partial z^1} \\ \dfrac{\partial x^1}{\partial z^2} & \dfrac{\partial x^2}{\partial z^2} \end{pmatrix}$ is the Jacobian matrix.

The Jacobian J is equal to the determinant of the Jacobian matrix $\hat{J}$.

1.9 The Theory of Surfaces. The Area of a Region on the Surface

Consider a region U in a plane with coordinates $x^1 = u, x^2 = v$, bounded by a curve Γ (Γ is the boundary of the region U). We assume the curve Γ to be continuous, and furthermore, to be piecewise smooth. This means that the curve consists of several smooth pieces $\gamma_1, \dots, \gamma_k$, the end of the piece γ_i coinciding with the origin of the piece γ_{i+1} (the end of γ_k coinciding with the origin of γ_1). (Figure 18).

The pieces γ_i can be given parametrically:

$$\gamma_i = \begin{cases} u = u_i(t), \\ v = v_i(t), \end{cases} \qquad a_i \le t \le b_i,$$

and $u_i(b_i) = u_{i+1}(a_{i+1}), v_i(b_i) = v_{i+1}(a_{i+1})$ (the origin of γ_{i+1} coincides with the end of γ_i). $\gamma_1(a_1) = \gamma_k(b_k)$.

EXAMPLE. (Figure 19).

$$\begin{aligned} &\gamma_1\colon\ v = a, \quad u_1 \le u \le u_2, \\ &\gamma_2\colon\ u = f(v), \quad a \le v \le b, \\ &\gamma_3\colon\ v = b, \quad u_4 \le u \le u_3, \\ &\gamma_4\colon\ u = g(v), \quad a \le v \le b, \\ &\Gamma = \gamma_1 \gamma_2 \gamma_3 \gamma_4, \end{aligned}$$

$f(v)$ and $g(v)$ are continuous functions, $f > g$.

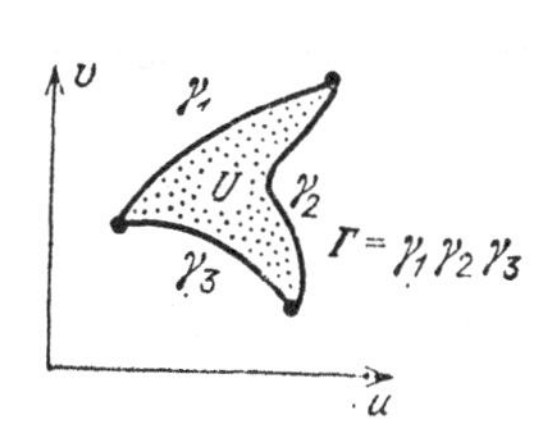

Figure 18.

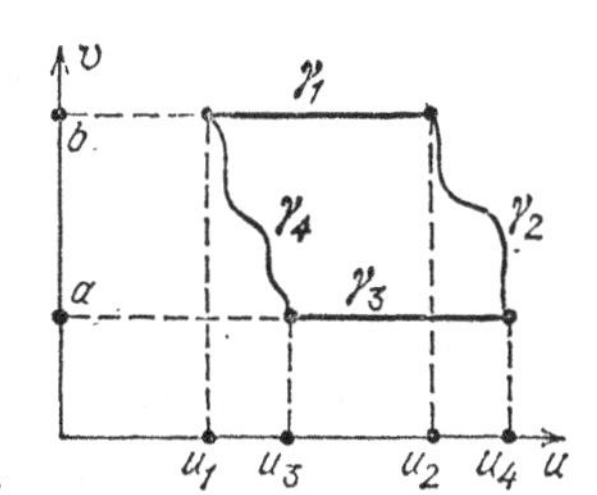

Figure 19.

Consider small numbrs Δu and Δv and partition the plane into rectangles with sides Δu and Δv.

We assume the umbers Δu and Δv to be sufficiently small and to tend to zero.

Obviously, we have: the area of the region U is greater than or equal to the sum of the areas of all interior rectangles. The area of U is equal to $\sigma(U) \sum_i S_i$, where S_i is the area of the interior rectangles indexed by the number i.

DEFINITION 1. The *area of the regioin U* is the limit of the sum of the areas of all interior rectangles as $\Delta u \to 0$ and $\Delta v \to 0$ if this limit exists.

Suppose next that we are given, in addition, a continuous function of two variables $f(u, v)$. We shall now recall the definition of the *integral* of the function $f(u, v)$ over the region U.

Consider all the interior (for the region) rectangles with sides Δu and Δv from a rectangular net. In a rectangle S_i we consider the value of $f(u_i, v_i)$ at the centre of the rectangle. Consider the integral sum:

$$S(f, U) = \sum_i f(u_i, v_i)(\Delta u)(\Delta v),$$

where the sum is taken over all the interior rectangles.

DEFINITION 2. The limit of the sums $S(U, f)$ as $\Delta u \to 0$, $\Delta v \to 0$ is called the *double integral* of the function $f(u, v)$ over the region U and is denoted by

$$\iint_U f(u, v)\, du\, dv.$$

We shall list the properties of the double integral (without justification).

1. If the boundary of a region is a continuous curve Γ without self-intersections, the region U is bounded and the function $f(u, v)$ is continuous, then the limit of the integral sumes exists and coincides also with the limit over all the rectangles intersecting the region U ("existence of the integral").

2. If $f(u, v) \equiv 1$ with (u, v) being Cartesian Euclidean coordinates, then the integral $\iint_U 1 \cdot du\, dv$ coincides with the area of the region U. recall that in Euclidean coordinates the square of the element (the differential) of the arc length has the form:

$$(dl)^2 = (du)^2 + dv)^2, \quad g_{ij} = \delta_{ij} = \begin{cases} 1, & i=j, \\ 0, & i \neq j. \end{cases}$$

In this case, the area of a rectangle with sides Δu and Δv is equal to $(\Delta u)(\Delta v)$. the curves $u=$ const. and $v =$ const. intersect at an angle which, in this case, is actually right.

3. If in the plane $u = x^1$, $v = x^2$ we are given a Riemannian metric:

$$(dl)^2 = \sum g_{ij}\, dx^i\, dx^j = g_{11}(du)^2 + 2g_{12}\, du\, dv + g_{22}(dv)^2,$$

then the area of the region U is equal to the integral:

$$\sigma(U) = \iint_U (g)^{1/2} du\, dv \qquad \text{(definition!)}.$$

Why is this definition natural?

If Δu and Δv are sufficiently small, the area of a small parallelogram centred at the point (u_α, v_α) with sides Δu and Δv is equal, approximately, to $S_i \cong (\Delta u)(\Delta v)$ $(g)^{1/2}$ where $(g)^{1/2} = \left(g_{11}g_{22} - g_{12}^2\right)^{1/2} = \left(EG - F^2\right)^{1/2}$, the numbers g_{ij} being calculated at the point (u_α, v_α).

For small Δu, Δv we have;

$$(\text{the area of } U) = \sigma(U) = \sum_\alpha S_\alpha = \sum_\alpha \left(g(u_\alpha, v_\alpha)\right)^{1/2} (\Delta u)(\Delta v).$$

The limit of these sums (as $(\Delta u) \to 0$, $(\Delta v) \to 0$) is called the *integral*

$$\iint_U (g)^{1/2} du\, dv\ .$$

As in the case of the curve length, the expression $\iint_U (g)^{1/2} du\, dv$ is taken, in effect, as the *definition of the area of a region.*

In the notation $g_{11} = E$, $g_{12} = F$, $g_{22} = G$ we have:

$$\sigma(U) = \iint_U (EG - F^2)^{1/2} du\, dv\ .$$

4. Under a one-to-one change of variables:

$$x^1 = u = u(z^1, z^2),$$

$$x^2 = v = v(z^1, z^2)$$

such that

$$J = u_{z^1}v_{z^2} - u_{z^2}v_{z^1} \neq 0,$$

$$J = J(z^1, z^2),$$

to the region U in the coordinates (u, v) there corresponds the region V in the coordinates (z^1, z^2) and we have the equality:

$$\iint_U f(u, v)\, du\, dv = \iint_V f(u(z^1, z^2), v(z^1, z^2))\, J(z^1, z^2)\, dz^1 dz^2$$

("the change of variables in the integral").

If the point z^1, z^2 belongs to the region V, then the point $u = x^1(z^1, z^2)$, $v = x^2(z^1, z^2)$ belongs to the region U.

EXAMPLE 1. The *Euclidean plane*.

$$u = r\cos\phi, \quad v = r\sin\phi,$$

$$(dl)^2 = (du)^2 + (dv)^2 = (dr)^2 + r^2(d\phi)^2,$$

$$J = u_r v_\phi - u_\phi v_r = r.$$

Conclusion. $$\iint_U du\, dv = \iint_V r\, dr\, d\phi.$$

Let U be a ring ($0 \le r \le R$, ϕ is arbitrary). We have:

$$\sigma(U) = \iint_{\substack{0\le r\le R\\ 0\le\phi\le 2\pi}} r\, dr\, d\phi = \int_0^R dr \left(\int_0^{2\pi} r\, d\phi\right) = \int_0^R 2\pi r\, dr = \pi R^2.$$

Conclusion. The area of the ring is equal to πR^2.

EXAMPLE 2. A *sphere of radius* r_0:

$$(dl)^2 = r_0^2((du)^2 + \sin^2 u\,(dv)^2), \quad 0 \le u \le \pi,$$

$$(g)^{1/2} = r_0^2 \sin u \quad (u = \theta,\ v = \phi).$$

A ring of radius R, where $R \le \pi r_0$, is a region, where:

$$U_R = \begin{cases} 0 \le u \le R/r_0, \\ \phi \text{ — arbitrary.} \end{cases}$$

The area

$$U_R = \sigma(U) = \iint\limits_{\substack{0 \le u \le R/r_0 \\ 0 \le v \le 2\pi}} r_0^2 \sin u\,du\,dv =$$

$$= r_0^2 \int_0^{R/r_0} du \left(\int_0^{2\pi} \sin u\,dv\right) = \int_0^{R/r_0} r_0^2\, 2\pi \sin u\,du =$$

$$= 2\pi r_0^2 \left(1 - \cos\frac{R}{r_0}\right), \quad R \le \pi r_0 \quad (\theta = u,\ \phi = v).$$

When $R = \pi r_0$ we have that U_R coincides with the whole of the sphere (Figure 20).

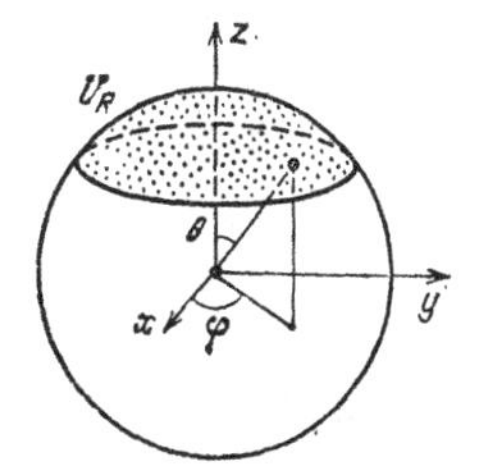

Figure 20.

Conclusion. The area of a sphere is equal to $4\pi r_0^2$ (since $\cos R/r_0 = \cos\pi = -1$).

EXAMPLE 3. The *Lobachevskian plane.*

$$(dl)^2 = r_0^2\,((du)^2 + \mathrm{sh}^2\,\mathrm{iu}\,(dv)^2),$$

$$(g)^{1/2} = r_0^2\,\mathrm{sh}\,u$$

($v = \phi$, where ϕ is the angle).

A ring of radius R is a region

$$U_R = \begin{cases} 0 \le u \le R/r_0, \\ 0 \le v \le 2\pi. \end{cases}$$

For the area we have:

$$\sigma(U) = \iint_U r_0^2\,\mathrm{sh}\,u\,du\,dv = \iint_{\substack{0\le u\le R/r_0 \\ 0\le v\le 2\pi}} (du\,dv\,)r_0^2\,\mathrm{sh}\,u =$$

$$= \int_0^{R/r_0} du\left(\int_0^{2\pi} r_0^2\;\mathrm{sh}\,u\,dv\right) = 2\pi r_0^2 \int_0^{R/r_0} \mathrm{sh}\,u\,du =$$

$$= 2\pi r_0^2\left(\mathrm{ch}\,\frac{R}{r_0} - 1\right).$$

Since $\mathrm{ch}\,(x) = (e^x + e^{-x})/2$, it follows that for large x we have $\mathrm{ch}\,x \cong e^x/2$, and therefore, for large R

$$(\text{the area of a ring of radius } R\,) \cong \pi r_0^2\, e^{R/r_0}.$$

Now we shall calculate the circumference of the ring, respectively, on a sphere and in a Lobachevskian plane.

1. The metric on the sphere: $(dl)^2 = r_0^2\left[(du)^2 + \sin^2 u(dv)^2\right]$.

Let the curve be given as follows: $u = R/r_0 = \text{const.}$, v is arbitrary, $v = t$ (circumference). Then the circumference is equal to:

$$\int_0^{2\pi} r_0 \sin u \, dt = 2\pi r_0 \sin r_0 \sin u = 2\pi r_0 \sin (R/r_0).$$

We see that when $R = \pi r_0/2$ the circumference is maximal (equator), while when $R = \pi r_0$ the circumference is equal to $2\pi r_0 \cdot \sin \pi = 0$.

1. The metric of the Lobachevskian sphere: $(dl)^2 = r_0^2 \left[(du)^2 + \text{sh}^2 u(dv)^2\right]$. Let the curve be given as follows: $u = R/r_0$ = const., v – arbitrary, $v = t$ (circumference). Then the circumference is equal to

$$\int_0^{2\pi} r_0 \,\text{sh}\, u \, dt = 2\pi r_0 \,\text{sh}\, u = 2\pi r_0 \frac{e^u - e^{-u}}{2}.$$

For large $u \to \infty$ the circumference is approximately equal to $\pi r_0 e^u = \pi r_0 e^{R/r_0}$. Conversely, for small $u \to 0$ we have

$$\text{sh} \frac{R}{r_0} \cong \frac{R}{r_0} \text{ and } \sin \frac{R}{r_0} \cong \frac{R}{r_0},$$

and therefore for small $u = R/r_0$ we have for the circumferences and areas of rings approximately the same formulae as in Euclidean geometry:

the circumference of a ring is approximately equal to

$$2\pi R \cong 2\pi r_0 \,\text{sh} \frac{R}{r_0} \cong 2\pi r_0 \sin \frac{R}{r_0},$$

the area of a ring is approximately equal to

$$\pi R^2 \cong \pi r_0^2 \left(\text{ch} \frac{R}{r_0} - 1\right) \cong \pi r_0^2 \left(1 - \cos \frac{R}{r_0}\right).$$

The parameter r_0 of the dimension of length is sometimes called "the radius of curvature" of the sphere and of the Lobachevskian plane, and the number r_0^{-2} coincides with the Gaussian curvature of the sphere.

For Riemannian metrics of the sphere and of the Lobachevskian plane (a pseudo-sphere or a "sphere of imaginary radius") we have the natural scale for measuring he length — this scale is the number r_0. Choosing it as the unit length, we assume $r_0 = 1$.

We obtain the dimnsionless metric of the unit sphere:

$$(d\chi)^2 = (du)^2 + \sin^2 u\,(dv)^2 \quad (u = \theta,\ v = \phi),$$

and of the (Lobachevskian) pseudo-sphere:

$$(d\chi)^2 = (du)^2 + \mathrm{sh}^2 u\,(dv)^2.$$

From school mathematics we remember the concept of a "solid angle".

The solid angle is, by definition, the area of a region on a sphere in dimnesionless metric:

$$(du)^2 + \sin^2 u\,(dv)^2, \quad u = \theta,\ v = \phi.$$

The solid angle is equal to $\iint_U \sin d\, du\, dv$ and correspondingly is equal to the solid angle of a bundle of rays coming from the origin in the direction of all points of the region U on the unit sphere.

The total solid angle is equal to

$$\int\limits_{\substack{\text{over all}\\ \text{the sphere}\\ (0\le u\le\pi)}} \sin u\, du\, dv = \int_0^{\pi} du \left(\int_0^{2\pi} \sin u\, dv\right) = 4\pi.$$

What prescriptions for calculating a double integral are naturally used? First, if a rectangle is given relative to the coordinates (u, v) (Figure 21)

$$U = \begin{cases} a \le u \le b, \\ c \le v \le d, \end{cases}$$

then we have the formula:

$$\iint_U f(u, v)\, du\, dv = \int_a^b du \left(\int_c^d f(u, v)\, dv\right) = \int_c^d dv \left(\int_a^b f(u, v)\, du\right).$$

The expression $\int_c^d f(u, v)\, dv$ is a function of u only:

$$\phi(u) = \int_c^d f(u, v)\, dv.$$

The value of the function $\phi(u_0)$ is the integral of the function $f(u_0, v)$ over the variable v.

For example, we had

$$\iint\limits_{\substack{0 \le r \le R \\ 0 \le \phi \le 2\pi}} r\, dr\, d\phi = \int_0^{2\pi} d\phi \left(\int_0^R r\, dr \right) = \int_0^{2\pi} \frac{R^2}{2}\, d\phi = \pi R^2$$

or

$$\iint\limits_{\substack{0 \le r \le R \\ 0 \le \phi \le 2\pi}} r\, dr\, d\phi = \int_0^R dr \left(\int_0^{2\pi} r\, d\phi \right) = \pi R^2.$$

Let us express this in a more general form. Suppose we are given a region U, relative to coordinates (u, v), between two curves $u = f(v)$ and $u = g(v)$, where $f > g$ over the distance between a and b (Figure 22).

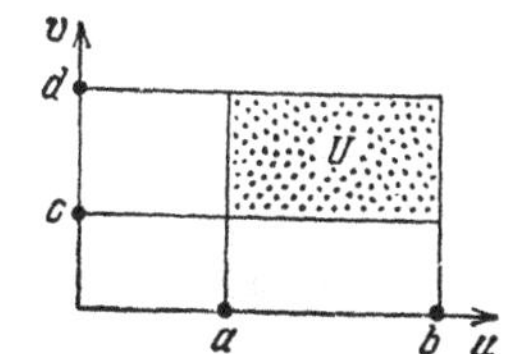

Figure 21.

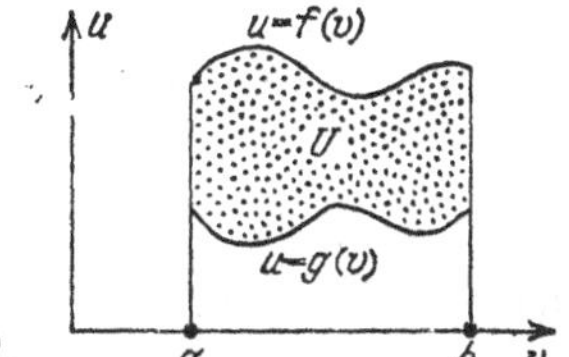

Figure 22.

Given a function $\phi(u, v)$, there holds the formula:

$$\iint_U \phi(u, v)\, du\, dv = \int_a^b \psi(v)\, dv .$$

Here $\psi(v) = \int_{g(v)}^{f(v)} \phi(u, v)\, du$ is the integral over the variable u, where the function ϕ itself and the limits of integration depend on the parameter v.

If $\phi(u, v) \equiv 1$ then the area of the region U is equal to

$$\sigma(U) = \iint_U du\, dv = \int_a^b dv \left(\int_{g(v)}^{f(v)} du \right) = \int_a^b (f(v) - g(v))\, dv.$$

We have given exact definitions of all the concepts referring to the double integral and have listed all its basic properties (the existnece for continuous functions, connection with the area in Euclidean and general Riemannian geometry of surfaces and the formula for the change of variables, the prescription for reduction to single integrals).

We have also pointed out the areas of rings in the simplest geometries (Euclidean, on the sphjere, on the Lobachevskian plane).

There holds:

THEOREM 1. *Let in Euclidean 3-space a surface be given in the form* $F(x, y, z) = C$, *where* $F \neq 0$.

I on the surface we are given a region U which is projected in a one-to-one fashion into a region V of the plane (x, y), *then the area of the region U on the surface* $F(x, y, z) = C$ *is calculated by the formula*

$$\sigma(U) = \iint_V \frac{|\text{grad } F|}{|F_z|}\, dx\, dy$$

(*we assume* $F_z \neq 0$ *at all points of the region*).

COROLLARY 1. *If the surface is given in the form* $z = f(x, y)$ *or* $F(x, y, z) = z - f(x, y) = 0$ (*where* $F(x, y, z) = z - f(x, y)$), *then the area of the region on the surface is calculated by the formula*

$$\sigma(U) = \iint\limits_V \left(1 + f_x^2 + f_y^2\right)^{1/2} dx\, dy.$$

THEOREM 2. *If the surface in a three-dimensional space is parametrized by the coordinates* $r = r(u, v)$, $r = (x, y, z)$, *where* $r_u = (x_u, y_u, z_u)$, $r_v = (x_v, y_v, z_v)$ *and the vectors* r_u *and* r_v *are linearly independent, then the area of the region on the surface is calculated by the formula*

$$\sigma(U) = \iint\limits_U |[r_u, r_v]|\, du\, dv,$$

where $[r_u, r_v]$ *is the vector product.*

We shall give a proof of Theorems 1 and 2. Both theorems are proved in a simple way. We should recall the general definition of area of a region in the case where we are given a Riemannian metric:

$$(dl)^2 = g_{ij}\, dx^i x^j; \quad x^1 = u, \ x^2 = v.$$

By definition:

$$\sigma(U) = \iint\limits_U (g)^{1/2} du\, dv, \quad g = g_{11}g_{22} - g_{12}^2 \ .$$

To prove Theorems 1 and 2 it is necessary to calculate $(g)^{1/2}$.

Proof of Theorem 1. Recall that for the surface $F(x, y, z) = C$ we have $g_{11} = 1 + F_x^2/F_z^2$, $g_{12} = F_x F_y/F_z^2$, $g_{22} = F_y^2/F_z^2 + 1$. If $F_z \neq 0$, then $u = x^1 = x$, $v = x^2 = y$ and $E = g_{11}$, $F = g_{12}$, $G = g_{22}$ by definition. From this we have:

$$g = g_{11}g_{22} - g_{12}^2 = \left(1 + \frac{F_x^2}{F_z^2}\right)\left(1 + \frac{F_y^2}{F_z^2}\right) - \frac{F_x^2 F_y^2}{F_z^4} = 1 + \frac{F_x^2}{F_z^2} + \frac{F_y^2}{F_z^2},$$

and therefore

$$(g)^{1/2} = \left(1 + \frac{F_x^2}{F_z^2} + \frac{F_y^2}{F_z^2}\right)^{1/2} = \frac{|\operatorname{grad} F|}{|F_z|}$$

which concludes the proof of Theorem 1.

Proof of Theorem 2. If $r = r(u, v)$, $u = x^1$, $v = x^2$, then $g_{ij} = r_{x^i} r_{x^j}$. Note that:

$$[r_u, r_v] = (x_u y_v - x_v y_u)e_3 + (z_u x_v - z_v x_u)e_2 + (y_u z_v - y_v z_u)e_1.$$

Hence $|[r_u, r_v]|^2 = g_{11}g_{22} - g_{12}^2$ and $(g)^{1/2} = |[r_u, r_v]|^2$. For the area we have, by definition:

$$\sigma(U) = \iint_U |[r_u, r_v]| \, du \, dv,$$

as required.

Thus, we have investigated area in the Riemannian geometry of surfaces and, in particular, made sure that the concept of area is defined, as is the concept of length, by giving the scalar product (g_{ij}) of the tangent vectors at each point.

1.10 The Theory of Surfaces. The Theory of Curvature and the Second Quadratic Form

In Section 1.9 we defined the double integral:

$$\iint_U f(u,v)\, f(u, v)\, du\, dv$$

over the region U in the (u, v)-plane and formulated the properties of the integral:

a) The existence theorem;

b) the change of variables $u = u(x, y)$, $v = v(x, y)$,

$$\iint_U f(u, v)\, du\, dv = \iint_V f(u, (x, y), v(x, y))\; J(x, y)\, dx\, dy,$$

where $J(x, y) = u_x v_y - u_y v_x > 0$;

c) the area $U = \iint_U du\, dv$ if $(dl)^2 = (du)^2 + (dv)^2$;

c') the area $U = \iint_U (g)^{1/2}\, du\, dv$,

if $(dl)^2 = g_{11}(du)^2 + 2g_{12}\, du\, dv + g_{22}(dv)^2$;

$$g = g_{11}g_{22} - g_{12}^2 = EG - F^2;$$

d) $\iint_U f(u, v)\, du\, dv = \int_c^d du \left(\int_{a(u)}^{b(u)} f(u, v)\, dv \right),$

where U is shown in Figure 23.

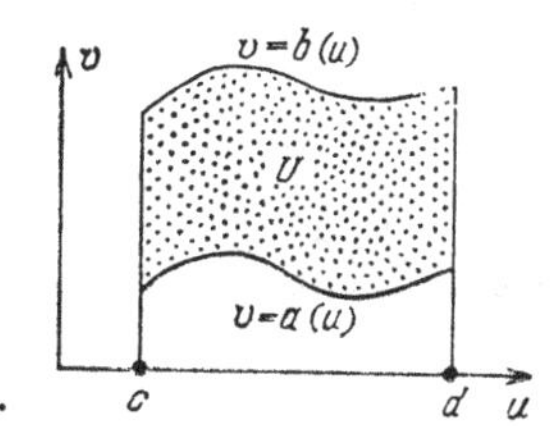

Figure 23.

As an example of calculation of the areas of the simplest figures, we have considered circles of radius R in the Euclidean plane, on the sphere and on the Lobachevskian plane.

1. The Euclidean metric:

$$(dl)^2 = (du)^2 + (dv)^2 = (dr)^2 + r^2 (d\phi)^2,$$

$$\sigma(U_R) = \iint\limits_{\substack{0 \ll r \ll R \\ 0 \ll \phi \ll 2\pi}} r\, dr\, d\phi = \pi R^2.$$

2. The metric of the sphere:

$$(dl)^2 = r_0^2 ((du)^2 + \sin^2 u (d\phi)^2),$$

$$\sigma(U_R) = \iint\limits_{\substack{0 \le u \le R/r_0 \\ 0 \le \phi \le 2\pi}} r_0^2 \sin u\, d\phi\, d\phi = 2\pi r_0^2 \left(1 - \cos \frac{R}{r_0}\right).$$

As $R \to 0$ we have $1 - \cos \frac{R}{r_0} \cong \frac{R^2}{r_0^2}$ and, therefore, $2\pi r_0^2 \left(1 - \cos \frac{R}{r_0}\right) \cong \pi R^2$.

We obtain the area of the entire sphere by setting $\cos \frac{R}{r_0} = -1$, or $R = \pi r_0$.

Then $2\pi r_0^2(1 - \cos \pi) = 4\pi r_0^2$. If $r_0 = 1$, then we have the total area of the sphere, equal to 4π (as has already been mentioned, this is the total solid angle).

3. The Lobachevskian metric:

$$(dl)^2 = r_0^2 ((du)^2 + \operatorname{sh}^2 u (d\phi)^2),$$

$$\sigma(U\ ) = \iint\limits_{\substack{0 \le u \le R/r_0 \\ 0 \le \phi \le 2\pi}} r_0^2 \operatorname{sh}\ u\, du\, d\phi = 2\pi r_0^2 \left(\operatorname{ch} \frac{R}{r_0} - 1\right).$$

If $R \to 0$, then $2\pi r_0^2 \left(\operatorname{ch} \frac{R}{r_0} - 1\right) \cong \pi R^2$ since $\operatorname{ch} x = \frac{e^x + e^{-x}}{2}$.

We have proved the following assertions for the calculation of the areas on surfaces situated in a three-dimensional space.

a) *If we are given a surface in the form* $z = f(x, y)$ *and on this surface a region* U *which is projected into a region* V *of a plane* (x, y), *then there holds the equality*:

$$\sigma(U) = \iint_V \left(1 + f_x^2 + f_y^2\right)^{1/2} dx\, dy,$$

where $f_x = \partial f/\partial x, f_y = \partial f/\partial y$.

b) *If a surface is given in the form* $F(x\, y, z) = C$ *and a region* U *on this surface is projected in a one-to-one manner into a region* V *in a plane parametrized by coordinates* (x, y), *then there holds the formula*:

$$\sigma(U) = \iint_V \frac{|\mathrm{grad}\, F|}{|F_z|} dx\, dy,$$

where $|F_z| = |\partial F/\partial z| \neq 0$ *for* (x, y, z) *lying in the region* U.

c) *If a surface is given parametrically in the form* $r = r(u, v)$ *or* $x = x(u, v)$, $y = y(u, v)$, $z = z(u, v)$, *then we have*:

$$\sigma(U) = \iint_U |[r_u, r_v]|\, du\, dv,$$

where U *is the region in the* (u, v)*-plane*, $[r_u, r_v]$ *is the vector product.*

The proof of this theorem consisted in the calculation of the quantitiy $(g)^{1/2}$ for three different ways in which the surface was given.

EXAMPLE 4. Let a surface be given as a rotation paraboloid $z = f(x, y) = \dfrac{x^2 + y^2}{a}$. We are going to calculate the area of a region $U_R = (r \leq R)$, where $r^2 = x^2 + y^2$:

$$\sigma(U_R) = \iint_{U_R} (g)^{1/2}\, dx\, dy =$$

$$= \iint_{U_R} \Big(1 + \frac{4x^2}{a^2} + \frac{4y^2}{a^2}\Big)^{1/2} dx\, dy = \iint_{U_R} \Big(1 + \frac{4r^2}{a^2}\Big)^{1/2} r\, dr\, d\phi =$$

$$= \int_0^R dr \Big(\int_0^{2\pi} \Big(1 + \frac{4r^2}{a^2}\Big)^{1/2} r\, d\phi\Big) = 2\pi \int_0^R \Big(1 + \frac{4r^2}{a^2}\Big)^{1/2} r\, dr =$$

$$= \int_0^{R^2/a^2} \pi a^2\, (1 + 4\rho)^{1/2}\, d\rho, \quad \rho = \frac{r^2}{a^2}.$$

Finally, $\sigma(U_R) = \dfrac{\pi a^2}{6}\Big[\Big(1 + 4\dfrac{R^2}{a^2}\Big)^{3/2} - 1\Big]$.

EXAMPLE 5. An ellipse in the Euclidean plane.

$$U_{a,b} = \Big(\frac{x^2}{a^2} + \frac{y^2}{b^2} \le 1\Big).$$

Let

$$x' = x/a, \quad y' = y/b,$$

$$\sigma(U_{a,b}) = \iint_{((x')^2+(y')^2)\le 1} ab\, dx'\, dy' = \pi ab.$$

The theory of curvature of curves on a surface. Suppose we are given a surface in the parametric form $r = r(u, v)$. Then $[r_u, r_v] = (g)^{1/2} \cdot m$, m is the normal vector and $|m| = 1$, $|[r_u, r_v]| = (g)^{1/2}$.

Consider the curve $r = r(u(t), v(t))$. We have:

$$\dot{r} = r_u \dot{u} + r_v \dot{v},$$

$$\ddot{r} = r_{uu}\dot{u}^2 + r_{uv}\dot{u}\dot{v} + r_{vu}\dot{u}\dot{v} + r_{vv}\dot{v}^2 + r_u \ddot{u} + r_v \ddot{v}.$$

Since $r_u \perp m$ and $r_v \perp m$, we obtain:

$$\ddot{r}m = (r_{uu}m)\dot{u}^2 + 2(r_{uv}m)\dot{u}\dot{v} + (r_{vv}m)\dot{v}^2 = b_{11}\dot{u}^2 + 2b_{12}\dot{u}\dot{v} + b_{22}\dot{v}^2.$$

This is the quadratic form of the velocity vector $(\dot{u}, \dot{v})$ in local coordinates u, v, $u = x^1$, $v = x^2$.

Let $b_{11} = L$, $b_{12} = M$, $b_{22} = N$. We have:

$$(\ddot{r}m)\, dt^2 = b_{ij}dx^i dx^j = L(du)^2 + 2M\, du\, dv + N(dv)^2.$$

The form $(\ddot{r}m)\, dt^2$ is called the *second quadratic form.*

Let the curve line $u(t)$, $v(t)$ be taken with respect to the natural parameter $t = l$.

According to the Frenet formulae, we obtain for the curve line:

$$r = r(t) = r(u(t), v(t)), \quad r = (x, y, z),$$

$\ddot{r} = \dfrac{d^2 r}{dt^2} = kn$, where n is the principal normal to the curve line, k is the curvature of the curve line.

Since $\ddot{r}m = k(nm) = k \cos\theta$, we obtain ($\theta$ is the angle between m and n):

$$k \cos\theta\, (dl)^2 = (\ddot{r}m)\,(dl)^2 = L(du)^2 + 2M\, du\, dv + N(dv)^2,$$

where

$$(dl)^2 = g_{ij}dx^i dx^j = g_{11}(du)^2 + 2g_{12}\, du\, dv + g_{22}(dv)^2.$$

Conclusion.

$$k\cos\theta = \frac{L(du)^2 + 2M\, du\, dv + N(dv)^2}{g_{11}(du)^2 + 2g_{12}\, du\, dv + g_{22}(dv)^2}.$$

Thus, we have obtained the following:

THEOREM. *The curvature of a curve on a surface in Euclidean 3-space, when multiplied by the cosine of the angle between the normal to the surface and the principal normal of the curve is the same (up to the sign) as the ratio of the second and first quadratic forms.*

COROLLARY 1. *If the curve is obtained by sectioning the surface with a plane normal to the surface, then:*

$$\cos\theta = \pm 1,$$

$$\pm k = \frac{b_{ij}\,dx^i dx^j}{g_{ij}\,dx^i dx^j}, \qquad x^1 = u,\ x^2 = v,\quad i = 1, 2:\quad j = 1, 2.$$

1.11 The Theory of Surfaces. Gaussian Curvature

In the preceding section we have defined the second quadratic form of the surface as follows: a surface is given in the parametric form $r = r(u, v)$; we consider a curve $r = (u(t), v(t))$ and the normal projection of its acceleration:

$$\ddot{r}m = (r_{uu}m)\dot{u}^2 + 2(r_{uv}m)\dot{u}\dot{v} + (r_{vv}m)\dot{v}^2,$$

where $m = \dfrac{[r_u, r_v]}{|[r_u, r_v]|}$ is the unit vector of the normal. Then the expression $(\ddot{r}m)$ is the quadratic form of the components of the velocity vector $(\dot{u}, \dot{v})$ and is called the *second quadratic form.*

For curves given in terms of the natural parameter $l = t$, we have derived the following formula:

$$\pm k \cos\theta = k(nm) = \frac{b_{ij}\, dx^i dx^j}{g_{ij}\, dx^i dx^j}, \qquad x^1 = u,\ x^2 = v,$$

where k is the curvature of the curve, n is the vector of the principal normal to the curve:

$$d^2r\backslash dl^2 = kn,\ L = b_{11},\ M = b_{12} = b_{21},\ n = b_{22}.$$

For the normal cross-section $\cos\theta = \pm 1$ since $\pm n = m$ by definition (at a point under investigation).

Thus, each point of the surface is associated with the pair of quadratic forms:

$$1)\ \ (dl)^2 = g_{ij}\, dx^i dx^j,$$

$$2)\ \ (\ddot{r}m)\, dt^2 = b_{ij}\, dx^i dx^j,$$

the form $(dl)^2$ being positive definite.

What are the known algebraic invariants of a pair of quadratic forms?

Consider any pair of quadratic forms, in a plane, of which one is positive definite. Let the matrices of the quadratic forms look like:

$$G = \begin{pmatrix} g_{11} & g_{12} \\ g_{21} & g_{22} \end{pmatrix} \text{ and } Q = \begin{pmatrix} b_{11} & b_{12} \\ b_{21} & b_{22} \end{pmatrix} = \begin{pmatrix} L & M \\ M & N \end{pmatrix}.$$

Next we write the equation det $(Q - \lambda G) = 0$, which, when written out in detail, becomes $(L - \lambda g_{11})(N - \lambda g_{22}) - (M - \lambda g_{12})^2 = 0$. We call the roots λ_1, λ_2 of these equations the *eigenvalues* of the pair of quadratic forms.

Let us solve the system of linear equations:

$$(L - \lambda_i g_{11})\, \xi_{1i} + (M - \lambda_i g_{12})\, \xi_{2i} = 0,$$
$$(M - \lambda_i g_{21})\, \xi_{1i} + (N - \lambda_i g_{22})\, \xi_{2i} = 0, \quad i = 1, 2,$$

where ξ_{1i}, ξ_{2i} are the unknowns.

If the roots λ_1, λ_2 are the eigenvalues, then the system of equations has non-trivial solutions (ξ_{11}, ξ_{21}) and (ξ_{12}, ξ_{22}), which are the vectors $\bar{e}_1 = (\xi_{11}, \xi_{21})$, $\bar{e}_2 = (\xi_{12}, \xi_{22})$.

The directions of the vectors $\bar{e}_1$ and $\bar{e}_2$ are called *principal directions* of the pair of quadratic forms, that of $\bar{e}_1$ corresponding to λ_1 and that of $\bar{e}_2$ to λ_2.

Recall that the scalar products of the basis vectors in the plane have the form:

$$e_i e_j = g_{ij}, \quad \begin{cases} i = 1, 2, \\ j = 1, 2. \end{cases}$$

(The Riemannian metric is given by the form g_{ij}).

PROPOSITION 1. *If the eigenvalues of a pair of quadratic forms are distinct, then the principal directions are orthogonal.*

We have two principal directions $\bar{e}_2, \bar{e}_2$:

$$\bar{e}_1 = \xi_{11} e_1 + \xi_{21} e_2, \quad \bar{e}_2 = \xi_{12} e_1 + \xi_{22} e_2.$$

By definition, their orthogonality implies that

$$\bar{e}_1 \bar{e}_2 = \xi_{11}\xi_{12} g_{11} + (\xi_{11}\xi_{22} + \xi_{21}\xi_{12}) g_{12} + \xi_{21}\xi_{22} g_{22} = 0.$$

Proof. Choose a pair of plane vectors d_1, d_2, such that

$$d_i d_j = \delta_{ij} = \begin{cases} 1, & i = j, \\ 0, & i \neq j. \end{cases}$$

This can be done by virtue of positive definiteness of the quadratic form with the matrix g_{ij} since it can be brought through a linear transformation into the sum of squares. The second quadratic form will now be considered in a new basis d_1, d_2, where:

$$e_1 = a_{11}d_1 + a_{12}d_2, \qquad A = \begin{pmatrix} a_{11} & a_{12} \\ a_{21} & a_{22} \end{pmatrix}$$
$$e_2 = a_{21}d_1 + a_{22}d_2,$$

For matrices of quadratic forms we have $d_i d_j = \delta_{ij}$ or $\bar{G} = \begin{pmatrix} 1 & 0 \\ 0 & 1 \end{pmatrix}$,

where $G = A \circ A^{\mathrm{T}}$ in the basis d_1, d_2, where $\bar{Q}$ is the matrix of the second quadratic form in the new basis d_1, d_2.

Since $G = A \circ A^{\mathrm{T}}, Q = A \circ \bar{Q} \circ A^{\mathrm{T}}$, it follows that:

$$Q - \lambda G = A \circ (\bar{Q} - \lambda \cdot E) \circ A^{\mathrm{T}}$$

and

$$\det(Q - \lambda G) = (\det A) \det(\bar{Q} - \lambda \cdot E) \cdot (\det A^{\mathrm{T}}) =$$
$$= (\det A)^2 \det(\bar{Q} - \lambda \cdot E), \quad E = \begin{pmatrix} 1 & 0 \\ 0 & 1 \end{pmatrix},$$

inasmuch as the determinant of the product of matrices is equal to the product of the determinants of the matrices.

It should be noted that $(\text{set } A^{\mathrm{T}}) = \det A = (g)^{1/2} = (\det G)^{1/2}$ and $g \neq 0$, and so the two equations for the eigenvalues:

$$\text{(I)} \qquad \det(Q - \lambda G) = 0,$$

$$\text{(II)} \qquad \det(\bar{Q} - \lambda \cdot E) = 0$$

are equivalent. The solution of either of these two equations yields the eigenvalues λ_1 and λ_2 to which there correspond the principal directions $\bar{e}_1$ and $\bar{e}_2$.

In the basis (d_1, d_2) the scalar product is Euclidean — it is given by a unit matrix $\begin{pmatrix} 1 & 0 \\ 0 & 1 \end{pmatrix} = \bar{G} = E = (\delta_{ij})$. It is a well-known fact in algebra that the quadratic form $\bar{Q}$, when rotated, can be brought into the diagonal form, and its eigenvectors $\bar{e}_1, \bar{e}_2$ are orthogonal in the usual Euclidean sense. Next, we choose $\bar{e}_1$, $\bar{e}_2$ in such a way that $|\bar{e}_1| = |\bar{e}_2| = 1$; then $\bar{e}_i \bar{e}_j = \delta_{ij}$.

Thus, our algebraic assertion is proved: it is a somewhat more extended version of the theorem which states that the quadratic form in the Euclidean plane can be brought to the diagonal form by rotation.

Now we shall turn again to the first and second quadratic forms on the surface in a three-dimensional space:

$$\text{(I)} \qquad (dl)^2 = g_{ij}\, dx^i dx^j,$$

$$\text{(II)} \qquad b_{ij}\, dx^i dx^j.$$

The ratio of these quadratic forms gives the curvature of the normal cross-section (up to the sign).

The eigenvalues of this pair of quadratic forms are called the *principal curvatures* of the surface at the point under investigation.

The product of the principal curvatures is called the *Gaussian curvature* of the surface, and their sum the *mean curvature* of the surface.

EXAMPLE. Let the surface be given in the form $z = f(x\ y)$, and at the point (x_0, y_0) that we are studying, we have $f_x = f_y = 0$, let $x = u$, $y = v$, $z = f(u, v)$. For the first and second quadratic forms we obtain (at the point x_0, y_0 under study):

$$\text{(I)} \qquad g_{11} = 1, \quad g_{12} = g_{21} = 0, \quad g_{22} = 1 \quad (g_{ij} = \delta_{ij}),$$

$$\text{(II)} \qquad L = b_{11} = r_{uu}m = f_{xx}|_{x_0,y_0};$$

$$M = b_{12} = r_{uv}m = f_{xy}|_{x_0,y_0};$$

$$N = b_{22} = r_{vv}m = f_{yy}|_{x_0,y_0}.$$

Here the vector m coincides with the unit vector along the z-axis.

So, at the point under consideration, the second quadratic form is represented by:

$$b_{ij}\, dx^i dx^j = f_{x^i x^j}\, dx^i dx^j = 2d^2 f.$$

The Gaussian curvature coincides in this case with the determinant of the matrix:

$$\begin{pmatrix} f_{xx} & f_{xy} \\ f_{yx} & f_{yy} \end{pmatrix};$$

the eigenvalues can be obtained from the equation $(f_{xx} - \lambda)(f_{yy} - \lambda) - (f_{xy})^2 = 0$ since $g_{ij} = \delta_{ij}$. The tangent plane to the surface at this point is parallel to the (x, y)-plane. The principal directions at this point can be obtained from the solution of the system of equations:

$$\begin{cases} (f_{xx} - \lambda_1)\,\xi_{11} + f_{xy}\,\xi_{21} = 0, \\ f_{xy}\,\xi_{11} + (f_{yy} - \lambda_1)\,\xi_{21} = 0, \end{cases} \qquad \text{for } \bar{e}_1,$$

$$\begin{cases} (f_{xx} - \lambda_2)\,\xi_{12} + f_{xy}\,\xi_{22} = 0, \\ f_{xy}\,\xi_{12} + (f_{yy} - \lambda_2)\,\xi_{22} = 0, \end{cases} \qquad \text{for } \bar{e}_2.$$

In so far as $\bar{e}_1 \perp \bar{e}_2$, we can take the unit vectors of the principal directions as the new coordinate axes x', y', obtained from the old system through rotation of the (x, y)-plane. It is only necessary that there hold the condition $\lambda_1 \neq \lambda_2$.

Relative to the new coordinates (z, x', y') we have:

$$z = f(x(x', y'), y(x', y')),$$

where $x = x' \cos\phi + y' \sin\phi$, $y = -x' \sin\phi + y' \cos\phi$, ϕ is the angle of rotation.

Relative to the new coordinates, the second quadratic form becomes (at the point under investigation, only):

$$\lambda_1 (dx')^2 + \lambda_2 (dy')^2.$$

Relative to the coordinates (x', y'), the curvature at our point of a normal cross-section is given by the formula:

$$k = \frac{\lambda_1 (dx')^2 + \lambda_2 (dy')^2}{(dx')^2 + (dy')^2}.$$

The tangent vector e to this normal cross-section of the surface has, at the point in question, the form $(\dot{x}', \dot{y}') = e$, where $dx' = \dot{x}'\,dt$, $dy' = \dot{y}'\,dt$. On this account,

$$\cos^2\alpha = \frac{(dx')^2}{(dx')^2 + (dy')^2}. \quad \sin^2\alpha = \frac{(dy')^2}{(dx')^2 + (dy')^2},$$

where α is the angle between the x'-axis and the tangent vector e to the normal cross-section.

We shall now derive *Euler's formula.*

THEOREM 1. *The curvature of the normal cross-section is given by the formula*:

$$k = \lambda_1 \cos^2\alpha + \lambda_2 \sin^2\alpha,$$

where λ_1, λ_2 *are the principal curvatures,* α *is the angle on the surface between the tangent vector to the normal cross-section and the principal direction corresponding to* λ_1.

Proof. We have derived Euler's formula in the case where the surface is given in the form $z = f(x, y)$, and at the point x_0, y_0 under investigation we have $f_x = f_y = 0$. However, since the result itself is independent of the choice of local coordinates, for any neighbourhood of the point we can always choose coordinates associated with it, such that the z-axis be normal to the surface at this point and the x- and y--axes be tangent to the surface and mutually orthogonal (we may even choose them to be the principal directions). Then the surface in the neighbourhood of this point is given in the form:

$$z = f(x, y), \quad \text{where } f_x = f_y = 0 \quad (x = x_0,\ y = y_0),$$

and, moreover, $f_{xy} = f_{yx} = 0$ provided that the axes are the principal directions at this point.

In this case, $\lambda_1 = f_{xx}, \lambda_2 = f_{yy}$ at this point.

Since we have already derived Euler's formula relative to such coordinates, we have completed the proof.

We shall now present the formulae useful in the case of the second quadratic form.

If the surface is given in the form $f(x, y) = z$, then for the coefficients of the second quadratic form we have (here $x = u, y = v$)

$$r_u = (1, 0\ f_x), \quad r_v = (0, 1, f_y),$$

$$[r_u, r_v] = (-f_x, -f_y, 1),$$

$$r_{uu} = (0, 0, f_{xx}), \quad r_{uv} = r_{vu} = (0, 0, f_{xy}), \quad r_{vv} = (0, 0, f_{yy}),$$

$$m = \frac{|r_u, r_v]}{|[r_u, r_v]|} = \frac{(-f_{x_1} - f_{y_1}, 1)}{\left(1 + f_x^2 + f_y^2\right)^{1/2}};$$

$$L = b_{11} = \frac{f_{xx}}{\left(1 + f_x^2 + f_y^2\right)^{1/2}}, \qquad M = b_{12} = b_{21} = \frac{f_{xx}}{\left(1 + f_x^2 + f_y^2\right)^{1/2}},$$

$$N = b_{22} = \frac{f_{yy}}{\left(1 + f_x^2 + f_y^2\right)^{1/2}}.$$

From this, we can obtain:

$$b_{ij}\, dx^i dx^j = \frac{1}{\left(1 + f_x^2 + f_y^2\right)^{1/2}} (f_{x^i x^j}\, dx^i dx^j),$$

$$x^1 = x = u, \quad x^2 = y = v.$$

Recall that for the coefficients g_{ij} we had the formulae

$$g_{11} = 1 + f_x^2, \quad g_{12} = g_{21} = f_x f_y, \quad g_{22} = 1 + f_y^2$$

and

$$g = g_{11} g_{22} - g_{12}^2 = 1 + f_x^2 + f_y^2.$$

As has already been defined above, the Gaussian curvature of the surface is the product of the principal curvatures (eigenvalues) $K = \lambda_1, \lambda_2$. The mean curvature is the sum $H = \lambda_1 + \lambda_2$.

THEOREM 2. *The Gaussian curvature of a surface is equal to the ratio of the determinants of the matrices of the second and first quadratic forms*:

$$K = \frac{b_{11} b_{22} - b_{12}^2}{g_{11} g_{22} - g_{12}^2}.$$

In particular, if the surface is given in the form $z = f(x, y)$, then there holds the formula:

$$K = \frac{f_{xx}f_{yy} - f_{xy}^2}{\left(1 + f_x^2 + f_y^2\right)^2}.$$

Proof. The eigenvalues λ_1 and λ_2 were determined form the equation:

$$\det (Q - \lambda G) = 0,$$

where $Q = \begin{pmatrix} L & M \\ M & N \end{pmatrix}$ is the matrix of the second quadratic form,

$$G = \begin{pmatrix} g_{11} & g_{12} \\ g_{21} & g_{22} \end{pmatrix}, \quad g_{12} = g_{21},$$

It should be noted that:

$$Q - \lambda G = \begin{pmatrix} b_{11} - \lambda g_{11} & b_{12} - \lambda g_{12} \\ b_{12} - \lambda g_{12} & b_{22} - \lambda g_{22} \end{pmatrix}$$

and

$$\det (Q - \lambda G) = (b_{11} - \lambda g_{11})\ (b_{12} - \lambda g_{22}) - (b_{12} - \lambda g_{12})^2.$$

The matrix$G = (g_{ij})$ is positive definite and is, therefore, non-degenerate. We shall denote the matrix inverse to this one by G^{-1}.

There holds the equality:

$$\det (Q - \lambda G) = (\det G) \det (G^{-1} Q - \lambda \cdot E),$$

where

$$E = \begin{pmatrix} 1 & 0 \\ 0 & 1 \end{pmatrix} = G \circ G^{-1}.$$

The eigenvalues λ_1 and λ_2 can be determined from the solution of the equation:

$$\det (G^{-1}Q - \lambda \cdot E) = 0,$$

as long as $g = \det G = g_{11}g_{22} - g_{12}^2 \neq 0$. We remind the reader of the well-known algebraic fact.

If there exists a matrix A and the equation $\det (A - \lambda \cdot E) = 0$ determines the eigenvalues, then for the second-order matrices the product of all the eigenvalues of the matrix is equal to its determinant $\det A = \lambda_1 \cdot \lambda_2$.

Assuming $A = G^{-1}Q$, we see that:

$$\lambda_1, \lambda_2 = \det (G^{-1} \circ Q) = \frac{\det Q}{\det G}.$$

This implies that the Gaussian curvature is equal to the ratio of the determinants of the matrices of the second and first quadratic forms.

Next, if a surface is given in the form $z = f(x, y)$, then we have the table of coefficients L, M, N, g_{ij} written above. Calculating the determinants, we use their ratio to deduce the formula for the Gaussian curvature.

COROLLARY 1. *If a surface is given in the form $z = f(x, y)$, then the sign of the Gaussian curvature κ is the same as the sign of the determinant $(f_{xx} f_{yy} - f_{xy}^2)$ because*

$$\kappa = \frac{f_{xx}f_{yy} - f_{xy}^2}{\left(1 + f_x^2 + f_y^2\right)^2}.$$

EXAMPLE. Suppose that we are given a surface in the form $z = f(x, y)$, where the function $f(x, y)$ satisfies the *Laplace equation*

$$f_{xx} + f_{yy} = 0.$$

Then we have $f_{xx}f_{yy} - f_{xy}^2 \leq 0$ since $f_{xx} = -f_{yy}$. Hence, at all points of the surface, where at least one of the partial derivatives f_{xx}, f_{yy}, f_{xy} is non-zero, we shall have that the Gaussian curvature $K < 0$.

Conversely: if the curvature K is positive at all points of the surface, then the surface is referred to as (strictly) *convex*.

The use of this terminology is due to representations of visual geometry. Suppose $z = f(x, y)$ and at a point (x_0, y_0) we have $f_x = f_y = 0$ (such coordinates can always be chosen). The graphs of the function $z = f(x, y)$ with poisitive and negative K are shown in Figures 24 to 26. In the case where K is positive, there may be either $\lambda_1 > 0, \lambda_2 > 0$, and then (x_0, y_0) is the point of the minimum of the function f, (Figure 24) or inversely, $\lambda_1 < 0, \lambda_2 < 0$, and then (x_0, y_0) is the point of the maximum of the function f, (Figure 25); in the case where K is negative, we have $\lambda_1 < 0, \lambda_2 > 0$, and the function has a saddle (pass) (Figure 26).

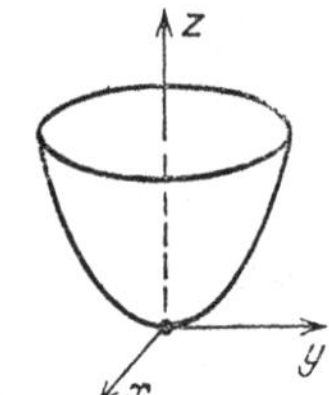

Figure 24.

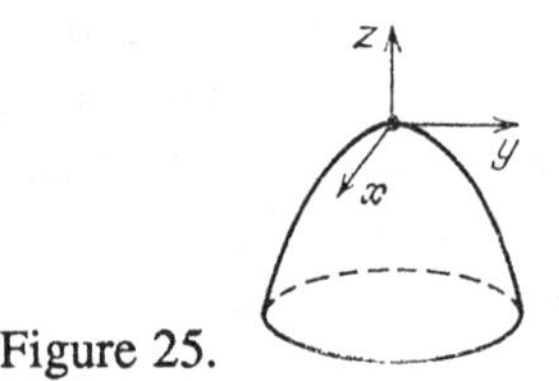

Figure 25.

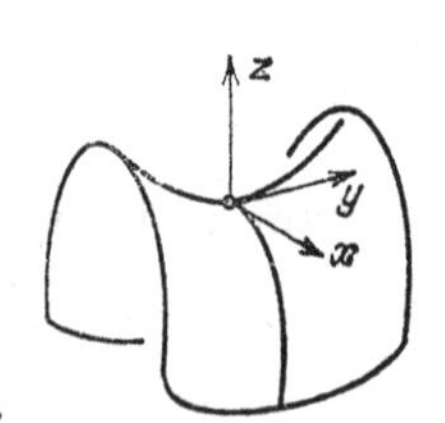

Figure 26.

1.12 The Theory of Surfaces. Invariants of a Pair of Quadratic Forms and Euler's Theorem

We shall systematize the facts from the theory of surfaces which have been discussed above.

I. Mathematical representation of the surface, non-singular points, equivalence of different ways of representing a surface (local equivalence).

a) $z = f(x, y), \quad x = x^1 = u, \quad y = x^2 = v,$

b) $F(x, y, z) = C, \quad \operatorname{grad} F \neq 0,$

c) $r = r(u, v), \quad r = (x, y, z), \quad [r_u, r_v] \neq 0,$

$$u = x^1, \quad v = x^2.$$

II. Riemannian metric on the surface (the first quadratic form) generated by the embedding into space.

$$(dl)^2 = g_{ij}\,dx^i dx^j = (dx)^2 + (dy)^2 + (dz)^2,$$

where

$$x^1 = u, \quad x^2 = v, \quad x = x(u, v), \quad y = y(u, v), \quad z = z(u, v),$$

$$g_{11} = E, \quad g_{12} = F, \quad g_{22} = G,$$

a) $g_{11} = 1 + f_x^2, \quad g_{12} = f_x . f_y, \quad g_{22} = 1 + f_y^2,$

c) $g_{ij} = r_{x^i x^j}, \quad i = 1, 2; \quad j = 1, 2.$

The Riemannian metric g_{ij} serves to determine the scalar product of the velocity vectors of curves on the surface $u(t)$, $v(t)$, $\xi = (\dot{u}, \dot{v})$ are the lengths and angles; for example,

$$|\xi|^2 = g_{11}\,\dot{u}^2 + 2g_{12}\,\dot{u}\dot{v} + g_{22}\,\dot{v}^2$$

the length of the curve segment is equal to

$$\int_a^b \sqrt{|\xi(t)|^2}\, dt.$$

III. The area of a region on the surface. The area of a region is calculated by the formula:

$$\sigma(U) = \iint_U (g)^{1/2} du\, dv = \iint_U (EG - F^2)^{1/2} du\, dv,$$

$$g = g_{11}g_{22} - g_{12}^2 = \det G, \quad G = (g_{ij}).$$

For the calculation of the quantity $(g)^{1/2}$ for different ways of representation, see above.

IV. The second quadratic form on tangent vectors (the normal projection of the acceleration vector of a curve on the surface). Given a curve $u(t)$, $v(t)$, $u = x^1$, $v = x^2$, the surface

$$r(t) = r(u(t), v(t)), \qquad m = \frac{[r_u, r_v]}{|[r_u, r_v]|},$$

where m is the unit vector of the normal

$$\ddot{r}m = b_{ij}\dot{x}^i\dot{x}^j,$$

$$L = b_{11},\ M = b_{12} = b_{21}, \quad N = b_{22}.$$

Let the surface be given in the form:

$$z = f(x, y),$$

$$x = x^1 = u, \quad t = x^2 = v,$$

$$z = f(u, v).$$

The table of quantities r_u, r_v, $[r_u, r_v]$, $(g)^{1/2} = |[r_u, r_v]|$, g_{ij}, b_{ij} is given above. From this, we have:

$$b_{ij} = \frac{f_{x^i x^j}}{(g)^{1/2}}.$$

V. The properties of the second quadratic form. a) if a curve is given in the form $u(t)$, $v(t)$ on the surface or in the form $r = r(u(t), v(t))$, $r = (x, y, z)$ in Euclidean 3-space, then there holds the formula:

$$\pm k \cos\theta = \frac{b_{ij}\, dx^i dx^j}{g_{ij}\, dx^i dx^j} = \frac{b_{ij}\, \dot{x}^i \dot{x}^j}{g_{ij}\, \dot{x}^i \dot{x}^j},$$

where k is the curvature of the curve, $\cos\theta = mn$ and n is the vector of the principal normal to the curve.

$$\frac{d^2 r}{dl^2} = kn \qquad \text{(the definition of } k \text{ and } n).$$

For normal cross-sections $\cos\theta = \pm 1$ ($\pm n = m$).

b) Algebraic invariants of a pair of quadratic forms (the first and the second) at a given point of the surface are two quadratic forms in a plane with basis vectors $r_u = e_1$, $r_v = e_2$

$$e = \xi^1 e_1 + \xi^2 e_2,$$

the first form $g_{ij}\, \xi^i \xi^j$; the second form $b_{ij}\, \xi^i \xi^j$,

the first quadratic form $g_{ij}\, \xi^i \xi^j$ being positive.

Let us consider the matrices of the forms:

$$Q = (b_{ij}) = \begin{pmatrix} L & M \\ M & N \end{pmatrix}, \qquad \mathfrak{G} = (g_{ij}) = \begin{pmatrix} E & F \\ F & G \end{pmatrix}.$$

We shall write the equation $\det(Q - \lambda \mathfrak{G}) = 0$ and find the roots λ_1, λ_2 of this equation.

The *Gaussian curvature* $K = \lambda_1 \lambda_2$. The *mean curvature* $H = \lambda_1 + \lambda_2$.

Suppose $\mathfrak{G}^{-1} = g^{ij}$ is the inverse matrix:

$$\sum_{j=1}^{2} g^{ij} g_{jk} = \delta_{ik} = \begin{cases} 1, & i = k, \\ 0, & i \neq k. \end{cases}$$

We have considered the matrix $A = \mathfrak{G}^{-1} \circ Q = (a_{ik})$, where $a_{ik} = \sum_{q=1}^{2} g^{iq} b_{qk}$. Then

$$K = \lambda_1\lambda_2 = \det A = \det(\mathfrak{G}^{-1} \circ Q) = \frac{\det Q}{\det \mathfrak{G}},$$

$$H = \lambda_1 + \lambda_2 = \operatorname{Sp} A = a_{11} + a_{22} = \sum_{i,j=1}^{2} g^{ij} b_{ji}.$$

In particular, we have

$$K = \frac{LN - M^2}{g_{11}g_{22} - g_{12}^2} = \frac{LN - M^2}{EG - F^2}.$$

If $z = f(x, y)$ parametrizes the surface, then

$$K = \frac{f_{xx}f_{yy} - f_{xy}^2}{\left(1 + f_x^2 + f_y^2\right)^2}.$$

Therefore, the sign of the Gaussian curvature is the same as the sign of the determinant.

$$\det \begin{pmatrix} f_{xx} & f_{xy} \\ f_{yx} & f_{yy} \end{pmatrix} = f_{xx}f_{yy} - f_{xy}^2.$$

For example, if $f_{xx} + f_{yy} = 0$, then K is always either less than or equal to zero.

VI. The geometrical meaning of the Gaussian curvature. Let us choose, for a given point of the surface, an orthogonal frame (x, y, z), where the z-axis is normal to the surface. Then locally the surface is written as $z = f(x, y)$, where $f_x = f_y = 0$ at this point.

At our point we get:

$$g_{ij} = \delta_{ij} = \begin{cases} 1, & i = j, \\ 0, & i \neq j. \end{cases}$$

since $g_{11} = f_x^2 + 1$, $g_{12} = f_x f_y$, $g_{22} = f_y^2 + 1$. Next, $L = b_{11} = f_{xx}$, $M = b_{12} = f_{xy}$, $N = b_{22} = f_{yy}$.

We shall consider the three cases depicted in Figure 27:

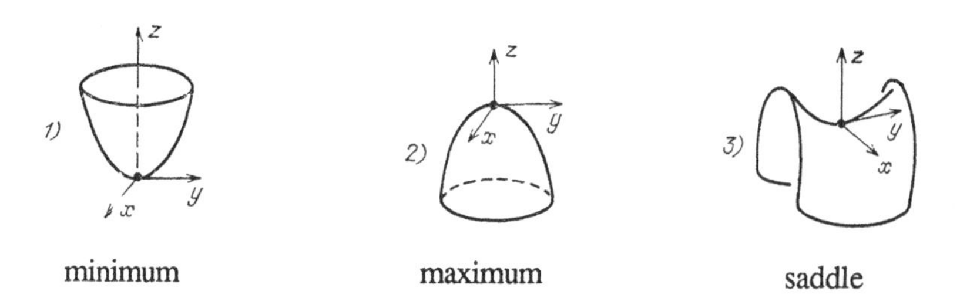

Figure 27.

1) $K > 0, \lambda_1 > 0, \lambda_2 > 0$ (the function $f(x, y)$ is minimal when $x = x_0, y = y_0$);

2) $K > 0, \lambda_1 < 0, \lambda_2 < 0$ (the function $f(x, y)$ is maximal when $x = x_0, y = y_0$);

3) $K < 0$, and therefore $\lambda_1 > 0, \lambda_2 < 0$ or vice versa (this is a saddle or a "path").

Conclusion. When K is locally positive, the surface lies on one side of the tangent plane in a neighbourhood of the point under study. When K is negative, the surface necessarily intersects the tangent plane.

If the Gaussian curvature is everywhere positive, then we may say that the surface is strictly positive, as, for instance, an ellipsoid.

Thus the property that the Gaussian curvature is positive in a neighbourhood of a given point is sufficient for the curvature to be *locally* convex. It should be recalled that a closed surface in a three-dimensional space is called *globally convex* if it bounds a convex region in $\mathbb{R}^3$, i.e. a region which contains, along with any two of its points, the whole of the straight line segment joining these points. As a visual geometrical exercise, we suggest that the reader prove the following : *any region with a locally convex boundary is globally convex.*

EXERCISE. Show that for the ellipsoid, and for the two-sheeted hyperboloid, K is positive and for the one-sheeted hyperboloid, K is negative.

For the cylinder, K is zero (whatever the base line of the cylinder). K is also zero for the cone.

An important class of surfaces of negative Gaussian curvature is that parametrized by $z = f(x, y)$, where $f_{xx} + f_{yy} = 0$. Such functions f are called *harmonic* (e.g. the reader can verify that $f(x, y) = \ln (x^2 + y^2)^{1/2}, f_{xx} + f_{yy} = 0$.

EXAMPLE. Suppose $w = x + iy$, where $i = (-1)^{1/2}$ and $f(x, y)$ is the real part of the polynomial $P(w) = a_0 w^n + \ldots + a_{n-1} w + a_n$; the reader may check that $f_{xx} + f_{yy} \equiv 0$.

Interesting geometric relations arise on the surface of negative curvature. Suppose the Gaussian curvature K is strictly negative. Then in the neighbourhood of each point on the surface we can introduce local regular coordinates p and q, such that relative to them, the second quadratic form $L\,dp^2 + 2M\,dp\,dq + N\,dq^2$ assumes the form: $2M\,dp\,dq$. If in addition the Gaussian curvature is constant, e.g. $K \equiv -1$, then it turns out that we may assume the coefficients of the first quadratic form $E + dp^2 + 2F\,dpdq + G\,dq^2$ to satisfy the relations $\partial E/\partial q = \partial G/\partial p = 0$. i.e. $E = E(p)$, $G = G(q)$. We shall now introduce on the surface new local coordinates u and v, putting $u = \int_{p_0}^{p} \sqrt{E(p)}\,dp$ $\int_{q_0}^{q} \sqrt{G(q)}\,dq$. Then relative to these new coordinates, the first and second quadratic forms become $du^2 + 2F(u, v)\,du\,dv + dv^2$ and $2M(u, v)\,du\,dv$. Consider on the surface the coordinate lines $u =$ const. and $v =$ const. These are usually called *asymptotic lines*. Consider on the surface the function $\omega(u, v)$ defined as $F = \cos\omega$. In other words, ω is the angle between asymptotic lines at a given point. When $K \equiv -1$, the function ω satisfies the following differential equation $\omega_{uv} = \sin\omega$ (the reader may prove it himself), occasionally referred to as "sine-Gordon" *equation.*

It should be emphasized that our definition of asymptotic lines has sense only for surfaces of strictly negative Gaussian curvature. When K is positive, these lines do not exist.

VII. Invariants of a pair of quadratic forms, principal directions and principal curvatures (eigenvalues). Euler's formulae. Suppose that we are given a pair of quadratic forms on the plane ($n = 2$) with basis vectors e_1 (for a surface $e_1 = r_u$) and e_2 (for a surface $e_2 = r_v$), the vector $e = \xi^1 e_1 + \xi^2 e_2$. The first quadratic form:

$$g_{ij}\xi^i\xi^j = |e|^2 = ee > 0,$$

the second quadratic form:

$$b_{ij}\xi^i\xi^j, \quad b_{11} = L, \quad b_{12} = b_{21} = M, \quad b_{22} = N.$$

The matrices have the form:

$$Q = \begin{pmatrix} L & M \\ M & N \end{pmatrix}, \quad G = \begin{pmatrix} g_{11} & g_{12} \\ g_{12} & g_{22} \end{pmatrix}.$$

Let us write the linear equations:

$$(L - \lambda g_{11})\,\xi^1 + (M - \lambda g_{12})\,\xi^2 = 0,$$

$$(M - \lambda g_{12})\,\xi^1 + (N - \lambda g_{22})\,\xi^2 = 0.$$

The solution of these equations exists only for $\lambda = \lambda_1$ or $\lambda = \lambda_2$ where λ_1, λ_2 are the roots of the equation $\det(Q - \lambda G) = 0$.

Let $\lambda_1 \neq \lambda_2$. Recall that λ_1 and λ_2 are the eigenvalues (principal curvatures).

Substitute $\lambda = \lambda_1$ and $\lambda = \lambda_2$ into the linear equations and find two non-zero solutions of these equations:

$$\lambda = \lambda_1 \text{ vector } \bar{e}_1 = (\xi^1, \xi^2),$$

$$\lambda = \lambda_2 \text{ vector } \bar{e}_2 = (\eta^1, \eta^2).$$

We have:

a) $$(L - \lambda_1 g_{11})\,\xi^1 + (M - \lambda_1 g_{12})\,\xi^2 = 0,$$

$$(M - \lambda_1 g_{12})\,\xi^1 + (N - \lambda_1 g_{22})\,\xi^2 = 0;$$

b) $$(L - \lambda_2 g_{11})\,\eta^1 + (M - \lambda_2 g_{12})\,\eta^2 = 0,$$

$$(M - \lambda_2 g_{12})\,\eta^1 + (N - \lambda_2 g_{22})\,\eta^2 = 0.$$

Normalize the vectors $\bar{e}_1 = (\xi^1, \xi^2)$ and $\bar{e}_2 = (\eta^1, \eta^2)$; let

$$|\bar{e}_1|^2 = \sum g_{ij}\xi^i\xi^j = 1, \quad |\bar{e}_2|^2 = \sum g_{ij}\eta^i\eta^j = 1.$$

An important property of the vectors $\bar{e}_1, \bar{e}_2$ is that they are orthogonal provided that $\lambda_1 \neq \lambda_2$, i.e.

$$\bar{e}_1\bar{e}_2 = \sum g_{ij}\xi^i\eta^j = 0.$$

We assume the vectors $\bar{e}_1, \bar{e}_2$ to be the new basis, and for any vector e we have:

$$e = \bar{x}^1\bar{e}_1 + \bar{x}^2\bar{e}_2 = x^1 e_1 + x^2 e_2.$$

Here $(\bar{x}^1, \bar{x}^2)$ and (x^1, x^2) denote the components of the vector relative to the bases $(\bar{e}_1, \bar{e}_2)$ and (e_1, e_2), respectively.

In the new basis $\bar{e}_1, \bar{e}_2$, the first and second quadratic forms are:

$$1)\ g_{ij} x^i x^j = \bar{g}_{ij}\, \bar{x}^i \bar{x}^j = (\bar{x}^1)^2 + (\bar{x}^2)^2,$$

(the first quadratic form), $\bar{g}_{ij} = \delta_{ij}$,

$$2)\ b_{ij} x^i x^j = \bar{b}_{ij}\, \bar{x}^i \bar{x}^j = \lambda_1 (\bar{x}^1)^2 + \lambda_2 (\bar{x}^2)^2,$$

(the second quadratic form) $\bar{e}_{12} = \lambda_1,\ \bar{e}_{22} = \lambda_2,\ \bar{b}_{12} = 0.$

The directions of the vectors $\bar{e}_1, \bar{e}_2$ are called *eigenvectors* or *principal* directions.

It was essential that $\lambda_1 \neq \lambda_2$.

The Euler formula (algebraic): the ratio of two quadratic forms is equal to (for the vector $e = \bar{x}^1\bar{e}_1 + \bar{x}^2\bar{e}_2$)

$$\frac{\bar{b}_{ij}\bar{x}^i\bar{x}^j}{\bar{g}_{ij}\bar{x}^i\bar{x}^j} = \lambda_1 \cos^2\phi + \lambda_2 \sin^2\phi,$$

where ϕ is the angle between the vector $e = \bar{x}^1\bar{e}_1 + \bar{x}^2\bar{e}_2$ and the eigenvector $\bar{e}_1$ where λ_1, λ_2 are the eigenvalues. Note that

$$\cos^2\phi = \frac{(e\bar{e}_1)^2}{|e|^2 \cdot |\bar{e}_1|^2} \qquad (|\bar{e}_1|^2 = 1).$$

By definition:

$$\sin^2\phi = \frac{(e\bar{e}_2)^2}{|e|^2 \cdot |\bar{e}_2|^2} \qquad (|\bar{e}_2|^2 = 1),$$

(since $\bar{e}_1 \perp \bar{e}_2$), and for any vector e by definition:

$$|e|^2 = \bar{g}_{ij}\, \bar{x}^i \bar{x}^j,$$

$$e = \bar{x}^1 \bar{e}_1 + \bar{x}^2 \bar{e}_2$$

($\bar{g}_{ij}$ is the Riemannian metric on the plane).

In application to geometry, we have the first quadratic form g_{ij} and the second quadratic form (b_{ij}) at a given point on the surface (on tangent vectors).

We know that the ratio of these two quadratic forms:

$$\frac{b_{ij}\dot{x}^i \dot{x}^j}{g_{ij}\dot{x}^i \dot{x}^j} = \frac{b_{ij}\, dx^i dx^j}{g_{ij}\, dx^i dx^j}$$

is equal (up to the sign) to the curvature of the normal cross-section with the tangent vector $(\dot{u}, \dot{v}) = e = \dot{x}^1 e_1 + \dot{x}^2 e_2$, where $x^1 = u$, $x^2 = v$, $dx^1 = \dot{x}^1\, dt$, $dx^2 = \dot{x}^2\, dt$.

Conclusion. The curvature of the normal cross-section is equal to $k_1 \cos^2 \phi + k_2 \sin^2 \phi$, where $k_1 = \lambda_1$, $k_2 = \lambda_2$ and ϕ is the angle between the tangent vector of the normal cross-section and the principal direction $\bar{e}_1$ at a given point on the surface (Euler's formula).

The proof has been given above. It is very simply deduced on the tangent plane in the basis $e_1\, e_2$ in which the matrices of the quadratic forms are diagonal:

$$\bar{g}_{ij} = \delta_{ij} \text{ and } (\bar{b}_{ij}) = \begin{pmatrix} \lambda_1 & 0 \\ 0 & \lambda_2 \end{pmatrix},$$

since

$$\frac{\lambda_1 (\dot{\bar{x}}^1)^2 + \lambda_2 (\dot{\bar{x}}^2)^2}{(\dot{\bar{x}}^1)^2 + (\dot{\bar{x}}^2)^2} = \lambda_1 \left[\frac{\dot{\bar{x}}^1}{|e|} \right]^2 + \lambda_2 \left[\frac{\dot{\bar{x}}^2}{|e|} \right]^2 = \lambda_1 \cos^2 \phi + \lambda_2 \sin^2 \phi .$$

(Recall that $|\bar{e}_1| = |\bar{e}_2| = 1$, $\bar{e}_1 \perp \bar{e}_2$, $e = \dot{\bar{x}}^1 \bar{e}_1 + \dot{\bar{x}}^2 \bar{e}_2 = \dot{x}^1 e_1 + \dot{x}^2 e_2$).

1.13 The Language of Complex Numbers in Geometry. Conformal Transformations. Isothermal Coordinates.

The majority of problems in geometry are most conveniently formulated in terms of the language of complex numbers. We, therefore, present here the simple facts which we shall need for our further purpose.

Suppose we are given an n-dimensional linear space over the field of complex numbers and that this space has the basis $e_1, e_2, \dots, e_n$, where any vector has the form $\xi = \xi^\alpha e_\alpha$, $\xi^\alpha = x^\alpha + iy^\alpha$ being complex coordinates. From the point of view of real numbers, this is a $2n$-dimensional space over the field of real numbers with the basis $\{e_j, ie_j\}$. In this $2n$-dimensional space, the scalar product in the complex language is given as (the real scalar square of the vector is Euclidean):

$$\langle \xi, \eta \rangle = \sum_{\alpha=1}^{n} \xi^\alpha \bar{\eta}^\alpha; \quad \langle \xi, \xi \rangle = \sum_{\alpha=1}^{n} |\xi^\alpha|^2 \tag{1}$$

and possesses the following properties:

$$\begin{aligned} \langle \lambda\xi, \eta \rangle &= \lambda \langle \xi, \eta \rangle, \\ \langle \xi, \lambda\eta \rangle &= \bar{\lambda} \langle \xi, \eta \rangle, \\ \langle \xi, \eta \rangle &= \overline{\langle \eta, \xi \rangle}. \end{aligned} \tag{2}$$

Any scalar product possessing the properties (2) is called *Hermitian*.

Complex linear transformations A that preserve the scalar product (1),

$$\langle A\xi, A\eta \rangle = \langle \xi, \eta \rangle,$$

are called *unitary*.

From the point of view of reals, the unitary transformation is simply an orthogonal transformation of a $2n$-dimensional real space, which is, at the same time, a complex linear transformation,

$$U_n = O_{2n} \cap GL(n, \mathbb{C}).$$

We use the following notation: $GL(n, \mathbb{C})$, $SL(n, \mathbb{C})$ is a group of complex linear transformations and its sub-group with the determinant equal to 1 (the determinant is complex). Next, U_n, SU_n is a group of unitary transformations and its sub-group with determinant 1.

SIMPLE EXAMPLES. In a one-dimensional space, the group $GL(1, \mathbb{C})$ consists of multiplications by a complex number a, and the sub-group U_1 is multiplication by numbers $a = e^{i\theta}$. In the two-dimensional case $n = 2$, the group U_2 consists of the matrices:

$$\begin{pmatrix} a & b \\ c & d \end{pmatrix}$$

in which $|a|^2 + |b|^2 = 1$, $|c|^2 + |d|^2 = 1$ and there holds the orthogonality condition for the rows: $a\bar{c} + b\bar{d} = 0$. Its sub-group SU_2 is specified by the condition $ad - bc = 1$. Thus, the group SU_2 is described by matrices of the form:

$$\begin{pmatrix} a & b \\ -\bar{b} & \bar{a} \end{pmatrix}$$

where $|a|^2 + |b|^2 = 1$. The reader may check this, as well as the fact that the group of motions of a real plane along the dilatations has, in the complex notation, the form of affine transformation of a complex one-dimensional space:

$$z \to az + b$$

(the sub-group of motions without dilatations is specified by the condition $|a| = 1$ or $a = e^{i\theta}$). The element of length in a complex Euclidean space $(z^1, \ldots, z^n)$ is written in the form:

$$dl^2 = \sum_{\alpha=1}^{n} |dz^\alpha|^2 = \sum_{\alpha=1}^{n} dz^\alpha \, d\bar{z}^\alpha,$$

where $dz^\alpha = dx^\alpha + idy^\alpha$, $d\bar{z}^\alpha = dx^\alpha - idy^\alpha$. Here $z^\alpha = x^\alpha + iy^\alpha$, and the set $(x^1, y^1, \ldots, x^n, y^n)$ represents real Euclidean coordinates in this space (considered here as a $2n$-dimensional real space). The length of any curve $x^\alpha(t)$, $y^\alpha(t)$ (or $z^\alpha = z^\alpha(t)$) is written in the form:

$$l = \int_a^b \Big(\sum_{\alpha=1}^{n} |\dot{z}^\alpha|^2 \Big)^{1/2} dt = \int_a^b \Big(\sum_{\alpha=1}^{n} \dot{z}^\alpha \dot{\bar{z}}^\alpha \Big)^{1/2} dt.$$

It appears convenient, if we pass over, purely formally, from the real variables (x^α, y^α) to the complex variables:

$$z^{\alpha} = x^{\alpha} + iy^{\alpha}, \qquad \bar{z}^{\alpha} = x^{\alpha} - iy^{\alpha},$$

$$x^{\alpha} = 1/2(z^{\alpha} + \bar{z}^{\alpha}), \quad y^{\alpha} = 1/2i(z^{\alpha} - \bar{z}^{\alpha}). \tag{3}$$

Let us introduce differential operators:

$$\frac{\partial}{\partial z^{\alpha}} = 1/2\left(\frac{\partial}{\partial x^{\alpha}} - i\frac{\partial}{\partial y^{\alpha}}\right), \qquad \frac{\partial}{\partial \bar{z}^{\alpha}} = 1/2\left(\frac{\partial}{\partial x^{\alpha}} + i\frac{\partial}{\partial y^{\alpha}}\right). \tag{4}$$

Note that there hold the identities:

$$\frac{\partial}{\partial \bar{z}^{\alpha}}(z^{\alpha}) = 0, \qquad \frac{\partial}{\partial z^{\alpha}}(\bar{z}^{\alpha}) = 0,$$

$$\frac{\partial}{\partial z^{\alpha}}(z^{\alpha}) = 1, \qquad \frac{\partial}{\partial \bar{z}^{\alpha}}(\bar{z}^{\alpha}) = 1. \tag{5}$$

The following assertion can be verified, immediately.

LEMMA 1 *The differential of an arbitrary complex-valued function* $f(x^1, y^1, \dots$ $\dots, x^n y^n)$ *has the form*:

$$df = \frac{\partial f}{\partial z^1} dz^1 + \dots + \frac{\partial f}{\partial z^n} dz^n + \frac{\partial f}{\partial \bar{z}^1} d\bar{z}^1 + \dots + \frac{\partial f}{\partial \bar{z}^n} d\bar{z}^n.$$

We can verify this by calculation.

We now turn to considering an arbitrary polynomial of a certain degree with complex coefficients $P(x^1, y^1, \dots, x^n, y^n)$ of the variables $(x^1, y^1, \dots, x^n, y^n)$.

After the change of variables (3) we obtain from the polynomial $P(x^1, y^1, \dots,$ $x^n, y^n)$, the polynomial:

$$Q(z^1, \bar{z}^1, \dots, z^n, \bar{z}^n) = P(x^1, y^1, \dots, x^n, y^n).$$

The following assertion holds.

THEOREM 1. *After the change of variables* (3), *any polynomial*

$$P(x^1, y^1, \dots, x^n, y^n),$$

depends on the variables $z^1, \ldots, z^n$ *only and does not depend on* $\bar{z}^1, \bar{z}^2, \ldots, \bar{z}^n$ *if and only if there hold the identities*:

$$\partial P/\partial \bar{z}^\alpha \equiv 0, \quad \alpha = 1, \ldots, n$$

(these are referred to as Cauchy-Riemann conditions or the complex analyticity conditions).

Proof. The operators $\partial/\partial z^\alpha$ and $\partial/\partial \bar{z}^\alpha$ possess the following property (Leibniz formula):

$$\frac{\partial}{\partial z^\alpha}(fg) = \frac{\partial f}{\partial z^\alpha} g + f \frac{\partial g}{\partial z^\alpha}, \quad \frac{\partial}{\partial \bar{z}^\alpha}(fg) = \frac{\partial f}{\partial \bar{z}^\alpha} g + f \frac{\partial g}{\partial \bar{z}^\alpha}.$$

Next, by virtue of the identities (5), $\partial/\partial \bar{z}^\alpha (z^\alpha) = 0$. From this we obtain, using the Leibniz formula, that:

$$\partial/\partial \bar{z}^\alpha (z^\alpha)^k \equiv 0, \quad \partial/\partial \bar{z}^\alpha (\bar{z}^\alpha)^k \equiv k(\bar{z}^\alpha)^{k-1}.$$

LEMMA 2. *If the polynomial* $Q(z^1, \bar{z}^1, \ldots, z^n, \bar{z}^n)$ *has at least one non-zero coefficient, then the polynomial* $P(x, y)$, *which corresponds to it after the change of variables* (3), *is non-zero.*

Lemma 2 follows from the fact that the change of variables (3) *has a non-zero determinant seeing that* $\bar{z}$, z *are independent.*

Conclusion of proof of Theorem 1. Let now the polynomial P depend on $\bar{z}^\alpha$ and let the variable $\bar{z}^\alpha$ enter in it in the (maximal) power n.

We shall show that $\partial P/\partial \bar{z}^\alpha \neq 0$. The polynomial P has the form:

$$P = A_0(\bar{z}^\alpha)^n + A_1(\bar{z}^\alpha)^{n-1} + \ldots + A_n,$$

where $A_0, A_1, \ldots, A_n$ are polynomials of all variables $z^1, \ldots, z^n$ and all $\bar{z}^j$ except $\bar{z}^\alpha$. Obviously, on the basis of the Leibniz formula, we have:

$$\partial/\partial \bar{z}^\alpha P = A_0 \cdot n(\bar{z}^\alpha)^{n-1} + A_1(n-1)(\bar{z}^\alpha)^{n-2} + \ldots + A_{n-1},$$

since $\partial A_i/\partial \bar{z}^\alpha \equiv 0$ (A_i does not depend on $\bar{z}^\alpha$). So long as $A_0 \not\equiv 0$, we have $\partial P/\partial \bar{z}^\alpha \not\equiv 0$, which concludes the proof.

REMARK. The theorem is applicable not only to polynomials, but also to convergent power series: independence of variables $\bar{z}^\alpha$ is equivalent to the conditions $\partial f/\bar{z}^\alpha \equiv 0$. Such complex-valued functions $f(x^1, y^1, \dots, x^n, y^n)$ for which there hold the identities $\partial f/\bar{z}^\alpha = 0$, $\alpha = 1, \dots, n$, are therefore referred to as *complex analytic* functions.

For functions of two real (one complex) variables, $f(x, y) = f(z, \bar{z})$, where $z = x + iy$, $\bar{z} = x - iy$, the analyticity condition is:

$$\frac{\partial f}{\partial \bar{z}} = \frac{\partial f}{\partial x} + i\frac{\partial f}{\partial y} \equiv 0,$$

or, if $f(x, y) = u(x, y) + iv(x, y)$, then:

$$\frac{\partial u}{\partial x} \equiv \frac{\partial v}{\partial y}, \quad \frac{\partial u}{\partial y} \equiv -\frac{\partial v}{\partial x}. \tag{6}$$

Equations (6) are called *Cauchy-Riemann equations*. From (6) it obviously follows that:

$$\left(\frac{\partial^2}{\partial x^2} + \frac{\partial^2}{\partial y^2}\right) u = 0, \quad \left(\frac{\partial^2}{\partial x^2} + \frac{\partial^2}{\partial y^2}\right) v = 0.$$

The operator $\dfrac{\partial^2}{\partial x^2} + \dfrac{\partial^2}{dy^2} = \dfrac{\partial^2}{\partial z\, \partial \bar{z}}$ is called the *Laplace operator*.

DEFINITION 1. The map $z \to w(z)$ of a complex plane is called *conformal* if there holds the complex analyticity or analyticity condition:

$$\partial w/\partial \bar{z} \equiv 0 \text{ or } \partial w/\partial z \equiv 0.$$

The simplest examples are:

1) affine transformations

$$z \to az + b = w(z);$$

2) linear fractional transformations:

$$z \to \frac{az+b}{cz+d} = w(z);$$

3) transformation given by a rational function:

$$z \to \frac{a_0 z^n + \ldots + a_n}{b_0 z^m + \ldots + b_m} + w(z);$$

4) transformations given by the exponent, trigonometrical functions, etc.:

$$w(z) = e^z, \quad w = \sin z, \quad w = \operatorname{sh} z, \ldots .$$

The differential of the function $f(x, y) = f(z, \bar{z})$ satisfies the equality:

$$df = \frac{\partial f}{\partial x} dx + \frac{\partial f}{\partial y} dy = \frac{\partial f}{\partial z} dz + \frac{\partial f}{\partial \bar{z}} d\bar{z},$$

where

$$f(x, y) = u(x, y) + iv(x, y).$$

For the complex analytic function $f(x, y) = f(z)$, this formula assumes the form $df = \partial f/dz\, dz$ since $\partial f/\partial \bar{z} \equiv 0$.

The differential of the complex-valued function $f(x^1, y^1, y^2\, x^2) = f(z^1, z^2, \bar{z}^1, \bar{z}^2)$ of two complex variables is equal to

$$df\ \sum_{\alpha} \frac{\partial f}{\partial z^{\alpha}} dz^{\alpha} + \sum_{\alpha} \frac{\partial f}{\partial \bar{z}^{\alpha}} d\bar{z}^{\alpha}.$$

Provided that there holds the condition $\partial f/\partial \bar{z}^2 \equiv 0$, $\alpha = 1, 2$, we have:

$$df\ \sum_{\alpha=1}^{2} \frac{\partial f}{\partial z^{\alpha}} dz^{\alpha}, \quad \text{where } f = u + iv.$$

A two-dimensional surface can be given by the equation (one complex equation):

$$f(z^1, z^2) = 0$$

(these are two real equations $u = 0$ and $v = 0$, where $f = u + iv$).

We can obviously introduce a complex gradient for complex analytic functions,

$$\nabla_c f = \left(\frac{\partial f}{\partial z^1}, \frac{\partial f}{\partial z^2}\right).$$

We shall later make use of a complex analogue of the implicit function theorem. This complex analogue of the implicit function theorem is as follows:

$$\text{if } \partial f/\partial \bar{z}^{\alpha} \equiv 0, \ \alpha = 1, 2, \text{ and } \nabla_c f \neq 0 \ \ (\text{let } \partial f/\partial z^1 \neq 0)$$

at a given point (z_0^1, z_0^2) of the surface $f = 0$, then in a sufficiently small neighbourhood of this point the equation $f(z^1, z^2) = 0$ has a unique solution which is, at the same time, complex analytic:

$$z^1 = \phi(z^2), \ f(\phi(z^2), z^2) = 0,$$

where $z_0^1 = \phi(z_0^2)$, $\partial\phi/\partial\bar{z}_2 = 0$. Such points are called "non-singular".

Let $f(z^1, z^2)$ be the polynomial $P(z^1, z^2)$. Then the totality of the solutions of the equation $P(z^1, z^2) = 0$ of the form $z^1 = \phi(z^2)$ is called a *multiple-valued algebraic function*, and the surface $P(z^1, z^2) = 0$ is called the *graph* or the *Riemann surface of this multiple-valued algebraic function.*

If $P(z^1, z^2) = (z^1)^q - P_n(z^2)$, where $P_n(z^2)$ is a polynomial of degree n, we obtain:

$$z^1 = \sqrt[q]{P_n(z^2)}, \ z^1 = w, \ z^2 = z.$$

Let us consider a complex gradient:

$$\nabla_c P(w, z) = \left(\frac{\partial P}{\partial w}, \frac{\partial P}{\partial z}\right) = \left(qw^{q-1}, \frac{\partial P_n}{\partial z}\right).$$

What are the zeros of this gradient? To answer this question for the case $q \geq 2$, it is necessary and sufficient to solve the equation

$$w = 0, \ \partial P/\partial z = 0.$$

The zeros of the gradient get onto the surface $w = \sqrt[q]{P_n(z)}$ provided that the equations $w = 0$ and $\partial P_n/\partial z = 0$ have a common solution with the equation $P_n(z) - w^q = 0$. Obviously, this is possible if and only if the polynomial $P_n(z)$ has at least one multiple root. This completes the proof of the theroem which follows.

THEOREM 2. *The set of all complex solutions of the equation* $0 = w^q - P_n(z)$, $q \geq 2$, *consists of non-singular points if and only if the polynomial* $P_n(z)$ *has no multiple roots.*

The complex implicit function theorem allows us now to introduce in a neighbourhood of the non-singular point (w_0, z_0) on the surface $P(w, z) = 0$ the local coordinate a) z in the case $\partial P/\partial w|_{(z_0,w_0)} \neq 0$. Then in the neighbourhood of this point, we have:

$$w = w(z), \quad P(w(z), z) \equiv 0, \quad \partial w/\partial \bar{z} \equiv 0;$$

or else, we may introduce the local coordinate b) w in the case $\partial P/\partial z|_{(z_0,w_0)} \neq 0$. Then we have

$$z = z(w), \quad P(w, z(w)) \equiv 0, \quad \partial z/\partial \bar{w} \equiv 0.$$

In the space (z, w) we are given the Hermitian metric:

$$dl^2 = |dz|^2 + |dw|^2 = dz\, d\bar{z} + dw\, d\bar{w}. \tag{7}$$

The surface $P(z, w) = 0$ (where $\nabla_c P \neq 0$ in a neighbourhood of the non-singular point) is given parametrically in the form:

$$z = z(t), \quad w = w(t), \quad t = (u + iv);$$

where t is the complex parameter and $\partial z/\partial \bar{t} \equiv 0$, $\partial w/\partial \bar{t} \equiv 0$. On the surface we obtain:

$$dl^2 = dz\, d\bar{z} + dw\, d\bar{w} = \left(\left|\frac{dz}{dt}\right|^2 + \left|\frac{dw}{dt}\right|^2\right) dt\, d\bar{t}.$$

In case a) we have $t = z$, and therefore

$$dz\, d\bar{z} + dw\, d\bar{w} = dl^2 = \left(1 + \left|\frac{dw}{dt}\right|^2\right) dz\, d\bar{z}$$

on the surface given by the equation

$$w = w(z), \quad \partial w/\partial \bar{z} \equiv 0.$$

Returning again to the real coordinates x, y, where $z = x + iy$, we obtain, from formula (7), in the region of local coordinates the following formula for the square of the length:

$$dl^2 = f^2(x, y)\,(dx^2 + dy^2). \tag{8}$$

Now let us go back to surfaces in $\mathbb{R}^3$.

DEFINITION 2. Local coordinates x, y in a neighbourhood of a certain point on the surface are called *conformal* if the Riemannian metric of the surface induced by embedding the surface into $\mathbb{R}^3$ has, in these coordinates, the form analagous to (8):

$$g_{12} = g_{21} = 0,$$

$$g_{11} = g_{22} = f^2(x, y),$$

$$dl^2 = f^2(x, y)\,(dx^2 + dy^2). \tag{9}$$

The form of the metric (9) is called *conformal Euclidean* and the coordinates are called isothermal. Consider other conformal coordinates u, v, in the neighbourhood of the same point (Figure 28). Let $t = u + iv$.

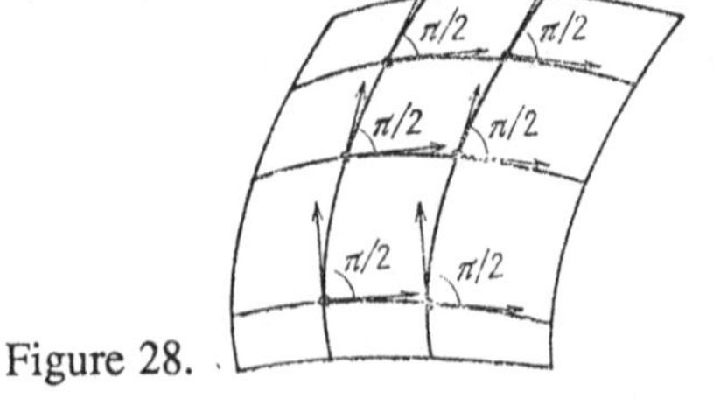

Figure 28.

THEOREM 3. *The transformation from one set of conformal coordinates to another set of conformal coordinates is called conformal transformation $x = x(u, v)$. This means that either $\partial z/\partial \bar{t} = 0$ or $\partial z/\partial t = 0$, where $z = z(t, \bar{t})$.*

Proof. Let us write the formula for the transformation of the metric g_{ij} under the change of coordinates, where g'_{kl} denotes the new metric in the coordinates u, v. Let $x^1 = x, x^2 = y, u^1 = u, u^2 = v$. Then:

$$g'_{kl}(u^1, u^2) = g_{ij} \frac{\partial x^i}{\partial u^k} \frac{\partial x^j}{\partial u^l} .$$

Assuming $z = x + iy$, $t = u + iv$, $z = z(t, \bar{t})$ and taking into account the condition $g_{12} = g_{21} = 0$, $g_{11} = g_{22} = f^2$ and also the condition $g'_{12} = g'_{21} = 0$, $g'_{11} = g'_{22} = (f')^2$, we obtain that $dz\, d\bar{z}$ is transformed into the metric:

$$\frac{\partial z}{\partial t} \frac{\partial \bar{z}}{\partial t} (dt)^2 + \frac{\partial z}{\partial t} \frac{\partial \bar{z}}{\partial \bar{t}} dt\, d\bar{t} + \frac{\partial z}{\partial \bar{t}} \frac{\partial \bar{z}}{\partial t} dt\, d\bar{t} +$$

$$+ \frac{\partial z}{\partial \bar{t}} \frac{\partial \bar{z}}{\partial \bar{t}} (d\bar{t})^2 = \left(\left| \frac{\partial z}{\partial t} \right|^2 + \left| \frac{\partial z}{\partial \bar{t}} \right|^2 \right) dt\, d\bar{t} +$$

$$+ \frac{\partial z}{\partial t} \frac{\partial \bar{z}}{\partial t} (dt)^2 + \overline{\left(\frac{\partial z}{\partial \bar{t}} \frac{\partial z}{\partial t} (dt)^2 \right)} .$$

This exactly implies that either $\partial z/\partial t = 0$ or $\partial z/\partial \bar{t} = 0$, as required.

Suppose that a two-dimensional surface $M^2 \subset \mathbb{R}^3$ is given parametrically, that is, $x = x(p, q)$, $y = y(p, q)$, $z = z(p, q)$, where (p, q) vary in a certain domain D (Figure 29).

On the surface M^2 we consider the Riemannian metric $ds^2 = E\, dp^2 + 2F\, dp\, dq + G\, dq^2$ induced by the embedding $M^2 \subset \mathbb{R}^3$ This means that $E = (r_p, r_p)$; $F = (r_p, r_q)$; $G = (r_q, r_q)$, where r is the radius vector of the surface M^2. Since $ds^2(M^2) = ds^2(\mathbb{R}^3)|_{M^2}$ and $ds^2(\mathbb{R}^3) = dx^2 + dy^2 + dz^2$ is a positive definite form, it follows that $ds^2(M^2)$ is also a positive definite quadratic form, that is, $g = EG - F^2 > 0$, and we can consider the real-valued function $(g)^{1/2} = (EG - F^2)^{1/2}$. Suppose that P_0 is a non-singular point on the surface.

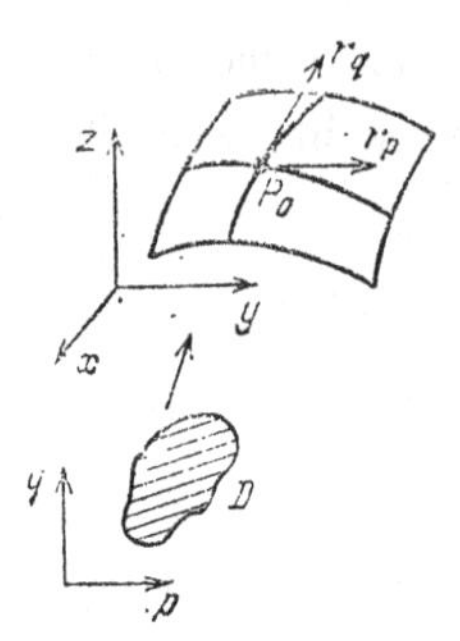

Figure 29.

If we change to the new coordinates u and v on M^2, then the form ds^2 will undergo the corresponding transformation, and it is, therefore, quite natural to ask the following: *what is the most simple form to which we can reduce the form $ds^2(M^2)$ by choosing different local systems of coordinates on the surface*? The theorem which follows answers this question.

THEOREM 4. (without proof). *Let $M^2 \subset \mathbb{R}^3$ be a surface given in certain parameters (local regular coordinates) p and q. Suppose that the metric $ds^2 = E\,dp^2 + 2F\,dp\,dq + G\,dq^2$ is smooth, that is, the coefficients of the form $E(p, q)$; $F(p, q)$; $G(p, q)$ are C^2 smooth functions of p and q. Then in a certain neighbourhood U of the point $P_0 \in M^2$ we can introduce new isothermal coordinates u and v, such that the metric is ds^2 in these coordinates assumes the following conformal Euclidean form*:

$$ds^2 = f(u, v)\,(du^2 + dv^2).$$

The existence of these coordinates is reduced to the solution of the so-called *Beltrami's equations* in the following way.

Let us expand the quadratic forms ds^2 into factors:

$$ds^2 = \left((E)^{1/2}dp + \frac{F + i(g)^{1/2}}{(E)^{1/2}}\,dq\right) \cdot \left((E)^{1/2}dp + \frac{F - i(g)^{1/2}}{(E)^{1/2}}\,dq\right).$$

We seek the new coordinates u and v as functions of p, q: $u = u(p, q)$, $v = v(p, q)$. We wish to represent ds^2 in the form $(du + idv)(du - idv) = du^2 + dv^2$. This can be achieved if we succeed in choosing the integration factor, that is, such a complex-valued function $\lambda = \lambda(p, q)$ that there hold two identities:

$$\lambda \cdot \left((E)^{1/2} dp + \frac{F + i(g)^{1/2}}{(E)^{1/2}} dq\right) = du + i\,dv,$$

$$\bar{\lambda}\left((E)^{1/2} dp + \frac{F - i(g)^{1/2}}{(E)^{1/2}} dq\right) = du - i\,dv.$$

It should be noted that the second of these identities is obtained from the first one by means of complex conjugation. Indeed, if such a function $\lambda(p, q)$ is found, then multplying the two identities we arrive at:

$$|\lambda|^2 ds^2 = du^2 + dv^2; \quad ds^2 = |\lambda|^{-2}(du^2 + dv^2)$$

and may set $f(u, v) = f(u(p, q), v(p, q)) = |\lambda|^{-2}$. Thus, the unknown functions are $u(p, q)$; $v(p, q)$; $\lambda(p, q)$. These functions must satisfy the equation:

$$\lambda\left((E)^{1/2} dp + \frac{F + i(g)^{1/2}}{(E)^{1/2}} dq\right) = du + i\,dv =$$

$$= \left(\frac{\partial u}{\partial p} + i\frac{\partial v}{\partial p}\right) dp + \left(\frac{\partial u}{\partial q} + i\frac{\partial v}{\partial q}\right) dq,$$

whence

$$\lambda(E)^{1/2} = \frac{\partial u}{\partial p} + i\frac{\partial v}{\partial p}; \quad \lambda \cdot \frac{F + i\,(g)^{1/2}}{(E)^{1/2}} = \frac{\partial u}{\partial q} + i\frac{\partial v}{\partial q}.$$

Elimination of λ gives:

$$\left(F + i\,(g)^{1/2}\right) \cdot \left(\frac{\partial u}{\partial p} + i\frac{\partial v}{\partial p}\right) = E\left(\frac{\partial u}{\partial q} + i\frac{\partial v}{\partial q}\right)$$

or

$$F\frac{\partial u}{\partial p} - (g)^{1/2}\frac{\partial v}{\partial p} = E\frac{\partial u}{\partial q}; \quad (g)^{1/2}\frac{\partial u}{\partial p} + F\frac{\partial v}{\partial p} = E\frac{\partial v}{\partial q}.$$

From this we have:

$$\frac{\partial v}{\partial p} = \frac{F\dfrac{\partial u}{\partial p} - E\dfrac{\partial u}{\partial q}}{(g)^{1/2}}; \qquad \frac{\partial v}{\partial q} = \frac{G\dfrac{\partial u}{\partial p} - F\dfrac{\partial v}{\partial q}}{(g)^{1/2}},$$

$$\frac{\partial u}{\partial p} = \frac{E\dfrac{\partial v}{\partial q} - F\dfrac{\partial v}{\partial p}}{(g)^{1/2}}; \qquad \frac{\partial u}{\partial q} = \frac{F\dfrac{\partial v}{\partial q} - G\dfrac{\partial v}{\partial p}}{(g)^{1/2}}.$$

Since $\dfrac{\partial^2}{\partial p\,\partial q} = \dfrac{\partial^2}{\partial q\,\partial p}$, we obtain the following equations: $L(u) = 0$, $L(v) = 0$, where the differential operator L has the form:

$$L = \frac{\partial}{\partial q}\left[\frac{F\dfrac{\partial}{\partial p} - E\dfrac{\partial}{\partial q}}{(EG - F^2)^{1/2}}\right] + \frac{\partial}{\partial p}\left[\frac{F\dfrac{\partial}{\partial q} - G\dfrac{\partial}{\partial p}}{(EG - F^2)^{1/2}}\right].$$

The equation $L(f) = 0$ is called *Beltrami's equation*, and the operator L, *Beltrami's* operator. Thus, we have found that the unknown functions u and v must satisfy Beltrami's equation. It is a well-known fact of the theory of differential equations that if the functions E, F, G are smooth, then the equation $L(f) = 0$ always has a solution. Since, in our case, the functions E, F, G are smooth by the assumption, all the unknown functions $u(p, q)$, $v(p, q)$, $\lambda(p, q)$ are determined.

Note that the coordinates u, v serve, generally speaking, only for a certain neigbourhood of the point P_0.

1.14 The Concept of a Manifold and the Simplest Examples

In cartography, there exist several ways of drawing maps of the earth's surface. All of them are necessarily reduced to one procedure, namely, to projecting the convex spherical surface of the globe onto a plane. It is more or less obvious that to make a one-to-one and continuous projection of the whole sphere into a certain region of a plane is impossible. Moreover, in attempting to project onto a flat map large enough pieces of the earth's surface, we inevitably introduce distortions. Therefore, cartographers resort to various contrivances to the effect that the sphere is cut into several sufficiently small pieces, each of which is projected separately into part of a plane. The original sphere is reconstructed from them by the reverse operation of glueing together according to the rules usually indicated on a flat map. Thus, a rather complicated object (sphere) is obtained from several simpler objects by glueing them together along their common part. Precisely this idea is an underlying one in the construction of a wide class of geometrical objects which are called manifolds.

The most clear formulation of the concept of a manifold is due to K.F. Gauss who porposed his definition in mathematical terms in connection with his studies in the field of geodesy and cartography of the earth's surface. In the practical mapping of sufficiently large regions of the earth's surface, these regions are sub-divided into smaller, partially overlapping, ones, each of which is assigned to a certain group of cartographers. They draw a map of each separate region endowed with reference points (landmarks, etc.) (Figure 30).

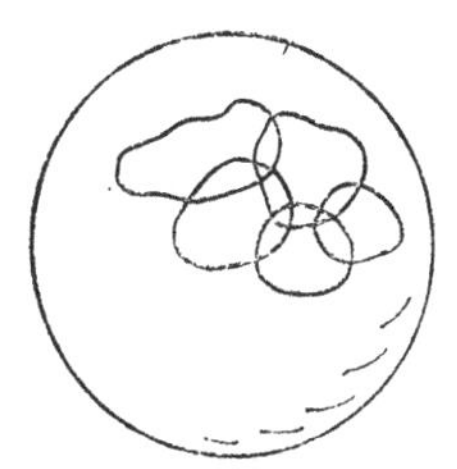

Figure 30.

In forming the total atlas, these maps are sewn or glued together. Those parts which were overlapping are reflected in several local maps. Adjusting individual local maps is realized by comparison and imposition of their common reference points. This procedure underlies the very important mathematical concept of a *manifold*. The simplest examples of manifolds are *surfaces* of certain dimension in Euclidean space.

If a surface in Euclidean 3-space is given by the equation:

$$f_1(x^1, \dots, x^n) = 0, \dots, f_{n-k}(x^1, \dots, x^n) = 0, \tag{1}$$

then in a neighbourhood of any *non-singular* point $x_0^1, \dots, x_0^n$, such that the rank of the matrix $(\partial f_q/\partial x^\alpha)$ is exactly equal to $n-k$, we can introduce local coordinates. Indeed, suppose the minor which is not equal to zero is $(\partial f_q/\partial x^\alpha)$, where $\alpha = i_1, \dots, i_{n-k}$; then in the neighbourhood of the point $x_0^1, \dots, x_0^n$ on the surface we choose, as local coordinates, the missing variables $x^{j_i} = z^1, \dots, x^{j_k} = z^k$ and solve equations (1) in the neighbourhood of this point by the implicit function theorem:

$$x^{i_1} = x^{i_1}(z^1, \dots, z^k),$$

$$\dots\dots\dots\dots$$

$$x^{i_{n-k}} = x^{i_{n-k}}(z^1, \dots, z^k). \tag{2}$$

We obtain the parametric representation of the surface relative to the variables $z^1, z^2, \dots, z^k$ in a neighbourhood of the point of interest. In a neighbourhood of each non-singular point of the surface we can, generally speaking, set special local coordinates. To calculate the length of any curve on the surface we can, in a domain near each non-singular point, use the local coordinates associated with the neighbourhood of this point; the length of any curve can be calculated by pieces positioned in each coordinate region. In general, by definition the length of a curve (and of any vector) does not depend on the choice of coordinates.

On this ground, we can give the general definition of a differentiable manifold.

DEFINITION 1. A *differentiable (smooth) manifold* is an arbitrary set M of points endowed with the following structure called the "atlas": the set M is covered with a collection of its sub-sets U_q called "local charts" (rather than "maps"), i.e. $M = \bigcup_q U_q$.

There exists a one-to-one correspondence ϕ_q between each set U_q and a certain open region V_q of Euclidean space $\mathbb{R}^n$ with coordinates $y^1, \dots, y^n$. This correspondence introduces into the set U_q a family of functions called local coordinates:

$$x_q^\alpha(P) = y^\alpha(\phi_q(P)).$$

One and the same point of the set M may belong to different local charts: $P \in U_p \cap U_q$. In the intersection of local charts $U_p \cup U_q$ there are already two systems of local coordinates x_p^α and x_q^α. It is required that each of such systems of local coordinates in all such intersections $U_p \cap U_q$ be smoothly expressed in terms of the other and, inversely, and that:

$$\det \left(\frac{\partial x_p^\alpha}{\partial x_q^\gamma} \right) \neq 0.$$

The general smoothness class of these coordinate changes $x_p^\alpha(x_q^1, \ldots, x_q^n)$ for all intersecting pairs of regions U_p, U_q is called the smoothness class of the manifold M, which, later on, we always assume equal to infinity.

This completes the definition of a smooth manifold endowed with an atlas of local charts.

In the following we shall specify in which case distinct atlases are equivalent, i.e. define identical manifolds.

We shall give the simplest examples of manifolds.

1. A Euclidean space or any region of it.

2. A surface in a space $f_1(x^1, \ldots, x^n) = 0, \ldots, f_{n-k}(x^1, \ldots, x^n) = 0$, where all points are non-singular; for example, a hypersurface $f(x^1, \ldots, x^n) = 0$, where $|\operatorname{grad} f| \neq 0$ on the surface.

3. Group manifolds (Lie groups):

a) a group of matrices with non-zero determinant over the field $\mathbb{R}$ of real numbers or over the field $\mathbb{C}$ of complex numbers, i.e. a region in space of dimension n^2 for $\mathbb{R}$ or $(2n)^2$ for $\mathbb{C}$, denoted by $GL(n, \mathbb{R})$ (or $GL(n, \mathbb{C})$);

b) a group of matrices with determinant 1, which is given by one equation (hypersurface) in the space of all matrices:

$$\det (a_{ij}) = 1.$$

It is denoted by $SL(n, \mathbb{R})$ (or $SL(n, \mathbb{C})$);

c) a group of orthogonal matrices O_n given by the system of equations:

$$AA^{\mathrm{T}} = E;$$

d) a group of unitary matrices U_n given in the group $GL(n, \mathbb{C})$ by the system of equations:

$$A\bar{A}^{\mathrm{T}} = E,$$

where the bar implies complex conjugation of all the coefficients of the matrix.

(We have not listed all of the well-known even matrix Lie groups.)

4. Projective spaces (real and complex): we are given a vector $\bar{y} = (y^0, \dots, y^n) \neq 0$, the vectors $\bar{y}$ and $\alpha\bar{y}$ for $\alpha \neq 0$ being assumed to define one and the same point of the projective space denoted by $\mathbb{R}P^n$ (or $\mathbb{C}P^n$).

Let us consider the region $U_q = (y^q \neq 0)$; in this region U_q we introduce local coordinates:

$$x_q^1 = y^0/y^q, \dots, x^q_q = y^{q-1}/y^q,$$

$$x_q^{q+1} = y^{q+1}/y^q, \dots, x^n_q = y^n/y^q.$$

EXERCISE. Find the changes of local coordinates in the intersection of regions $U_q \cap U_p$ for a projectve space of dimension $n = 1, 2$. What is meant by the real and complex projective straight lines?

The simplest example of a complex projective space is the Riemannian sphere, i.e. a "projective straight line" which is a z-plane with an extra infinitely remote point. The reader is no doubt familiar with the real projective plane $\mathbb{R}P^2$ from the course in analytical geometry. It turns out that the three-dimensional real projective space $\mathbb{R}P^3$ coincides with the matrix group SO_3 (orthogonal matrices with determinant + 1).

In what follows, the reader will find a number of other examples of manifolds. It should be noted that the general concept of a manifold that we have introduced is too wide from the logical point of view, and we shall restrict it. It is required from the very beginning that a manifold, by definition, be situated as a smooth non-singular surface in a Euclidean space of a certain (perhaps, large) dimension.

Let us introduce an important concept of a *smooth sub-manifold in Euclidean space*. Suppose that we are given an arbitrary covering of the Euclidean space $\mathbb{R}^n$ by open domains W_q. A smooth sub-manifold N^k, in $\mathbb{R}^n$, of dimension k is given by a system of local equations in the domains W_q:

$$(*) \quad f_q^1(y^1, \dots, y^n) = 0, \dots, f_q^{n-k}(y^1, \dots, Y^n) = 0,$$

where the functions f_q^α of class C^∞ are defined only in the domain W_q. It is required that the rank of each matrix $(\partial f_q^\alpha/\partial y^j)$ be equal to $n - k$ at all points of the sub-manifold N^k.

It is also required that the systems of local equations $f_q = 0$ and $f_p = 0$ be equivalent in the intersections $W_p \cap W_q$.

We shall investigate, in detail, two-dimensional surfaces in three-dimensional Euclidean space.

We shall now construct an atlas of local charts U_q on the sub-manifold N^k, where the indices q will be in a natural one-to-one correspondence with points of the sub-manifold N^k, i.e. $q = Q \in N^k$.

Let $Q \in N^k$. By the definition of a sub-manifold there exists a set of numbers $i_1, \dots, i_{n-k}$, such that:

$$\det\left(\frac{\partial f^{\alpha}_{i_p}}{\partial y^j}\right) \neq 0.$$

According to the implicit function theorem, in a certain neighbourhood of the point Q, equations (*) can be solved in the form:

$$(**) \quad y^{i_p} = \phi_Q^{i_p}(y^{j_1}, \dots, y^{j_k}),$$

where $1 \leq p \leq n-k$, and the numbers $j_1, \dots, j_k$ make up a complementary set to $i_1, \dots, i_{n-k}$.

Let us fix in the space $\mathbb{R}^k$ a certain sufficiently small region where the expression (**) holds. We denote this region by V_Q. The coordinates $y^{j_1}, \dots, y^{j_k}$ in this region will be denoted by $x_Q^1, \dots, x_Q^k$. By U_p we denote the set of points on the sub-manifold N^k corresponding to points of the region V_Q by virtue of (**).

THEOREM. *The set N^k, along with the atlas of local charts U_Q in which local coordinates x_Q^{α} are introduced as shown above, is a smooth manifold.*

Proof. By definition, the totality of the regions U_Q yields the covering of the set N^k by virtue of the non-degeneracy of equations (*). Suppose P and Q are two points of the sub-manifold N^k, such that the regions U_P and U_Q have a non-empty intersection. According to the implicit function theorem, the mapping (**) is infinitely differentiable. Therefore, the expression of local coordinates x_P^{α} in terms of the set $x_Q^1, \dots, x_Q^k$ is infinitely differentiable by the definition of these coordinates, and inversely. This implies that:

$$(***) \quad \det\left(\frac{\partial x_Q^{\alpha}}{\partial x_P^{\beta}}\right) \neq 0.$$

Indeed, if for any pair of indices P and Q the Jacobian (***) were equal to zero, then the Jacobian of the inverse mapping would not exist, as required.

REMARK. A particular case of the argument for $n = 3$, $k = 2$ has, in fact, been considered in the proof of equivalence of the concept of non-singular points for different ways of defining the surface (see Section 1.7).

Hereafter, we restrict our consideration only to such differentiable manifolds which are equivalent to smooth sub-manifolds in Euclidean space, although we shall not prove this equivalence for particular cases.

The differentiable manifolds equivalent to sub-manifolds of Euclidean space are called *Hausdorff*. Any two non-coincident points x and x' of such a manifold can be "separated" from one another, that is, their open neighbourhoods $U(x)$ and $U(x')$ can be so constructed that they do not intersect.

We now discuss the concept of equivalence of manifolds: we have not yet said when two manifolds are thought of as identical.

Suppose two manifolds are given:

$$M = \bigcup_p U_v, \quad N = \bigcup_q V_q$$

(coordinates x_p^α and y_q^β).

DEFINITION 2. An arbitrary transformation

$$f: M \to N$$

is called *smooth of smoothness class k* if all the functions $y_q^\beta(x_p^1, \dots, x_p^n)$ for all pairs (q, p), when defined, are smooth functions of smoothness class k in the regions where they are defined.

By definition, the smoothness class of a transformation (or of a mapping) is assumed to be not higher than the smoothness class of any of the manifolds M, N. In the case N is a real straight line $N = \mathbb{R}$ or a complex straight line $N = \mathbb{C}$ pthe mapping $f: M \to \mathbb{R}$ or $f: M \to \mathbb{C}$ is naturally called the numerical function $f(x)$, where x is a point of the manifold M.

The situation is possible when a smooth map (or a numerical function) is defined not on the entire manifold M, but only on a part of it.

Such a situation can be illustrated on an example of the local coordinates x_p^α themselves which are numerical functions for any α and are defined only in the region U_p already by their meaning.

DEFINITION 3. Two manifolds M and N are called *smoothly equivalent* if there exists a one-to-one and smooth onto map of some smoothness class:

$$f: M \to N, \ f^{-1}: N \to M.$$

In particular, the Jacobian of the local coordinates $J_{pq} = \det(\partial y_q^\beta / \partial x_p^\alpha)$ is non-zero wherever these functions$y_q^\beta(x_p^1, \dots, x_p^n)$ are defined.

Later, we shall everywhere assume that the smoothness class of the manifolds and mappings between them is precisely the one that we need for our particular pupose (always not less than 1, and if we need second derivatives then it is not less than two, etc.).

Suppose on a manifold M we are given a curve $x = x(\tau)$, $a \leq \tau \leq b$, where x is a poit on the manifold. Whenever the curve is coordinatized by the local system of coordinates (x_p^α) of the region U_p, it can be represented in the form:

$$x_p^\alpha = x_p^\alpha(\tau), \quad \alpha = 1, \dots, n.$$

The velocity vector in these coordinates has the form:

$$\dot{x} = (\dot{x}_p^1, \dots, \dot{x}_p^n).$$

In regions $U_p \cap U_q$, where two coordinate systems apply, we have two representations:

$$x_p^\alpha(\tau) \text{ and } x_q^\gamma(\tau),$$

where $x_p^\alpha(x_q^1(\tau), \dots, x_q^n(\tau)) \equiv x_p^\alpha(\tau)$.

For the velocity we obtain:

$$\dot{x}_p^\alpha = \frac{\partial x_p^\alpha}{\partial x_q^\gamma} \dot{x}_q^\gamma.$$

As for Euclidean space, this formula provides the basis for the following definition.

DEFINITION 4. A *tangent vector* to a manifold M at an arbitrary point x is a vector represented in terms of a system of local coordinates (x_p^α) by a set of numbers (ξ_p^α); the representations of one and the same vector in terms of distinct local coordinates containing this point are related by the formula:

$$\xi_p^\alpha = \left(\frac{\partial x_p^\alpha}{\partial x_q^\gamma}\right)_x \xi_p^\gamma .$$

The tangent vectors form an n-dimensional linear space. In particular, the velocity vector of any smooth curve is a tangent vector.

DEFINITION 5. A *Riemannian (pseudo-Riemannian) metric* on a manifold M is a positive definite (indefinite) quadratic form given on tangent vectors at each point of the manifold and smoothly depending on all the local coordinates, pointed out in the definition, in the region where they apply. In each region U_p coordinatized by the local coordinates $(x_p^1, \ldots, x_p^n)$ the metric is given by the matrix:

$$g_{\alpha\beta}(x_p^1, \ldots, x_p^n) \quad \text{and} \quad |\xi|^2 = g_{\alpha\beta}\, \xi_p^\alpha\, \xi_p^\beta$$

for any vector ξ at the point x.

A Riemannian (pseudo-Riemannian) metric determines a symmetric scalar product of two vectors at one and the same point by the usual formula:

$$\xi\eta = g_{\alpha\beta}\, \xi_p^\alpha\, \eta_p^\beta = \eta\xi.$$

Here $\xi\eta = \langle \xi, \eta \rangle$. In the mathematical literature both the notations are used. In this notation, the modulus squared, $|\xi|^2$, does not depend on the choice of the system of coordinates:

$$g_{\alpha\beta}\, \xi_p^\alpha\, \xi_p^\beta = g'_{\gamma\delta}\, \xi_q^\gamma\, \xi_q^\delta$$

$$\left(\text{or} \quad g'_{\gamma\delta} = \frac{\partial x_p^\alpha}{\partial x_q^\gamma}\, g_{\alpha\beta}\, \frac{\partial x_p^\beta}{\partial x_q^\delta} \right).$$

The length of any smooth curve on a manifold is determined by the usual formula:

$$l = \int_a^b |\dot{x}(\tau)|\, d\tau .$$

In manifolds with a pseudo-Riemannian metric the class of space-like curves $x(\tau)$, such that $|\dot{x}(\tau)| > 0$, is naturally distinguishable.

The concept of a manifold might, at first glance, seem excessively abstract. In fact, however, even in Euclidean space or in its regions we often have to make coordinate changes and, consequently, to discover and apply the transformation rule for one quantity or another. Furthermore, it is often convenient to solve one problem relative to distinct coordinate systems and then to see how the solutions are "sewn" together in the region of intersection of theses distinct systems of coordinates. In addition, not all surfaces can be coordinatized by a single system of coordinates without singular points (e.g. the sphere has no such coordinate system).

Continuous transformation groups of a space are also manifolds.

Of particular interest is "space-time continuum". The generally accepted hypothesis suggests that this space-time continuum is a four-dimensional differentiable manifold. This means that if an observer is at an arbitrary point of a space-time continuum, then the space-time region U_p which is surrounding this space-time continuum admits an introduction of the coordinates $x_p^0, x_p^1, x_p^2, x_p^3$. Given this, the coordinates x_p^α and x_q^α introduced by different observers positioned at distinct points are expressed in terms of each other in a smooth invertible way in the region where both the coordinate systems apply:

$$x_p^\alpha = x_p^\alpha(x_q^0, \dots, x_q^3)$$

(in the region $U_p \cap U_q$). This hypothesis is the most convenient and simple one, although in the neighbourhood of an observer there of course exist ponts which he cannot observe at a given instant of time.

In the special theory of relativity it is assumed, in addition, that the space is a pseudo-Euclidean Minkowski space which admits the introduction of a unique coordinate system, $ct = x^0, x^1, x^2, x^3$ and possesses a pseudo-Euclidean metric:

$$(dl)^2 = -\sum_{\alpha=1}^{3} (dx^\alpha)^2 + (dx^0)^2 .$$

The underlying (Einstein's) hypothesis of the general theory of relativity suggests that the space-time possesses a pseudo-Riemannian metric which in its physical meaning is identical to the gravitational field. In each local system of coordinates $(x_p^0, x_p^1, x_p^2, x_p^3)$ this metric is given by:

$$dl^2 = g_{\alpha\beta}^{(p)}\, dx_p^\alpha\, dx_p^\beta .$$

The gravitational field is said to be *weak* in the case where this metric is close to the pseudo-Euclidean one in some coordinates x^0, x^1, x^2, x^3, where $x^0 = ct$; given this, the components $g_{\alpha\beta}$ are small when $\alpha \neq \beta$, g_{00} is close to unity and $g_{\alpha\alpha}$, $\alpha \neq 0$, is close to minus unity. In classical mechanics, the gravitational field was descibed by the potential $\phi(x)$ (force is grad ϕ); in comparison with mechanics, the metric should approximately be given by:

$$g_{00} = 1 - 2\phi/c^2 + o(1/c^2), \quad g_{\beta\alpha} = g_{\alpha\beta} = o(1/c^2),$$

$$g_{\alpha\alpha} = -\ 1 - 2\phi/c^2 + o(1/c^2), \quad (\alpha = 1, 2, 3; \quad \beta \neq \alpha),$$

where c is the speed of light in a vacuum, and its value is high.

As far as positive definite (Riemannian) metrics are concerned, they have occurred, of course, due to the geometry of three-dimensional space. At the same time, the concept of a positive Riemannian metric is often a convenient tool for investigating various essential manifolds, for example, group manifolds (Lie groups). We shall give a number of useful examples of the Euclidean metric.

For example, in a linear space of skew-symmetric matrices

$$A = (a_{ij}), \quad a_{ij} = -\ a_{ji}$$

there exists a positive scalar product

$$\langle A, B \rangle = -\operatorname{trace}(A \cdot B),$$

where the trace $(A \cdot A) > 0$.

Another example: in an infinite-dimensional linear space consisting of all continuous real functions (for instance, on a segment $[a, b]$) there also exists a positive scalar product:

$$\langle f, g \rangle = \int_a^b f(x)\, g(x)\, dx,$$

$$\|f\|^2 = \int_a^b f(x)^2\, dx.$$

Let us now sum up the results. On defining the concept of a manifold, we considered the basic examples.

1. k-dimensional surfaces, in an n-dmensional Euclidean space, given by a system of equations

$$f_i(x^1, \dots, x^n) = 0, \; i = 1, \dots, n - k,$$

the rank $(\partial f_i/\partial x^j) = n - k$.

A more general case is a sub-manifold of a Euclidean space the whole of which cannot be given by a non-degenerate system of equation, for example, a projective plane.

2. The basic groups given by equations in an n^2-dimensional space of real matrices:

$$GL\,(n, \mathbb{R})\;(\det \neq 0), \quad SL\,(n, \mathbb{R})\;\det A = 1);$$

O_n is a group of matrices whose rows make up an orthonormal basis of vectors;

SO_n is the part of the group O_n, for which

$$\det A = 1;$$

Sp_n is the transformation group preserving the skew-symmetric scalar product in a $2n$-dimensional space.

Analogous groups are defined over the field of complex numbers $GL\,(n, \mathbb{C})$, $SL\,(n, \mathbb{C})$, $O_n^{\mathbb{C}}$, $Sp_n^{\mathbb{C}}$. In addition to these groups, in the complex case there appear other transformation groups preserving the Hermitian positive scalar product:

$$U_n; \;\; SU_n \;\; (\text{where } \det = 1);$$

$$U_n = O_{2n}\,(\mathbb{R}) \cap GL\,(n, \mathbb{C}).$$

There exist some other groups of linear transformations preserving the pseudo-Euclidean real scalar product which has p positive and q negative squares $O_{p,q}$; $SO_{p,q}$ (where det = 1); $O_{0,n} = O_n$, $SO_{0,n} = SO_n$.

Analogously, in the Hermitian complex case we determine the groups $U_{p,q}$; $SU_{p,q}$ $(U_{0,n} = U_n, SO_{0,n} = SU_n)$.

3. Projective spaces (real and complex); points of these spaces are non-zero vectors considered up to a factor

$$y = (y^0, \dots, y^n); \;\; y \sim \alpha y, \;\; \alpha \neq 0.$$

The coordinate regions U_j are distinguished here, where $y^j \neq 0$, with coodinates x_j^α:

$$x_j^\alpha = \begin{cases} y^\alpha/y^j, & \alpha \geq j, \ \alpha = 1, \dots, n, \\ y^{\alpha-1}/y^j, & \alpha < j, \ j = 0, \dots, n. \end{cases}$$

In particular, the region U_0, where $y^0 \neq 0$ is called *finite part* with coordinates $x_0^\alpha = y^\alpha/y^0$, $\alpha = 1, \dots, n$.

4, Riemannian surfaces of multi-valued functions are defined as follows: in the space of two complex variables (w, z) for any analytic function $f(z, w)$ (e.g. a polynomial) we take the surface of its zeros

$$f(z, w) = 0. \tag{3}$$

Here the function f is complex-valued, $f = u + iv$, and analytic,

$$\frac{\partial f}{\partial \overline{w}} = 1/2\left(\frac{\partial f}{\partial h} + i\frac{\partial f}{\partial g}\right) \equiv 0, \quad \frac{\partial f}{\partial \bar{z}} = 1/2\left(\frac{\partial f}{\partial x} + i\frac{\partial f}{\partial y}\right) \equiv 0,$$

$$z = x + iy, \quad w = h + ig,$$

(or, which is equivalent, in a neighbourhood of any point z_0, w_0 it is expanded into a series $f(z, w) = f(z_0, w_0) + Q_{m,n}(z - z_0)^m(w - w_0)^n$). The set of solutions of equation (3)

$$w = w(z), \quad f(z, w(z)) \equiv 0 \tag{4}$$

may appear to be a multi-valued function.

For example: a) $w = (P_n(z))^{1/2}$, where $P_n(z)$ is a polynomial;

b) $w = \ln z = \ln |z| + i \arg z + 2\pi i n$, $f(w, z) = e^w - z$.

Then the surface (3) is called the *Riemannian surface of the multi-valued function* (4).

That the function $w(z)$ is multi-valued means that the projection of the surface (3) into the z-plane along w is not one-to-one.

Suppose the function $f(z, w)$ is a polynomial of degree n in the set of variables. We make the substitution $z = y^1/y^0$, $w = y^2/y^0$. Then $f(z, w) = (1/(y^0)^n)Q_n(y^0, y^1, y^2)$, where Q_n is a homogeneous polynomial. To the projective plane $\mathbb{C}P^2$ the equation $f(z, w) = 0$ is continued in the form

$$Q_n(y^0, y^1, y^2) = 0. \tag{5}$$

Those points of the surface (5), where $y^0 = 0$, are called infinitely remote points of the Riemannian surface (3).

What is meant by two-dimensional manifolds? To which of the surfaces known to us are they equivalent as manifolds?

The reader is already acquainted with the following surfaces in two-dimensional and three-dimensional spaces.

A. Regions in a plane with k holes (Figure 31).

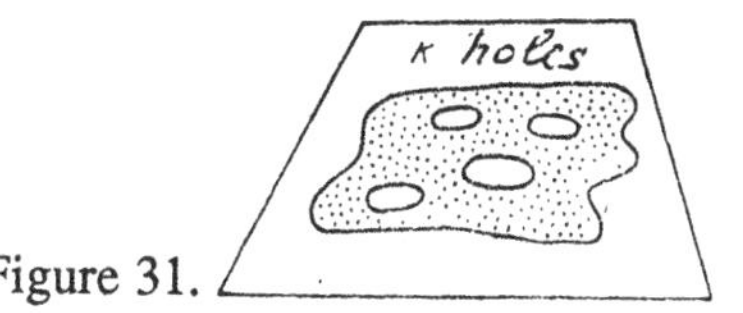

Figure 31.

B. A surface in a three-dimensional space with g handles (Figure 32).

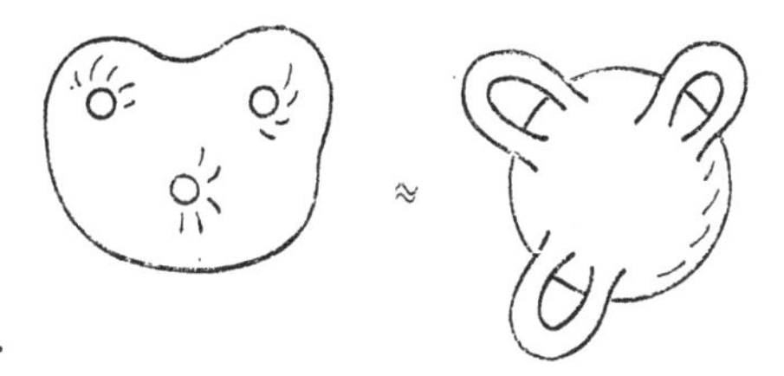

Figure 32.

C. Arbitrary regions on surfaces with g handles.

Now consider the following examples.

EXAMPLE 0. Let

$$f(w, z) = w^2 - z, \quad Q_2(y^0, y^1, y^2) = (y^2)^2 - y^1y^0 = 0.$$

Consider points $(z = \infty)$ and $(z = 0)$ and join them with a straight line a (Figure 33).

Figure 33. $z = 0$ a ∞

On the sphere S^2 equivalent to $\mathbb{C}P_1$ this line looks as shown in Figure 34.

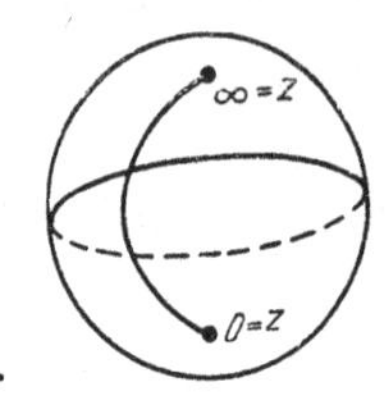

Figure 34.

Outside this line, the graph of the Riemannian surface $f(z, w) = 0$ falls into two disjoint parts (two "branches") each of which is equivalent (can be projected in a one-to-one manner) to the appearance of the line a in a z-plane (Figure 35).

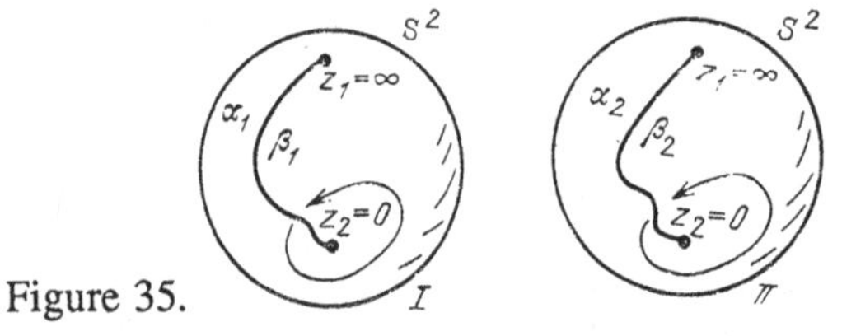

Figure 35.

At the points 0 and ∞ the values of these two branches of the function $w(z) = (z)^{1/2}$ merge.

To obtain a surface, it is necessary that the piece of a boundary α_1 of region I be identified with the piece of boundary β_2 of region II and that the piece of boundary β_1 of region I be identified with the piece of boundary S^2 of region II.

It can be readily seen that after glueing we again obtain a surface equivalent to the sphere S^2 (this can be done with scissors).

EXAMPLE I. $f(z, w) = w^2 - P_2(z)$, where $P_2(z)$ is a polynomial of degree 2 with simple (aliquant) roots $z = z_0$, $z = z_1$, $z_0 \neq z_1$.

Join the roots z_0 and z_1 with a segment a. Outside the segment a, the graph $f(z, w) = 0$ falls into two parts which are disjoint. Over the sphere $\mathbb{C}P^1$ this set looks exactly as the one in Example 0 (Figure 36).

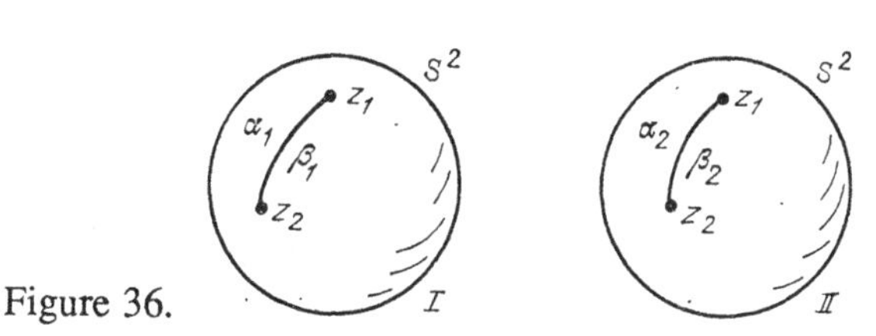

Figure 36.

The only difference is that here $z_1 \neq \infty$. By analogy with example 0, identifying the curves

$$\alpha_1 \sim \beta_2,$$
$$\beta_1 \sim \alpha_2,$$

we shall obtain the sphere S^2.

EXAMPLE II. $f(z, w) = w^2 - P_3(z)$, where the roots z_0, z_1, z_2 of the polynomial $P_3(z)$ are not pairwise equal. Let us make cutsa_1 and a_2 (Figure 37).

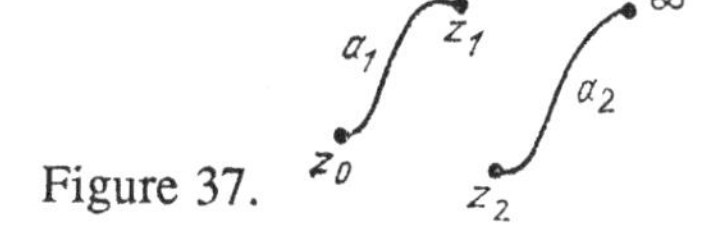

Figure 37.

Outside these cuts the surface falls into two disjoint parts (Figure 38).

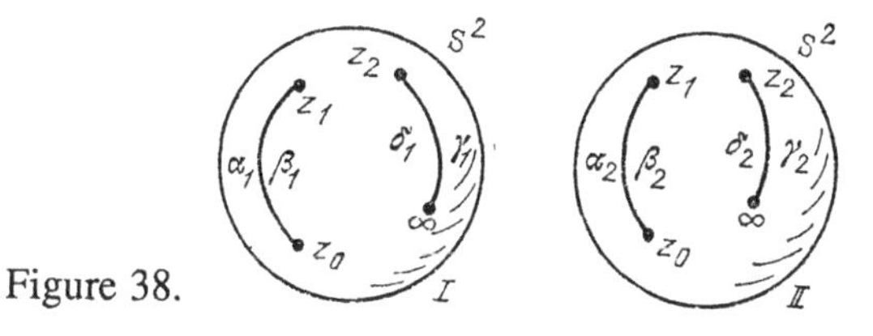

Figure 38.

Identifying the curves

$$\alpha_1 \sim \beta_2, \quad \gamma_1 \sim \delta_2,$$
$$\alpha_2 \sim \beta_1, \quad \gamma_2 \sim \delta_1,$$

we shall obtain a torus ($g = 1$) (Figure 39).

EXAMPLE III. $f(z, w) = w^2 - P_4(z)$, where the roots of the polynomial $P_4(z)$ are not pairwise equal. Let z_0, z_1, z_2, z_3 be the roots of the polynomial $P_4(z)$. Using arguments similar to those of Example II, we also arrive at a torus here.

For the Riemannian surfaces of muliple-valued functions of the form $w = (P_{2n}(z))^{1/2}$ or $w = (P_{2n-1}(z))^{1/2}$ (the polynomials $P(z)$ have aliquant roots) we obtain, as the Riemannian surface, a surface with $(n - 1)$ handles (Figure 40).

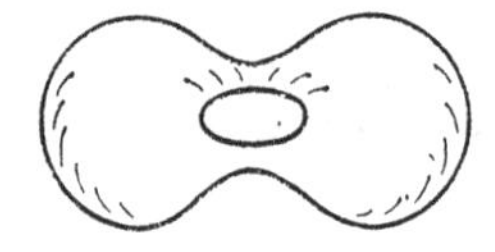
Figure 39.

Figure 40.

We can see that quite different definitions of the surface lead to equivalent results.

Let us discuss some more examples.

1. A special role is played by the torus which can be obtained as follows.

Let (x, y) be points in a plane. We shall assume the points (x, y) and $(x + mx_0, y + n y_0)$, where m,n are atbitrary integers, $x_0 \neq 0$, $y_0, \neq 0$, to be equivalent.

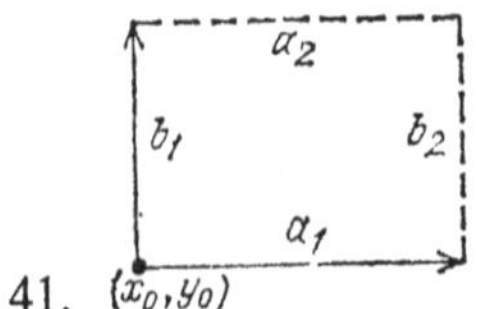

Figure 41.

Within the rectangle in Figure 41 there are no equivalent points, while the segments a_1 and a_2 on the boundary are equivalent, and so are the segments b_1 and b_2. Glueing them, respectively, together we obtain a torus. From this it is obvious that the doubly periodic functions in the plane are functions on the torus. Figure 42 illustrates several two-dimensional manifolds.

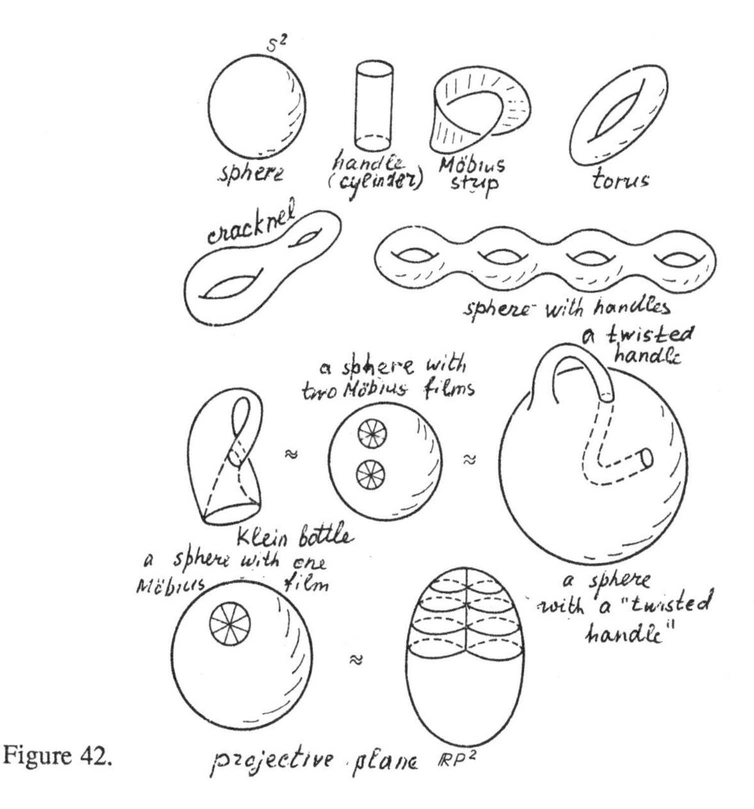

Figure 42.

2. We shall point out that even functions on the sphere are functions in the projective plane (similarly for the n-dimensional sphere $x_0^2, + x_1^2 + \dots + x_n^2 = 1$). An even function on the sphere is a function $f(x^0, \dots, x^n)$, where $\sum (x^i)^2 = 1$, such that $f(x) = f(-x)$.

Among group manifolds, the simplest are the following:

1) the rotation group of a plane as a manifold is equivalent to the circumference $U_1 = SO_2 \cong S^1$;

2) the group of motions of a plane is described by three coordinates:

$$(\phi, x, y), \quad \phi \sim \phi + 2\pi n,$$

x nd y assuming any values; this manifold is equivalent to a region in the space $\mathbb{R}^3$ from which one straight line is removed (this straight line may be, say, the z-axis);

3) the group $SL(2, \mathbb{R})$ as a manifold is equivalent to the preceding manifold (as the group $SL(2, \mathbb{R})/-E$ is equivalent to the group of motions of a Lobachevskian plane or to the group $Q^R_{2,1}$). Here $SL(2, \mathbb{R})/-E$ is the factor group with respect to the sub-group $(E, -E)$;

4) the group SO_3 as a manifold is equivalent to the real projective space $\mathbb{R}P^3$;

5) the group SU_2 as a manifold is equivalent to the three-dimensional sphere S^3

$$(x^0)^2 + (x^1)^2 + (x^2)^2 + (x^3)^2 = 1$$

and consists of matrices of the form $\begin{pmatrix} a & b \\ -\bar{b} & \bar{a} \end{pmatrix}$, where $|a|^2 + |b|^2 = 1$, a and b being complex numbers.

Note that as a group, $SU_2/-E$ is isomorphic to the rotation group SO_3.

The notation $SU_2/-E$ or $SL(2, \mathbb{R})/-E$ implies the factor-group with respect to the sub-group which consists of matrices $E = \begin{pmatrix} 1 & 0 \\ 0 & 1 \end{pmatrix}$ and $-E = \begin{pmatrix} -1 & 0 \\ 0 & -1 \end{pmatrix}$.

If $\phi(z) = w$ is a multiple-valued function given by the equation $f(z, w) = 0$, then the graph of this surface is the Riemannian surface of the function $w = \phi(z)$.

The function $f(z, w)$ is analytic, i.e. near any point (z_0, w_0) it is expanded into a power series

$$f(z, w) = f(z_0, w_0) + \sum_{m,n} a_{mn}(z - z_0)^n (w - w_0)^m .$$

For example, this is the case if $f(z, w)$ is a polynomial (the variables z and w are complex numbers).

What can be said about the multiple-valued function $w = \phi(z)$?

First, at each point z (except at the discrete set z_α) this set has a certain number of values ("branches") $w_i = \phi_i(z)$ which are not equal to each other and are continuous in a neighbourhood of the point z.

For example, $w_\pm = \phi_\pm(z) = \pm (z)^{1/2}$, $z \neq 0, \infty$.

Second, at the points $z = z_\alpha$ the number of values of the function is smaller (these are the "branching points").

For example, $+ (z)^{1/2} = - (z)^{1/2}$ when $z = 0, \infty$. Two or more branches are said to *merge* at branching points.

Third, if we move continuously along one branch and pass round the branching point z_α, then we can go over from one branch to another.

For example, from the value $+ (z)^{1/2}$ we shall go over, by passing round the point $z = 0$, to the value $- (z)^{1/2}$.

How can we imagine the geometrical structure of the Riemannian surface?

If branching points are removed, the local coordinates on the surface can be given as follows: suppose U is an arbitrary region in the branching plane. It is necesssary for us that the region U contain no closed paths moving along which we could, on returing to the same point z, go over to another value of the function $w = \phi(z)$. If the region U is such, then over it the graph of the function $w = \phi(z)$ falls into disjoint parts (branches) which can be somehow indexed by the subscript i: $w_i = \phi_i(z)$ in the region U.

On each branch we can introduce the same coordinates as in the region U. Therefore, we obtain charts (or "maps") of the coordinate atlas (U, i). Such regions may, in principle, cover the entire Riemannian surface except thc branching points.

We have already discussed above the example $w = (P_n(z))^{1/2}$, where $P_n(z)$ is a polynomial with aliquant roots $z = z_1, \dots, z_n$ of degree n. Let us represent the result as a theorem.

THEOREM 1. *The Riemannian surface of the function* $w = (P_n(z))^{1/2}$ *is equivalent to the surface of the sphere with g handles, where* $n = 2g + 1$ *or* $n = 2_g + 2$.

We shall explain this once again. Suppose n is even. Let us divide the roots of the polynomial $P_n(z)$ into pairs and then join each pair with a curve $a_1, \dots, a_{n/2}$ which is disjoint with the other curves.

$$\underset{1 \quad 2}{\overset{a_1}{\text{—}}} \quad \underset{3 \quad 4}{\overset{a_2}{\text{—}}} \quad \cdots \quad \underset{n-1 \quad n}{\overset{a_{n/2}}{\text{—}}}$$

Now cut the z-plane along the segments a_i. We have made sure that the Riemannian surface falls into two disjoint parts U_1 and U_2 (going round two roots at a time does not change the branch).

The edges of the cuts we denote by α_i and β_i; they lie, respectively, on the pieces U_1 and U_2 of the Riemannian surface.

After this, we glue the edges together by the rule

$$(U_1, \alpha_i) \sim (U_2, \beta_i) \quad (U_1, \beta_i) \sim (U_2, \alpha_i).$$

This glueing reflects the fact that when approaching the edge α_i we must go over from the piece U_1 onto the piece U_2 (the edge β_i).

We can make sure (as in the case $w = z^{1/2}$) that near any root $z = z_q$ the function has the form:

$$w = \left(z - z_q\right)^{1/2} \cdot \left((z - z_1) \dots (z - z_n)\right)^{1/2} = \left((z - z_q)\right)^{1/2} \cdot \left(Q_{n-1}(z)\right)^{1/2}$$

where $Q_{n-1}(z_q) \neq 0$.

From this, we can see that going round the point $z = z_q$ along the small path changes the values of two branches as in the case $w = (z)^{1/2}$.

If the closed path is large, it can be continuously deformed to the one consisting of small paths that cover the branching points (Figure 43). For odd n the construction is much the same, but we take $z_{n+1} = \infty$ for one of the branching points and, after this, repeat the whole procedure.

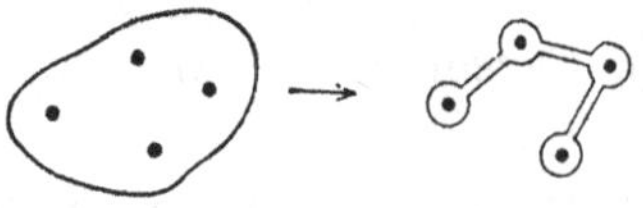

Figure 43.

The crucial point which we have left without rigorous proof is that after we remove an appropriately chosen collection of paths, the Riemannian surface falls into two disjoint pieces.

This can also be represented as follows: choose a point P in a z-plane aside from the branching points.

Consider all possible closed directed paths which start and end at the point P and leave aside the branching points (Figure 44).

Suppose γ is such a path. By γ^{-1} we denote the same path in the opposite direction.

If there exist paths γ_1 and γ_2, then we can first move along the path γ_1 and then γ_2. Then we shall cover the path $\gamma = \gamma_2 \cdot \gamma_1$ (the "product" of paths). It is obvious that generally $\gamma_1 \cdot \gamma_2 \neq \gamma_2 \cdot \gamma_1$ (the order of taking paths is different).

Moving along the path γ, we shall return to the same point. The values of the multi-valued function at the point P will be somehow permuted (if the function has n branches):

$$\gamma \to \sigma(\gamma) = \begin{pmatrix} 1 & 2 & \dots & n \\ i_1 & i_2 & \dots & i_n \end{pmatrix},$$

here $\sigma(\gamma)$ is the permutation of the values of the function induced by the motion along the path γ.

Obviously, we have:

1) $\sigma(\gamma^{-1}) = \sigma(\gamma)^{-1}$,

2) $\sigma(\gamma_1 \cdot \gamma_2) = \sigma(\gamma_1)\, \sigma(\gamma_2)$.

A surface $w = (P_n(z))^{1/2}$ always has two branches; for a path γ_i (Figure 45) which embraces one point, we obtain:

$$\sigma(\gamma_i) = \begin{pmatrix} 1 & 2 \\ 2 & 1 \end{pmatrix}.$$

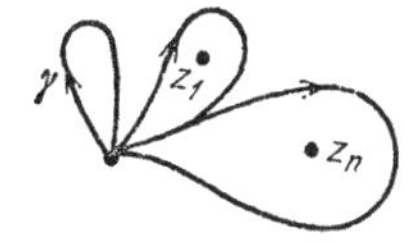

Figure 44.

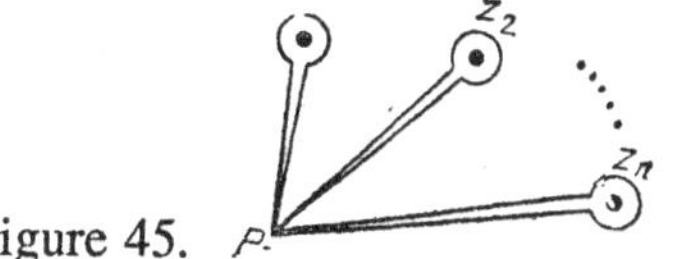

Figure 45.

We can easily reconstruct the Riemannian surface if we know the permutations $\sigma(\gamma)$ for all closed paths γ which start and end at one point P. It suffices to know only $\sigma(\gamma_i)$ for paths γ_i embracing one branching point.

As an example, investigate the Riemannian surface of the function $w = (P_m(z))^{1/n}$.

The specific feature of the Riemannian surface is the complexity of these two-dimensional manifolds:

1) each coordinate region U is a region in a plane with coordinate $z(\alpha)$,

2) in the intersection of regions $U_\alpha \cap U_\beta$ the change of coordinates

$$z(\beta) = w = f(z(\alpha)), \quad z(\alpha) = z = x(\alpha) + iy(\alpha),$$

is analytic (or conformal),

$$\frac{\partial f}{\partial \bar{z}(\alpha)} = 1/2\left(\frac{\partial f}{\partial x(\alpha)} + i\frac{\partial f}{\partial y(\alpha)}\right) \equiv 0,$$

$$z(\beta) = x(\beta) + iy(\beta) = u + iv,$$

$$z(\alpha) = x(\alpha) + iy(\alpha) = u + iy.$$

The Jacobian matrix

$$\begin{pmatrix} u_x & u_y \\ v_x & v_y \end{pmatrix} = J$$

with account taken of the identities:

$$u_x - v_y = 0, \quad u_y + v_x = 0,$$

has the form $J = \begin{pmatrix} u_x & u_y \\ -u_x & u_y \end{pmatrix}$, and its determinant is positive: $\det J = u_x^2 + u_y^2 > 0$.

The transformations $w = f(z)$, where $\partial f/\partial \bar{z} = 0$ (or the complex conjugate ones) are called *conformal*.

By definition, $\frac{\partial f}{\partial z} = 1/2\left(\frac{\partial f}{\partial x} - i\frac{\partial f}{\partial y}\right)$. Obviously, $\left|\frac{\partial f}{\partial z}\right|^2 = u_x^2 + v_y^2$.

It would be of interest to pay attention to the following circumstance. In this sub-section we have imposed the requirement that all the manifolds we are dealing with should be smoothly equivalent to sub-manifolds in a Euclidean space of some dimension or in its regions. We call such manifolds *Hausdorff*. Among the simplest examples of manifolds considered above, all except projective spaces are, by definition, as follows. For projective spaces this fact does hold, but requires a special proof which we do not give here. The general smooth manifolds that we have introduced need not necessarily be Hausdorff. The simplest example: consider two copies of a real straight line with coordinates x and y respectively. Identify the points $x = y$ for $x < 0$, $y < 0$. When $x \geq 0$, $y \geq 0$, we assume the points to be distinct. Obviously, we obtain a smooth one-dimensional manifold. Prove that it is not realized as a smooth sub-manifold in a Euclidean space. (The points $x = 0$ and $y = 0$ are not identified!)

The requirement that manifolds be Hausdorff may also be formulated in some other equivalent ways (without proof).

1. On the manifold there exists a Riemannian metric, such that there exists not a single pair of infinitely close points. A pair of points P and Q is called infinitely close if, for any $\varepsilon > 0$, these points are joined in this manifold by a piecewise smooth curve γ, such that its length is less then ε.

2. There exists a "sufficiently small" partition of unity, so-called, i.e. an atlas of a finite or countable number of maps (charts) $M^n \bigcup_q U_q$, where each point belongs only to a finite number of regions. Given this, each region is homeomorphic to a region of Euclidean space and endowed with local coordinates x_q^α, and there exist smooth real functions $\phi_q \geq 0$ on the manifold M^n, such that $\sum_q \phi_q \equiv 1$, $\phi_q \equiv 0$ outside the region U_q.

3. For any pair of points P and Q of the manifold M there exist continuous functions ϕ_P and ϕ_Q on the manifold M, such that $\phi_P(Q) = 0$, $\phi_P(P) = 1$ and $\phi_Q(P) = 1$, $\phi_Q(P) = 0$.

(The last definition extends to all topological spaces.)

For instance, with the partition of unity in mind, we construct the Riemannian metric as follows. We introduce the tensor relative to the coordinates x_q of the point $P \in U_q$: $\overset{(q)}{g}_{ij}(x_q) = \delta_{ij} \cdot \phi_q(x_q)$, where $\overset{(q)}{g}_{ij} \equiv 0$ outside the region U_q. This tensor is defined on the entire manifold and is non-negative. We assume $g_{ij}(P) = \sum_q \overset{(q)}{g}_{ij}(P)$.

EXERCISE. Prove that the tensor $g_{ij}(P)$ determines a positive definite Riemannian metric on M. Prove that there exist no infinitely close points.

As has been mentioned above, that the manifold M^n is Hausdorff is a consequence of a rather small-sized partition of unity. Non-negative functions ϕ_q form partition of unity if at each point of the manifold the identity $\sum_q \phi_q(x) \equiv 1$ holds. For a sufficiently small-sized partition of unity, the support $|\phi_q|$ of the functions ϕ_q (i.e. the closure of the set of points Int $|\phi_q|$ at which the function $\phi_q$is positive) is obviously embedded into the Euclidean space $\mathbb{R}^n$. Suppose that the points P and Q are inseparable (i.e. assume the manifold not to be Hausdorff). Let U_j and U'_j be contracting sequences of neighbourhoods of these points. Choose points $x_j \in U_j \cap U'_j$. Choose a function ϕ_q such that $\phi_q(P) > 0$. Then $P \in$ Int $|\phi_q|$. Since the support $|\phi_q|$ is embedded into the Euclidean space $\mathbb{R}^n$, it follows that $\phi_q(Q) = 0$. Hence $Q \notin$ Int $|\phi_q|$. But continuity of the function ϕ_q implies that the condition $\phi_q(Q) \lim_j \phi_q(x_j) = \phi_q(P) \neq 0$ must hold. The contradiction obtained shows that hte manifold M cannot contain inseparable points.

It should be emphasized that the existence on a manifold of an indefinite non-degenerate metric does not necessasrily require that the manifold be Hausdorff. Therefore, the requirement that a manifold should necessarily be Hausdorff is unnatural from the point of view of the general theory of relativity.

For real analytic manifolds, the existence of an analytic embedding into Euclidean space is an exceedingly sophisticated and not at all elementary theorem.

The requirement that the manifold be Hausdorff is not natural in a number of other fields of mathematics, namely, in algebraic geometry, in the theory of invariants and others. It is precisely for this reason that we think it most fundamental only to determine a manifold as an atlas of local coordinate regions with certain requirements imposed upon the class of functions of the change from one local coordinate to others.

In the definition of the basic concepts of analysis on manifolds, the requirement that these manifolds be Hausdorff in the sense of items 2 and 3 (see above) is important for the manifold as the domain of definition of functions and is not so important for the manifold as the domain of values of functions.

1.15 Geodesics

Suppose we are given a certain Riemannian manifold M with a positive definite metric g_{ij}. It is natural to define an important class of curves on a manifold, which are called *geodesics* and posses the property that they are *locally minimal*. i.e. they minimize the length between any of its two *sufficiently close* points. In Euclidean space such curves coincide with *straight lines*. If points on a Riemannian manifold are situated *far* from each other, then the geodesic joining them may turn out to be not a minimal trajectory. In other words, there may exist another curve of smaller length between these points.

Thus, geodesic lines are (at least locally) the shortest trajectories, i.e. their length does not exceed that of any other curve joining the same two points sufficiently close to one another. Let us consider this question in more detail.

It is instructive to approach this question from a more general point of view. Suppose $L(z, \xi)$ is a function of the point $z = (z^1, z^2, z^3)$ and of the tangent vector $\xi = (\xi^i)$ at this point. Consider a fixed pair of points $P = (z_1^1, z_1^2, z_1^3)$ and $Q = (z_2^1, z_2^2, z_2^3)$, as well as various smooth curves $\gamma : z^i = z^i(t)$ joining these two points

$$z^i(a) = z_1^i, \quad z^i(b) = z_2^i, \quad a \le t \le b.$$

Consider the quantity

$$S(\gamma) = \int_P^Q L(z(t), \dot{z}(t))\, dt .$$

On which curve γ will the quantity S be minimal? The quantity $S(\gamma)$ will be called the *action*.

EXAMPLE 1. Let $L(z, \xi) = g_{ij}\, \xi^i \xi^j$, then

$$S(\gamma) = \int_P^Q L(z, \dot{z})\, dt = \int_P^Q g_{ij}\, \dot{z}^i \dot{z}^j\, dt = \int_P^Q |\dot{z}|^2\, dt.$$

On which curve $\gamma = \{z(t)\}$ will the function $S(\gamma)$ be minimal?

EXAMPLE 2 Let $L(z, \xi) = \left(g_{ij}\, \xi^i \xi^j\right)^{1/2}$, then

$$S(\gamma) = \int_P^Q (g_{ij}\dot{z}^i\dot{z}^j)^{1/2}\, dt = \text{(the length of the curve } \gamma).$$

On which curve γ is the length minimal?

EXAMPLE 3. Let the metric be Euclidean and let $L = \frac{m}{2}\delta_{ij}\dot{\xi}^i\dot{\xi}^j - U(z)$. Then

$$S(\gamma) = \int_P^Q \Big[\sum_{i=1}^{S} \frac{m}{2}(\dot{z}^i)^2 - U(z)\Big]\, dt.$$

The curves γ along which $S(\gamma)$ is minimal are the trajectories of motion of the point of mass m in the field of forces $f_i = -\partial U/\partial z^i$.

Then a simple theorem holds.

THEOREM 1. *If the quantity* $S(\gamma) = \int_P^Q L(z, \dot{z})\, dt$ *reaches its minimum on a certain curve* $\gamma: \{z^i = z^i(t)\}$ *among all the smooth curves going from P into Q, then along the curve there hold the equations*

$$\left(\frac{\partial L}{\partial \dot{z}^i}\right)^{\cdot} = \frac{\partial L}{\partial z^i}, \quad i = 1, 2, 3$$

where

$$\frac{\partial L}{\partial \dot{z}^i} = \frac{\partial L}{\partial \xi^i}(z, \dot{z}), \quad \left(\frac{\partial L}{\partial \dot{z}^i}\right)^{\cdot} = \frac{\partial^2 L}{\partial \xi^i \partial \xi^j}\ddot{z}^j + \frac{\partial^2 L}{\partial \xi^i \partial z^j}\dot{z}^j$$

(it is assumed that $L = L(z^1, z^2, z^3, \xi^1, \xi^2, \xi^3)$*, where z and* ξ *are independent variables, but then* $\xi^i = dz^i/dt$ *is substituted along a given curve* $\gamma: z^i = z^i(t)$*).*

Proof. Let $\eta^i = \eta^i(t)$ be any smooth function, such that $\eta^i(a) = 0$ and $\eta^i(b) = 0$, $a \le t \le b$. Let ε be a small number. Consider the equation

$$\lim_{\varepsilon \to 0} \frac{S(\gamma + \varepsilon\eta) - S(\gamma)}{\varepsilon} = \frac{d}{d\varepsilon} S(\gamma + \varepsilon\eta)\Big|_{\varepsilon=0}.$$

Here $\gamma + \varepsilon\eta$ is the curve $\{z^i = z^i(t) + \varepsilon\eta^i(t)\}$ close to the curve $\gamma(t)$ as $\varepsilon \to 0$.

LEMMA 1. *If $S(\gamma)$ is minimal, then for any smooth vector function $\eta(t)$ which vanishes at the ends of the time interval (the curve $\gamma+\varepsilon\eta$ also going from P into Q) we have*

$$\lim_{\varepsilon\to 0}\frac{S(\gamma+\varepsilon\eta)-S(\gamma)}{\varepsilon} = \frac{d}{d\varepsilon}S(\gamma+\varepsilon\eta)\Big|_{\varepsilon=0} = 0.$$

The proof of this lemma is obvious.

We now proceed to the theorem. For the expression $d/d\varepsilon\, S(\gamma+\xi\eta)|_{\varepsilon=0}$, we have

$$d/d\varepsilon\, S(\gamma+\varepsilon\eta)\Big|_{\varepsilon=0} = \int_a^b \{\frac{\partial L}{\partial z^i}\eta^i(t)+\frac{\partial L}{\partial \xi^i}\dot{\eta}^i\}\, dt = 0 \tag{1}$$

where the integral, by definition, is calculated along the curve γ:

$$z^i = z^i(t),\quad \xi^i = \dot{z}^i(t).$$

This equality holds for any smooth vector function $\eta(t)$ which becomes zero at the ends of the time interval.

Note that there holds the identity

$$\int_a^b \frac{\partial L}{\partial \xi^i}\dot{\eta}^i\, dt = \left(\frac{\partial L}{\partial \xi^i}\eta^i\right)_{t=b} - \left(\frac{\partial L}{\partial \xi^i}\eta^i\right)_{t=a} - \int_a^b \eta^i\left(\frac{\partial L}{\partial \xi^i}\right)^{\cdot} dt$$

(integration by parts).

Since $\eta^i(a)=\eta^i(b)=0$, we obtain

$$\int_a^b \frac{\partial L}{\partial \xi^i}\dot{\eta}^i\, dt = -\int_a^b \left(\frac{\partial L}{\partial \xi^i}\right)^{\cdot}\eta^i\, dt.$$

Substituting this expression into formula (1) we see that for any smooth vector function $\eta^i(t)$ which vanishes at the ends of the time interval, there holds the equality

$$\int_a^b \left[\frac{\partial L}{\partial z^i} - \left(\frac{\partial L}{\partial \xi^i}\right)^{\cdot}\right]\eta^i\, dt = 0,$$

if $\gamma : z^i = z^i(t)$ gives the minimum of the function $S = S(\gamma)$ among all the smooth curves joining the points P and Q, whence

$$\psi^i(t) = \frac{\partial L}{\partial z^i} - \left(\frac{\partial L}{\partial \xi^i}\right)^{\cdot} = 0, \quad i = 1, 2, 3.$$

Indeed, if $\psi^i(t) \neq 0$ for some i and $t = t_0$ between a and b, then we can easily choose a function $\eta^i(t)$ such that the integral is not equal to zero (e.g. for $\eta^i = \psi^i \cdot f(t)$ we have in the integrand a positive number if $f(t)$ is greater than or equal to zero and if it vanishes at the ends). This completes the proof.

REMARK. The solutions of the equations $\left(\frac{\partial L}{\partial \xi^i}\right)^{\cdot} = \frac{\partial L}{\partial z^i}$ are called *extremals*.

Now, we shall give some definitions.

1. The *energy* is the expression:

$$E = E(z, \dot{z}) = E(z, \xi) = \xi^i \frac{\partial L}{\partial \xi^i} - L = \dot{z}^i \frac{\partial L}{\partial \dot{z}^i} - L.$$

2. The *momentum* is described by the expression:

$$\frac{\partial L}{\partial \dot{z}^i} = p_i = \frac{\partial L}{\partial \xi^i} \quad \text{(covector).}$$

3. The *Lagrangian* is an original integrand:

$$L = L(z, \xi) = L(z, \dot{z}).$$

4. The *force* is given by the expression:

$$f_i = \frac{\partial L}{\partial z^i}.$$

5. The *Euler-Lagrange equations* are those of Theorem 1 (equations for extremals)

$$\left(\frac{\partial L}{\partial \dot{z}^i}\right)^{\cdot} = \frac{\partial L}{\partial z^i} \quad \text{or} \quad \dot{p}_i = f_i.$$

EXAMPLE 1. If $L = 1/2 g_{ij}\xi^i\xi^j$, then $p_i = g_{ij}\xi^j$,

$$f^k = 1/2\frac{\partial g_{ij}\xi^i\xi^j}{\partial z^k}.$$

We obtain the equations for the extremals $\frac{dp_k}{dt} = 1/2\frac{\partial g_{ij}}{\partial z_k}\cdot \dot z^i\cdot \dot z^j$ where

$$\frac{\partial p^k}{\partial t} = \ddot z^j g_{jk} + \dot z^j\frac{\partial g_{jk}}{\partial z^i}\dot z^j = 1/2\frac{\partial g_{ij}}{\partial z^k}\dot z^i\dot z^j.$$

Since $g^{km}g_{jk} = \delta^m_j$, we obtain:

$$\ddot z^m + g^{km}\left(\frac{\partial g_{jk}}{\partial z^i} - 1/2\frac{\partial g_{ij}}{\partial z^k}\right)\dot z^i\dot z^j = 0.$$

Now we shall point to the identity:

$$g^{km}\frac{\partial g_{jk}}{\partial z^k}\dot z^i\dot z^j = 1/2\,\dot z^i\dot z^j g^{km}\left(\frac{\partial g_{jh}}{\partial z^i} + \frac{\partial g_{ik}}{\partial z^j}\right).$$

Substituting it into the preceding equation we obtain

$$\ddot z^m + \Gamma^m_{ij}\dot z^i\dot z^j = 0, \tag{2}$$

where

$$\Gamma^m_{ij} = 1/2\, g^{km}\left(\frac{\partial g_{jk}}{\partial z^i} + \frac{\partial g_{ik}}{\partial z^j} - \frac{\partial g_{ij}}{\partial z^k}\right).$$

Equations (2) are called *equations for geodesics* (relative to a given Riemannian metric). The functions Γ^m_{ij} are called *Christoffel coefficients* (or Christoffel symbols) for a symmetric connection compatible with the metric g_{ij}. We shall again deal with these equations in Part II, where they are derived in another way.

Thus we have obtained

THEOREM 2. *The Euler-Lagrange equations for extremals* (*in particular, for minima*) *coincide with the equations for geodesics provided that*

$$L = g_{ij}\dot{z}^i\dot{z}^j = |\dot{z}|^2, \quad S(\gamma) = \int_P^Q |\dot{z}|^2\, dt.$$

EXAMPLE 2. If $L = \left(g_{ij}\dot{z}^i\dot{z}^j\right)^{1/2} = |\dot{z}|$, then the expression $S = (\text{length}) = \int_P^Q |\dot{z}|\, dt$ does not depend on the introduction of the parameter t. The Euler-Lagrange equations have the form:

$$\left(\frac{\partial L}{\partial \dot{z}^k}\right)^{\cdot} = \frac{\partial L}{\partial z^k}$$

or

$$\left(\frac{g_{kj}\dot{z}^j}{\left(g_{ij}\dot{z}^i\dot{z}^j\right)^{1/2}}\right)^{\cdot} = \left(\frac{\partial g_{ij}}{\partial z^k}\dot{z}^i\dot{z}^j\right)\Big/ 2\left(g_{ij}\dot{z}^i\dot{z}^j\right)^{1/2}$$

If we associate a curve with the natural parameter $l = t$, where $\left(g_{ij}\dot{z}^i\dot{z}^j\right)^{1/2} \equiv 1$, we obtain $\left(g_{kj}\dot{z}^j\right)^{\cdot} = \frac{\partial g_{ij}}{\partial z^k}\dot{z}^i\dot{z}^j$. This is the same equation as in Example 1, but only for the curves associated with the natural parameter.

We have arrived at the following:

THEOREM 3. *The Euler-Lagrange equations for extremals (in particular, for minimal curves) in the sense of length* $\left(L = \left(g_{ij}\xi^i\xi^j\right)^{1/2}\right)$ *coincide for the natural parameter with the equations for geodesics. Therefore, any smooth curve which is the shortest between two points P and Q satisfies the equation for geodesics if it is run through with the natural parameter of time (proportional to the length).*

Note some general properties of energy and momentum, for any Lagrangian $L = L(z, \dot{z})$.

Property 1. The total derivative of the energy along an extremal is always equal to zero (it is assumed that $L = L(z, \xi)$ and does not depend explicitly on t):

$$\frac{dE}{dt} = \frac{d}{dt}\left(\dot{z}^i \frac{\partial L}{\partial \dot{z}^i} - L\right) \equiv 0.$$

(This can be checked by direct calculation!)

Property 2. If coordinates $(z^1, z^2 z^3)$ are so chosen that $\partial L/\partial z^1 \equiv 0$, then there holds the equality

$$\left(\frac{\partial L}{\partial \dot{z}^1}\right)^{\bullet} = \dot{p}_1 = 0 \quad \text{(along the extremal).}$$

Indeed, $\dot{p}_1 = \left(\frac{\partial L}{\partial \dot{z}^1}\right) \equiv 0.$

Properties 1 and 2 are the laws of conservation of energy and momentum.

EXAMPLES.

1. If $L = g_{ij}\dot{z}^i\dot{z}^j$, then $E = L = |\dot{z}|^2$. From the energy conservation law we have $dE/dt \equiv 0$ along the extremal of this functional $S = \int L\, dt$. Thus, extremals are always geodesics, and the velocity of their motion is constant (the natural parameter).

2. If a surface in three-dimensional Euclidean space is given relative to cylindrical coordinates z, r, ϕ by the equation $f(z, r) = 0$ (the surface of rotation), then one of the coordinates on the surface may be an angle, and the other will be r or z (locally).

Then the components of the first quadratic form do not depend on ϕ:

$$\partial g_{ij}/\partial\phi \equiv 0; \quad i, j = 1, 2;$$

for coordinates z^1 and z^2 on the surface, we shall take $z^1 = \phi$ and $z^2 = (r \text{ or } z)$. This implies momentum conservation:

$$0 = \partial L/\partial\phi = \dot{p}_\phi = 0, \quad p_\phi = \partial L/\partial\dot{\phi},$$

$$L = 1/2\left(g_{\phi\phi}\dot{\phi}^2 + 2g_{r\phi}\dot{r}\dot{\phi} + g_{rr}\dot{r}^2\right),$$

$$p_\phi = g_{\phi\phi}\dot{\phi} + g_{r\phi}\dot{r} = \text{const}$$

(along the geodesic curve). However, for surfaces of rotation we always have:

$$g_{r\phi} \equiv 0, \quad g_{\phi\phi} = r^2, \quad p_\phi = r^2 \dot{\phi}.$$

Thus, there holds

THEOREM 4. *For surfaces of rotation $f(r, z) = 0$ in Euclidean space with cylindrical coordinates (z, r, ϕ) and Euclidean metric $dl^2 = dz^2 + dr^2 + r^2(d\phi)^2$*, the quantity $r^2\dot{\phi}$ *is constant along any geodesic curve.*

(Recall that the parameter is natural!)

Indeed, if (r, ϕ) are coordinates on the surface, we always have $g_{\phi r} = 0$, $g_{\phi\phi} = r^2$. Therefore $p'_\phi = r^2\dot{\phi}$. Since $\dot{p}_\phi = \partial L/\partial\phi = 0$, the theorem is proved.

PART II

TENSORS. RIEMANNIAN GEOMETRY

2.1 Rank-One and Rank-Two Tensors

We have already got used to the fact that many quantities are given as numerical functions of a point in space. For example, the distance from a point to a certain fixed centre, etc. If we have several such quantities at our disposal, we already have several functions of a point (or, so to say, the vector function of this point). In three-dimensional space, for a complete characteristic of the poisition of a point in space it is nceecessary, as is well-known, to know the values of at least three numerical functions called coordinates of the point (x^1, x^2, x^3): each of the coordinates x^i is a function of the point, and the set (x^1, x^2, x^3) completely determines the point. The reader has already met with different types of coordinates, for example, in a plane there exist Cartesian coordinates x^1, x^2 and polar coordinates r, ϕ, where $x^1 = r \cos \phi, x^2 = r \sin \phi$; in space there exist Cartesian, cylindrical r, z, ϕ or spherical r, θ, ϕ coordinates.

Thus, coordinates make up the set of numerical functions of a point which determine completely the position of this point in space. In precisely the same way, the coordinates of any physical system make up such a set of numerical functions of the state ot this system which completely determines this state. The state of a system is a point in "the space of all possible states" of the system. For instance, the state of a moving material point is determined by six numbers: three coordinates and three components of the velocity vector; here we deal with a six-dimensional state space.

It turns out, however, that the numerical function of a point, or the set of such functions, is insufficient for the investigation of many problems. The point is that many geometrical and physical quantities can be described as a set of numerical functions only after a certain set of coordinates (x^1, x^2, x^3) in space is already given; the numerical representation of these quantities may change significantly if we assume some other coordinates z^1, z^2, z^3, where

$$x^i = x^i(z^1, z^2, z^3), \quad i = 1, 2, 3.$$

To clarify this, we shall consider the concept of a vector, for example, the velocity vector for motion along a certain curve:

$$z^j = z^j(t), \quad j = 1, 2, 3;$$

$$x^i = x^i(t) = x^i(z^1(t), z^2(t), z^3(t)), \quad i = 1, 2, 3.$$

In the coordinates z^1, z^2, z^3 we have the components of the velocity vector in the form:

$$\left(\frac{dz^1}{dt}, \frac{dz^2}{dt}, \frac{dz^3}{dt}\right)_{t=t_0} = (\xi^1, \xi^2, \xi^3).$$

Representing the same curve in the other coordinates x^1, x^2, x^3, we obtain other components of the same vector:

$$\left(\frac{dx^1}{dt}, \frac{dx^2}{dt}, \frac{dx^3}{dt}\right)_{t=t_0} = (\eta^1, \eta^2, \eta^3),$$

where

$$\frac{dx^i}{dt} = \frac{\partial x^i}{\partial z^j}\frac{dz^j}{dt}, \qquad j, i = 1, 2, 3.$$

Thus, for the components of the vector, we have the formula of their transformation under the change of coordinates:

$$\xi^i = \eta^j \frac{\partial x^i}{\partial z^j}, \qquad j, i = 1, 2, 3,$$

$$x^i = x^i(z^1, z^2, z^3);$$

(ξ^1, ξ^2, ξ^3) are the components of the vector in the coordinates (z^1, z^2, z^3) at a given point;

(η^1, η^2, η^3) are the components of the vector in the coordinates (x^1, x^2, x^3) at the same point.

Tensors are the most important class of quantities whose numerical representation changes under the change of coordinates. The vector is the simplest and most visual example of the tensor. A trivial example of the tensor is the scalar which does not change under the change of coordinates.

Before we introduce mathematically the exact concept to a tensor, we shall consider some other examples which we have already encountered repeatedly.

1. The gradient of a numerical function. It is normal to say that the gradient of a numerical function $f(x^1, x^2, x^3)$ in Cartesian coordinates x^1, x^2, x^3 is a vector with components:

$$\nabla j = \operatorname{grad} f = \left(\frac{\partial f}{\partial x^1}, \frac{\partial f}{\partial x^2}, \frac{\partial f}{\partial x^3}\right) = (\xi_1, \xi_2, \xi_3).$$

Let us see how the gradient of the same function looks in coordinates z^1, z^2, z^3, where

$$x^i = x^i(z^1, z^2, z^3), \quad i = 1, 2, 3.$$

We have

$$\operatorname{grad} f(x^1(z), x^2(z), x^3(z)) = \left(\frac{\partial f}{\partial z^1}, \frac{\partial f}{\partial z^2}, \frac{\partial f}{\partial z^3}\right) = (\eta_1, \eta_2, \eta_3),$$

$$\frac{\partial f}{\partial z^i} = \frac{\partial f}{\partial x^j}\frac{\partial x^j}{\partial z^i}, \quad i = 1, 2, 3.$$

Hence

$$\eta_i = \frac{\partial x^j}{\partial z^i}\xi_j,$$

where (ξ_1, ξ_2, ξ_3) are the components of the gradient in the coordinates x^1, x^2, x^3; (η_1, η_2, η_3) are the components of the gradient in the coordinates z^1, z^2, z^3.

Now compare the formlae of transformation of the numerical representation of the velocity vector of a curve and that of the gradient of a function.

The velocity vector

$$\xi^i = \eta^i \frac{\partial x^j}{\partial z^i}. \tag{1}$$

The gradient

$$\eta_i = \xi_j \frac{\partial x^j}{\partial z^i}. \tag{2}$$

These formulae are distinct!

To compare these formulae, we shall introduce the matrix $A = (a^i_j)$, where $a^i_j = \partial x^i/\partial z^j$, and a transposed matrix $A^T = (b^j_k)$, where $b^j_k = a^k_j$.

For the vectors ξ and η we rewrite formulae (1) and (2) in the form

$$(x) \leftarrow (z)$$

$$\xi = A\eta \quad \text{(for the velocity vector),}$$

$$\eta = A^T\xi \quad \text{(for the gradient)}$$

$$(z) \leftarrow (x).$$

In case the matrix A^T has the inverse $(A^T)^{-1}$, formula (2) can be rewritten to become

$$(A^T)^{-1}\eta = (A^T)^{-1} A^T\xi = \xi,$$

$$\xi = (A^T)^{-1}\eta \quad \left(\text{or } \xi_i = \eta_j \frac{\partial z^j}{\partial x^i}\right). \tag{2'}$$

In which case will the transformation laws for the velocity vectors and gradients of functions coincide under the change from the coordinate system (x) to the coordinate system (z)?

From formulae (1), (2) and (2') we obtain

$$(x) \rightarrow (z)$$

$$\xi = A\eta \qquad \text{(for the velocity vector)}$$

$$\xi = (A^T)^{-1}\eta \quad \text{(for the gradient)}$$

The final conclusion is that for the transformation formulae (1) and (2) to coincide, it is necessary that we have equality of the matrices

$$A = (A^T)^{-1} \text{ or } A \circ A^T = E,$$

where

$$A = (a^i_j) = \frac{\partial x^i}{\partial z^j}.$$

Such matrices A for which $A^T = A^{-1}$ are called *orthogonal.* Note that the change of coordinates $\lambda x = x(z)$, such that at each point the Jacobian matrix $(\partial x^i/\partial z^j) = A$ is orthogonal, is a linear orthogonal transformation A = const., that is, the matrix A does not depend on the point.

Thus, the gradient of a function under the change of coordinates transforms in a way *different* from that of the velocity vector. This is *another form of the tensor* which is occasionally referred to as "*covector*" as distinct from velocity vectors.

2. Riemannian metric. As has already been said, given the coordinates x^1, x^2, x^3, in three-dimensional space or in a region of space, the metric concepts (such as lengths and angles) are determined by the set of functions $(g_{ij}(x))$, $j, i = 1, 2, 3$. For the length of a curve, by definition

$$l = \int_a^b \left(g_{ij}(x(t))\,\dot{x}^i\,\dot{x}^j\right)^{1/2} dt,$$

where $\dot{x} = dx/dt$ and the quadratic form $\sum g_{ij}\xi^i\xi^j$ is positive. This is the quadratic form determined on vectors of the "velocity vector" type at each particular point $x = (x^1, x^2, x^3)$ and dependent on the point. We have called (g_{ij}) the Riemannian metric. Under the change of coordinates

$$x^i = x^i(z^1, z^2, z^3), \quad i = 1, 2, 3,$$

the formula for the lentth of the curve assumes the form

$$l = \int_a^b \left(g'_{ij}\,(z(t))\,\dot{z}^i\,\dot{z}^j\right)^{1/2} dt,$$

where $x^i(t) = x^i(z^1(t), z^2(t), z^3(t))$, the transformation law for metric components being of the form

$$g'_{ij}\,(z) = g_{kl}(x(z)) \cdot \frac{\partial x^k}{\partial z^i} \cdot \frac{\partial x^l}{\partial z^j}. \tag{3}$$

Hence the quadratic forms on the vectors are transformed by the law (3). This is one more type of tensor (called tensor of rank two).

Thus, we have already pointed out several types of tensors:

a) scalar (does not transform);
b) vector (transforms by the law (1));
c) covector (transforms by the law (2));
d) Riemannian metric (transforms by the law (3)).

It should be recalled that the Riemannian metric (g_{ij}) in coordinates x^1, x^2, x^3 was needed to define the concept of the length of the vector at a given popint (x).

Given the vector $\xi = (\xi^1, \xi^2, \xi^3)$ at the point (x^1, x^2, x^3), we have

$$|\xi|^2 = \text{the length squared} = g_{ij}(x)\, \xi^i \xi^j.$$

In particular, this rule has been applied to the velocity vectors of parametrized curves to determine the length of a curve as the integral of the length of the velocity vector.

The transformation law (3) for the components of the metric under the change of coordinates follows unambiguously from the law (1) for the components of the vector and from the obvious requirement that the length of the curve should not depend on the choice of coordinates relative to which it is calculated. The lenth of a curve is, in fact, the time integral of the lenth of the velocity vector. It is, therefore, necessary that the square of the length of the velocity vector

$$|\xi|^2 = g_{ij}(x)\, \xi^i \xi^j$$

should not depend on the choice of coordinates. This requirement and formula (1) for the components of the vector imply the transformation law (3) for the components of the metric (g_{ij}).

3. If we wish to define the invariant concept of the square of the length of the covector which transforms by the law (2) or (2'), we have to introduce the components $(g^{ij}(x))$ and put

$$|\xi|^2 = g^{ij}(x)\, \xi_i \xi_j$$

$$\xi = (\xi_1, \xi_2, \xi_3), \ \text{(at the point } x).$$

Under the change $x^i = x^i(z)$, $i = 1, 2, 3$ we obtain the transformation law

$$\eta_i = \xi_j \frac{\partial x^j}{\partial z^i}. \tag{2}$$

$$g^{ij}(z) = g^{kl}(x(z)) \cdot \frac{\partial z^i}{\partial x^k} \cdot \frac{\partial z^j}{\partial x^l}. \tag{4}$$

Then the length will not depend on the choice of coordinates:

$$|\eta|^2 = |\xi|^2 = g'^{ij}\eta_i\eta_j = g^{ij}\xi_i\xi_j,$$

where $\xi = (\xi_1, \xi_2, \xi_3)$ is a covector in the coordinates x^1, x^2, x^3 at the point (x), $\eta = (\eta_1, \eta_2, \eta_3)$ is the same covector at the same point but relative to the coordinates z^1, z^2, z^3.

The transformation law (4) yields one more type of tensor (of rank two).

4. Finally, we have to examine the last type of tensor of rank two, namely, linear operators on vectors.

Suppose that at each point of a space with coordinates (x^1, x^2, x^3) we are given a matrix $(a^i_j(x)) = A$, which determines the linear transformation of vectors at each point $x = (x^1, x^2, x^3)$. This linear transformation $A(x)$ has the form $\eta = A\xi$, where

$$\eta^i = a^i_j(x)\, \xi^j. \tag{5}$$

Here $\xi = (\xi^1, \xi^2, \xi^3)$ is the vector at the point x.

The same matrix will determine the linear transformation of covectors by the formula $\eta = A\xi$, where

$$\eta_j = a^i_j(x)\, \xi_i. \tag{6}$$

Under the change of coordinates $x^i = x^i(z^1, z^2, z^3)$, from formulae (1) and (2) we can deduce that the components of the matrix A are transformed by the law $A \to \widetilde{A} = (\widetilde{a}^i_j)$:

$$\widetilde{a}^i_j = \frac{\partial z^i}{\partial x^k} a^k_l \frac{\partial x^l}{\partial z_j},$$

where $x^i = x^i(z)$, $z^j = z^j(x)$ and $z^i(x(z)) = z^i$, and

$$\frac{\partial z^i}{\partial x^k} \cdot \frac{\partial x^k}{\partial z^j} = \begin{cases} 1, & i = j, \\ 0, & i \neq j. \end{cases}$$

For the covector, formula (2) can now become

$$\xi_i = \eta_j \frac{\partial x^j}{\partial z^i}, \tag{2'}$$

since

$$\frac{\partial z^j}{\partial x^i} \cdot \frac{\partial x^i}{\partial z^k} = \delta^j_k .$$

Now we shall tabulate the transformation laws for the scalar, vector, covector and all three types of rank-two tensors:

The laws of transformation.

1. The scalar (rank-0 tensor) is not transformed.
Tensors of rank one:
2. The vector $\xi = (\xi^i)_x \to \acute{\xi} = (\acute{\xi}^j)_z$ (of the type of velocity vector):

$$\acute{\xi}^j = \xi^i \frac{\partial z^j}{\partial x^i} .$$

3. The covector $\xi = (\xi_i)_x \to \acute{\xi} = (\acute{\xi}_j)_z = \acute{\xi}$ (of the type of the gradient of a function):

$$\acute{\xi}_j = \xi_i \frac{\partial x^i}{\partial z^j} .$$

Tensors of rank two:
4. The scalar product of vectors $(g_{ij}) \to (\acute{g}_{ij})$:

$$\acute{g}_{ij} = g_{kl} \frac{\partial x^k}{\partial z^i} \cdot \frac{\partial x^l}{\partial z^j} ; \quad (x) \to (z).$$

5. The scalar product of covectors $(g^{ij}) \to (\acute{g}^{ij})$:

$$\acute{g}^{ij} = g^{kl} \frac{\partial z^i}{\partial x^k} \cdot \frac{\partial z^j}{\partial x^l} .$$

6. The linear operator on vectors (covectors) $A = (a^i_j) \to A' = (a'^i_j)$:

$$a'^i_j = a^k_l \, \frac{\partial x^l}{\partial z^j} \cdot \frac{\partial z^i}{\partial x^k} \, .$$

Here $x^i = x^i(z^1, z^2, z^3)$, $z^i = z^i(x^1, x^2, x^3)$, $i = 1, 2, 3$, and

$$z^i(x^1(z), x^2(z), x^3(z)) \equiv z^i,$$

$$\frac{\partial z^i}{\partial x^k} \cdot \frac{\partial x^k}{\partial z^j} = \delta^i_j \, .$$

2.2 Tensors of General Form. Examples

In the preceding section we have considered tensors of rank one (vectors and covectors) and tensors of rank two (quadratic forms on vectors g_{ij}, quadratic forms on covectors g^{ij} and linear transformations — operators or affiners — a^i_j). It should be recalled that in each coordinate system x^1, x^2, x^3 ... the tensor was given by the set of numbers at a given point x

(ξ^i) — vector	(g_{ij})	
	(g^{ij})	rank-2 tensors of all the three types.
(ξ_i) — covector	(d^i_j)	

Under the change of coordinates $x^i = x^i(z^1, z^2, z^3, \dots)$, $i = 1, 2, 3, \dots, k$, the tensor in the coordinates z at the same point was given by a set (a different one) of numbers

$$(\xi'^i),\ (\xi'_i),\ (g'_{ij}),\ (g'^{ij}),\ (a'^i_j).$$

Given this, there hold the relations

$$\xi^i = \xi'^j \frac{\partial x^i}{\partial z^j}, \tag{1}$$

$$\xi_i = \xi'_j \frac{\partial z^j}{\partial x^i}, \tag{2}$$

$$g_{ij} = g'_{kl} \frac{\partial z^k}{\partial x^i} \frac{\partial z^l}{\partial x^j}, \tag{3}$$

$$g^{ij} = g'^{kl} \frac{\partial x^i}{\partial z^k} \frac{\partial x^j}{\partial z^l}, \tag{4}$$

$$d^i_j = a'^k_l \frac{\partial z^l}{\partial x^j} \frac{\partial x^i}{\partial z^k}. \tag{5}$$

By definition

$$x^i = x^i(z^1, z^2, \dots, z^k) \text{ and } z^j = z^j(x^1, x^2, \dots, x^k),$$

where $z^i = z^i(x(z))$, $\dfrac{\partial z^i}{\partial z^k} = \delta^i_k = \dfrac{\partial z^i}{\partial x^j}\dfrac{\partial x^j}{\partial z^k}$. Hence, the matrices $A = \left(\dfrac{\partial x^j}{\partial z^k}\right)$ and $B = \left(\dfrac{\partial z^j}{\partial x^q}\right)$ are mutually inverse: $B = A^{-1}$. We can now define tensors of general form.

DEFINITION 1. *A tensor of type* (m, n) and *rank* $m + n$ is an object which is given an arbitrary system of coordinates $(x^1, x^2, \dots, x^k)$ by a set of numbers $(T^{i_1 i_2 \dots i_m}_{j_1 j_2 \dots j_n})$ and whose numerical representation depends on the coordinate system obeying the following law:

$$\text{if } x^i = x^i(z^1, z^2, \dots, z^k),\ z^j = z^j(x^1, x^2, \dots, x^k),$$

then there holds the formula

$$T^{i_1 i_2 \dots i_m}_{j_1 j_2 \dots j_n} = \acute{T}^{k_1 k_2 \dots k_m}_{l_1 l_2 \dots l_n} \frac{\partial x^{i_1}}{\partial z^{k_1}} \cdots \frac{\partial x^{i_m}}{\partial z^{k_m}} \frac{\partial z^{l_1}}{\partial x^{j_1}} \cdots \frac{\partial z^{l_n}}{\partial x^{j_n}}; \tag{6}$$

Here $\acute{T}^k_l$ is the numerical representation of the tensor in the coordinates (z) and T^i_j is the numerical representation of the tensor in the coordinates (x). The indices $(i_1, \dots, i_m, j_1, \dots, j_n)$ and $(k_1, \dots, k_m, l_1, \dots, l_n)$ vary from 1 to 3 for tensors in a three-dimensional space, and, in k dimensions, all these indices vary from 1 to k.

The velocity vector is a tensor of type (1, 0).
The covector is a tensor of type (0, 1).
The quadratic form on vectors is a tensor of type (0, 2).
The quadratic form on covectors is a tensor of type (2, 0).
The linear operator on vectors is a tensor of type (1, 1).

THEOREM 1. *The components of the tensor* $\acute{T}^{k_1 \dots k_m}_{l_1 \dots l_n}$ *can be expressed in terms of* $T^{i_1 \dots i_m}_{j_1 \dots j_n}$ *by the formula*

$$\acute{T}^{k_1 \dots k_m}_{l_1 \dots l_n} = T^{i_1 \dots i_m}_{j_1 \dots j_n} \frac{\partial z^{k_1}}{\partial x^{i_1}} \dots \frac{\partial z^{k_m}}{\partial x^{i_m}} \frac{\partial x^{j_1}}{\partial z^{l_1}} \dots \frac{\partial x^{j_n}}{\partial z^{l_n}} . \tag{7}$$

LEMMA 1. *There holds the relations*

$$\frac{\partial x^i}{\partial z^j} \frac{\partial z^j}{\partial x^k} = \delta^i_k ; \qquad \frac{\partial z^k}{\partial x^j} \frac{\partial x^j}{\partial z^q} = \delta^k_q . \tag{8}$$

Proof. From the fact that the transformations $x = x(z)$ and $z = z(x)$ are inverse to each other, we have

$$x^i(z(x)) = x^i ; \quad z^q(x(z)) = z^q .$$

Therefore, from the formulae for differentiation of composite functions and from the fact that $\frac{\partial x^i}{\partial x^k} = \delta^i_k$, $\frac{\partial z^i}{\partial z^q} = \delta^i_q$, we obtain formulae (8). Indeed,

$$\delta^i_j = \frac{\partial x^i}{\partial x^j} = \frac{\partial x^i}{\partial z^l} \frac{\partial z^l}{\partial x^j} ,$$

$$\delta^k_q = \frac{\partial z^k}{\partial z^q} = \frac{\partial z^k}{\partial x^j} \frac{\partial x^j}{\partial z^q} .$$

Formulae (8) are thus derived.

Let us now prove formula (7).

By the definition of a tensor, we have the relation (6). Consider the relation (6) as a linear equation with the right-hand sides T^i_j and with the unknowns $\acute{T}^k_l$. Solving this equation, we must derive (7).

By virtue of (6) there holds the formula

$$T^{i}_{j} \frac{\partial z'^{k_1}}{\partial x^{i_1}} \cdots \frac{\partial z'^{k_m}}{\partial x^{i_m}} \frac{\partial x^{j_1}}{\partial z'^{l_1}} \cdots \frac{\partial x^{j_n}}{\partial z'^{l_n}} =$$

$$= \Big(\sum_{p,q} T'^{q_1 \dots q_m}_{p_1 \dots p_n} \frac{\partial x^{i_1}}{\partial z'^{q_1}} \cdots \frac{\partial x^{i_m}}{\partial z'^{q_m}} \frac{\partial z'^{p_1}}{\partial x^{j_1}} \cdots \frac{\partial z'^{p_n}}{\partial x^{j_n}}\Big) \frac{\partial z'^{k_1}}{\partial x^{i_1}} \cdots$$

$$\cdots \frac{\partial z'^{k_m}}{\partial x^{i_m}} \frac{\partial x^{j_1}}{\partial z'^{l_1}} \cdots \frac{\partial x^{j_n}}{\partial z'^{l_n}} =$$

$$= T'^{q}_{p} \frac{\partial x^{i}}{\partial z'^{q}} \frac{\partial z'^{p}}{\partial x^{j}} \frac{\partial z'^{k}}{\partial x^{i}} \frac{\partial x^{j}}{\partial z'^{l}} \equiv T'^{q}_{p} \delta^{p}_{l} \delta^{k}_{q} \equiv T'^{k}_{l} ,$$

where

$$i = (i_1, \dots, i_m), \ j = (j_1, \dots, j_n),. \ k = (k_1, \dots, k_m),$$

$$l = (l_1, \dots, l_n), \ p = (p_1, \dots, p_n), \ q = (q_1, \dots, q_m).$$

Thus we come to the relations (7). This completes the proof.

Now we shall point out the simplest properties of tensors.

At any arbitrarily given point of space, the tensors form a linear space

a) if $T = \left(T^{i_1 \dots i_m}_{j_1 \dots j_n}\right)$, $S = \left(S^{k_1 \dots k_m}_{l_1 \dots l_n}\right)$ are tensors of type (m, n), then their linear combination

$$\lambda T + \mu S = U \text{ with components } U^{i_1 \dots i_m}_{j_1 \dots j_n} = \lambda T^{i_1 \dots i_m}_{j_1 \dots j_n} + \mu S^{i_1 \dots i_m}_{j_1 \dots j_n}$$

is also a tensor of type (m, n);

b) it is important to note that a tensor is an object fixed to a point, and there exists no rule for summation of tensors fixed to different points;

c) the dimension of the linear space of tensors of type (m, n) in a k-dimensional space is calculated as k^{m+n}. If the basis coordinate vectors in a k-dimensional space coordinatized by a system of coordinates $x^1, x^2, \dots, x^k$ are expressed in terms of $e_1, e_2, \dots, e_k$ and the basis vectors in terms of $e^1, e^2, \dots, e^k$, then any tensor will be conveniently represented in the form

$$\text{vector } \xi = \xi^i e_i \ \left(\text{e.g. } \frac{dx}{dt} = \frac{dx^i}{dt} e_i\right),$$

$$\text{covector } \xi = \xi_i e^i \ \left(\text{e.g. grad} f = \frac{\partial f}{\partial x^i} e^i\right),$$

$$\text{quadratic forms } (g) = g_{ij}\, e^i \circ e^j \ \text{(on vectors)},$$

$$\text{quadratic forms } (g) = g^{ij}\, e_i \circ e_j \ \text{(on covectors)},$$

$$\text{linear operators } A = a^i_j\, e_i \circ e^j.$$

$$\text{Any tensor } T = \left(T^{i_1 \dots i_m}_{j_1 \dots j_n}\right) \text{ will be written as}$$

$$T = T^{i_1 \dots i_m}_{j_1 \dots j_n}\, e_{i_1} \circ \dots \circ e_{i_m} \circ e^{j_1} \dots \circ e^{j_n}.$$

It is essential to note that in this notation the order of indices is of importance — e_{i_1} and e_{i_2}, for example, should not , generally speaking, exchange places.

Thus, in the linear space of tensors of type (m, n) at a given point (x) of the space, the basis has the form

$$e_{i_1} \circ \dots \circ e_{i_m} \circ e^{j_1} \dots \circ e^{j_n} \quad (\text{altogether } k^{m+n})$$

where i, j independently take on values $1, 2, \dots, k$. Making the change of coordinates $x^i = x^i(z^1, z^2, \dots, z^k)$, we go over to another basis in the linear space of tensors fixed at a given point to the basis connected with the coordinate vectors of the system $z^1, z^2, \dots, z^k$ at this point.

The mutual expression of these bases in terms of each other proceeds, according to formulae (6) and (7), at a given point of space. We shall consider several examples.

1. *The stress tensor*. In a continuous medium in $\mathbb{R}^3$, at each point $x = (x^1, x^2, x^3)$, the pressure upon a small element of area ΔS orthogonal to the unit vector n is given by $(\Delta S)P(n)$, where P is a linear operator $P = P^i_j$. If $n = n^j e_j$, then $P_n = (n^j P^i_{\ j})e_i$

or $\{P(n)\}^i = n^j P^i_j$, for instance, if the medium satisfies Pascal's law $P^i_j = \delta^i_j P$, where the quantity P is called the *pressure* at this point.

2. *The strain tensor.* If in a continuous medium with coordinates x^1, x^2, x^3 each point is displaced

$$x^i \to x^i + u^i,$$

then the medium is said to *have undergone deformation* (or strain). If originally the distance between two close points of the medium, for example, in Euclidean coordinates x^1, x^2, x^3 was

$$(\Delta l)^2 = \Sigma(\Delta x)^2 = \sum_{i=1}^{3} (x^i - x'^i)^2,$$

after the deformation the distance between the same points will be different:

$$(\overline{\Delta l})^2 = \Sigma\, (x^i + u^i(x) - x'^i - u^i(x'))^2.$$

Obviously, we have

$$(\overline{\Delta l})^2 = (\Delta l)^2 - 2\sum_{i=1}^{3} \Delta x^i \Delta u^i + \sum_{i=1}^{3} (\Delta u^i)^2,$$

$$\Delta x^i = x^i - x'^i,$$

$$\Delta u^i = u^i(x) - u^i(x') \cong \frac{\partial u^i}{\partial x^j} \Delta x^j .$$

Therefore as $\Delta x^i \to 0$,

$$(\overline{dl})^2 = (dl)^2 - 2dx^i dx^j \frac{\partial u^i}{\partial x^j} + \frac{\partial u^i}{\partial x^k} \frac{\partial u^i}{\partial x^l} dx^k dx^l .$$

Given this, there holds the equality

$$2\frac{\partial u^i}{\partial x^j} dx^i dx^j = \left(\frac{\partial u^i}{\partial x^j} + \frac{\partial u^j}{\partial x^i}\right) dx^i dx^j,$$

since $$\frac{\partial u^i}{\partial x^j} dx^i dx^j = \frac{\partial u^j}{\partial x^i} dx^i dx^j.$$

DEFINITION 2. The *strain tensor* η_{ij} *of the medium* is given by the difference

$$(dl)^2 - (\overline{dl})^2 = \eta_{ij} dx^i dx^j - 1/2 \frac{\partial u^i}{\partial x^k} \frac{\partial u^i}{\partial x^l} dx^k dx^l$$

where $\eta_{ij} = (\frac{\partial u^i}{\partial x^j} + \frac{\partial u^j}{\partial x^i}) = \eta_{ij}$. In the case when u^i are small displacements, the quadratic terms in u_i are ignored, yielding the *strain tensor for small deformations*

$$(\eta_{ij}) = (\frac{\partial u^i}{\partial x^j} + \frac{\partial u^j}{\partial x^i}) .$$

According to *Hooke's law*, such small deformations induce stresses in this medium, which depend *linearly* on the deformation. Therefore, the stress tensor and the strain tensor must be *related linearly* as

$$P = U(\eta).$$

This relation is a tensor of rank four. In index notation it is given by

$$P^i_j = U_j^{ikl} \eta_{kl} ,$$

where

$$P = (P^i_j), \quad \eta = (\eta_{kl}), \quad (U = U_j^{ikl}).$$

The tensor U_j^{ikl} of rank four is described by 81 components. Hooke's law in a continuous medium does not actually require 81 components for its specification and can manage with a much smaller number.

In the case when the coordinates are Euclidean, we need not (under orthogonal transformations) distinguish between vectors and covectors, and may generally do without distinguishing between upper and lower indices of a tensor, since they transform in a similar way. The general tensor $U = (U_j^{ikl})$ in Euclidean coordinates is specified by 81 parameters, but the medium is hypothesized to be isotropic. This hypothesis means that the tensor U at each point should be such that its numerical notation remains invariant in all coordinates that differ by rotation around this point, i.e. under orthogonal transformations. We may write either U or (U^{kl}_{ij}) since we do not distinguish between the types of tensors U_j^{ikl} and U^{kl}_{ij} so long as they are equivalent under orthogonal transformations.

We can, therefore, write $P_{ij} = P^j_i = P^{ij}$ and $\eta_{ij} = \eta^j_i = \eta^{ij}$.
There holds

THEOREM 2 (without proof). *The class of tensors of rank four which, when written numerically, are invariant under all rotations of Euclidean space is specified by three parameters* λ, μ, ν: *this class consists of the tensors*

$$P_{ij} = \lambda\eta_{ij} + \mu(\mathrm{Sp}\,\eta)\delta_{ij} + \nu\eta_{ji},$$

where

$$\mathrm{Sp}\,\eta = \sum_{i=1}^{3}\eta_{ii} = \sum_{i=1}^{3}\eta^i_i,$$

(here we do not distinguish between upper and lower indices under rotations), $P_{ij} = U^{ijkl}\eta_{kl}$.

In the theory of elasticity, we should take into account the symmetry

$$\eta_{ij} = \eta_{ji}, \quad P_{ij} = P_{ji}.$$

Therefore, Hooke's tensor is described by two parameters only.

It is obvious here only that the tensors U of the indicated form are actually invariant under all rotations.

The condition that the medium be isotropic is fulfilled in many liquids. In a solid, this hypothesis is far from being always valid. Of course, in an isotropic substance any linear law relating two symmetric physical tensors of rank two, which is described by a tensor of rank four, depends only on two constants at each given point of the medium.

It should be emphasized that the condition of isotropy suggests the presence of the Riemannian metric (we have formulated it for the Euclidean one), whereas the concept of a tensor is not associated with a metric, and the Riemannian metric itself is simply a special type of tensor of rank two (g_{ij}).

It is natural to ask a simpler question: what form may be assumed by tensors of rank one and rank two?

Obviously, there exist no non-zero vectors (covectors) whose numerical notation would not change under all rotations.

As for tensors of rank two (η_{ij}), the only invariants of all the rotations are the eigenvalues of the matrix (η_{ij}), namely, the solutions of the equations $\det(\eta_{ij} - \lambda\delta_{ij}) = 0$.

The tensor $\lambda\delta_{ij}$ is invariant under all rotations (the eigenvalues are all the same). If there exists a pair of distinct eigenvalues, then the tensor of rank two is already not invariant under rotation. Therefore, $\lambda\delta_{ij}$ is the only isotropic tensor of rank two. It can be shown that there exist no isotropic tensors of rank three.

Now we shall consider the class of tensors of rank two invariant not only under rotations but also under all linear transformations (it is already needless here to assume the space to be Euclidean and generally to introduce the Riemannian metric).

For rank two tensors we shall have a single invariant tensor of type (1, 1):

$$(d^i_j) = \lambda\delta^j_i .$$

It can be verified that there exist no tensors of type (0, 2) and (2, 0) invariant under all linear transformations.

For fourth rank tensors we obtain (without proof)

$$P^i_j = \lambda\eta^i_j + \mu(\eta^k_k)\,\delta^i_j = U^{i\,q}_{ij}\,\eta^p_q,$$

$$P_{ij} = \lambda\eta_{ij} + \nu\eta_{ji} = U^{k\,\,l}_{\,\,ij}\,\eta_{kl} .$$

We can see once again that the result is different for tensors of type (1, 1) and (0, 2).

Thus in the absence of the Riemannian metric, the properties of tensors of different types are distinct.

2.3 Algebraic Operations on Tensors

In the preceding section we have discussed the concept of a tensor of type (m, n)

$$T^{i_1 \dots i_m}_{j_1 \dots j_n}(x) = T'^{l_1 \dots l_m}_{k_1 \dots k_n}(z) \frac{\partial x^{i_1}}{\partial z^{l_1}} \dots \frac{\partial z^{k_1}}{\partial x^{j_1}} \dots$$

(the sum over k, l).

The inverse transformation from T to T' has been shown to have the form

$$T'^{k_1 \dots k_m}_{l_1 \dots l_n} = T^{i_1 \dots i_m}_{j_1 \dots j_n} \frac{\partial z^{k_1}}{\partial x^{i_1}} \dots \frac{\partial x^{j_1}}{\partial z^{l_1}} \dots$$

(the sum over i, j).

We have considered several examples:

1) the stress tensor $P^i_j = \hat{P}$,

2) the strain tensor $\eta_{ij} = \hat{\eta}$,

3) Hooke's law — the relation between $\hat{\eta}$ and $\hat{P}$,

4) the isotropy principle — the restriction upon the relation between strain and stress tensors which follows from the rotational invariance of Hooke's law. We are now in a position to proceed to algebraic operations on tensors.

1. Permutation of indices. We shall say that two tensors of the same type $T^{i_1 \dots i_m}_{j_1 \dots j_n}$ and $\tilde{T}^{i_1 \dots i_m}_{j_1 \dots j_n}$ can be obtained one from the other *by means of a permutation of the upper indices* if there exists a permutation of indices $\begin{pmatrix} i_1 \dots i_n \\ q_1 \dots q_n \end{pmatrix}$, where $i_k \to q_k$, such that for all $i_1, \dots, i_n, j_1, \dots, j_m, q_1, \dots, q_n$ there holds the equality

$$\tilde{T}^{i_1 \dots i_m}_{j_1 \dots j_n} = T^{q_1 \dots q_m}_{j_1 \dots j_n}.$$

Thus, from the tensor T we have obtained a new tensor $\tilde{T}$. In a similar way we can make a *permutation of the lower indices* and obtain a new tensor. *We cannot interchange the lower and upper indices* (this operation is not preserved under the change of coordinates).

EXAMPLE 1. From a second-rank tensor (a_{ij}) we can obtain a tensor $(b_{ij}) = (a_{ji})$ using the permutation. Similarly, from a tensor (a^{ij}) we can obtain a tensor $(b^{ij}) = (a^{ji})$.

2. Contraction (trace). The contraction of a tensor T of type (m, n) with respect to the indices (i_k, j_l) is the sum

$$\sum_{i=1}^{p} T_{j_1 \dots i=j_l, \dots, j_n}^{i_1 \dots i=i_k, \dots, i_m} = \tilde{T}_{j_1 \dots j_{n-1}}^{i_1 \dots i_{m-1}},$$

which is a tensor of type $(m-1, n-1)$.

For a tensor of type T^i_j (a linear operator), the trace Sp $T = T^i_i$ is invariant (a scalar).

3. Product of tensors. Given two tensors $T_{j_1 \dots j_n}^{i_1 \dots i_m}$ and $P_{\beta_1 \dots \beta_l}^{\alpha_1 \dots \alpha_k}$ of ranks (m, n) and (k, l) respectively, we define their product to be a tensor of type $(m+k, n+l)$ with components

$$A_{j_1 \dots j_n \beta_1 \dots \beta_l}^{i_1 \dots i_m \alpha_1 \dots \alpha_k} = T_{j_1 \dots j_n}^{i_1 \dots i_m} \cdot P_{\beta_1 \dots \beta_l}^{\alpha_1 \dots \alpha_k} .$$

So, we have three invariant operations on tensors, namely, permutation, contraction (tracing) and product. Now let us consider several examples.

1. A vector (ξ^i) and a covector (η_j). Consider their product which is a second-rank tensor $T^i_j = (\xi^i \eta_j)$ and its trace

$$T^i_i = \xi^i \eta_i.$$

We obtain the scalar $\xi^i \eta_i$ from the vector and covector, which is their scalar product $\lambda = \xi^i \eta_i$.

2. A vector (ξ^i) and a linear operator (A^k_l). Consider their product $(T_l^{ik}) = A^k_l \xi^i$. This is a tensor of type (2, 1). The trace (contraction) of this product

$$\eta^k = T^i_l \eta^l = A^k_l \eta^l$$

is again a vector — the result of application to the initial vector (ξ^i) of the operator (A^k_l). To justify the definitions proposed, we have to prove the following assertion.

LEMMA 1. *The contraction of any tensor of type (m, n) with respect to any pair of indices (upper and lower) is again a tensor of type $(m-1, n-1)$. The product of tensors respectively of type (k, l) and (m, n) is a tensor of type $(k+m, l+n)$ which depends on the order of the cofactors.*

The proof of the lemma involves immediate verification of the tensor transformation law as applied to the results of contraction or product.

EXAMPLE 2. Suppose we are given two vectors (ξ^i) and (η^i) and a quadratic form $(g_{\alpha\beta})$, i.e. a tensor of type (0, 2). Then we can consider the triple product $T^{ij}_{\alpha\beta} = \xi^i\eta^i g_{\alpha\beta}$ and after this the double contraction

$$\langle \xi, \eta \rangle = T^{\alpha\beta}_{\alpha\beta} = \xi^\alpha \eta^\beta g_{\alpha\beta}.$$

Thus, any tensor of type (0, 2) determines the scalar product of vectors.

EXAMPLE 3. Suppose we are given two covectors (ξ_i) and (η_i) and a tensor $(g^{\alpha\beta})$ of type (2, 0). Consider the product

$$T^{\alpha\beta}_{ij} = g^{\alpha\beta}\xi_i\eta_j$$

and then the double contraction

$$T^{\alpha\beta}_{\alpha\beta} = g^{\alpha\beta}\xi_\alpha\eta_\beta = \langle \xi, \eta \rangle.$$

As a result we have obtained the scalar product of two covectors using a tensor of type (2, 0). One of the most important operations of tensorial calculus is the operation of raising (lowering) indices.

4. Lowering indices. If in our space we are given a Riemannian metric (g_{ij}) and an arbitrary tensor $T^{i_1 \dots i_m}_{j_1 \dots j_n}$ relative to some system of coordinates $(x^1, x^2, \dots, x^k)$, then we may consider a new tensor, for example,

$$T^{i_1 \dots i_m}_{ij_1 \dots j_n} = g_{ik} T^{ki_1 \dots i_m}_{j_1 \dots j_n}.$$

We can readily see that this is again a tensor (the composition of operations of the product by the tensor (g_{ij}) and contraction). The result of this operation is called *lowering the index* i_1 using the Riemannian metric (g_{ij}).

Thus, all the indices can be lowered if there exists a Riemannian metric (g_{ij}). For instance, in the Euclidean metric given in Euclidean coordinates, $g_{ij} = \delta_{ij}$. Hence $T_{i_1 j_1 \dots j_n}^{\ \ i_2 \dots i_m} = T_{j_1 \dots j_n}^{i_1 \dots i_m}$. Thus, in Euclidean coordinates, we may assume all indices to be lower if we lower them using the metric $g_{ij} = \delta_{ij}$.

5. Raising indices. If we have the Riemannian metric (g_{ij}), then to raise lower indices we should necessarily consider an inverse matrix (g^{ij}), such that

$$g^{ij} g_{jk} = \delta^i_k = \begin{cases} 1, & i = k, \\ 0, & i \neq k, \end{cases}$$

By definition,

$$T^{j_1 i_1 \dots i_m}_{\ \ j_2 \dots j_n} = g^{j_1 k} T_{k j_1 \dots j_n}^{\ \ i_1 \dots i_m}$$

(the operation of *raising an index* using a Riemannian metric).

Let us now fix the following formal (generally accepted) rule for handling tensors: the summation sign in performing the operation of contraction is omitted, but the indices that undergo summation (it is always one upper and one lower index) are marked by identical symbols implying summation. For example:

1) $\langle \xi, \eta \rangle = g^{\alpha\beta} \xi^\alpha \xi^\beta$ (vectors);

2) $\langle \xi, \eta \rangle = g^{\alpha\beta} \xi_\alpha \eta_\beta$ (covectors);

3) $(A\xi)^\alpha = A^\alpha_\beta \xi^\beta$ (operator on vectors);

4) $(A\xi)_\beta = A^\alpha_\beta \eta_\alpha$ (operator on covectors);

5) if $A = (A^\alpha_\beta)$ (operator), then $\operatorname{Sp} A = A^\alpha_\alpha$;

6) if $e_1, e_2, \dots, e_k$ are basis vectors, then any vector has the form $\xi = \xi^i e_i$.

2.4 Symmetric and Skew-Symmetric Tensors

In Section 2.3 we introduced the basic algebraic operations on tensors:

1) permutation of indices (only among upper or only among lower indices),
2) product of tensors (non-commutative),
3) operation of tracing (contraction) one upper and one lower index.

If a space comes endowed with a Riemannian metric (g_{ij}) relative to some coordinates $x^1, x^2, \dots, x^n$, then the trace of the tensor T_{jk} is given by

$$\mathrm{Sp}(T_{jk}) = \mathrm{Sp}(g^{ji}T_{jk}) = g^{ij}T_{ij}.$$

In the right-hand side of the formula, $g^{ij}T_{ij}$ is, by definition, the sum over all the values of the indices (i, j), that is, a double trace.

Obviously, the trace $\mathrm{Sp}(T_{jk})$ is a *metric invariant*, and its definition requires the Riemannian metric on which its value depends. Recall, as an example, that on a surface in space $x = x(u, v)$, $y = y(u, v)$, $z = z(u, v)$, where $u = x^1$, $v = x^2$, there occurred two quadratic forms:

1) the metric $g_{ij}\, dx^i dx^j$ gives a tensor (g_{ij}),
2) the second quadratic form $b_{ij}dx^i dx^j$ gives a tensor (b_{ij}).

By definition we assume that:

the Gaussian curvature

$$K = \frac{\det (b_{ij})}{\det (g_{ij})},$$

the mean curvature

$$H = g^{ij}\, b_{ij} = \mathrm{Sp}(b^i_k),$$

where $b^i_k = g^{ij}b_{jk}$. This is the trace of the tensor (b_{ij}). Thus, the mean curvature is the trace of the two-dimensional tensor (b_{ij}) provided that we have the metric (g_{ij}).

We have pointed out all the invariant algebraic operations on tensors (permutation of indices, product, trace, sum, product by a number, raising and lowering indices using the metric g_{ij}).

EXERCISE. In the two-dimensional case $n = 2$, express $K = \dfrac{\det b_{ij}}{\det g_{ij}}$ in terms of invariant tensor operations.

Undoubtedly, K is an invariant! The determinant det (T_{ij}) is generally not an invariant. On the contrary, the determinant of a linear operator (T^i_j) is an invariant, for example, for $n = 2$ we have $2\det(T^i_j) = T^i_i T^k_k - T^k_i T^i_k$ (in a plane, for $n = 2$, where all $i, j, k, l = 1, 2$).

There exist two especially important operations on tensors of type (2, 0) or (0, 2) associated with the operation of permutation of indices:

1) alternation $b_{ij} = T_{[ij]} = 1/2(T_{ij} - T_{ji})$. The symbol $[ij]$ implies that $b_{ij} = -b_{ji}$;

2) symmetrization $q_{ij} = 1/2(T_{ij} + T_{ji}) = T_{(ij)}$. The symbol (ij) implies that $q_{ij} = q_{ji}$.

We always have $T_{ij} = b_{ij} + q_{ij}$, and this separation is preserved under all changes of coordinates.

Given a linear operator T^i_j, it is useless to speak of symmetry or skew-symmetry of this tensor if we have no Riemannian metric.

Given a Riemannian metric g_{ij}, we can omit the index $T_{ij} = g_{ik}T^k_j$.

DEFINITION 1. A linear operator T^i_j in a space endowed with a Riemannian metric (g_{ij}) is said to be *symmetric* (*skew-symmetric*) if for the tensor $T_{ij} = g_{ik}T^k_j$ there holds the symmetry $T_{ij} = T_{ji}$ (or the skew-symmetry $T_{ij} = -T_{ji}$).

We can make the following simple assertion.

THEOREM 1. *The linear operator in a space endowed with a Riemannian metric is symmetric (skew-symmetric) if and only if for any vectors* $\xi = (\xi^i)$, $\eta = (\eta^i)$ *there holds the equality*

$$\langle T\xi, \eta\rangle = \langle \xi, T\eta\rangle \qquad (symmetry)$$

$$\langle T\xi, \eta\rangle = -\langle \xi, T\eta\rangle \qquad (skew\text{-}symmetry).$$

Proof. Since $(T\xi)^i = T^i_k\xi^k$, it follows that always

$$\langle T\xi, \eta\rangle = g_{ij}(T\xi)^i\eta^j = g_{ij}T^i{}_k\xi^k\eta^j = T_{jk}\xi^k\eta^j$$

where $g_{ik}T^k_j = T_{jk}$ since the summations over distinct indices are independent. If $T_{ik} = +T_{ki}$, then $T_{ik}\xi^k\eta^i = T_{ik}\eta^k\xi^i$. Inversely, if $T_{ik}\xi^k\eta^i = T_{ik}\eta^k\xi^i$ for all vectors

ξ and η, then $T_{ik} = T_{ki}$. For the symmetric case, this completes the proof. For the skew-symmetric case, the proof is identical, and the theorem follows.

Let us touch upon another simple fact. Given a Riemannian metric (g_{ij}), we are given the scalar product of vectors $\langle \xi, \eta \rangle = \xi\eta = g_{ij}\xi^i\eta^j$, $\xi = (\xi^i)$, $\eta = (\eta^i)$, and that of covectors $\langle \xi, \eta \rangle = \xi\eta = g^{ij}\xi_i\eta_j$, $\xi = (\xi_i)$, $\eta = (\eta_i)$, where $g^{ij}g_{jk} = \delta^i{}_k$ by definition.

There exists the operation of "raising and lowering indices":

$$(\xi^i) \to (\xi_i) = (g_{ij}\xi^j)$$

$$(\xi_i) \to (\xi^i) = (g^{ij}\xi_j).$$

THEOREM 2. *The following equality holds:*

for any pair of vectors $\xi = (\xi^i)$, $\eta = (\eta^i)$ *and a corresponding pair of covectors* $(\hat{\xi}_i) = \hat{\xi} = (g_{ij}\xi^j)$, $\hat{\eta} = (\hat{\eta}_i) = (g_{ij}\eta^j)$ *the scalar products coincide*: $\hat{\xi}\hat{\eta} = \xi\eta$.

Proof. Since $\xi\eta = g_{ij}\xi^i\eta^j$ and $\hat{\xi}\hat{\eta} = g^{ij}\hat{\xi}_i\hat{\eta}_j$, then we have

$$\hat{\xi}\hat{\eta} = g^{ij}\hat{\xi}_i\hat{\eta}_j = g^{ij}g_{ik}\xi^k g_{jl}\eta^l = \delta^j{}_k\xi^k g_{jl}\eta^l = \xi^i\eta^l g_{il} = \xi\eta$$

as required. Thus, the scalar product on covectors has been introduced proceeding from the requirement that after lowering indices we obtain the same scalar product as for vectors. Concluding the purely algebraic theory of tensors, we think it is instructive to make the following remark. An important role is played by special types of tensors which possess additional symmetry properties under permutation of one-type (upper or lower) indices.

For example, for rank-two tensors we had two classes:

$$T_{ij} = -T_{ji} \quad \text{(skew-symmetric)},$$

$$T_{ij} = T_{ji} \quad \text{(symmetric)}.$$

Symmetric tensors of rank two have already been repeatedly encountered in the form of quadratic forms on vectors. Of independent interest are any rank skew-symmetric tensors of type $(0, k)$ or $(k, 0)$:

$$T_{i_1 \dots i_k} \text{ or } T^{i_1 \dots i_k}.$$

DEFINITION 1. A *skew-symmetric tensor* $T_{i_1 \dots i_k}$ or $T^{i_1 \dots i_k}$ is a tensor that changes sign under an odd permutation of indices and preserves its value under any even permutation of indices.

For example,

$$T_{i_1 i_2 \dots} = -T_{i_2 i_1 \dots}; \quad T^{i_1 i_2 \dots} = -T^{i_2 i_1 \dots}.$$

In a three-dimensional space with coordinates x^1, x^2, x^3, a skew-symmetric tensor of rank three is specified by one number

$$T_{123} = T_{312} = T_{231} = -T_{213} = -T_{321} = -T_{132}$$

(the other are equal to zero).

Similarly, in the two-dimensional case, a skew-symmetric tensor of rank two is given by one number (the coordinates being x^1, x^2):

$$T_{12} = -T_{21}, \; T_{11} = T_{22} = 0 = -T_{11} = -T_{22}.$$

Conclusion. Skew-symmetric tensors of rank equal to the dimension of the underlying space are specified by one number, while those of higher rank, are equal to zero (since at least one pair of indices necessarily coincides).

How will skew-symmetric tensors transform under the change of coordinates?

THEOREM 3. *Skew-symmetric tensors of rank equal to the dimension of the underlying space are transformed under the change of coordinates $x = x(z)$, $x = (x^1, x^2, \dots)$, $z = (z^1, z^2, \dots)$ by the following law:*

$$T'_{12\dots} = T_{12\dots} \cdot J, \quad T'^{12\dots} = T^{12\dots} J^{-1},$$

where $J = \det\left(\frac{\partial x^i}{\partial z^j}\right)$ is the Jacobian of the change of coordinates, $T'_{12\dots}, T^{12\dots}$ is the component of the tensor in coordinates $(z^1, z^2, \dots)$.

Proof. (To make the notation shorter, we shall restrict ourselves to the case of three dimensions). From the general transformation rule, we have

$$T'_{123} = T_{ijk} \frac{\partial x^i}{\partial z^1} \frac{\partial x^j}{\partial z^2} \frac{\partial x^k}{\partial z^3}.$$

However, by virtue of skew symmetry we always have $i \neq j \neq k$, and all the T_{ijk} will, in this case, coincide with T_{123} up to the sign of permutation:

$$\begin{pmatrix} 1 & 2 & 3 \\ i & j & k \end{pmatrix}.$$

Recall that the determinant of any matrix $a^i_j = A$ is algebraically defined as

$$\sum_{i_1, i_2, \ldots, i_n} (-1)^\alpha a_1^{i_1} \ldots a_n^{i_n}$$

where $(-1)^\alpha$ is the sign of the permutation $\begin{pmatrix} 1 & 2 & \ldots & n \\ i_1 & i_2 & \ldots & i_n \end{pmatrix}$.

Obviously, this implies the theorem. For tensors of type (3, 0) we use the same arguments with the same result.

Thus, the quantity $f(x^1, x^2, x^3)$, which under the change of coordinates $x = x(z)$ is multiplied by the Jacobian $f'(z^1, z^2, z^3) = f(x(z)) \cdot J$ is a skew-symmetric tensor of type (0, 3). We arrive at the following conclusion: in fact, in the analysis we determine the integral over the region $\iiint_u f\, dx^1 dx^2 dx^3$ of a skew-symmetric tensor which in the analysis is denoted by the sign $[f\, dx^1 dx^2 dx^3]$ since under the change of coordinates it is multiplied by the Jacobian. In the sequel we shall return to the theory of integration of skew-symmetric tensors of type (0, k).

Here we shall mention, in addition, the concept of the element of volume associated with the Riemannian metric: on a surface with coordinates x^1, x^2 and a metric g_{ij} there is introduced the quantity $(g)^{1/2}\, dx^1 dx^2$, and the area of the region U is equal to $\iint_U (g)^{1/2} dx^1 dx^2$. The quantity g is equal to $\det(g_{ij})$. We can easily see that under the change of coordinates the quantity $g = \det(g_{ij})$ transforms as

$$x = x(z), \quad x = (x^1, x^2, \ldots, x^n), \quad z = (z^1, z^2, \ldots, z^n),$$

$$g' = \det(g'_{ij}) = J^2 \det(g_{ij}),$$

where J is the Jacobian of the coordinate change $J = \det\left(\frac{\partial x^i}{\partial z^j}\right)$.

Therefore the quantity $g^{1/2}$ transforms by the law

$$(g^{1/2})' = |J|\, g^{1/2}.$$

Conclusion. Under coordinate changes, where $J > 0$, the element of the volume behaves as a skew-symmetric tensor of rank n, where n is the dimension of the underlying space (this case is of importance for $n = 2, 3$). The tensor here is of type $(0, n)$.

The volume of a region in space of arbitrary dimension n endowed with a Riemannian metric g_{ij} is defined, as in the case of two dimensions, by the formula $\sigma(U) = \int\!\!\int_U (g)^{1/2} dx^1 \ldots dx^n$. The element of a volume will sometimes be written as $(g)^{1/2}\, dx^1 \wedge \ldots \wedge dx^n$, where $dx^i \wedge dx^j + dx^j \wedge dx^i = 0$. For more details see Section 2.10.

For convenience in our further calculations it is instructive to scrutinize the tensor $\varepsilon_{i_1 \ldots i_n}$ defined as follows: the component $\varepsilon_{i_1 \ldots i_n}$ is other than zero if and only if there are no repeated indices among $i_1, \ldots, i_n$; given this, we have

$$\varepsilon_{i_1 \ldots i_n} = \begin{cases} +1, & \operatorname{sgn}(i_1, \ldots, i_n) = +1 \text{ (even permutation)} \\ -1, & \operatorname{sgn}(i_1, \ldots, i_n) = -1 \text{ (odd permutation)} \end{cases}$$

(Of course, $\varepsilon_{i_1 \ldots i_n}$ is a tensor only under coordinate transformations with Jacobian $J \equiv 1$.)

Clearly, for any skew-symmetric tensor $T_{i_1 \ldots i_n}$ in a space of dimension n the equality $T_{i_1 \ldots i_n} = T_{12 \ldots n}\, \varepsilon_{i_1 \ldots i_n}$ holds.

Suppose in a region of space $\mathbb{R}^n$ we are given a metric g_{ij}. Then we may define the essential operation * which permits identification of skew-symmetric tensors of type $(0, k)$ with those of type $(0, n-k)$, that is, tensors of complementary ranks.

DEFINITION 2. If $T_{i_1 \ldots i_k}$ is a skew-symmetric tensor, then by $*T$ we denote a skew-symmetric tensor of a complementary rank $n - k$ given by the formula

$$(*T)_{i_{k+1} \ldots i_n} = (k!)^{-1}\, g^{1/2}\, \varepsilon_{i_1 \ldots i_n} T^{i_1 \ldots i_k}.$$

Given this, $T^{i_1 \dots i_k}$ is a tensor obtained form the tensor $T_{i_1 \dots i_k}$ through raising indices, i.e. $T^{i_1 \dots i_k} = g^{i_1 j_1} \dots g^{i_k j_k} T_{j_1 \dots j_k}$.

The expression $g^{1/2} \, \varepsilon_{i_1 \dots i_n}$ is a tensor already under all regular coordinate changes (with a positive Jacobian). Thus, the expression $*T$ is a skew-symmetric tensor.

We can easily verify the formula:

$$*(*T) = (-1)^{k(n-k)} T.$$

2.5 Differential Calculus of Skew-Symmetric Tensors of Type $(0, k)$

Most of the physical laws are represented as differential relations between physical quantities. Many of these quantitites are tensor fields (in particular, vector fields) in a space or in a region of space. It is, therefore, of interest to us which differential operations on tensors generally exist that, in a certain sense (specified below), do not depend on the system of coordinates. For example, the simplest of the operations is as follows: if a function $f(x, \alpha)$ or a tensor field $T^{i_1 \dots i_n}_{k_1 \dots k_n}(x, \alpha)$ depends on a point of space $x = (x^1, x^2, x^3)$ and on a certain parameter α not associated with the space, then we can take the partial derivative with respect to the parameter

$$\frac{\partial f}{\partial \alpha}(x, \alpha) \quad \text{or} \quad \frac{\partial T^{i_1 \dots i_n}_{k_1 \dots k_n}(x, \alpha)}{\partial \alpha}$$

at each given point. In classical mechanics, this parameter is time $t = \alpha$. This operation is not connected with the geometry of space (x^1, x^2, x^3) and is performed separately at each point. Another well-known differential operation not connected with the Riemannian metric is the gradient of a function (scalar field):

$$\nabla^s f = \left(\frac{\partial f}{\partial x^1}, \frac{\partial f}{\partial x^2}, \frac{\partial f}{\partial x^3}\right) = \text{grad}\, f.$$

This is a covector constructed in an invariant manner from the function f in the sense that under coordinate changes its numerical notation changes according to the tensor law

$$x = x(z), \quad \frac{\partial f}{\partial x^i} = \frac{\partial f}{\partial x^i} \cdot \frac{\partial x^i}{\partial x^j}.$$

A frequently encountered case is a multi-dimensional extension of the gradient to skew-symmetric tensors.

DEFINITION 1. If $T_{i_1 \dots i_k}$ is a tensor which is skew-symmetric with respect to all indices in an n-dimensional space with coordinates $(x^1, \dots, x^n)$, $i_q = 1, \dots, n$, then its gradient $(\nabla^s T)_{j_1 \dots j_k, j_{k+1}}$ is a skew-symmetric rank-$(k + 1)$ tensor of type $(0, k + 1)$ with the components

$$(\nabla^s T)_{j_1 \ldots j_{k+1}} = \sum_{q=1}^{k+1} \frac{\partial T_{j_1, \ldots, \hat{j}_q, \ldots, j_{k+1}}}{\partial x^{j_q}} (-1)^q$$

(the hat over j_q in the numerator $j_1, \ldots, \hat{j}_q, \ldots, j_{k+1}$ implies here that the index j_q is omitted).

Before turning to verification of the fact that $\nabla^s T$ is a tensor, we shall consider some examples.

1. If $k + 1 = 1$ and $T = f(x)$ is a function, then, by definition,

$$(\nabla^s T)_i = \frac{\partial T}{\partial x^i},$$

i.e. this is the usual gradient.

2. If $T = (T_i)$ is a covector, then

$$(\nabla^s T)_i = \frac{\partial T_i}{\partial x^j} - \frac{\partial T_j}{\partial x^i} = -(\nabla^s T)_{ji}.$$

This tensor is often defined as the *curl of the covector field*, $(\nabla^s T) = \text{rot}\, T$ if T is a covector. (The alternative term for the curl is "rotation", which is responsible for the notation rot T). The curl is a skew-symmetric rank-two tensor of type (0, 2).

REMARK. If $n = 3$, i.e. the space and the coordinates x^1, x^2, x^3 are Euclidean, then it is customary to associate the tensor $(\nabla^s T)_{ij}$ with the vector $(\eta^k) = \text{rot}\, T$, where

$$\eta^1 = \frac{\partial T_2}{\partial x^3} - \frac{\partial T_3}{\partial x^2} = (\nabla^s T)_{23};$$

$$\eta^2 = \frac{\partial T_3}{\partial x^1} - \frac{\partial T_1}{\partial x^3} = (\nabla^s T)_{31} = -(\nabla^s T)_{13};$$

$$\eta^3 = \frac{\partial T_1}{\partial x^2} - \frac{\partial T_2}{\partial x^1} = (\nabla^s T)_{12}.$$

3. Given a skew-symmetric tensor $T_{ij} = -T_{ji}$ in a Euclidean 3-space, the third-rank skew-symmetric tensor $\nabla^s T$ has the form

$$(\nabla^s T)_{123} = \frac{\partial T_{12}}{\partial x^3} - \frac{\partial T_{13}}{\partial x^2} + \frac{\partial T_{23}}{\partial x^1}.$$

REMARK. If the coordinates (x^1, x^2, x^3) are Euclidean and if according to the above mentioned rule of association of the skew-symmetric tensor with a vector $\eta^1 = T_{23}, \eta^2 = -T_{13}, \eta^3 = T_{12}$, then we have

$$(\nabla^s T)_{123} = \frac{\partial \eta^1}{\partial x^1} + \frac{\partial \eta^2}{\partial x^2} + \frac{\partial \eta^3}{\partial x^3} = \frac{\partial \eta^i}{\partial x^i}.$$

In Euclidean coordinates, the operation associating a vector field $(\eta^i) = \eta$ with a number $\operatorname{div} \eta = \dfrac{\partial \eta^i}{\partial x^i}$ is called *divergence*. There holds

THEOREM 1. *The gradient* $\nabla^s T$ *of a skew-symmetric rank-k tensor of type* $(0, k)$ *is a skew-symmetric* $k+1$*-rank tensor of type* $(0, k+1)$.

Proof. Suppose we are given the coordinate change

$$x^i = x^i(z^1, \dots, z^n), \quad i = 1, \dots, n.$$

By definition

$$(\nabla^s T)_{i_1 \dots i_{k+1}} = \sum_q (-1)^q \frac{\partial T_{i_1, \dots, \hat{i}_q, \dots, i_{k+1}}}{\partial x^{i_q}}$$

in any coordinate system.

Let $T_{i_1 \dots i_k}$ be components of the tensor in coordinates (x) and let $\tilde{T}_{j_1 \dots j_k}$ be those in coordinates (z).

By definition we have

$$\tilde{T}_{j_1 \dots j_k} = T_{i_1 \dots i_k} \frac{\partial x^{i_1}}{\partial z^{j_1}} \cdots \frac{\partial x^{i_k}}{\partial z^{j_k}}. \tag{1}$$

Next, by the definition of the tensor gradient,

$$(\nabla^s\tilde{T})_{j_1 \dots j_{k+1}} = \sum(-1)^q \frac{\partial T_{\dots \hat{j}_q \dots}}{\partial z^{j_q}},$$

$$(\nabla^s T)_{i_1 \dots i_k} = \sum(-1)^p \frac{\partial T_{\dots \hat{i}_p \dots}}{\partial x^{i_p}}. \tag{2}$$

To prove the theorem, it is necessary to substitute formula (2) into (1) and make sure that the gradient $\nabla^s\tilde{T}$ is expressed in terms of $\nabla^s T$ by the tensor law. Since the corresponding calculations are cumbersome, we shall present a complete proof for $k = 1$, $k + 1 = 2$.

If T_i is a covector and $(\nabla^s T)_{ij} = \frac{\partial T_i}{\partial x^j} - \frac{\partial T_j}{\partial x^i}$, then

$$\hat{T}_k = T_i \frac{\partial x^i}{\partial z^k}; \quad (\nabla^s\tilde{T})_{kl} = \frac{\partial \tilde{T}_k}{\partial z^l} - \frac{\partial \tilde{T}_l}{\partial z^k},$$

and we have

$$(\nabla^s\tilde{T})_{kl} = \frac{\partial \tilde{T}_k}{\partial z^l} - \frac{\partial \tilde{T}_l}{\partial z^k} = \frac{\partial}{\partial z^l}\left(T_i \frac{\partial x^i}{\partial z^k}\right) - \frac{\partial}{\partial z^k}\left(T_j \frac{\partial x^j}{\partial z^l}\right) =$$

$$= \frac{\partial T_i}{\partial z^l}\frac{\partial x^i}{\partial z^k} - T_i \frac{\partial^2 x^i}{\partial z^l \partial z^k} - \frac{\partial T_j}{\partial z^k}\frac{\partial x^j}{\partial z^l} + T_j \frac{\partial^2 x^j}{\partial z^k \partial z^l} =$$

$$= \left(\frac{\partial T_i}{\partial x^p}\frac{\partial z^p}{\partial z^l}\right)\frac{\partial x^i}{\partial z^k} - \left(\frac{\partial T_j}{\partial x^q}\frac{\partial x^q}{\partial z^k}\right)\frac{\partial x^j}{\partial z^l}$$

(only indices k, l are not summed). Let us denote p in terms of j in the first summand and q in terms of i in the second summand. Then we obtain

$$\frac{\partial T_i}{\partial x^j}\frac{\partial x^j}{\partial z^l}\frac{\partial x^i}{\partial z^k} - \frac{\partial T_j}{\partial z^i}\frac{\partial x^l}{\partial z^k}\frac{\partial x^j}{\partial z^l} = \left(\frac{\partial T_i}{\partial x^j} - \frac{\partial T_j}{\partial x^i}\right)\frac{\partial x^j}{\partial x^i} = (\nabla^s T)_{ij}\frac{\partial x^i}{\partial z^k}\frac{\partial x^j}{\partial z^l}.$$

This completes the proof for $k = 1$, $k + 1 = 2$.

In the three-dimensional case there also exists the case $k = 2, k + 1 = 3$. The third-rank tensor $(\nabla^s T)_{ijk}$ in a three-dimensional space has the form

$$(\nabla^s T)_{123} = \frac{\partial T_{12}}{\partial x^3} - \frac{\partial T_{13}}{\partial x^2} + \frac{\partial T_{23}}{\partial x^1}. \tag{I}$$

On the basis of Theorem 3 of Section 2.4, we shall prove that under the coordinate change

$$x = x(z): (\nabla^s T)_{123} = J(\nabla^s T)_{123}, \tag{II}$$

where

$$J = \det\left(\frac{\partial x^i}{\partial z^j}\right).$$

It is useful to consider

LEMMA 1. *Under the change of coordinates $x = x(z)$ the components of a skew-symmetric tensor of rank two are transformed by the law*

$$\tilde{T}_{ij} = \sum_{k<l} J_{ij}^{kl} T_{kl},$$

where

$$J_{ij}^{kl} = \frac{\partial x^k}{\partial z^i}\frac{\partial x^l}{\partial z^j} - \frac{\partial x^l}{\partial z^i}\frac{\partial x^k}{\partial z^j}$$

is the minor of the matrix $\left(\frac{\partial x^\alpha}{\partial z^\beta}\right)$.

Proof. Since

$$\tilde{T}_{ij} = T_{kl}\frac{\partial x^k}{\partial z^i}\frac{\partial x^l}{\partial z^j} \quad \text{and} \quad \tilde{T}_{ij} = -\tilde{T}_{ji} = -T_{lk}\frac{\partial x^l}{\partial z^i}\frac{\partial x^k}{\partial z^j},$$

it follows that

$$\tilde{T}_{ij} = \sum_{k<l} T_{kl}\left[\frac{\partial x^k}{\partial z^i}\frac{\partial x^l}{\partial z^j} - \frac{\partial x^l}{\partial z^i}\frac{\partial x^k}{\partial z^j}\right] = \sum_{k<l} T_{kl} J_{ij}^{kl}. \tag{III}$$

as required.

Substituting formula (III) into (I) and making use of the fact that the determinant of a matrix is equal to the sum of the elements of the row multiplied by the additional minors, we obtain formula (II). We shall not present a detailed derivation of this formula. We have, in fact, carried out a complete proof of the theorem for covector field gradients ($k = 1, k + 1 = 2$) and also pointed out the useful formula for transformation of skew-symmetric tensors of rank two.

We should like to draw the reader's attention to another useful property of the gradient of skew-symmetric tensors, namely, to the property that the square of the operation ∇^s is equal to zero.

THEOREM 2. *In an n-dimensional space (n is arbitrary), two successive applications of the gradient operation to a skew-symmetric tensor yield identical zero*: $\nabla^s(\nabla^s T) = 0$.

We shall prove this identity for a plane $n = 2$ and for a space $n = 3$.

1. $n = 2$. We should show that for any function $f(x^1, x^2)$ there holds the identity $\nabla^s\nabla^s f = 0$.

Since $(\nabla^s f)_i = \dfrac{\partial f}{\partial x^i}$ and $(\nabla^s T^s)_{ij} = \dfrac{\partial T_i}{\partial x^j} - \dfrac{\partial T_j}{\partial x^i}$, we should substitute $T_i = \dfrac{\partial f}{\partial x^i}$ and verify the identity

$$\frac{\partial}{\partial x^j}\left(\frac{\partial f}{\partial x^i}\right) - \frac{\partial}{\partial x^i}\left(\frac{\partial f}{\partial x^j}\right) \equiv 0.$$

This relation is familiar from the analysis and completes the proof for the case $n = 2$.

2. $n = 3$. Here, two cases exist:

a) $\nabla^s\nabla^s f = 0$ for the function $f(x^1, x^2, x^3)$ or rot grad $f \equiv 0$.

Indeed

$$\frac{\partial}{\partial x^i}\left(\frac{\partial f}{\partial x^j}\right) - \frac{\partial}{\partial x^j}\left(\frac{\partial f}{\partial x^i}\right) \equiv 0;$$

b) if (T_i) is a vector field, then

$$\nabla^s\nabla^s T \equiv 0 \quad (\text{div rot } (T) \equiv 0).$$

Indeed,

$$(\nabla^s\nabla^s T)_{123} \equiv \frac{\partial}{\partial x^1}(\nabla^s T)_{23} - \frac{\partial}{\partial x^2}(\nabla^s T)_{13} + \frac{\partial}{\partial x^3}(\nabla^s T)_{12} =$$

$$= \frac{\partial}{\partial x^1}\left(\frac{\partial T_2}{\partial x^3} - \frac{\partial T_3}{\partial x^2}\right) - \frac{\partial}{\partial x^2}\left(\frac{\partial T_1}{\partial x^3} - \frac{\partial T_3}{\partial x^1}\right) + \frac{\partial}{\partial x^3}\left(\frac{\partial T_1}{\partial x^2} - \frac{\partial T_2}{\partial x^1}\right) \equiv 0.$$

This proves the case $n = 3$.

(For all $n > 3$ the proof is similar.)

We have, in fact, included in our consideration, all the differential operations on tensors which are not in any way related to any space geometry, in particular, metric.

We shall point out, in addition, a frequently exploited essential fact for tensors in $\mathbb{R}^n$: in Euclidean coordinates of a Euclidean space for any tensor the *upper and lower indices are indistinguishable*: $T^{i_1 \dots i_m}_{j_1 \dots j_n} = T^{i_1 \dots i_m j_1 \dots j_n} = T_{i_1 \dots i_m j_1 \dots j_n}$ since $g_{ij} = \delta_{ij}$. There naturally exists a partial derivative

$$T_{i_1 \dots i_m j_1 \dots j_n, k} = \frac{\partial T_{i_1 \dots i_m j_1 \dots j_n}}{\partial x^k}$$

and the divergence of the tensor

$$\frac{\partial}{\partial x^k} T^{i_1 \dots i_q = k, \dots, i_m}_{j_1 \dots j_n} = \operatorname{div} T$$

(with respect to the index i_q). Note that these operations are carried out in Euclidean coordinates only.

For example, the divergence of the tensor T_{ij} in Euclidean coordinates has the form

$$\operatorname{div} T = \frac{\partial}{\partial x^j} T_{ij} \quad \text{(the sum over } j\text{)}.$$

We shall, very soon, proceed to a more detailed investigation of the differential operations on tensors connected with the geometry of space (in particular, with metric).

The gradient operation upon a skew-symmetric tensor, which we have considered above, has the following properties.

1. The result of the operation is again a tensor.

2. This operation applies exactly the same formulae relative to any coordinate system not related to any additional geometrical structure of space. The skew-symmetric gradient proves to be the only operation possessing such properties in the sense that all the rest may be derived from this and from the above-mentioned purely algebraic operations on tensors.

As an example we shall consider a four-dimensional space-time, coordinatized by $x^0 = ct, x^1, x^2, x^3$ (c is the speed of light), and endowed with a pseudo-Euclidean metric

$$-(dx^0)^2 + \sum_1^3 (dx^i)^2 = (ds)^2 \quad \text{or} \quad g_{ij} \begin{pmatrix} -1 & 0 & 0 \\ 0 & 1 & 0 \\ 0 & 0 & 1 \end{pmatrix} = g^{ij}.$$

It turns out that the electro-magnetic field is a skew-symmetric tensor of rank two, type (0, 2), i.e. the field is equal to (F_{ij}), $i, j = 0, 1, 2, 3$.

$$F_{ij} = -F_{ji}.$$

The components $F_{0i} = E_i$ are called the *vector of an electric field*, $E = (E_1, E_2, E_3)$.

The components $F_{ij} = (H_{ij})$, $i, j = 1, 2, 3$, are called the *skew-symmetric tensor (axial vector) of a magnetic field*, and $H_1 = H_{23}, H_2 = H_{31}, H_3 = H_{12}$; $H = (H_1, H_2, H_3)$.

Under the coordinate changes $x = x(z)$ with the time unchanged $x^0 = z^0, x^i = x^i(z^1, z^2, z^3)$, the electrical field E_i and the magnetic field $H = (H_{ij})$ behave as a vector and a skew-symmetric tensor.

The first pair of Maxwell's equations has the form $\nabla^s F_{ij} = 0$ or

$$(\nabla^s F)_{ijk} = \frac{\partial F_{jk}}{\partial x^i} - \frac{\partial F_{ik}}{\partial x^j} + \frac{\partial F_{ij}}{\partial x^k} \equiv 0.$$

In components we have

$$1) \quad \nabla^s(H_{ij}) = 0, \qquad \frac{\partial H_{12}}{\partial x^3} - \frac{\partial H_{13}}{\partial x^2} + \frac{\partial H_{23}}{\partial x^1} = 0$$

or

$$\frac{\partial H_i}{\partial x^i} = 0, \quad \operatorname{div} H = 0,$$

2)

$$\nabla E + \frac{\partial H}{\partial x^0} = 0, \quad \operatorname{rot} E = -\frac{1}{c}\frac{\partial H}{\partial t}.$$

We see that the first pair of Maxwell's equations has no relation to pseudo-Euclidean geometry and always has the form $\nabla^s F = 0$. As for the second pair of Maxwell's equations, it is related, unlike the first pair, to pseudo-Euclidean divergence

$$\operatorname{div}_{(4)} F_{ij} = \sum_{j=1}^{3} \frac{\partial F_{ij}}{\partial x^j} - \frac{\partial F_{i0}}{\partial x^0} = g^{jl}\frac{\partial F_{ij}}{\partial x^l}$$

and a four-dimensional vector of electric current $j = (j_0, j_1, j_2, j_3)$, where $j_0 = \rho c$, ρ is the charge density and $j' = (j_1, j_2, j_3)$ is the electric current vector in the usual three-dimensional sense.

The equation has the form

$$\operatorname{div}_{(4)} F_{ij} = \frac{4\pi}{c} j.$$

Expressing the operator in terms of E and H we have

$$\operatorname{div} E = \sum_{i=1}^{3} \frac{\partial E_i}{\partial x} = 4\pi\rho$$

(here div implies the usual Euclidean divergence),

$$\operatorname{rot} H + \frac{1}{c}\frac{\partial E}{\partial t} = \frac{4\pi}{c} j'.$$

Thus, the concept of the divergence of a tensor depends essentially on the metric, whereas the concept of the gradient does not. On this account, the first pair of Maxwell's equations is equally written in any coordinates as $\nabla^s F \equiv 0$, while the second pair requires that the space be endowed with a metric (moreover, it requires Euclidean coordinates in order that the skew-symmetric tensor H_{ij} could be identified with the vector H). Recall that the Lorentz force f acting upon a charge in an electro-magnetic field $F_{ij} = (E, H)$ is calculated as $f = eE + e[v/c, H]$, where v is the velocity and [ ] is the vector product. It is just this formula that implies that E is a vector and H is a skew-symmetric tensor, which allows them to be united in a single four-dimensional skew-symmetric tensor.

REMARK. The vector product of two vectors (covectors) of an n-dimensional space

$$\begin{aligned} \eta &= (\eta^i), \quad \xi = (\xi^j), \\ & \qquad\qquad\qquad\qquad i, j = 1, 2, \dots, n, \\ \hat{\eta} &= (\hat{\eta}_i), \quad \hat{\xi} = (\xi_j), \end{aligned}$$

is given by

$$[\eta, \xi]^{ij} = \eta^i \xi^j - \xi^i \eta^j = -[\eta, \xi]^{ji},$$

$$[\hat{\eta}, \hat{\xi}]_{ij} = \eta_i \hat{\xi}_j - \hat{\xi}_i \eta_j = -[\hat{\eta}, \hat{\xi}]_{ji}.$$

This is a second-rank skew-symmetric tensor. It can be associated with a vector in Euclidean 3-space only.

Given a skew-symmetric tensor $T_{ij} = -T_{ji}$ of rank two in Euclidean space (relative to Euclidean coordinates), the vector product of the vector (v_i) by the tensor T_{ij} $(= T^i_j)$ has the form

$$[T, v]_j = T^i_j v_i = T_{ij} v_i = -T_{ji} v_i = -[v, T]_j$$

i.e. the product $[T, v]$ is again a vector, which is the result of an application of the operator T to the vector v. For example, we had the Lorentz force $f = eE + e/c\ [v, H]$ understood as a vector, $v = (c, v_1, v_2, v_3)$, $T = F_{ij} c^{-1}$.

2.6 Covariant Differentiation. Euclidean and General Connections

In Section 2.5 we have examined the gradient operation on a skew-symmetric tensor (tensor field) leading to a skew-symmetric tensor of rank higher by unity than the rank of the initial tensor. This operation had the form (in components)

$$(\nabla^s T)_{i_1 \dots i_{k+1}} = \sum_{q=1}^{k+1} (-1)^q \frac{\partial T_{i_1 \dots \hat{i}_q \dots i_{k+1}}}{\partial x^{i_q}} .$$

In particular, for $k = 1$ we had

$$(\nabla^s T)_{ij} = \frac{\partial T_i}{\partial x^j} - \frac{\partial T_j}{\partial x^i} .$$

It was pointed out that $\nabla^s T$ is again a tensor (this was derived rigorously for $k = 0, 1$). It was also emphasized that the operation ∇^s is the only one not related to any geometry. The differential operations on tensors are reduced to this one and the purely algebraic operations discussed above (permutation of indices, summation, product, trace).

Concerning the usual extension of the gradient of a function of tensors

$$T^{i_1 \dots i_m}_{j_1 \dots j_n} = \frac{\partial T^{i_1 \dots i_m}_{j_1 \dots j_n}}{\partial x^k}$$

in a space with Cartesian coordinates (x^j), we have already said that the resultant "tensor" is really not a tensor. Since this operation is used rather frequently, we shall point out the class of transformations under which its result transforms as a tensor. These are linear transformations of coordinates.

THEOREM 1. *If in a space we are given coordinates and a tensor field* $T = T^{i_1 \dots i i_m}_{j_1 \dots j j_n}$,

then the field $T^{i_1 \dots i i_m}_{j_1 \dots j j_n , k} = \dfrac{\partial T^{i_1 \dots i_m}_{j_1 \dots j_n}}{\partial x^k}$ *transforms as a tensor under all linear coordinate changes*

$$x^i = a^i_j z^j, \quad a^i_j = \text{const.},$$

$$z^i = b^i_j x^j, \quad b^i_j a^j_k = \delta^i_k .$$

Proof. For linear transformations, we have

$$\frac{\partial x^i}{\partial z^j} = a^i_j = \text{const. and } \frac{\partial^2 x^i}{\partial z^j \partial z^k} = 0,$$

$$\frac{\partial z^i}{\partial x^j} = b^i_j = \text{const.,} \quad a^i_j b^j_k = \delta^i_k.$$

By the definition of a tensor, we have

$$\tilde{T}^{k_1 \dots k_m}_{l_1 \dots l_n} = T^{i_1 \dots i_m}_{j_1 \dots j_n} \frac{\partial x^{j_1}}{\partial z^{l_1}} \dots \frac{\partial z^{k_1}}{\partial x^{i_1}} \dots = T^{(i)}_{(j)} a^{(j)}_{(l)} b^{(k)}_{(i)},$$

where

$$(i) = (i_1, \dots, i_m), \quad (k) = (k_1, \dots, k_m),$$

$$(j) = (j_1, \dots, j_n), \quad (l) = (l_1, \dots, l_n). \tag{1}$$

Since $a^i_j = \text{const.}$, $b^j_k = \text{const.}$, differentiating formula (1), we obtain

$$\tilde{T}^{(k)}_{(l)q} = \frac{\partial \tilde{T}^{(k)}_{(l)}}{\partial z^q} = \frac{\partial T^{(i)}_{(j)}}{\partial z^q} a^{(j)}_{(l)} b^{(k)}_{(i)} = \frac{\partial T^{(i)}_{(j)}}{\partial x^p} \frac{\partial x^p}{\partial z^q} a^{(j)}_{(l)} b^{(k)}_{(i)} = T^{(i)}_{(j)} a^{(j)}_{(l)} a^p_q b^{(k)}_{(i)}.$$

This is the transformation law of a tensor, which implies the theorem.

In the proof we have essentially used the fact that $\dfrac{\partial^2 x^i}{\partial z^k \partial z^l} \equiv 0$. Consider, for example, tensors of type (0, 1) or (1, 0):

$$\frac{\partial T_j}{\partial x^k} = T_{jk}, \quad \frac{\partial T^i}{\partial x^k} = T^i_k.$$

By virtue of the theorem just proved, T_{ik} and T^i_k transform as tensors under linear coordinate changes. Under general changes of coordinates $x^i = x^i(z^1, \dots, z^n)$, $i = 1,$

2, ... , n, where $\frac{\partial^2 x^i}{\partial z^k \partial z^p} = 0$, we obtain

$$\tilde{T}_{jq} = \frac{\partial \tilde{T}_j}{\partial z^q} = \frac{\partial}{\partial z^q}\left(T_i \frac{\partial x^i}{\partial z^j}\right) = \frac{\partial T_i \partial x^i}{\partial z^q \partial z^j} + T_i \frac{\partial^2 x^i}{\partial z^q \partial z^j} =$$

$$= \frac{\partial T_i}{\partial x^p} \frac{\partial x^p}{\partial z^q} \frac{\partial x^i}{\partial z^j} + T_i \frac{\partial^2 x^i}{\partial z^q \partial z^j} =$$

$$= T_{iq} \frac{\partial x^p}{\partial z^q} \frac{\partial x^i}{\partial z^j} + T_i \frac{\partial^2 x^i}{\partial z^q \partial z^j} = \tilde{T}_{jq} .$$

Here, as always $\tilde{T}$ are components in the coordinate system (z) and T are components in the coordinate system (x). Thus, the general transformation formula has the form

$$\frac{\partial \tilde{T}_j}{\partial z^q} = \tilde{T}_{jq} = T_{ip} \frac{\partial x^p}{\partial z^q} \frac{\partial x^i}{\partial z^j} + T_i \frac{\partial^2 x^i}{\partial z^q \partial z^j} . \tag{2}$$

The summand $T_i \frac{\partial^2 x^i}{\partial z^q \partial z^j}$ is not of tensor character. As deduced in Section 2.5, the expression

$$(\nabla^s \tilde{T})_{iq} = \tilde{T}_{jq} - \tilde{T}_{qj} = (T_{ip} - T_{pi}) \frac{\partial x^p}{\partial z^q} \frac{\partial x^i}{\partial z^j}$$

is a tensor. But the symmetrized part

$$(\tilde{T}_{jq} + \tilde{T}_{qj}) = (T_{jq} + T_{qj}) \frac{\partial x^p}{\partial z^q} \frac{\partial x^i}{\partial z^j} + 2T_i \frac{\partial^2 x^i}{\partial z^q \partial z^j}$$

is already not a tensor relative to arbitrary coordinate systems.

Similarly, for T^i_k we have

$$\tilde{T}^j_l = \frac{\partial \tilde{T}^j}{\partial z^l} = \frac{\partial}{\partial z^l}\left(T^i \frac{\partial z^j}{\partial x^i}\right) = \frac{\partial T^i}{\partial z^l}\frac{\partial z^j}{\partial x^i} + T^i \frac{\partial}{\partial z^l}\left(\frac{\partial z^j}{\partial x^i}\right) =$$

$$= \frac{\partial T^i}{\partial x^p}\frac{\partial x^p}{\partial z^l}\frac{\partial z^j}{\partial x^i} + T^i \frac{\partial^2 z^j}{\partial x^i \partial x^q}\frac{\partial x^q}{\partial z^l}. \tag{3}$$

As we can readily see, this is not a tensor because of the second summand.

Formula (3) implies

$$\tilde{T}^j_j = \frac{\partial \tilde{T}^j}{\partial z^j} = \frac{\partial T^i}{\partial x^p}\frac{\partial x^p}{\partial z^l}\frac{\partial z^l}{\partial x^i} + T^i \frac{\partial^2 z^j}{\partial x^i \partial x^q}\frac{\partial x^q}{\partial z^j} =$$

$$= T^i_p \delta^p_i + T^i \frac{\partial^2 z^j}{\partial x^i \partial x^q}\frac{\partial x^q}{\partial z^j} = T^i_i + T^i \frac{\partial x^q}{\partial z^j}\frac{\partial^2 z^j}{\partial x^i \partial x^q}. \tag{4}$$

REMARK. The expression $\frac{\partial T^i}{\partial x^i} = T^i_i$ is often called (in Euclidean coordinates) the *divergence* of a vector field. We can see that the expression T^i_i is not a scalar if the change of coordinates is non-linear.

EXAMPLE. It is customary to use this formula to calculate the divergency in terms of Euclidean coordinates (x^1, x^2, x^3) only. The meaning of the divergence, as is well known, is as follows: given small displacements of points in a space

$$x^i \to x^i + T^i(x^1, x^2, x^3) = \bar{x}^i,$$

the element of a Euclidean volume $dx^1 \wedge dx^2 \wedge dx^3$, after the displacement of the region, takes on an additional term $T^i_i \wedge x^1 \wedge x^2 \wedge x^3$. Indeed, the new volume is equal to

$$d\bar{x}_1 \wedge d\bar{x}_2 \wedge d\bar{x}_3 =$$

$$= \left(dx^1 + \frac{\partial T^1}{\partial x^j} dx^j\right) \wedge \left(dx^2 + \frac{\partial T^2}{\partial x^j} dx^j\right) \wedge \left(dx^3 + \frac{\partial T^3}{\partial x^j} dx^j\right).$$

We shall recall here that by definition $dx^i \wedge dx^j = -dx^j \wedge dx^i$, and, in particular $dx^i \wedge dx^i = 0$, as is natural in the formulae for the volume element. By virtue of this,

$$\left(dx^1 + \frac{\partial T^1}{\partial x^j}\, dx^j\right) \wedge \left(dx^2 + \frac{\partial T^2}{\partial x^j}\, dx^j\right) \wedge \left(dx^3 + \frac{\partial T^3}{\partial x^j}\, dx^j\right) =$$

$$= dx^1 \wedge dx^2 \wedge dx^3 + T^i{}_i\, dx^1 \wedge dx^2 \wedge dx^3 + dx^1 \wedge dx^2 \wedge dx^3,$$

(quadratic and cubic expressions of the components T^i_k).

In the the case where $T^i(x^1, x^2, x^3)$ and $T^i{}_k$ are small, we have approximately

$$d\bar{x}_1 \wedge d\bar{x}_2 \wedge d\bar{x}_3 \cong (1 + T^i_i)\, dx^1 \wedge dx^2 \wedge dx^3,$$

or, more precisely, in the case $x^i \to x^i + tT^i(x^1, x^2, x^3) = \bar{x}^i$, where t is a numerical parameter, we introduce the "volume element distortion" function

$$\frac{\partial \bar{x}^1 \wedge d\bar{x}^2 \wedge d\bar{x}^3}{\partial x^1 \wedge dx^2 \wedge dx^3} = f(t).$$

Then the following equality holds

$$\frac{df}{dt}\Big|_{t=0} = T^i_i(x^1, x^2, x^3) = \text{div}\,(T^i),$$

where

$$T^i = \left[\frac{d\bar{x}^i}{dt}\right]_{t=0}$$

and (x^1, x^2, x^3) are Euclidean coordinates. The reader is no doubt familiar with this from the analysis. We have dwelt on this remark specially just to recall the concept of divergence of a vector field in Euclidean geometry.

Let us now return to our subject connected with the gradient transformation law

$$T^{i_1 \dots i_m}_{j_1 \dots j_n, k} = \frac{\partial T^{i_1 \dots i_m}_{j_1 \dots j_n}}{\partial x^k}$$

Let us agree, on the basis of the theorem proved above, to apply this operation only in terms of Euclidean coordinates (x^j) and any other coordinates differing from Euclidean by the linear change

$$x^i = a^i_j z^j, \quad a^i_j = \text{const.}$$

We have already said that applying this operation by means of the same formulae in terms of another coordinate system which differs from (x) in a non-linear change, we obtain the expression $\tilde{T}^{k_1 \dots k_m}_{l_1 \dots l_n, q}$ related to $T^{(i)}_{(j)}$ through a non-tensor transformation law.

Now let us approach this question from another point of view. How do we know that the gradient operation should always be applied using one and the same formula? We may assume that

a) this operation is essentially related to Euclidean geometry;

b) it is applicable by this formula in terms of Euclidean coordinates (x) only;

c) the result of this operation is a tensor.

What are the consequences of these hypotheses? What are the formulae to be used to apply this operation in terms of other systems of coordinates related to a Euclidean non-linear change? To deduce corollaries from these hypotheses, we should first calculate the result of the applicaton of this operation to a tensor field T in Euclidean coordinates (x^i) and only after that transform this result, using the tensor law, into another coordinate system: $x^i = x^i(z^1, \dots, z^n)$, $i = 1, 2, \dots, n$.

Let us do so:

$$T^{(i)}_{(j)p} = \frac{\partial T^{(i)}_{(j)}}{\partial x^p} \qquad \begin{aligned} (i) &= (i_1, \dots, i_m), \\ (j) &= (j_1, \dots, j_n). \end{aligned}$$

By definition $T^{(i)}_{(j)p}$ is assumed to be a tensor. Therefore,

$$\tilde{T}^{(k)}_{(l)q} = T^{(k)}_{(j)p} \frac{\partial x^{(j)}}{\partial z^{(l)}} \frac{\partial z^{(k)}}{\partial x^{(i)}} \frac{\partial x^p}{\partial z^q}, \tag{5}$$

where

$$\frac{\partial x^{(j)}}{\partial z^{(l)}} = \frac{\partial x^{j_1}}{\partial z^{l_1}} \cdots \frac{\partial x^{j_m}}{\partial z^{l_m}} = \frac{\partial z^{k_1}}{\partial x^{i_1}} \cdots \frac{\partial z^{k_n}}{\partial x^{i_n}}.$$

The question is, what operation in the coordinate system $z^{(j)}$ is used to transform the components $\tilde{T}\{^{(k)}_{(l}\}_q$ into $\tilde{T}\{^{(k)}_{(j}\}_q$.

Consider, for the sake of simplicity, vector fields (T^i) and covector fields (T_j). In this case

$$\tilde{T}^k_p = T^i_p \frac{\partial z^k}{\partial x^i} \frac{\partial x^p}{\partial z^q},$$

$$\tilde{T}_{lq} = T_{jp} \frac{\partial x^j}{\partial z^l} \frac{\partial x^p}{\partial z^q}. \tag{6}$$

Since $T^i_p = \dfrac{\partial T^i}{\partial x^p}$, from formula (6) it follows that

$$\tilde{T}^k_q = \frac{\partial T^i}{\partial x^p} \frac{\partial x^p}{\partial z^q} \frac{\partial x^k}{\partial x^i} = \frac{\partial T^i}{\partial z^q} \frac{\partial z^k}{\partial x^i}. \tag{7}$$

Recall that $T^i = \tilde{T}^p \dfrac{\partial x^i}{\partial z^p}$. Formula (7) implies the equality

$$\tilde{T}^k_q = \frac{\partial T^i}{\partial z^q} \frac{\partial z^k}{\partial x^i} = \frac{\partial}{\partial z^q}(\tilde{T}^k) - T^i \frac{\partial}{\partial z^q}\left(\frac{\partial z^k}{\partial x^i}\right). \tag{8}$$

Since $T^i = \tilde{T}^p \dfrac{\partial x^i}{\partial z^p}$, we obtain the final equality

$$\tilde{T}^k_q = \frac{\partial \tilde{T}^k}{\partial z^q} - \tilde{T}^p \frac{\partial x^i}{\partial z^p} \frac{\partial^2 z^k}{\partial x^i \partial x^m} \frac{\partial x^m}{\partial z^q}. \tag{9}$$

Now let us introduce the notation

$$\Gamma^k_{pq} = -\frac{\partial x^i}{\partial z^p} \frac{\partial x^m}{\partial z^q} \frac{\partial^2 z^k}{\partial x^i \partial x^m} \equiv \frac{\partial z^k}{\partial x^m} \frac{\partial^2 x^m}{\partial z^p \partial z^q}. \tag{10}$$

Formula (9) asumes the form

$$\tilde{T}^k_q = \frac{\partial \tilde{T}^k}{\partial z^q} + \Gamma^k_{pq}\tilde{T}^p.$$

We have proved

THEOREM 2. *If the gradient of a vector field* (T^i) *transforms as a tensor and is applied in a natural way in terms of Euclidean coordinates* (x)*:*

$$T^i_k = \frac{\partial T^i}{\partial x^k},$$

then in terms of any other coordinate system (z) *the gradient is given by*

$$\tilde{T}^k_q = \frac{\partial \tilde{T}^k}{\partial z^q} + \Gamma^k_{pq}\tilde{T}^p,$$

where the coefficients Γ^k_{pq} *are calculated by formula* (10).

COROLLARY 1. *The divergence of a vector field,* div (T^i) *is defined as contraction of its gradient, and in terms of any coordinat system is given by the formula*

$$\operatorname{div}(\tilde{T}) = \tilde{T}^k_k = \frac{\partial \tilde{T}^k}{\partial z^k} + \Gamma^k_{pq}\tilde{T}^p,$$

where

$$\Gamma^k_{pq} = -\frac{\partial x^i}{\partial z^p}\frac{\partial x^m}{\partial z^q}\frac{\partial^2 z^k}{\partial x^i \partial x^m}. \tag{11}$$

Here $\tilde{T}$ *are components in the coordinate system* (z), and $x^i(z)$ *are Euclidean coordinats as functions of the coordinates* (z).

The corollary follows immediately from the theorem by means of the substitution $k = q$.

We can similarly transform expression (6) for a covector field:

$$T_{lq} = T_{jp} \frac{\partial x^j}{\partial z^l} \frac{\partial x^p}{\partial z^q} = \frac{\partial T_j}{\partial x^p} \frac{\partial x^p}{\partial z^q} \frac{\partial x^j}{\partial z^l} = \frac{\partial T_j}{\partial z^q} \frac{\partial x^j}{\partial z^l} =$$

$$= \frac{\partial}{\partial z^q} (\tilde{T}_k \frac{\partial z^k}{\partial x^j}) \frac{\partial x^j}{\partial z^l} = \frac{\partial \tilde{T}_k}{\partial z^q} (\frac{\partial z^k}{\partial x^j} \frac{\partial x^j}{\partial z^l}) + \tilde{T}_k \frac{\partial x^j}{\partial z^l} \frac{\partial}{\partial z^q} (\frac{\partial x^k}{\partial x^j}) =$$

$$= \frac{\partial \tilde{T}_k}{\partial z^q} \delta^k_l + \tilde{T}_k \frac{\partial x^j}{\partial z^l} \frac{\partial^2 z^k}{\partial x^j \partial x^p} \frac{\partial x^p}{\partial z^q} =$$

$$= \frac{\partial \tilde{T}_l}{\partial z^q} + \tilde{T}_k (\frac{\partial x^j}{\partial z^l} \frac{\partial x^p}{\partial z^q} \frac{\partial^2 z^k}{\partial x^i \partial x^p}) = \frac{\partial \tilde{T}_l}{\partial z^q} - \Gamma^k_{lq} \tilde{T}_k,$$

$$\Gamma^k_{lq} = - \frac{\partial x^j}{\partial z^l} \frac{\partial x^p}{\partial z^q} \frac{\partial^2 z^k}{\partial x^j \partial x^p} .$$

So we have come to

THEOREM 3. *If the gradient of a covector field* (T_i) *transforms as a tensor and if in the Euclidean coordinate system it is calculated in the usual manner:* $T_{ik} = \frac{\partial T_i}{\partial x^k}$, *then any other coordinate system* (z) *it is given by the formula*

$$\tilde{T}_{ik} = \frac{\partial \tilde{T}_i}{\partial z^k} - \tilde{T}_q \Gamma^q_{ik}$$

where the set Γ^q_{ik} *is the same as for the vectors* (T^i) *in Theorem* 1 *and is calculated using formula* (10).

Thus, the application of the gradient operation based on the fact that its result behaves as a tensor under any coordinate changes $x = x(z)$ yields distinct formulae for vectors and covectors:

$$\tilde{T}_{ik} = \frac{\partial \tilde{T}_i}{\partial x^k} + \Gamma^q_{ik} \tilde{T}_q \qquad \text{(for a covector)},$$

$$\tilde{T}^i_k = \frac{\partial \tilde{T}^i}{\partial x^k} + \Gamma^i_{qk} \tilde{T}^q \qquad \text{(for a vector).}$$

However, the set Γ^i_{qk} is common for them.

We shall not carry out detailed calculations for any tensors of type (m, n) but only give the result.

THEOREM 4 (without proof). *If the gradient $T^{(i)}_{(j);q}$ of any tensor field $T^{(i)}_{(j)}$ of type (m, n) behaves as a tensor under any coordinate changes and if, in a Euclidean coordinate system it is determined by the formula*

$$T^{(i)}_{(j);q} = \frac{\partial T^{(i)}_{(j)}}{\partial x^q},$$

then in any other coordinate system $x = x(z)$, it is calculated by the formula

$$\tilde{T}^{(k)}_{(l);q} = \frac{\partial \tilde{T}^{(k)}_{(l)}}{\partial x^q} + \sum_{p=1}^{m} \tilde{T}^{k_1 \dots k_j = p \dots k_m}_{l_1 \dots l_n} \cdot \Gamma^{k_j}_{pq} - \sum_{p=1}^{m} \tilde{T}^{k_1 \dots k_m}_{l_1 \dots l_j = p \dots l_n} \cdot \Gamma^p_{l_j q}, \tag{12}$$

where the set of functions Γ^p_{kq} is calculated by the same formula (10).

For example, for tensors or rank two

1) $$\tilde{T}^i_{jk} = \frac{\partial \tilde{T}^i_j}{\partial x^k} + \tilde{T}^p_j \Gamma^i_{pk} - \tilde{T}^i_p \Gamma^p_{jk},$$

2) $$\tilde{T}_{ij;k} = \frac{\partial T_{ij}}{\partial x^k} - T_{pj} \Gamma^p_{ik} - T_{ip} \Gamma^p_{jk},$$

3) $$\tilde{T}^{ij}_k = \frac{\partial T^{ij}}{\partial x^k} + T^{pj} \Gamma^i_{pk} + T^{ip} \Gamma^j_{pk}.$$

The operation of calculation of the gradient of a tensor $T^{(i)}_{(j)}$ is always denoted by

$$\text{gradient } T^{(i)}_{(j)} = T^{(i)}_{(j);k} \qquad (i) = (i_1, \dots, i_m), \quad (j) = (j_1, \dots, j_m).$$

We should like to emphasize that the operation introduced is essentially related to Euclidean geometry. The point is that we have defined this operation proceeding from two requirements:

a) the result of the operation is a tensor,

2) in Euclidean coordinates it is calculated by the formula

$$\frac{\partial T^{(i)}_{(j)}}{\partial x^k} = T^{(i)}_{(j);k} .$$

From the point of view of this operation we can say that we call affine such coordinates in which the gradient of any tensor is calculated by the formula

$$T^{(i)}_{(j);k} = \frac{\partial T^{(i)}_{(j)}}{\partial x^k} .$$

These coordinates differ from Euclidean coordinates by an affine transformation.

We should find out how the set $\Gamma^k_{ij}(z)$ changes in a given coordinate system (z) under the change $z^i = z^i(y)$, $i = 1, 2, 3$.

If there exist Euclidean coordinates $(x^1 x^2 x^3)$

$$x^i = x^i(z) = x^i(z(y)),$$

then, according to formulae (10), we assume

$$\Gamma^k_{pq} = -\frac{\partial x^i}{\partial z^p}\frac{\partial x^j}{\partial z^q}\frac{\partial^2 z^k}{\partial x^i \partial x^j} = \frac{\partial^2 x^m}{\partial z^p \partial z^q}\frac{\partial z^k}{\partial x^m} .$$

In the coordinate system (y) we shall have

$$\tilde{\Gamma}^k_{pq} = -\frac{\partial x^i}{\partial y^p}\frac{\partial x^j}{\partial y^q}\frac{\partial^2 y^k}{\partial x^i \partial x^j} = \frac{\partial^2 x^m}{\partial y^p \partial y^q}\frac{\partial y^k}{\partial x^m} .$$

From formulae (10) and (11) we obtain

$$\Gamma^k_{pq}\frac{\partial z^p}{\partial y^m}\frac{\partial z^q}{\partial y^n} = -\frac{\partial^2 z^k}{\partial x^i \partial x^j}\frac{\partial x^i}{\partial z^p}\frac{\partial x^j}{\partial z^q}\frac{\partial z^p}{\partial y^m}\frac{\partial z^q}{\partial y^n} =$$

$$= -\frac{\partial^2 z^k}{\partial x^i \partial x^j}\frac{\partial x^i}{\partial y^m}\frac{\partial x^j}{\partial y^n} = -\frac{\partial^2 z^k}{\partial y^m \partial y^n} + \frac{\partial z^k}{\partial x^j}\frac{\partial^2 x^j}{\partial y^m \partial y^n} ,$$

since

$$\frac{\partial^2 (z^k(x(y)))}{\partial y^n \partial y^m} = \frac{\partial}{\partial y^n}\left(\frac{\partial z^k}{\partial x^i}\frac{\partial x^i}{\partial y^m}\right) \frac{\partial^2 x^i}{\partial y^m \partial y^n}\frac{\partial z^k}{\partial x^i} + \frac{\partial x^i}{\partial y^m}\frac{\partial^2 z^k}{\partial x^i \partial x^j}\frac{\partial x^j}{\partial y^n}.$$

Therefore

$$\Gamma^k_{pq}\frac{\partial z^p}{\partial y^m}\frac{\partial z^q}{\partial y^n} + \frac{\partial^2 z^k}{\partial y^m \partial y^n} = \frac{\partial z^k}{\partial x^i}\frac{\partial x^i}{\partial y^m \partial y^n},$$

whence there follows the equality

$$\frac{\partial y^s}{\partial z^k}\left(\Gamma^k_{pq}\frac{\partial z^p}{\partial y^m}\frac{\partial z^q}{\partial y^n} + \frac{\partial^2 z^k}{\partial y^m \partial y^n}\right) =$$

$$= \frac{\partial y^s}{\partial z^k}\frac{\partial z^k}{\partial x^j}\frac{\partial^2 x^j}{\partial y^m \partial y^n} = \frac{\partial^2 x^j}{\partial y^m \partial y^n}\frac{\partial y^s}{\partial x^j} = \tilde{\Gamma}^s_{mn}.$$

Finally we arrive at the transformation formula

$$\Gamma^s_{mn} = \frac{\partial y^s}{\partial z^k}\left(\Gamma^k_{pq}\frac{\partial z^p}{\partial y^m}\frac{\partial z^q}{\partial y^n} + \frac{\partial^2 z^k}{\partial y^m \partial y^n}\right). \tag{13}$$

The covariant differentiation which we have introduced in Euclidean space is symmetric, that is, $\Gamma^i_{jk} = \Gamma^i_{kj}$. This follows from the explicit formulae for the Christoffel symbols expressing these symbols in terms of first and second partial derivatives of the new coordinates with respect to the old coordinates. It turns out that the concept of covariant differentiation can be introduced on an arbitrary smooth manifold. This operation can be defined using formulae (12) and (13). It should be noted now that the Christoffel symbols do not necessarily have the form (10) and (11). We shall conclude this section with the following definition.

DEFINITION 1. A *general operation of covariant differentiation* (taking the gradient) of tensors of arbitrary type is said to be defined if we are given, in terms of any system of coordinates z^1, z^2, z^3, a family of functions $\Gamma^k_{pq}(z)$ which transform under arbitrary coordinate changes $z = z(y)$ according to formula (13).

It should be emphasized that in going over to the definition of the general operation of covariant differentiation, we have taken, as the basic one, only the transformation formula (13) and renounced the requirement of the existnece of "affine coordinates", in terms of which the Christoffel symbols Γ^i_{jk} are equal to zero. Such coordinates may not exist for general connections.

For vectors and covectors, the operation of covariant differentiation of the gradient is specified by the formulae

$$T^i_k = \frac{\partial T^i}{\partial x^k} + \Gamma^i_{jk} T^j,$$

$$T_{i;k} = \frac{\partial T_i}{\partial x^k} - \Gamma^j_{ik} T_j,$$

and by formula (12) for general tensors. Given this, the transformation law (13) for components Γ^k_{ij} is determined from the requirement that the covariant gradient of a tensor be again a tensor. (In spite of the fact that the components Γ^k_{ij} themselves do not form a tensor.)

REMARK 1. An operation of covariant differentiation (of a gradient) is often called a *differential-geometric connection*, or *affine connection.*

REMARK 2. A connection is said to be *Euclidean* if there exist coordinates (x^j) in terms of which $\Gamma^k_{ij} = 0$, i.e. such that $T^{(i)}_{(j);k} = \dfrac{\partial T^{(i)}_{(j)}}{\partial x^k}$. These coordinates are called *affine* since they are defined up to an affine transformation.

If an affine connection is given beforehand, it may so happen that for it there exist no affine coordinates. This will be the case, for example, if a connection is non-symmetric. Indeed, if for such a connection there existed affine coordinates, we should have, in terms of these coordinates, the equality $\Gamma^i_{jk} = 0$, and since the difference $\Gamma^i_{jk} - \Gamma^i_{kj} = T^i_{jk}$ always forms a tensor (and this can be verified), it follows that the symbols T^i_{jk} will become identical zeros in any regular coordinate system, which would mean that the connection is symmetric.

2.7 Basic Properties of Covariant Differentiation

In the preceding section we gave (without proof) the formula for covariant differentiation of tensors of arbitrary rank.

In some expressions for the Euclidean case we obtain for Γ^i_{jq}

$$\Gamma^i_{jq} = \frac{\partial^2 x^i}{\partial z^j \partial z^m} \frac{\partial z^m}{\partial x^q} = \frac{\partial}{\partial x^q}\left(\frac{\partial x^i}{\partial z^j}\right), \tag{1}$$

in other expressions, as we have seen in the previous section, we have

$$\Gamma^i_{jq} = -\frac{\partial^2 z^i}{\partial z^\alpha \partial x^\beta} \frac{\partial x^\alpha}{\partial z^j} \frac{\partial x^\beta}{\partial z^q}, \tag{2}$$

LEMMA 1. *There holds the identity*

$$\Gamma^i_{jq} = -\frac{\partial^2 z^i}{\partial z^\alpha \partial x^\beta} \frac{\partial x^\alpha}{\partial z^j} \frac{\partial x^\beta}{\partial z^q} = \frac{\partial^2 x^i}{\partial z^j \partial z^m} \frac{\partial z^m}{\partial x^q} \tag{3}$$

for an arbitrary transformation

$$x^i = x^i(z^1, \dots, z^n),$$

$$z^j = z^j(x^1, \dots, x^n),$$

where $x(z(x)) = x$, $z(x(z)) = z$ *and the matrix* $\left(\frac{\partial x^\alpha}{\partial z_\beta}\right)$ *has a non-zero determinant (the Jacobian is not equal to zero).*

Proof. Since $\frac{\partial z^i}{\partial x^\alpha} \frac{\partial x^\alpha}{\partial z^j} = \delta^i_j$ = const., we have

$$0 \equiv \frac{\partial}{\partial z^q}(\delta^i_j) = \frac{\partial}{\partial z^q}\left(\frac{\partial z^i}{\partial x^\alpha} \frac{\partial x^\alpha}{\partial z^j}\right) = \frac{\partial^2 z^i}{\partial z^\alpha \partial x^\beta} \frac{\partial x^\beta}{\partial z^q} \frac{\partial x^\alpha}{\partial z^j} + \frac{\partial z^i}{\partial x^\alpha} \frac{\partial^2 x^\alpha}{\partial z^j \partial z^q},$$

which proves the lemma.

Thus, we have two distinct expressions for Christoffel symbols Γ^i_{jk} of Euclidean connection

$$\Gamma^i_{jq} = \frac{\partial^2 x^\alpha}{\partial z^j \partial z^q} \frac{\partial z^i}{\partial x^\alpha} = - \frac{\partial^2 z^i}{\partial z^\alpha \partial x^\beta} \frac{\partial x^\alpha}{\partial z^j} \frac{\partial x^\beta}{\partial z^q} . \tag{4}$$

We shall recall the definition of the general concept of covariant differentiation (already not Euclidean).

DEFINITION 1. *Covariant differentiation* of vector (covector) fields is an operation which in each coordinate system (z^1, z^2, z^3) is given by the formula

$$T^i_{;q} = \frac{\partial T^i}{\partial z^q} + \Gamma^i_{jq} T^j \quad \text{(for vectors)} \tag{5}$$

$$T_{i;q} = \frac{\partial T_i}{\partial z^q} - \Gamma^j_{iq} T_j \quad \text{(for covectors)}, \tag{6}$$

where Γ^i_{jq} are some functions in a given coordinate system. Given this, the transformation law for the quantity (Γ^i_{jq}) under coordinate changes $z^i = z^i(y^1, \dots, y^n)$ is specified proceeding from the requirement that the result of the operation of covariant differentiation be a tensor.

REMARK 1. *Affine coordinates* for the operation of covariant differentiation (if these coordinates do exist) are (x^j), where $x^i = x^i(z)$, such that, in terms of these coordinates, the following formula holds

$$T^i_{;q} = \frac{\partial T^i}{\partial x^q} \quad \text{or} \quad \Gamma^i_{jq} = 0.$$

REMARK 2. The operation of covariant differentiation is often denoted by the symbol ∇

$$\nabla_q T^i \equiv T^i_{;q} \quad \text{(by definition).}$$

The first point to be clarified is the transformation law for the symbols Γ^i_{jq} under the change $z = z(y)$. There holds

THEOREM 1. *Under the change of coordinates $z = x(y)$ the quantities Γ^i_{jq} transform by the formula*

$$\widetilde{\Gamma}^k_{l\,p} = \frac{\partial y^k}{\partial z^i}\left(\Gamma^i_{jq}\,\frac{\partial z^j}{\partial y^l}\,\frac{\partial z^q}{\partial y^p} + \frac{\partial^2 z^i}{\partial y^l \partial y^p}\right). \tag{7}$$

Proof. Since the expression

$$\frac{\partial T^i}{\partial z^q} + \Gamma^i_{jq} T^j = T^i_{;q} = \nabla_q T^i$$

is a tensor under the change $z = z(y)$, we have (using the equality $T^i = \widetilde{T}^\alpha \frac{\partial z^i}{\partial y^\alpha}$);

$$\widetilde{T}^l_{;p} = \frac{\partial \widetilde{T}^l}{\partial y^p} + \widetilde{\Gamma}^l_{jp}\widetilde{T}^j = \left(\frac{\partial T^i}{\partial z^q} + \Gamma^i_{mq}T^m\right)\frac{\partial z^q}{\partial y^p}\,\frac{\partial y^l}{\partial z^i} =$$

$$= \frac{\partial T^i}{\partial y^p}\,\frac{\partial y^l}{\partial z^i} + \Gamma^i_{mq}T^m\,\frac{\partial z^q}{\partial y^p}\,\frac{\partial y^l}{\partial z^i} =$$

$$= \frac{\partial}{\partial y^p}\left(\widetilde{T}^\alpha \frac{\partial z^i}{\partial y^\alpha}\right)\frac{\partial y^l}{\partial z^i} + \Gamma^i_{mq}\left(\widetilde{T}^\alpha \frac{\partial z^m}{\partial y^\alpha}\right)\frac{\partial z^q}{\partial y^p}\,\frac{\partial y^l}{\partial z^i} =$$

$$= \frac{\partial \widetilde{T}^\alpha}{\partial y^p}\delta^l_\alpha + \widetilde{T}^\alpha\left[\frac{\partial^2 z^i}{\partial y^\alpha \partial y^p}\,\frac{\partial y^l}{\partial z^i} + \frac{\partial z^m}{\partial y^\alpha}\,\frac{\partial z^q}{\partial y^p}\,\frac{\partial y^l}{\partial z^i}\Gamma^i_{mq}\right] =$$

$$= \frac{\partial \widetilde{T}^l}{\partial y^p} + \widetilde{T}^\alpha \widetilde{\Gamma}^l_{\alpha p}.$$

Thus, we obtain

$$\widetilde{\Gamma}^l_{\alpha p} = \frac{\partial y^l}{\partial z^i}\left(\frac{\partial^2 z^i}{\partial y^\alpha \partial y^p} + \Gamma^i_{mq}\,\frac{\partial z^m}{\partial y^\alpha}\,\frac{\partial z^q}{\partial y^p}\right),$$

and the theorem follows.

COROLLARY 1. *Symbols $\Gamma^l_{\alpha p}$ transform as tensors only under linear or affine tranadormations of coordinates $z = z(y)$, where* $\dfrac{\partial^2 z^i}{\partial y^\alpha y^p} \equiv 0$ *for all i, α, p.*

COROLLARY 2. *The alternative expression*

$$\Gamma^k_{lp} - \Gamma^k_{pl} = \Gamma^k_{[lp]}$$

is a tensor (the "torsion").

Proof. From formulae (7) we can see that under permutation of indices l and p, the summand $\dfrac{\partial y^k}{\partial z^i} \dfrac{\partial^2 z^i}{\partial y^l \partial y^p}$ remains unchanged. Therefore, the transformation law for $\Gamma^k_{lp} - \Gamma^k_{pl}$ will not contain this summand. Hence, this is a tensor (called "torsion"). On the basis of the result of Corollary 2, we introduce

DEFINITION 2. A covariant differentiation of Γ^k_{ij} is called *symmetric* if the torsion tensor $\Gamma^k_{ij} - \Gamma^k_{ji}$ is identical zero in each coordinate system or $\Gamma^k_{ij} = \Gamma^k_{ji}$.

EXAMPLE. If there exist affine coordinates $(x^1, x^2, \ldots, x^n)$, where $\Gamma^i_{kj} \equiv 0$, then relative to all coordinate systems (z), the torsion tensor is equal to zero and we have

$$\tilde{\Gamma}^i_{kj} = \tilde{\Gamma}^i_{jk}.$$

Indeed, in the coordinate system (z) we had, earlier, the formula

$$\Gamma^i_{kj} = -\frac{\partial^2 z^i}{\partial x^\alpha \partial x^\beta} \cdot \frac{\partial x^\alpha}{\partial z^k} \cdot \frac{\partial x^\beta}{\partial z^j}.$$

This expression is symmetric with respect to (k, j).

Next, the operation of covariant differentiation of a vector field enables the divergence of a vector field to be defined by the formula

$$\operatorname{div}(T^i) = T^i_{;i} = \frac{\partial T^i}{\partial z^i} + \Gamma^i_{ji} T^j. \tag{8}$$

For the Euclidean covariant differentiation, where (x^j) are affine coordinates, we have

$$\Gamma^i_{jq} = \frac{\partial^2 x^\alpha}{\partial x^j \partial z^q} \frac{\partial z^i}{\partial x^\alpha}.$$

Therefore,

$$\Gamma^m_{jm} = \frac{\partial^2 x^i}{\partial z^j \partial z^m} \frac{\partial z^m}{\partial x^i}.$$

We shall now present formulae which define covariant differentions of second-rank tensors:

$$T_{ij;k} = \nabla_k T_{ij} = \frac{\partial T_{ij}}{\partial z^k} - T_{qj}\Gamma^q_{ik} - T_{iq}\Gamma^q_{jk},$$

$$T^i_{j;k} = \nabla_k T^i_j = \frac{\partial T^i_j}{\partial z^k} + T^q_j \Gamma^i_{qk} - T^i_q \Gamma^q_{kj},$$

$$T^{ij}_{;k} = \nabla_k T^{ij} = \frac{\partial T^{ij}}{\partial z^k} + T^{qj}\Gamma^i_{qk} + T^{iq}\Gamma^j_{qk}, \tag{9}$$

where Γ^i_{jq}, the Christoffel symbols, are the same as for vector (covariant) fields.

What is, generally, the relation of a metric tensor to the manner of covariant differentiation defined by (Γ^k_{ij})?

These two entities have been introduced for different purposes. The metric g_{ij} has been introduced to determine metric relations in space — first of all the lengths of curve segments and the angles between them at the points where they intersect. The symbols (Γ^k_{ij}) have come as the only possible way to construct the differential calculus of tensor fields (in particular, vector fields). We have seen that the formula of covariant differentiation involving Γ^k_{ij} appears already for vector fields:

$$\nabla_k T^i = T^i_{;k} = \frac{\partial T^i}{\partial z^k} + \Gamma^i_{qk} T^q.$$

In fact, for functions (zero-rank tensors) we had the gradient operation $\nabla_k f = \dfrac{\partial f}{\partial z^k_i}$ and the derivative with respect to direction: given the vector $(\xi^i) = \xi$ at a point P and a function $f(z^1, \dots, z^n)$, its derivative with respect to the direction ξ was the expresssion

$$\xi_i \frac{\partial f}{\partial z^i} = \xi^i(\nabla_i f)$$

at the point P. Given a covariant differentiation we can determine the derivative with respect to direction for vector fields T^i.

DEFINITION 3. The *covariant derivative of a vector field* (T^i) *(or a covector field* (T_i)*) with respect to the direction of the vector* (ξ^k) at a certain point $P = (z^1, \dots, z^n)$ is the expression $\nabla_\xi (T^i) = \xi^k \nabla_k T^i$ at the point P (or the expression $\nabla_\xi T_i = \xi^k \nabla_k T_i$ at the point P for a covector field). The result of covariant differentiation of a vector field with respect to the direction ξ at a certain point P is the vector at this point P.

REMARK. Similarly, for an arbitrary tensor field, the covariant directional derivative at a point P in the direction ξ, is given by the expression $\xi^k \nabla_k T^{(i)}_{(j)}$, calculated at the point P, and is again a tensor of the same type at the same point P.

When the derivative of a function with respect to a certain coordinate is identically zero, we know that the function does not depend on this coordinate: when moving in the direction of this coordinate in such a manner that the other coordinates remain unchanged, we shall see that the functions remain constant. To put it more generally, if we are in motion along some curve in space

$$z^i = z^i(t), \quad i = 1, 2, \dots, n$$

and if the directional derivative of a function f in the direction of the velocity vector of that curve is zero, then the function is constant along the curve, i.e. if

$$\frac{\partial f(z^1(t), \dots, z^n(t))}{\partial t} = \xi^k \frac{\partial f}{\partial z^k} \equiv 0,$$

where $\xi^k = \dfrac{\partial z^k}{dt}$ is the velocity vector, then $f(z(t)) = \text{const}$.

Is the situation similar with vector and, in general, with tensor fields? The difficulty we come across, in this case, is that a vector, and generally a tensor, has different components in different coordinate systems; it is therefore rather difficult to compare two vectors, or two tensors, determined at distinct points of a space. The operation at least requires some additional definition and additional geometric structure in the space, namely, the structure of covariant differentiation.

Suppose, in a space, we are given some coordinates $(z^1, \dots, z^n)$, a covariant differentiation ∇ determined by Christoffel symbols (Γ^i_{kq}) and an arbitrary smooth curve $z^i(t)$, $a \le t \le b$.

DEFINITION 4. We shall say that a vector (or, more generally, tensor) field T is *covariantly constant* or *parallel along* the segment $a \le t \le b$ of a *curve* $z^i = z^i(t)$ if the directional covariant derivative of the field T at points of the curve in the direction of the velocity vector of the curve is equal to zero:

$$\nabla_\xi T = \xi^k \nabla_k T = 0, \quad a \le t \le b,$$

$$\xi^k = \frac{dz^k}{dt}.$$

For vector fields we have

$$\nabla_\xi T^i = \xi^k \nabla_k T^i = \xi_k \left(\frac{\partial T^i}{\partial z^k} + \Gamma^i_{jk} T^j \right) \equiv 0.$$

It should be emphasized that the concept of parallelism depends, generally speaking, on the curve. Only Euclidean geometry is an exception to this rule: in Euclidean coordinates $x^1, x^2, \dots, x^n$ we define parallel vector fields as fields possessing constant components in these (Euclidean) coordinates. These fields are, obviously, parallel along any curve. Since the result of covariant differentiation is independent of the choice of coordinates, the same fields will be parallel in terms of any coordinate system $(z^1, \dots, z^n)$, although in a new coordinate system the components of these fields will depend on the point.

Thus, we see that the concept of parallelism of vectors attached to distinct points depends both on the way the covariant differentiation (or, to put it differently, on the differential geometric connection) and on the path joining these two points. In the section following this, we shall investigate this question in more detail. Here, we shall only ask, once again, the question which was formulated earlier: what is the relation between covariant differentiation and the Riemannian metric?

DEFINITION. 5. A *covariant differentiation* (Γ^k_{ij}) *is said to be compatible with a Riemannian metric* (g_{ij}) if the covariant derivative of the tensor field g_{ij} at any point and in any direction is identically zero:

$$g_{ij;k} = \nabla_k g_{ij} = 0$$

(the tensor g_{ij} is covariantly constant or parallel along any curve).

In Euclidean geometry and in Euclidean coordinates, we had

$$g_{ij} = \text{const.}, \quad \nabla_k g_{ik} = \frac{\partial g_{ij}}{\partial x^k} = 0,$$

if $\Gamma^k_{ij} = 0$.

In the section which follows we shall show that the symmetric covariant differentiation $\Gamma^k_{ij} = \Gamma^k_{ji}$ compatible with the metric g_{ij} is uniquely determined by this metric.

The next item is supplementary.

Gauge fields. The most general concept of (affine) differential geometric connection is defined locally as a linear operation of "covariant" differentiation of N-component vector functions in a certain region U, coordinatized by coordinates $(z^1, \ldots, z^n)$, which is given by the formula

$$(\nabla_i \eta)^k = \frac{\partial \eta^k}{\partial z^i} + \Gamma^k_{li}\eta^l(z), \qquad i = 1, 2, \ldots, n; \quad l, k = 1, 2, \ldots, N, \tag{10}$$

or in vector notation

$$\nabla_i \eta = \frac{\partial \eta}{\partial z^i} + \hat{\Gamma}_i \eta \tag{11}$$

where the matrix $\hat{\Gamma}_i$ acts in an N-dimensional space of the values of the fields which, probably, have no relation to the tensors in the z-space. With respect to coordinate changes in the z-space, the set (Γ_i) transforms as a covector. Given this, the basis in the space of values is assumed to be constant.

Suppose we are given a non-degenerate linear transformation $a(x) = (a^k_p(z))$, $\det a^k_p \neq 0$ in an N-dimensional space. Let us change the basis of this space: $\eta^k = a^k_l \eta'^l$. Then the following simple lemma holds:

LEMMA 2. *In the new basis, the operation* (10) *is given by the formula*

$$(\nabla_i \eta')^q = \frac{\partial \eta'^q}{\partial z^i} + \Gamma'^q_{li}\eta'^l(z), \quad \hat{\Gamma}'_i = a^{-1}\hat{\Gamma}_i a + a^{-1}\frac{\partial a}{\partial z^i}. \tag{12}$$

Proof. In vector notation we have

$$\nabla_i \eta = \nabla_i (a \acute{\eta}) = \frac{\partial}{\partial z^i} (a\acute{\eta}) + \hat{\Gamma}_i a \acute{\eta} =$$

$$= a \left(\frac{\partial \acute{\eta}}{\partial z^i} + a^{-1} \Gamma_i a \acute{\eta} + a^{-1} \frac{\partial a}{\partial z^i} \acute{\eta} \right) = a \nabla'_i \acute{\eta}$$

and the lemma follows immediately.

DEFINITION 5. A *gauge field* or *affine connection* is a family of matrix functions $\hat{\Gamma}_i$ with the values within an N-dimensional space, which under the change of basis in this space transforms by the formula

$$\eta = a\acute{\eta}, \quad \eta^k(z) \equiv a^k_l(z)\, \eta^l(z),$$

$$\Gamma'_i = a^{-1} \Gamma_i a(z) + a^{-1} \frac{\partial a}{\partial z} \tag{13}$$

(these are "gauge transformations").

EXAMPLE 1. If $N = n$ and the matrix $a(z)$ is the Jacobi matrix of the change of coordinates in the z-space, then formula (13) coincides with the transformation formula for Christoffel symbols under tha change of coordinates.

EXAMPLE 2. If $N = 1$, then $\eta(z)$ is a scalar function. Suppose that $a(z) = \exp(\phi(z))$. Then we have

$$\hat{\Gamma}'_i = \hat{\Gamma}_i + \frac{\partial \phi}{\partial z^i} . \tag{14}$$

DEFINITION 6. The *curve* is the commutator of covariant derivatives

$$[\nabla_i, \nabla_j] = \nabla_i \nabla_j - \nabla_j \nabla_i = \hat{R}_{ij}, \quad ([\nabla_i, \nabla_j]\, \eta)^k = R^k{}_{ijl}\, \eta^l.$$

LEMMA 3. *This commutator is a zero-order linear operator. Under gauge transformations it transforms by the formula* $R'_{ij} = a^{-1} R_{ij}\, a(z)$. *This commutator determines the "curvature tensor"* R^k_{ijl}, *where* $i, j = 1, \dots, n$; $k, l = 1, 2, \dots, N$.

The following formula holds

$$\hat{R}_{ij} = \frac{\partial\hat{\Gamma}_j}{\partial z^i} - \frac{\partial\hat{\Gamma}_i}{\partial z^j} + [\hat{\Gamma}_i, \hat{\Gamma}_j].$$

The proof is obtained by means of the obvious substitution. In the language of differential forms we have

$$d\hat{\Gamma} = \hat{R} + [\hat{\Gamma}, \hat{\Gamma}]/2,$$

where $\hat{\Gamma} = \hat{\Gamma}_i\, dz^i$, $\hat{R} = \sum_{i<j} \hat{R}_{ij}\, dz^i \wedge dz^j$. The quantities $\hat{\Gamma}$, $\hat{R}$ are respectively called "the *form of connection*" and "the *form of curvature*". These are differential forms with the values in the matrices. For more details concerning the differential forms see Section 1.10. The commutator of two 1-forms $[\hat{\Gamma}, \hat{\Gamma}]$ is the 2-form $[\hat{\Gamma}, \hat{\Gamma}] = [\hat{\Gamma}_i, \hat{\Gamma}_i]\, dz^i \wedge dz^j$.

EXAMPLE. For $N = 1$ we have $[\hat{\Gamma}, \hat{\Gamma}] = 0$,

$$\hat{R} = d\hat{\Gamma}.$$

The form of curvature may be an arbitrary closed 2-form.

General connections which extend to the electro-magnetic field are of great importance in the mathematical apparatus of the modern theory of elementary particles ("gauge fields"). The case $N = 1$ corresponds to an electro-magnetic field where $\hat{R}_{ij}$ is the *field strength* and $\hat{\Gamma}$ is the *vector-potential*.

2.8 Covariant Differentiation and the Riemannian Metric. Parallel Transport of Vectors along Curves. Geodesics

The concept of covariant differentiation of vector (covector) fields was defined in the preceding sections. With respect to local coordinates it has the form

$$T^i_{;k} = \nabla_k T^i = \frac{\partial T^i}{\partial z^k} + \Gamma^i_{qk} T^q,$$

$$T_{i;k} = \nabla_k T_i = \frac{\partial T_i}{\partial z^k} - \Gamma^q_{ik} T_q. \tag{1}$$

Under the change of coordinates $z = z(y)$ the transformation law for Γ^k_{ij} is defined proceeding from the requirement that $T^i_{;k}$ be a tensor

$$\tilde{T}^j_{;k} = T^i_{;q} \frac{\partial z^q}{\partial y^k} \frac{\partial y^j}{\partial z^i} .$$

Therefore

$$\underset{(y)}{\tilde{\Gamma}^m_{np}} = \frac{\partial y^m}{\partial z^\alpha} \left(\underset{(z)}{\Gamma^\alpha_{qt}} \frac{\partial x^q}{\partial y^n} \frac{\partial z^t}{\partial y^p} + \frac{\partial^2 z^\alpha}{\partial y^n \partial y^p} \right). \tag{2}$$

In this case, when $\Gamma^\alpha_{qt} \equiv 0$ (or when z are affine), we obtain from (2) our old formula for Euclidean connection

$$\tilde{\Gamma}^m_{np} = \frac{\partial y^m}{\partial z^\alpha} \frac{\partial^2 z^\alpha}{\partial y^n \partial y^p} .$$

Inversely, under linear transformations, where $\dfrac{\partial^2 z^\alpha}{\partial y^\beta \partial y^\gamma} \equiv 0$, the transformation law (2) becomes tensorial.

The "torsion tensor" $\Gamma^\alpha_{qt} - \Gamma^\alpha_{tq} = T^\alpha_{qt}$ always transforms as a tensor; for the symmetric case, this tensor is equal to zero by definition: $T^\alpha_{qt} \equiv 0$. Covariant differentiation of tensors of any rank (connection) is determined by the following requirements:

a) a covariant differentiation should be a linear operation (where the derivative of the sum is equal to the sum of the derivatives) and should commute with contraction;

b) the covariant derivative of a zero-rank tensor (of a function) should be the ordinary derivative

$$\nabla_k f = \left(\frac{\partial f}{\partial z^k}\right);$$

c) the covariant derivative of a vector (covector) field should be given by formulae (1);

d) the covariant derivative of a product of tensors should be calculated by the Leibniz formula for differentiation of a product:

$$\nabla_k T^{(i),(j)}_{(k),(l)} = (\nabla_q \tilde{T}^{(i)}_{(k)} \cdot \tilde{\tilde{T}}^{(j)}_{(l)} + \tilde{T}^{(i)}_{(k)} \cdot (\nabla_q \tilde{\tilde{T}}^{(j)}_{(l)})$$

where $T^{(i),(j)}_{(k),(l)} = \tilde{T}^{(i)}_{(k)} \cdot \tilde{\tilde{T}}^{(j)}_{(l)}$ is the product of tensors.

As basic examples, we consider second-rank tensors T^{ij}; T^i_j; T_{ij}. There holds

THEOREM 1. *If in a space we are given a connection (or the way of covariant differentiation of vector (covector) fields) and if the differentiation of second-rank scalars and tensors is determined by the requirements* a), b), c) *and* d) *(listed above), then the differentiation in an arbitrary coordinate sysem* (z) *is given by the formulae*

$$\nabla_k T^{ij} = \frac{\partial T^{ij}}{\partial z^k} + \Gamma^i_{qk} T^{qj} + \Gamma^j_{qk} T^{iq},$$

$$\nabla_k T^i_j = \frac{\partial T^i_j}{\partial z^k} + \Gamma^i_{qk} T^q_j - \Gamma^q_{jk} T^i_q,$$

$$\nabla_k T_{ij} = \frac{\partial T_{ij}}{\partial z^k} - \Gamma^q_{ik} T_{qj} - \Gamma^q_{jk} T_{iq}. \tag{3}$$

Proof. We shall carry out the proof for tensors of the form T_{ij}; it is exactly the same for the other cases.

LEMMA 1. *Any tensor field* $T^{(i)}_{(j)}$ *can be represented as a linear combination of products of the first-rank tensors.*

Proof. Let us choose a convenient basis in the space of tensors.

Let e_i be basis vector fields and let e^i be basis covector fields. Here the fields e_i are unit vectors of the coordinate system (z^j), where any vector field has the form $T^i = T^i e_i$. (T^i are components). The covector fields e^i are specified by the formulae $e^i \cdot e_j = \delta^i{}_j$, and any covector field $T = T_j e^j$, where T_j are components. Let us consider the product

$$e_{i_1} \circ e_{i_2} \circ \dots \circ e_{i_k} \circ e^{j_1} \circ \dots \circ e^{j_l}$$

for all sets $(i_1, \dots, i_k)$, $(j_1, \dots, j_k)$. These are basis tensors of type (k, l); any tensor of type (k, l) has the form

$$T = T^{i_1 \dots i_k}_{j1 \dots jl} e_{i_1} \circ e_{i_2} \circ \dots \circ e_{i_k} \circ e^{j_1} \circ \dots \circ e^{j_l}, \tag{4}$$

where $T = T^{i_1 \dots i_k}_{j1 \dots jl}$ are components in the coordinate system (z). Hence, any tensor field is of the form

$$T^{(i)}_{(j)} = T^{i_1 \dots i_k}_{j1 \dots jl} e_{i_1} \circ e_{i_2} \circ \dots \circ e_{i_k} \circ e^{j_1} \circ \dots \circ e^{j_l},$$

which is a linear combination of products, as required.

On this account, a tensor field T_{ij} in a sufficiently small neighbourhood of an arbitrary point can be represented in the form of (4), where $T_{ij}(z)$ are functions (numerical) of the point.

It suffices, therefore, to prove the theorem only for products of the form

$$T_{ij} = \alpha(z) \tilde{T}_i \approx{T}_j.$$

By definition, according to requirement d),

$$\nabla_k(\alpha(z)\,\tilde{T}_i\tilde{\tilde{T}}_j) = \nabla_k(\alpha\tilde{T}_i)\tilde{\tilde{T}}_j + \alpha T_i(\nabla_k\tilde{\tilde{T}}_j) =$$

$$= \frac{\partial\alpha}{\partial z^k}\,\tilde{T}_i\tilde{\tilde{T}}_j + \alpha(\nabla_k\tilde{T}_i)\tilde{\tilde{T}}_j + \alpha\tilde{T}_i(\nabla_k\tilde{\tilde{T}}_j) =$$

$$= \frac{\partial\alpha}{\partial z^k}\,(\alpha\tilde{T}_i\tilde{\tilde{T}}_j) - \alpha\tilde{T}_p\tilde{\tilde{T}}_j\Gamma^p_{ik} - \alpha\tilde{T}_i\tilde{\tilde{T}}_p\Gamma^p_{jk} =$$

$$= \frac{\partial T_{ij}}{\partial z^k} - \Gamma^p_{ik}T_{pj} - \Gamma^p{}_{jk}T_{jp}.$$

which implies the theorem.

EXAMPLE 1. If T^i is a vector field and T_j is a covector field, then the scalar field $T^{\cdot\, i}T_j$ (the trace of the product of tensors) is determined. To meet the requirements a) to d), there holds the formula

$$\frac{\partial}{\partial z^k}\,(T^iT_i) = (\nabla_kT^i)T_i + T^i(\nabla_kT^i) =$$

$$= (\frac{\partial T^i}{\partial z^k} + \Gamma^i_{qk}T^q)T_i + (\frac{\partial T_i}{\partial z^k} - \Gamma^q_{ik}T_q)T^i =$$

$$= \frac{\partial}{\partial z^k}\,(T^iT_i) + \Gamma^i{}_{qk}T^q\,T_i - \Gamma^q_{ik}T_qT^i.$$

From this formula we can see that the components Γ^i_{kl} of covariant differentiation of covector fields must have opposite signs (and coincide in the absolute value) with the components of differentiation of vector fields, so that for the scalar T^iT_i the following formula holds

$$\frac{\partial}{\partial z^k}\,(T^iT_i) = (\nabla_kT^i)T_i + T^i(\nabla_kT_i).$$

EXAMPLE. 2. If in a space, a Riemannian metric g_{ij} is given, then the connection is said to be compatible with the metric provided that the following formula holds (by definition)

$$\nabla_k g_{ij} \equiv 0. \tag{5}$$

Therefore, for any tensor field $T^{(i)}_{(j)}$ we have

$$\nabla_q\left(g_{kl} T^{(i)}_{(j)}\right) = g_{kl}\left(\nabla_q T^{(i)}_{(j)}\right). \tag{6}$$

This formula follows from the requirement d) and from formula (5) since $\nabla_q g_{kl} \equiv 0$. Since the differentiation operation is linear, we have

COROLLARY 1. *If a given connection (a way of covariant differentiation) is compatible with the Riemannian metric g_{ij}, then the operation of lowering any tensor index commutes with the covariant differentiation.*

Indeed, we have

$$\nabla_k\left(g_{il} T^{l,(i)}_{(j)}\right) = g_{kl}\left(\nabla_q T^{l,(i)}_{(j)}\right).$$

Finally, we are led to an important

THEOREM 2. *If the metric g_{ij} is non-degenerate (i.e. if* $\det(g_{ij}) = g \neq 0$*), then there exists a unique symmetric connection which is compatible with the metric g_{ij}. Whatever the coordinate system (z), this connection is given by the formula*

$$\Gamma^q_{ik} = 1/2\, g^{qj}\left(\frac{\partial g_{jk}}{\partial z^i} + \frac{\partial g_{ij}}{\partial z^k} + \frac{\partial g_{ik}}{\partial z^j}\right). \tag{7}$$

Proof. By definition

1) $\Gamma^q_{ik} = \Gamma^q_{ki}$,

2) $\nabla_k g_{ij} = \dfrac{\partial g_{ij}}{\partial z^k} - \Gamma^q_{ik} g_{qj} - \Gamma^q_{jk} g_{iq} = 0.$

We shall attempt the solution of the equation $\nabla_k g_{ij} = 0$ with respect to Γ^q_{ik}. By the definition of lowering indices, we have

$$\Gamma_{j;ik} = \Gamma^q_{ik} g_{qj}.$$

The equations have the form

$$\Gamma_{j;ik} + \Gamma_{i;jk} = \frac{\partial g_{ij}}{\partial z^k}.$$

Given this, $\Gamma_{j;ik} = \Gamma_{j;ki}$; $\Gamma_{i;jk} = \Gamma_{i;kj}$. Permuting the indices i, j, k we obtain

$$\frac{\partial g_{ij}}{\partial z^k} = (\Gamma_{j;ik} + \Gamma_{i;jk}), \qquad [a]$$

$$\frac{\partial g_{ki}}{\partial z^j} = (\Gamma_{k;ij} + \Gamma_{i;kj}), \qquad [b]$$

$$\frac{\partial g_{kj}}{\partial z^i} = (\Gamma_{j;ki} + \Gamma_{k;ji}). \qquad [c]$$

Obviously, by virtue of the connection symmetry, there holds the equality

$$[b] + [c] - [a] = 2\Gamma_{k;ij}.$$

Hence

$$\Gamma_{k;ij} = 1/2\left(\frac{\partial g_{ki}}{\partial z^j} + \frac{\partial g_{kj}}{\partial z^i} + \frac{\partial g_{ij}}{\partial z^k}\right) = g_{kq}\Gamma^q_{ij}.$$

Since $g^{ij}g_{jq} = \delta^i_q$, we have

$$\Gamma^q_{ij} = g^{kq}\Gamma_{k;ij} = \frac{g^{kq}}{2}\left(\frac{\partial g_{ki}}{\partial z^j} + \frac{\partial g_{kj}}{\partial z^i} + \frac{\partial g_{ij}}{\partial z^k}\right).$$

This completes the proof.

From the theorem there follows

COROLLARY 2. *If the coordinates are so chosen that at a given point all the first-order derivatives of g_{ij} are equal to zero, then at this point the components Γ^q_{ij} are equal to zero (for a symmetric connection compatible with the metric).*

EXAMPLE 1. Let us consider the case of a surface situated in a three-dimensional Euclidean space with coordinates $x^1, = x, x^2 = y, x^3 = z$ (Euclidean coordinates):

$$x^1 = x^1(z^1, z^2), \quad x^2 = x^2(z^1, z^2), \quad x^3 = x^3(z^1, z^2).$$

At a given point of the surface, let the x^3-axis be orthogonal to the surface and let the x^1- and x^2-axes be tangent to the surface. In a neighbourhood of this point the surface is given by the equation

$$x^3 = f(z^1, z^2), \quad z^1 = u = x^1, \quad z^3 = v = x^3,$$

where $z^1 = x^1, z^2 = x^2$; moreover, since the x^3-axis is orthogonal to the surface at the point $P = (0, 0)$, we obtain

$$\frac{\partial f}{\partial z^1}\Big|_{0,0} = 0, \quad \frac{\partial f}{\partial z^2}\Big|_{0,0} = 0 \quad \text{or} \quad \operatorname{grad} f|_{0,0} = 0$$

at the point $P = (0, 0)$.

For the metric g_{ij} we have

$$g_{ij} = \delta_{ij} + \frac{\partial f}{\partial z^i}\frac{\partial f}{\partial z^j}.$$

It should be recalled that $g_{ij}\, dz^i dz^j = (dx)^2 + (dy)^2 + (dz)^2 = dx^2 + dy^2 + (df)^2$. At the point P, where $\frac{\partial f}{\partial z^i} = 0$, we have $g_{ij} = \delta_{ij}$ and

$$\frac{\partial f_{ij}}{\partial z^k} = \frac{\partial}{\partial z^k}\left(\frac{\partial f}{\partial z^i}\frac{\partial f}{\partial z^j}\right) = \frac{\partial^2 f}{\partial z^i \partial z^k}\frac{\partial f}{\partial z^j} + \frac{\partial^2 f}{\partial z^j \partial z^k}\frac{\partial f}{\partial z^i} = 0.$$

Therefore with respect to these coordinates, at the point P all the components $\Gamma^q_{ik} = 0$ for $q, i, k = 1, 2$. These coordinates at the point P were chosen in Part I; they were convenient for different purposes. The axis $z = x^3$ is orthogonal to the surface $x^1 = u = z^1, x^2 = v = z^2$.

EXAMPLE 2. The divergence of a vector field (T^i) has been defined as

$$\operatorname{div} T^i = \nabla_i T^i = T^i_{;i}.$$

For a symmetric connection compatible with some Riemannian metric we have

$$\nabla_i T^i = \frac{\partial T^i}{\partial z^i} + \Gamma^i_{qi} T^q,$$

where

$$\Gamma^i_{qi} = 1/2\, g^{it} \left(\frac{\partial g_{tq}}{\partial z^i} + \frac{\partial g_{ti}}{\partial z^q} - \frac{\partial g_{qi}}{\partial z^t}\right) = 1/2\, g^{it} \frac{\partial g_{it}}{\partial z^q} = 1/2 g \frac{\partial g}{\partial z^q}.$$

We can verify that $g = \det(g_{ij})$.

Question. What is the formula for the divergence in general coordinates?

Answer.

$$\nabla_i T^i = \frac{\partial T^i}{\partial z^i} + 1/2 g \frac{\partial g}{\partial z^q} T^q = \frac{\partial T^i}{\partial z^i} + \frac{T^q}{(|g|)^{1/2}} \cdot \frac{\ln (|g|)^{1/2}}{\partial z^q}.$$

A more convenient form is

$$\nabla_i T^i = \frac{\partial T^i}{\partial z^i} + 1/2 \frac{\partial g}{\partial z^q} T^q = \frac{1}{(|g|)^{1/2}} \frac{\partial}{\partial z^i} \left((|g|)^{1/2} T^i\right).$$

Conclusion: the divergence of a vector field $\nabla_i T^i$ has the usual form

$$\nabla_i T^i = \frac{\partial T^i}{\partial z^i},$$

if and only if the volume element $(|g|)^{1/2}\, dz^1 \wedge dz^2 \wedge dz$ coincides with the Euclidean one: $(|g|)^{1/2} = 1$, where $g = \det(g_{ij})$.

Now we already have a definite relation between the connection (the way of covariant differentiation) and the Riemannian metric g_{ij}, which can be interpreted as follows: any Riemannian geometry gives rise to a certain symmetric way of tensor differentiation with respect to which the Riemannian metric itself is a constant.

How can we distinguish between Euclidean and non-Eudlicean geometries? Can we find coordinates (x^j), where $g_{ij} = \delta_{ij}$ and $\Gamma^q_{ij} \equiv 0$? What geometrical properties distinguish Euclidean geometry from non-Euclidean?

In order to link up our present geometrical ideas with the most basic concepts from school geometry, we recall the so-called fifth postulate of Euclid: "Given a line through a point P, and a point Q not on the line, there is exactly one line through Q parallel to the given line".

The reader may know from school mathematics that the appearance of Lobachevski geometry is due to distrust of precisely this postulate, i.e. to its validity in a real space on a large scale.

We shall formulate this postulate in a (perhaps, not quite formal) way convenient for our further purposes: given a vector $(T^i)_P$ at a point P in Euclidean geometry, at an arbitrary point Q there exists only one vector $(\tilde{T}^i)_Q$ which is parallel to and has the same length as the given vector.

It is relevant no to ask: what exactly do we mean by parallel vectors attached to different points P and Q? By definition, a vector (like any tensor) is attached to a given point.

Recall that for any tensor field (T^i) and any vector ξ^α at a point P we have determined the directional derivative

$$\nabla_\xi T^i = \xi^\alpha \nabla_\alpha T^i.$$

This is the vector at the point P.

If we are given a curve $z^i = z^i(t)$ and $\xi^\alpha = \dfrac{dz^\alpha}{dt}$, then the derivative of the vector field T^i along the curve takes the form

$$\nabla_{\xi(t)} T^i = \frac{dz^\alpha}{dt} \nabla_\alpha T^i.$$

By definition, the vector field is parallel along the curve if and only if $\nabla_{\xi(t)} T^i \equiv 0$ at all points of the curve.

We shall now give the definition of the important concept of *parallel transport* of a vector (T^i) from a point P into a point Q along a curve $z^i(t)$, where $z^i(0) = z_0^i(P)$ and $z^i(1) = z_1^i(Q)$.

DEFINITION 1. *Parallel transport* of a vector T_P^i from a point $P = (z_0^1, \ldots, z_0^n)$ to a point $Q = (z_1^1, \ldots, z_1^n)$ along a curve $z^i = z^i(t)$ joining P and Q is the vector field T^i given at all points of the curve and parallel to itself along this curve: $\dfrac{dz^\alpha}{dt} \nabla_\alpha T^i = 0$

for all $0 \le t \le 1$. When $t = 0$, the vector field (T^i) at the point P must coincide with the initial vector $(T^i)_P$. When $t = 1$, the vector field T^i at the point Q is the vector $(T^i)_Q$, and is called the result of parallel transport of the vector $(T^i)_P$ along the given curve $z^i = z^i(t)$ from P to Q. We can see that *parallel transport of a vector along a curve depends on the connection* (Γ^q_{ij}) and thereby on the metric g_{ij} provided that the connection is symmetric and compatible with the metric.

With respect to coordinates $z^1, \ldots, z^n$ we obtain

$$0 = \frac{dz^\alpha}{dt} \nabla_\alpha T^i = \frac{\partial T^i}{\partial z^\alpha} \frac{dz^\alpha}{dt} + \Gamma^i_{p\alpha} \frac{dz^\alpha}{dt} =$$

$$= \frac{dT^i(t)}{dt} + \left(\frac{dz^\alpha}{dt} \Gamma^i_{p\alpha}\right) T^p = 0. \tag{8}$$

This is the equation of *parallel transport*. The initial data (for $t = 0$) has the form

$$T^i(0) = T^i_P.$$

Equation (8) is linear in T^i. From the existence and uniqueness theorem, for any smooth curve $z^i = z^i(t)$ we obtain the following result.

THEOREM 3. *The result of parallel transport along any smooth fixed curve exists, is uniquely determined by the initial vector* T^i_z *and depends linearly on the initial vector* T^i_z.

For connections compatible with the metric g_{ij}, where $\nabla_\alpha g_{ij} \equiv 0$, there holds

THEOREM 4. *If* $\tilde{\tilde{T}}^i(t)$ *and* $\tilde{\tilde{T}}^j(t)$ *are parallel vector fields along a curve* $z^i = z^i(t)$, *their scalar product is constant:* $\frac{d}{dt}(g_{ij}\tilde{T}^i(t)\,\tilde{\tilde{T}}^j(t)) = 0$ *providing* $\nabla_\alpha g_{ij} = 0$.

Proof. Since $\nabla_\alpha g_{ij} \equiv 0$, it follows from the definition of covariant differentiation for the product of tensors that

$$\frac{d}{dt}\left(g_{ij}\tilde{T}^i\tilde{\tilde{T}}^j\right) = \frac{dz^\alpha}{dt}\nabla_\alpha\left(g_{ij}\tilde{T}^i\tilde{\tilde{T}}^j\right) = \frac{dz^\alpha}{dt}\, g_{ij}\,\nabla_\alpha\left(\tilde{T}^i\tilde{\tilde{T}}^j\right) =$$

$$= g_{ij}\left(\frac{dz^\alpha}{dt}\ \nabla_\alpha\tilde{T}^i\right)\tilde{\tilde{T}}^j + g_{ij}\,\tilde{T}^i\ \left(\frac{dz^\alpha}{dt}\nabla_\alpha\nabla_\alpha\tilde{\tilde{T}}^j\right) = 0,$$

and this implies the theorem.

Thus, the concept of connection compatible with the metric has appeared for the property that parallel transport preserves the scalar product of vectors, i.e. is an orthogonal transformation.

If the geometry is Euclidean or $\Gamma^q_{ik} \equiv 0$, we are led to

COROLLARY 3. *In Euclidean geometry and in Euclidean coordinates, vectors attached to different points and having identical components are parallel along any curve. In any coordinates, the result of parallel transport of a vector along a curve does not depend on the curve provided that the geometry is Euclidean.*

The difference between Euclidean and non-Euclidean geometries is already intuitively clear now: parallel transport of one and the same vector form P to Q along different curcves (if curvature exists) yields distinct results.

In a real space the geometry is determined by the gravitational field, but this is the geometry of a four-dimensional space-time.

How should curvature be measured numerically? This will be our concern in the next section.

What lines are straight? They are called "geodesics" of a given connection. Given a line $z^i = z(t)$, the tangent vector field $T^i = \frac{dz^i}{dt} = T$ along this line is defined.

DEFINITION 2. The *geodesic line* of a given connection, $z^i = z^i(t)$, is such a line that the covariant derivative of the vector field $T^i = \frac{dz^i}{dt}$ along this curve is equal to zero:

$$0 = \nabla_T(T) = \frac{dz_i}{dt}\nabla_i\left(\frac{dz^j}{dt}\right) = \frac{dz^i}{dt}\left(\frac{\partial}{\partial z^i}\frac{dz^j}{dt} + \Gamma^j_{\alpha i}\frac{\partial z^\alpha}{\partial t}\right) =$$

$$= \frac{d^2 z^j}{dt^2} + \Gamma^j_{\alpha i}\frac{dz^\alpha}{dt}\frac{dz^i}{dt} = K^j_g(t) = 0.$$

The equation for geodesics has the form

$$\frac{d^2 z^j}{dt^2} + \Gamma^j_{\alpha i}\frac{dz^\alpha}{dt}\frac{dz^i}{dt} = 0. \tag{9}$$

Parallel transport of the velocity vector of a geodesic along the geodesic itself is again a velocity vecor (this is an alternative definition of geodesics).

Geodesics are given by equation (9). If $\Gamma \equiv 0$, these are ordinary straight lines, as they should be when the geometry is Euclidean.

REMARK. The vector $\nabla_T(T^j) = \frac{d^2 z^j}{dt^2} + \Gamma^j_{\alpha i}\frac{dz^\alpha}{dt}\frac{dz^i}{dt} = K^j_g(t)$ is frequently called the *vector of geodesic curvature* of this line, $T^i = \frac{dz^i}{dt}$. Given the metric g_{ij}, the curve can be determined in terms of the natural parameter l, where

$$dl = \left|\frac{dz^\alpha}{dt}\right| dt, \quad \left|\frac{dz^\alpha}{dt}\right|^2 = g_{ij}\frac{dz^i}{dt}\frac{dz^j}{dt};$$

the vector of geodesic curvature is often determined only with respect to the natural parameter:

$$K^j_g(l) = \frac{d^2 z^j}{dl^2} + \Gamma^j_{\alpha i}\frac{dz^\alpha}{dl}\frac{dz^i}{dl} = \nabla_{\tilde{T}}(\tilde{T}),$$

$$\tilde{T}^\alpha = \frac{dz^\alpha}{dl}, \quad |\tilde{T}| = 1.$$

Geodesic curvature is the length of the vector $K^i_g(l)$:

$$k_g(l) = |K_g(l)| = \left(g_{ij}K^i_g(l)\,K^j_g(l)\right)^{1/2}$$

where l is the natural parameter.

In this language, we can say that geodesics are lines whose geodesic curvature is equal to zero

$$K^j_g(t) = 0.$$

Let us compare the equation for geodesics (9) (in a given connection) which we derived in this section with the one obtained in Section 1.15 proceeding from the variational principle. Recall that in Section 1.15 we defined geodesics as locally the shortest curve segments for a given Riemannian metric. To compare these equations, we have to consider on a manifold a connection compatible with a Riemannian metric, i.e. a connection specified by the condition $\nabla_\alpha g_{ij} = 0$. Then, we arrive at the conclusion that equation (9) derived in this section and equation (1) of Section 1.15 coincide. In the case of general connections, their geodesics are not already obliged to be locally the shortest in a Riemannian metric if the connection is not compatible with this metric.

2.9 Riemannian Curvature Tensor. Gaussian Curvature as an Intrinsic Invariant of the Surface

In the preceding section we have explained that in a non-Euclidean space, the parallel transport of a vector is path-dependent. At the same time, the result of parallel transport of a vector T along a path $z(t)$ is given by the transport equation

$$\frac{dT^i}{dt} + \Gamma^i_{jk} T^k \frac{dz^j}{dt} = 0,$$

the solution of which requires some skill.

It would be much more convenient if, instead of solving this equation, we could find the local characteristic of the departure of the given connection (Γ^i_{jk}) from being Euclidean. What characteristic is this? What is the way to find whether or not there exist coordinates $(x^1, x^2, \dots, x^n)$ in terms of which the Γ^i_{jk} vanish identically? Of course, if the connection is not symmetric, then $T^i_{jk} = \Gamma^i_{jk} - \Gamma^i_{kj}$ is a non-zero tensor, and therefore we cannot introduce coordinates with respect to which the Γ^i_{jk} vanish identically. In this case we may understand Euclidean coordinates as such coordinates (x^j) that $\Gamma^i_{jk} = 1/2T^i_{jk}$, that is, the Γ^i_{jk} is skew-symmetric with respect to the lower indices (the symmetric part $\Gamma^i_{kj} + \Gamma^i_{jk} \equiv 0$).

How shall we find out whether or not there exist Euclidean coordinates? We shall attempt the answer in respect of symmetric connections. We are acquainted with the important property of partial derivatives in the usual analysis:

$$\frac{\partial}{\partial x^j}\frac{\partial f}{\partial x^i} = \frac{\partial}{\partial x^i}\frac{\partial f}{\partial x^j} = \frac{\partial^2 f}{\partial x^i \partial x^j}.$$

If the connection admits Euclidean coordinates $x^1, x^2, x^3, \dots, x^n$, then, in terms of these coordinates, the tensors are differentiated using ordinary formulae

$$\nabla_k T^{(i)}_{(j)} = \frac{\partial T^{(i)}_{(j)}}{\partial x_k}.$$

Therefore,

$$T^{(i)}_{(j);k;l} = T^{(i)}_{(j);l;k}$$

or, equivalently,

$$(\nabla_k \nabla_l - \nabla_l \nabla_k) T^{(i)}_{(j)} \equiv 0.$$

This property is valid relative to any coordinates, as long as $T^i_{j;k;l}$ is a tensor. Let us examine the general connections.

Let a manifold M^n be endowed with local coordinates $x^1, \ldots, x^n$ in a neighbourhood of an arbitrary point. Consider vector fields T^i on M and consider the differential expression $\nabla_k\nabla_l - \nabla_l\nabla_k$, where ∇ is some differential-geometric symmetric connection. We apply this expression to the vector field T^i, to obtain

$$\nabla_l T^i = \frac{\partial T^i}{\partial x^l} + T^p\Gamma^i_{pl},$$

$$\nabla_k\nabla_l(T^i) = \frac{\partial^2 T^i}{\partial x^k \partial x^l} + \frac{\partial T^p}{\partial x^k}\Gamma^i_{pl} +$$

$$+ T^p\frac{\partial}{\partial x^k}(\Gamma^i_{pl}) + \nabla_l(T^p)\,\Gamma^i_{pk} - \nabla_p(T^i)\,\Gamma^p_{kl} =$$

$$= \frac{\partial T^p}{\partial x^l}\Gamma^i_{pk} + T^q\,\Gamma^p_{ql}\,\Gamma^i_{pk} - \frac{\partial T^i}{\partial x^p}\Gamma^p_{kl} - T^q\Gamma^i_{qp}\Gamma^p_{kl}$$

and so

$$(\nabla_k\nabla_l - \nabla_l\,\nabla_k)T^i = T^p\left(\frac{\partial}{\partial x^k}\Gamma^i_{pl} - \frac{\partial}{\partial x^l}\Gamma^i_{pk}\right) -$$

$$-(\Gamma^p_{kl} - \Gamma^p_{lk})\frac{\partial T^i}{\partial x^p} + T^q(\Gamma^p_{ql}\Gamma^i_{pk} - \Gamma^p_{lk}\Gamma^i_{pl} - \Gamma^i_{qp}\Gamma^p_{lk} + \Gamma^i_{qp}\Gamma^p_{kl}).$$

In the case of symmetric connection $\Gamma^i_{jk} = \Gamma^i_{kj}$ we have

$$(\nabla_k\nabla_l - \nabla_l\nabla_k)T^i =$$

$$= T^p\left(\frac{\partial}{\partial x^l}\Gamma^i_{ql} - \frac{\partial}{\partial x^l}\Gamma^i_{qk} + \Gamma^p_{ql}\Gamma^i_{pk} - \Gamma^p_{qk}\Gamma^i_{pl}\right) = -T^qR^i_{q,kl},$$

where

$$-R^i_{q,kl} = \frac{\partial\Gamma^i_{ql}}{\partial x^k} - \frac{\partial\Gamma^i_{qk}}{\partial x^l} + \Gamma^p_{ql}\Gamma^i_{pk} - \Gamma^p_{qk}\Gamma^i_{pl}.$$

Thus, for symmetric connections we are led to

$$(\nabla_k \nabla_l - \nabla_l \nabla_k) T^i = -T^q R^i{}_{q,kl} \,.$$

In the case of general (non-symmetric) connections

$$(\nabla_k \nabla_l - \nabla_l \nabla_k) T^i = -T^q R^i{}_{q,kl} - (\Gamma^p_{kl} - \Gamma^p_{lk}) \left(\frac{\partial T^i}{\partial x^p} + T^q \Gamma^i_{qp}\right) =$$

$$= -T^q R^i_{q,kl} - T^p_{kl} (\nabla_p T^i).$$

Here $T^p_{kl} = \Gamma^p_{kl} - \Gamma^p_{lk}$ is a tensor called the *torsion tensor*; $R^i_{q,kl}$ is a tensor called the *Riemannian curvature tensor*. So, ultimately, we have

$$(\nabla_k \nabla_l - \nabla_l \nabla_k) T^i = -T^q R^i{}_{q,kl} - T^p_{kl} (\nabla_p T^i),$$

and for symmetric connections

$$(\nabla_k \nabla_l - \nabla_l \nabla_k) T^i = -T^q R^i{}_{q,kl}.$$

It turns out that $R^i_{q,kl}$ is a (Riemannian) tensor. It is called the Riemannian curvature tensor. For symmetric connections $T^p_{kl} \equiv 0$, and, therefore, in the symmetric case, there holds

THEOREM 1. *For symmetric connections and for any vector field T the expression* $(\nabla_k \nabla_l - \nabla_l \nabla_k) T^i$ *takes the form* $-R^i{}_{q,kl} T^q$, *where* $R^i_{q,kl}$ *is a Riemannian tensor, and the following formula holds*

$$-R^i_{q,kl} = \frac{\partial \Gamma^i_{ql}}{\partial z^k} - \frac{\partial \Gamma^i_{qk}}{\partial z^l} + \Gamma^i_{pk} \Gamma^p_{ql} - \Gamma^i_{pl} \Gamma^p_{qk}.$$

If the connection is Euclidean, then the tensor $R^i_{q,kl}$ is identically zero; at points where $\Gamma^i_{pq} = 0$ we have

$$-R^i_{q,kl} = \frac{\partial \Gamma^i_{ql}}{\partial z^k} - \frac{\partial \Gamma^i_{qk}}{\partial z^i}.$$

What are the properties of the curvature tensor?

1) Obviously $R^i_{q,kl} = -R^i_{q,lk}$ in all cases;

2) Let $R_{iq,kl} = g_{ip}R^p_{q,kl}$, where g_{ip} is a Riemannian metric.

There following theorem holds

THEOREM 2. *If a connection is symmetric and compatible with a Riemannian metric g_{ij}, we have the symmetries*

$$R_{iq,kl} = -R_{qi,kl}, \quad R_{iq,kl} = R_{kl,iq},$$

$$R^i_{q,kl} + R^i_{l,qk} + R^i_{k,lq} = 0.$$

Given this, the Riemannian metric determines the curvature tensor by the formula

$$R_{iq,kl} = 1/2\left(\frac{\partial^2 g_{il}}{\partial z^q \partial z^k} + \frac{\partial^2 g_{qk}}{\partial z^i \partial z^l} - \frac{\partial^2 g_{ik}}{\partial z^q \partial z^l} - \frac{\partial^2 g_{ql}}{\partial z^i \partial z^k}\right) +$$

$$+ g_{mp}\left(\Gamma^m_{qk}\Gamma^p_{il} - \Gamma^m_{ql}\Gamma^p_{ik}\right).$$

The proof of Theorem 2 for symmetric connections follows from the formulae which express Γ^k_{iq} in terms of the metric (g_{ij}). We omit this calculation.

Theorems 1 and 2 imply

COROLLARY 1. *If a Riemannian tensor does not vanish, we cannot introduce Euclidean coordinates with respect to which $g_{ij} \equiv \delta_{ij}$ and $\Gamma^k_{ij} \equiv 0$.*

REMARK. We can arrive at this conclusion in a different way. Consider the tansformation law for components Γ^k_{ij}:

in the case $z = z(y)$, the following formula holds

$$\Gamma'^m_{np} = \frac{\partial y^m}{\partial z^\alpha}\left(\Gamma^\alpha_{qt}\frac{\partial z^q}{\partial y^n}\frac{\partial z^t}{\partial y^p} + \frac{\partial^2 z^\alpha}{\partial y^n \partial y^p}\right).$$

Suppose the connection is symmetric $\Gamma^\alpha_{qt} = \Gamma^\alpha_{tq}$. We shall seek (Euclidean coordinates) y, such that $\Gamma^m_{np} = 0$. We obtain the equation for $z^\alpha = z^\alpha(y^1, y^2, \dots, y^n)$

$$\frac{\partial^2 z^\alpha}{\partial y^n \partial y^p} = -\Gamma^\alpha_{qt} \frac{\partial z^q}{\partial y^n} \frac{\partial z^t}{\partial y^p}, \quad \Gamma^\alpha_{qt} = \Gamma^\alpha_{qt}(z).$$

Are these equations soluble? If they are, we have

$$\frac{\partial}{\partial y^\beta}\left(\frac{\partial^2 z^\alpha}{\partial y^n \partial y^p}\right) - \frac{\partial}{\partial y^n}\left(\frac{\partial^2 z^\alpha}{\partial y^\beta \partial y^p}\right) \equiv 0.$$

We can verify that this is the condition on the right-hand side of the equations. Its validity is equivalent to

$$R^i_{q,kl} \equiv 0.$$

The curvature tensor is a tensor of rank four. It is naturally obtained as an operator on vector fields which depends on the pair (k, l) in a skew-symmetric manner:

$$-R^i_{q,kl}T^q = R^i_{q,lk}T^q = (\nabla_k\nabla_l - \nabla_l\nabla_k)T^i - T^p{}_{kl}\frac{\partial T^i}{\partial z^p},$$

where $T^p_{kl} = \Gamma^p_{kl} - \Gamma^p_{lk}$ is the torsion tensor.

In the symmetric case $T^p_{kl} \equiv 0$. If the connection is symmetric and compatible with the metric g_{ij}, then the components Γ^k_l and $R_{q;kl}$ are expressed in terms of g_{ij} and their derivatives, and the symmetries hold

1) $R^i_{q,kl} = R^i_{q,lk}$,

2) $R_{iq,kl} = g_{im}R^m{}_{q,kl} = -R_{qi,kl}$,

3) $R_{iq,kl} = R_{kl,iq}$,

5) $R^i_{q,kl} + R^i{}_{l,qk} + R^i{}_{k,lq} = 0.$

Theorem 2 also implies an important

COROLLARY 2. *The curvature tensor $R_{iq,kl}$ determines the symmetric bi-linear form on the vector space whose elements are skew-symmetric tensors of rank two with upper indices.*

Elements of this space are customarily called *bi-vectors*. *Simpe bi-vectors* are those of the form

$$T^{ij} = -T^{ji} = \xi^i \eta^j - \eta^i \xi^j$$

(that is, and exterior product of two vectors: $T = \xi \wedge \eta$).

The scalar product of two bi-vectors $\overset{\prime}{T}$ and $\overset{\prime\prime}{T}$, by virtue of the Riemannian tensor, is given by the formula

$$\langle \overset{\prime}{T}, \overset{\prime\prime}{T} \rangle_R = \overset{\prime}{T}{}^{ij} \overset{\prime\prime}{T}{}^{kl} R_{ij,kl} = \langle \overset{\prime\prime}{T}, \overset{\prime}{T} \rangle_R .$$

A Riemannian tensor is called *strictly positive* (*strictly negative*) providing $\langle T, T \rangle_R$ is strictly positive (strictly negative) for a non-zero T.

The *curvature of two-dimensional direction* generated by tangent vectors ξ and η is the quantity

$$R(\xi, \eta) = \frac{\langle T, T \rangle_R}{\langle T, T \rangle},$$

where $T = \xi \wedge \eta$ and $\langle T, T \rangle$ is a usual scalar square.

Riemannian space is called the space of *positive* (*negative*) *sectional curvature* if the curvature of all its two-dimensional directions is positive (negative).

EXERCISE. Prove that the curvatures of all two-dimensional directions in a Euclidean space, in a sphere and in a Lobachevskian plane, are constant and are a zero, a positive and a negative constant (number). Prove that any Riemannian metric with this property is locally isometric to one of these three metrics.

How many components are there in a Riemannian tensor?
Let us consider the two-dimensional and three-dimensional cases.

I. *The two-dimensional case.* From the symmetry condition

$$R_{iq,kl} = -R_{iq,lk} = -R_{qi,kl} = R_{kl,iq}$$

it follows that the Riemannian tensor is determined by the single non-zero component $R_{12,12}$. All the other components are either obtained from it by permuting the indices or are equal to zero.

DEFINITION 1. The *Ricci tensor* is the expression $R_{ql} = R^i{}_{q,il}$ — the trace of the Riemannian tensor.

DEFINITION 2. The *scalar curvature* is the trace of the Ricci tensor:

$$R = g^{lq}R_{ql} = g^{lq}R^{i}_{q,il}.$$

An important theorem holds.

THEOREM 3. *For two-dimensional surfaces in a three-dimensional space, the scalar curvature R is twice the Gaussian curvature. It follows that, unlike the mean curvature of a surface, the Gaussian curvature is expressible in terms of the Riemannian metric of the surface itself, i.e. is an intrinsic invariant of the surface.*

Proof. Let the surface be parametrized by coordinates $x = x(u, v)$, $y = y(u, v)$, $z = z(u, v)$, where (x, y, z) are Euclidean coordinates of the space and $(u, v) = (z^1, z^2)$ are coordinates on the surface. We choose at a given point P, where the z-axis is normal to the surface, $u = z^1 = x$, $v = z^2 = y$. Then in a neighbourhood of the point P the surface is given by the equation

$$z = f(x, y), \quad \text{where } \operatorname{grad} f|_P = 0, \quad P = (0. 0).$$

For the components of the metric on the surface, we obtain

$$g_{ij} = \delta_{ij} + \frac{\partial f}{\partial z^i}\frac{\partial f}{\partial z^j}, \quad z^1 = x,\ z^2 = y.$$

In particular, at the point $P = (0, 0)$ all $\partial g_{ij}/\partial z^k = 0$. Hence, at this point, $\Gamma^{k}_{ji} = 0$. So, according to Theorem 1 of this section, we have at this point

$$-R^{i}_{q,kl} = \frac{\partial \Gamma^{i}_{ql}}{\partial z^k} - \frac{\partial \Gamma^{i}_{q,k}}{\partial z^l},$$

$$R_{iq,kl} = 1/2\left(\frac{\partial^2 g_{il}}{\partial z^q \partial z^k} + \frac{\partial^2 g_{qk}}{\partial z^i \partial z^l} - \frac{\partial^2 g_{ik}}{\partial z^q \partial z^l} - \frac{\partial^2 g_{ql}}{\partial z^i \partial z^k}\right)$$

(Theorem 2 of this section).

From this we obtain ($z^1 = x$, $x^2 = y$):

$$R_{12,12} = 1/2\left(\frac{\partial^2 g_{12}}{\partial x \partial y} + \frac{\partial^2 g_{12}}{\partial x \partial y} - \frac{\partial g_{11}}{\partial y^2} - \frac{\partial g_{22}}{\partial x^2}\right).$$

Given this, the following equalities hold

$$g_{11} = (f_x)^2 + 1, \quad g_{22} = (f_y)^2 + 1, \quad g_{12} = f_x f_y,$$

$$\frac{\partial^2 g_{11}}{\partial y^2}\Big|_P = 2(f_{xy})^2, \quad \frac{\partial^2 g_{22}}{\partial x^2}\Big|_P = 2(f_{xy})^2, \quad \frac{\partial^2 g_{12}}{\partial x \partial y}\Big|_P = f_{xx} f_{yy} + f_{xy}^2 .$$

Finally, we are led (at the point P) to

$$R_{12,12} = (f_{xx} f_{yy} + f_{xy}^2 - f_{xy}^2 - f_{yy}^2) =$$

$$= f_{xx} f_{yy} - f_{xy}^2 = \det \begin{pmatrix} f_{xx} & f_{xy} \\ f_{yx} & f_{yy} \end{pmatrix} = K.$$

By definition, $K = \det \begin{pmatrix} f_{xx} & f_{xy} \\ f_{yx} & f_{yy} \end{pmatrix}$ at the point P, where $\delta_{ij} = g_{ij}$ relative to the chosen coordinates. However, the Gaussian curvature K is a scalar and $R_{12,12}$ is a component of the tensor. They are equal only in the particular given coordinate system, where $g_{ij} = \delta_{ij}$, $\det (g_{ij}) = 1 = g$. From the definition of R, where $R = g^{ql} R^i{}_{q,il}$, we can readily see that

$$R = 2 \det (g^{ql}) R_{12,12} = \frac{2}{\det (g_{ij})} \cdot R_{12,12} = \frac{2}{g} R_{12,12} = R.$$

In our coordinate system $g = 1$ and $R_{12,12} = K$. Therefore, relative to this system the equality $R = 2K$ holds; since R and K are both scalars, this equality is always valid, and we have reached the desired conclusion.

REMARK. The conclusion concerning the Riemannian tensor components implies

$$\frac{R}{2} (g_{11} g_{22} - g_{12}^2) = R_{12,12} = \frac{R \cdot g}{2} K \cdot g, \quad g = \det (g_{ij}).$$

Hence, the Gaussian curvature K is invariant equal to $R/2$, where $R = g^{ql} R^i{}_{l,il}$ for $n = 2$.

We shall consider several examples:

1) The metric of the Euclidean plane $dl^2 = dx^2 + dy^2$;

$$R^i_{q,il} \equiv 0, \quad K = R/2 = 0;$$

2) The metric on the sphere $dl^2 = dr^2 + \sin^2 \frac{r}{r_0} d\phi^2$; here

$$K = \frac{R}{2} = \frac{1}{r_0^2} > 0$$

that is, we have constant positive curvature.

3) The metric in the Lobachevskian plane $dl^2 = dr^2 + \mathrm{sh}^2 \frac{r}{r_0} d\phi^2$,

$$K = \frac{R}{2} = -\frac{1}{r_0^2} < 0$$

and we have constant negative curvature.

4) The conformal Euclidean metric

$$dl^2 = g(x, y)\,(dx^2 + dy^2), \quad g > 0,$$

$$g_{ij} = \delta_{ij} \cdot g(x, y).$$

Such coordinates x, y are called *conformal* (see Section 1.13). In these coordinates, the Gaussian curvature is given by the simple formula

$$K = -\frac{1}{2g} \Delta (\ln g) = -\frac{1}{2g} \frac{\partial^2}{\partial x \partial \bar{z}} (\ln g);$$

$$\Delta = \frac{\partial^2}{\partial z \partial \bar{z}} = \frac{\partial^2}{\partial x^2} + \frac{\partial^2}{\partial y^2};$$

$$\frac{\partial}{\partial z} = 1/2 \left(\frac{\partial}{\partial x} - i\frac{\partial}{\partial y}\right), \quad \frac{\partial}{\partial \bar{z}} = 1/2 \left(\frac{\partial}{\partial x} + i\frac{\partial}{\partial y}\right).$$

We invite the reader to derive this formula. In Part I we have shown the visual meaning of curvature when it is positive or negative.

We shall present some more fundamental facts from the theory of surfaces which we have not proved here.

1) If on a surface with coordinates z^1, z^2 we are given a closed curve $z^i(t)$, $i = 1, 2$, which bounds a region U, then the following theorem holds:

$$\Delta\phi = \iint_U K\,(\bar{g})^{1/2} dz^1 \wedge dz^2 = \text{ the angle of rotation of the vector under parallel}$$

enclosure along the curve $z^i(t)$, $i = 1, 2$.

2) If this curve consists of three geodesic arcs, and if the curvature of the surface is constant, then

$$\Delta\phi = \iint_U K\,(\bar{g})^{1/2} dz^1 \wedge dz^2 = K\sigma,$$

where σ is the area of U. We invite the reader to derive, from this, the following relations:

a) The sum of the angles of a geodesic traingle is eqaul to $\pi + K\sigma < \pi$, where σ is the area of the triangle, K is the curvature, $K = \text{const.}$ (a Lobachevskian plane).

b) For the sphere, the sum of the angles of a geodesic triangle is equal to $\pi + K\sigma > \pi$, where σ is the area of the triangle (sphere).

II. *The three-dimensional case.* Here the situation is more complicated. Let us consider a metric g_{ij} in a neighbourhood of a point P, and construct a tensor $R_{iq.kl} = q_{ip} R^p_{q,kl}$. The Riemannian tensor

$$R_{iq.kl} = -R_{qi,kl}\,; \quad R_{iq,lk} = R_{kl,iq}$$

at each point may, by virtue of these symmetry relations, be regarded as a quadratic form on the three-dimensional linear space of all skew-symmetric rank-two tensors.

If we denote the pair $[k, l] = -[l, k]$ by A, B, then

$$R_{[iq]\,[kl]} = R_{AB} = R_{BA}.$$

The Riemannian tensor is thus determined by the six numbers. Consider the Ricci tensor $R^i_{q,il} = R_{ql} = R_{lq}$. This is a symmetric tensor of rank two which is also determined by six numbers R_{ql} with $q \geq l$. The scalar curvature R is one number $R = g^{ql} R_{ql} = g^{ql} R^i_{q,il}$. In contrast to the two-dimensional case, the scalar R does not determine the whole of the tensor $R^i_{q,kl}$. However, in the three-dimensional case it suffices to know the Ricci tensor since the following formula holds

$$R_{\alpha\beta\gamma\delta} = R_{\alpha\gamma} g_{\beta\delta} - R_{\beta\delta} g_{\alpha\gamma} - R_{\beta\gamma} g_{\alpha\delta} + R/2(g_{\alpha\delta} g_{\beta\gamma} - g_{\alpha\gamma} g_{\beta\delta})$$

(prove this!).

The scalar curvature is the trace of the Ricci tensor $\mathrm{Sp}(R_{ql}) = g^{ql}R_{ql}$. There also exist invariants $\lambda_1, \lambda_2, \lambda_3$ which are the eigenvalues of the Ricci tensor. These invariants are given by the equation

$$\det (R_{ql} - \lambda g_{ql}) = 0,$$

and we have $\lambda_1 + \lambda_2 + \lambda_3 = R$.

When we say "a space of positive or negative curvature", we mean that the Riemannian tensor R_{AB} is a positive definite quadratic form on skew-symmetric rank-two tensors.

III. *The four-dimensional case.* The Riemann tensor is not determined here by the Ricci tensor which, nonetheless, remains to be of great importance. For example, in four-dimensional space-time the gravitational field is taken to be the metric (g_{ij}), i, $j = 0, 1, 2, 3$, and all the other properties of matter are thought of as involved in the "energy-momentum tensor" λT_{ij}, where λ is a dimensional constant.

The Einstein equations for the metric of space-time have the form

$$R_{ij} - 1/2Rg_{ij} = \lambda T_{ij}; \quad \Delta_j T^j{}_i = 0.$$

The same equations in the absence of matter become

$$R_{ij} - 1/2Rg_{ij} = 0$$

(or $R_{ij} = 0$ — check it!). The determinant of the metric g_{ij} is not equal to zero, but the metric is indefinitie (in the diagonal form it has three minuses and one plus).

2.10 Skew-Symmetric Tensors and the Theory of Integration

The reader is already acquainted with the usual concept of multiple integral: given a region U and a function $f(z^1, z^2, \dots, z^n)$ in an n-dimensional space, we have the definition of the integral of the function over the region
If, in addition, we are given the coordinate change

$$\iint_U \dots \int f(z)\, dz^1 \dots dz^n \equiv \iint_U \dots \int f\, dz^1 \wedge \dots \wedge dz^n.$$

$$z = z(y),$$

then the formula for the change of variables is
We have already mentioned that the integrand is, in fact, a skew-symmetric

$$\iint_U \dots \int f(z)\, dz^1 \dots dz^n \equiv \iint_V \dots \int f(z(y))\, dy^1 \dots dy^n J,$$

where $J = \det \left(\frac{\partial z^i}{\partial y^j}\right)$ is the Jacobian.

tensor of rank n. In the coordinate system $z^1, \dots, z^n$ the component $T_{1 \dots n}$ of this tensor is, by definition, $f(z) = T_{1 \dots n}$.

Recall that under the change of coordinates $z = z(y)$ we have proved the formula $T'_{1 \dots n} = JT_{1 \dots n}$, where
for skew-symmetric rank-n tensors in an n-dimensional space.

$$J = \det \left(\frac{\partial z^i}{\partial y^j}\right)$$

The formula for the change of variables for a multiple integral, as is well-known from analysis, has the form
where V is the same region as U but written in the coordinates $y^1, \dots, y^n$.

$$\iint_U \dots \int f(z)\, dz^1 \dots dz^n \equiv \iint_V \dots \int f(z(y))\, dy^1 \dots dy^n J,$$

We can see that $T'_{1 \dots n} = f(z) \cdot J = T_{1 \dots n} \cdot J$. Thus the integrand is a skew-symmetric tensor.

EXAMPLE 1. Given a Riemannian metric (g_{ij}), the determinant $g = \det(g_{ij})$, under the coordinate changes $z = z(y)$, behaves as

$$g' = \det(g'_{ij}) = \det\left(g_{kl}\frac{\partial z^k}{\partial y^i}\frac{\partial z^l}{\partial y^j}\right) = J^2 \cdot g.$$

Therefore, under coordinate changes with positive Jacobian, the expression $(|g|)^{1/2}$ behaves as a skew-symmetric tensor. Recall that the area of a region on the surface is

$$\sigma(U) = \iint_U (|g|)^{1/2}\, du\, dv, \quad u = z^1, \quad v = z^2, \quad n = 2.$$

Suppose we are given a surface $x^i = x^i(z)$, $i = 1, 2, 3$, $z = (z^1, z^2)$ in a space with Euclidean coordinates(x^1, x^2, x^3). If we now wish to take the surface integral of some scalar functions $f(x(z))$ whose definition is essentially related to this surface (let it be, say, its Gaussian curvature), this integral will be defined as follows: the integral over a region U on the surface is equal to

$$\iint_U f(x(z))\, (|g|)^{1/2} dz^1 \wedge dz^2,$$

where $\iint_U \phi(z)\, dz^1 dz^2$ is the usual multiple integral. The expression $(|g|)^{1/2} dz^1 dz^2$ is sometimes called the *measure* (the *element of volume*) *on the surface* and denoted by $d\sigma$.

EXAMPLE 2. Let K be a Gaussian curvature and let $\iint_U K\, (|g|)^{1/2} dz^1 dz^2$ be its integral over a region on a surface; the surface is given by $x^i = x^i(z)$. The region U in the coordinates(z^1, z^2) is bounded to a closed curve $\Gamma = \{z^j = z^j(t), j = 1, 2, z^j(0) = z^j(1)$ (Figure 46).

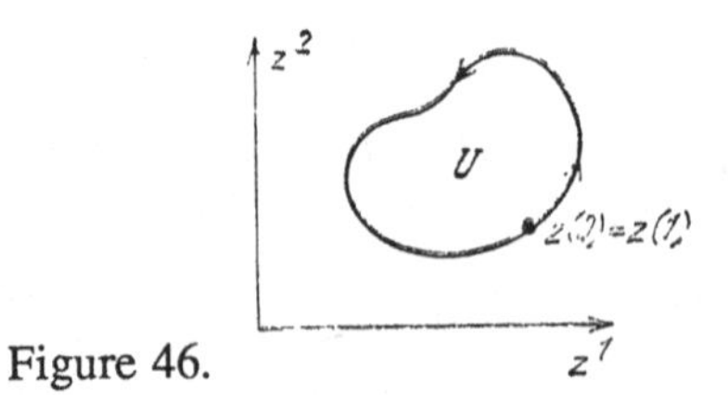

Figure 46.

The value of the expression $\iint K\, d\sigma$ is given by the following theorem (without proof).

THEOREM 1. *The angle of rotation of a vector under parallel enclosure along a closed curve* Γ *from the beginning* $t = 0$ *to the end* $t = 1$ *is calculated by the formula*

$$\Delta\phi = \iint_U K\,(|g|)^{1/2} dz^1 dz^2$$

(the connection is symmetric and compatible with the metric).

So, we arrive at the following conclusions:

1) In an n-dimensional space for any bounded region U the integral

$$\iint \underset{U}{\cdots} \int T,$$

is defined, where T is a skew-symmetric rank-n tensor of type $(0, n)$, $T = (T_{i_1 \dots i_n})$.

2) In coordinate notation this tensor is written as

$$T = T_{1, \dots, n}\, dz^1 \wedge \dots \wedge dz^n$$

(or $T_{1, \dots, n}\, dz^1 \dots dz^n$ if we omit the sign $\wedge$).

3) If $T_{1, \dots, n} = f(z)$ is a scalar function of a point, then under the change of coordinates $z = z(y)$ we have

$$\iint \underset{U}{\cdots} \int f(z)\, dz^1 \wedge \dots \wedge dz^n = \iint \underset{U}{\cdots} \int f(z(y))\, J\, dy^1 \wedge \dots \wedge dy^n;$$

4) If we wish to integrate a function $\phi(z)$ over the space, it is necessary that we have a given and marked skew-symmetric tensor T in the space (such a tensor is called the volume element or the measure); then, by definition, the integral of the function $\phi(z)$ is the integral of the tensor $\phi(z)T$

$$\iint \underset{U}{\cdots} \int \phi(z)\, T = \iint \underset{U}{\cdots} \int \phi(z)\, T_{1, \dots, n}\, dz^1 \wedge \dots \wedge dz^n;$$

5) Given a Riemannian metric (g_{ij}), such a marked skew-symmetric tensor $T = T_{1, \dots, n}\, dz^1 \wedge \dots \wedge dz^n$ (under changes with a positive Jacobian) can be represented in the form

$$T = d\sigma = (|g|)^{1/2}\, dz^1 \wedge \ldots \wedge dz^n;$$

and the integral of the function $\phi(z)$ is given by

$$\iint_U \ldots \int \phi(z)\, T = \iint_U \ldots \int \phi(z)\, (|g|)^{1/2}\, dz^1 \wedge \ldots \wedge dz^n;$$

6) The sign $\wedge$ implies that $dz^i \wedge dz^j = -\, dz^j \wedge dz^i$; under the changes of variables $z^i = z^i(y)$; if $dz^i = \dfrac{\partial z^i}{\partial y^j}\, dy^j$, then from the equality $dy^i \wedge dy^j = -\, dy^j \wedge dy^i$ we obtain

$$dz^{i_1} \wedge \ldots dz^{i_k} = \sum_{j_1<\ldots<j_k} J^{i_1 \ldots i_k}_{j_1 \ldots j_k}\, dy^{i_1} \wedge \ldots \wedge dy^{i_k},$$

where $J^{(i)}_{(j)}$ is the minor of the Jacobian matrix $\dfrac{\partial z^i}{\partial y^j}$; in particular for $k = n$ we have

$$dz^1 \wedge \ldots \wedge dz^n = J\, dy^1 \wedge \ldots \wedge dy^n,$$

where J is the Jacobian;

7) In Euclidean coordinates we usually have $(|g|)^{1/2} \equiv 1$, and therefore $d = dx^1 \wedge \ldots \wedge dx^n$.

We should distinguish between two expressions:

a) "the integral of a skew-symmetric rank-n tensor over a region". This integral always has sense. We call it "the integral of the second kind";

b) "the integral (of the first kind) of a function over a region": for this integral we should know over which volume element (measure) the integration is carried out — we should first multiply the function by this volume element (which is a skew-symmetric tensor) and then integrate. Obviously, this integral is reduced to the first one.

Now let us turn to arbitrary skew-symmetric rank-k tensors of type $(0, k)$ in an n-dimensional space. To begin with, we choose convenient symbols for co-vectors $(k = 1)$ and vectors. We have already mentioned the convenient basis in the tensor space provided we are given coordinates $z^1, z^2, z^3, \ldots, z^n$.

The basis fields for vectors are $e_1, \ldots, e_n$, $T = T^i e_i$, $e_i = \delta^j_i e_j$ (the components e_i are equal to δ^j_i).

The basis fields for covectors are $e^1, \dots, e^n$, $T = T_i e^i$, $e^i = \delta^i{}_j e^j$ (the compoonents e^i are equal to δ^i_j).

The basis fields for tensors of any rank are $e_{i_1} \circ \dots \circ e_{i_k} \circ e^{j_1} \circ \dots \circ e^{j_l}$,

$$T = T^{i_1 \dots i_k}_{j_1 \dots j_k} \; e_{i_1} \circ \dots \circ e_{i_k} \circ e^{j_1} \circ \dots \circ e^{j_l}.$$

Let us recall what we understood by the term "vector".

In each coordinate system the vector at a given point is given by a set of numbers T^i, i.e. by the set of its components in the basis e^i:

$$T = T^i e_i.$$

Under the change $z = z(y)$, this set of numbers varies by the law

$$T'^j = T^i \frac{\partial y^j}{\partial z^i} \text{ or } T^i = T^i = T'^j \frac{\partial z^i}{\partial y^j}.$$

In the case where $e'_1, \dots, e'_n$ is the basis in the system (y), we have

$$T' = (T'^j) = T'^j e'_j, \quad \text{where } e_i = e'_q \frac{\partial y^q}{\partial z^i}.$$

Indeed

$$T^i e_i = T'^k \frac{\partial z^i}{\partial y^k} e_i = T'^k \frac{\partial z^i}{\partial y^k} \left(e'_q \frac{\partial y^q}{\partial z^i}\right) = T'^k \delta^q_k e'_q = T'^k e'_k.$$

We can see that $T^i e_i = T'^k e'_k$ is the same vector provided that

$$e_i = e'_q \frac{\partial y^q}{\partial z^i}.$$

It should be noted that differentiating any function $f(z)$, we had, under the coordinate change $z = z(y)$, the equation

$$\frac{\partial f}{\partial z^q} = \frac{\partial f}{\partial y^\alpha} \frac{\partial y^\alpha}{\partial z^q}.$$

In other words, the operators $\frac{\partial}{\partial z^q}$ satisfied the relation

$$\frac{\partial}{\partial z^q} = \frac{\partial y^\alpha}{\partial z^q}\frac{\partial}{\partial y^\alpha}.$$

Let us associate the basis field e_i in the system (z) with the operator $\frac{\partial}{\partial z^i}$ and the field e'_i in the system (y) with the operator $\frac{\partial}{\partial y^j}$.

We obtain two relations

$$e_i = \frac{\partial y^q}{\partial z^i} e'_q,$$

$$\frac{\partial}{\partial z^i} = \frac{\partial y^\alpha}{\partial z^i}\frac{\partial}{\partial y^\alpha}.$$

This is one and the same transformation law! Hence, the operators $\partial/\partial z^i$ are, in fact, the basis vector fields e_i.

Recall that for any vector ξ the operator $\nabla_\xi = \xi^\alpha \frac{\partial}{\partial z^\alpha}$ (on scalar functions) is called the *directional derivative* of this vector.

We have the assignment $\partial/\partial z^i \leftrightarrow e_i$, $\nabla_\xi \leftrightarrow \xi^\alpha e_\alpha = \xi$. Vectors are often said to be differential operators of the form $\xi^\alpha \frac{\partial}{\partial z^\alpha}$ on scalar functions.

Now we shall return to the covectors of interest: $T = (T_i) = T_i e^i$. What did we begin with when we introduced the concept of a covector? The components of the gradient of a function $T_i = \frac{\partial f}{\partial z^i}$ do not form a vector, they form a covector. We know from analysis that the differential of a function is the expression

$$df = \frac{\partial f}{\partial z^\alpha} dz^\alpha.$$

Given the change $z^\alpha = z^\alpha(y)$, we have $dz^\alpha = \frac{\partial z^\alpha}{\partial y^j} dy^j$

$$df = \frac{\partial f}{\partial z^\alpha} dz^\alpha = \left(\frac{\partial f}{\partial z^\alpha} \frac{\partial z^\alpha}{\partial y^j}\right) dy^j = \frac{\partial f}{\partial y^j} dy^j.$$

What is the meaning of the symbols dx^α and dy^j? The basis covectors e^i, which we introduced before, were transformed by the law

$$e^\alpha = \frac{\partial z^\alpha}{\partial y^j} e'^j , \quad T_\alpha e^\alpha = T'_j e'^j,$$

where

$$T'_j = T_\alpha \frac{\partial z^\alpha}{\partial y^j}, \quad z^\alpha = z^\alpha(y).$$

We can see, again, that the basis covectors e^α transform by the same law as dz^α:

$$e^\alpha = \frac{\partial z^\alpha}{\partial y^j} e'^j \leftrightarrow dz^\alpha = \frac{\partial z^\alpha}{\partial y^j} dy^j ,$$

$$e^\alpha \leftrightarrow dz^\alpha, e' \leftrightarrow dy^j.$$

So, we can say that the symbols dx^α are the basis covectors e^α. The covector $(T_\alpha) = T$ can be written in two ways

$$(T_\alpha) = T_\alpha e^\alpha = T_\alpha dz^\alpha.$$

The covectors e^α or dz^α are determined from the identities $e^\alpha e_j = \delta^\alpha_j = (dx^\alpha, e_j)$. To say it in a different way, $dz^\alpha = e^\alpha$ are basis linear forms on vectors; any covector is a linear form (on vectors) expanded in the basis $e^\alpha = dz^\alpha$. Under the change $z = z(y)$ we simply go over to another basis e'_α and $e'_\alpha = dy^\alpha$ at each point of the space.

REMARK The value of the linear form $\frac{\partial f}{\partial z^\alpha} dz^\alpha$ on the vector $(\Delta\xi) = (\Delta z)^\alpha e_\alpha$ is equal, by definition, to $\left(\frac{\partial f}{\partial z^\alpha} dz^\alpha, \Delta\xi\right) = \frac{\partial f}{\partial z^\alpha} (\Delta z^\alpha)$. As is well-known, this is called the principal linear part of the increment of the function f due to the shift along the vector.

By virtue of the correspondence between the covectors e^α and the "differentials" dz^α, the covector fields $(T_\alpha) = T_\alpha e^\alpha = T_\alpha dx^\alpha$ are called "differential forms" of the first degree (first-rank tensors of type (0, 1)).

Why did the term "differential form" appear? It turns out that the expression $T_\alpha dx^\alpha$ can be integrated along any curve:

$$z^i = z^i(t), \quad a \le t \le b.$$

Indeed, let us consider the expression

$$\int_a^b T_\alpha \frac{dz^\alpha}{dt} dt = \int_a^b T_\alpha \xi^\alpha dt,$$

where $\xi^\alpha = \dot{z}^\alpha$ is the velocity vector.

This expression is termed the integral of a differential form along an arc (called in analysis "the integral of the second kind").

The differential form $T = T_\alpha dz^\alpha$ is given in a space, and it can be integrated along any arc in this space.

The situation is different when we are given a curve (arc) $z^i = z^i(t)$ and a certain scalar function $f(z(t))$, for example its torsion curvature, which is essentially related to this particular curve. Then we introduce the measure on the curve — the element of its length $dl = |\dot{z}|\, dt$, and the integral of the function $f(z(t))$ along the curve is the expression $\int_a^b f(z(t))\, dl$ (in analysis this is the integral of the first kind). The element of length on the curve $dl = |\dot{z}|\, dt$ is, in fact, the one-dimensional version of the general "volume element" $d\sigma = (|g|)^{1/2} dz^1 \wedge \ldots \wedge dz^n$ already introduced above, since, for $n = 1$, we have

$$|g| = |g_n| \quad \text{and} \quad (|g_{11}|)^{1/2} d = dl,$$

where $g_{11} = |\dot{z}|^2$ (for $n = 1$, where t is the only coordinate on the straight line).

Concerning the integral of the second kind, that is, the integral of a covector field (of a differential form) along any arc, it posseses the following properties:

1) it does not depend on the choice of the curve on which the parameter is introduced:

$$\int_a^b T_\alpha \frac{dz^\alpha}{dt} dt = \int_{a'}^{b'} T_\alpha \frac{dz^\alpha}{d\tau} d\tau,$$

where $t = t(\tau)$ and τ varies from a' to b' with t varying from a to b;

2) The result of the integration does not depend on the coordinates in space; in the case $z = z(y)$, $T'_i = T_\alpha \dfrac{\partial x^\alpha}{\partial y^i}$ and $z^i(t) = z^i(y(t))$, the following equality holds

$$\int_a^b T_\alpha \frac{dz^\alpha}{dt}\, dt = \int_a^b T'_\beta \frac{dy^\beta}{dt}\, dt .$$

Indeed, $T_\alpha\, dz^\alpha \equiv T_\beta\, dy^\beta$, and therefore both the integrals coincide along one and the same curve $z^\alpha = z^\alpha(t)$ or $y^\beta = y^\beta(t)$, where $z(t) = z(y(t))$.

So, we already have the integration of a skew-symmetric rank-n tensor over an n-dimensional region in an n-dimensional space and the integration of a covector field along any curve.

It turns out that skew-symmetric rank-k tensors of type $(0, k)$ are integrable over k-dimensional surfaces in an n-dimensional space. Suppose a k-dimensional surface is given parametrically as

$$x^i = x^i(z^1, \dots, z^k), \quad i = 1, \dots, n.$$

Suppose, also, that we are given a region U in a k-dimensional space with coordinates $z^1, \dots, z^k$.

How should we introduce the integral of a skew-symmetric tensor $T = (T_{i_1 \dots i_k})$ in n-space with coordinates $x^1, \dots, x^n$ over a region U in k-space with coordinates $z^1, \dots, z^k$, if we are given an embedding (a surface) $x^i = x^i(z^1, \dots, z^k)$?

To begin with, for the sake of convenience, we shall use the language of differential forms for skew-symmetric tensors. We have already introduced the symbols $dx^1, \dots, dx^n$ and $dz^1, \dots, dz^n$ which are basis covectors in corresponding coordinates. Covector fields (T_i) are written in terms of $T_i\, dx^i$.

We shall associate with a skew-symmetric tensor $(T_{i_1 \dots i_k})$ in any coordinates $(x^1, \dots, x^n)$ the formal expression

$$(T_{i_1 \dots i_k}) = T_{i_1 \dots i_k} e^{i_1} \circ \dots \circ e^{i_k} \leftrightarrow \sum_{i_1 < \dots < i_k} T_{i_1 \dots i_k} dx^{i_1} \wedge \dots \wedge dx^{i_k},$$

and assume, by definition, that $dx^i \wedge dx^j = -dx^j \wedge dx^i$. We shall define the operation of restriction of a tensor of type $(0, k)$ to the surface. For a surface $x^i = x^i(z^1, \dots, z^k)$ consider the expression (on the surface)

$$\sum_{i_1<\ldots<i_k} T_{i_1\cdots i_k} dx^{i_1} \wedge \ldots \wedge dx^{i_k},$$

where $T_{i_1\cdots i_k}(x(z))$ is expressed in terms of $z^1, \ldots, z^k$ at points of the surface and

$dx^j = \frac{\partial x^j}{\partial z^k} dz^k$, by definition. At points of the surface $x = x(z)$ we have

$$\sum_{i_1<\ldots<i_k} T_{i_1\cdots i_k} dx^{i_1} \wedge \ldots \wedge dx^{i_k} =$$

$$= \sum_{i_1<\ldots<i_k} I^{i_1 \cdots i_k}_{12 \ldots k} T_{i_1\cdots i_k} dz^1 \wedge \ldots \wedge dz^k,$$

$J^{i_1 \cdots i_k}_{1 \ldots k}$ being the minor of the matrix $(\frac{\partial x^i}{\partial z^j})$. This expression is called the *restriction of the skew-symmetric tensor* $T_{i_1\cdots i_k}$ to the surface $x = x(z)$. This is already a tensor of rank k in a k-dimensional space.

DEFINITION 1. The usual multiple integral of the restriction

$$\int \underset{U}{\cdots} \int \Big(\sum_{i_1<\ldots<i_k} T_{i_1\ldots i_k} J^{i_1\ldots i_k}_{1\ldots k} \Big) dz^1 \wedge \ldots \wedge dz^k$$

of the tensor $T_{i_1\cdots i_k}$ over a region U on the surface is called the *integral of the skew-symmetric tensor field* $T_{i_1\cdots i_k}$, in an n-dimensional space, over the region U on any surface

$$x^i = x^i(z^1, \ldots, z^k), \quad i = 1, \ldots, n.$$

REMARK The expression

$$\sum_{i_1<\ldots<i_k} T_{i_1\ldots i_k} dx^{i_1} \wedge \ldots \wedge dx^{i_1}$$

itself is called the *differential form of degree k*; we are already familiar with one of the *forms of writing skew-symmetric tensors of type*$(0, k)$.

The integral of the form (of the tensor) over a region on the surface possesses, as before, two properties:

1. The integral does not change under the change of variables on the surface $z^q = z^q(z')$, $q = 1, \ldots, k$. Indeed, the restriction

$$\Big(\sum_{i_1<\ldots<i_k} T_{i_1\ldots i_k} J^{i_1\ldots i_k}_{1\ldots k}\Big)\, dz^1 \wedge \ldots \wedge dz^k$$

is a tensor of rank k in a k-dimensional space $z^1, \ldots, z^k$; under the change of variables $z^q = z^q(z'^1, \ldots, z'^k)$ we have the usual change of variables in the multiple integral over a region U in a k-dimensional space.

2. The integral does not change under the change of coordinates $x^i = x^i(x'^1, \ldots, x'^n)$ in the n-dimensional space itself, where $x^i = x^i(z^1, \ldots, z^k)$ and $x^i(z^1, \ldots, z^k) \equiv x^i(x'^1(z), \ldots, x'^n(z))$. This result is immediate, as it is also for $k = 1$, from the fact that under the change $x = x(x')$ there holds the identity

$$\sum_{i_1<\ldots<i_k} T_{i_1\ldots i_k}\, dx^{i_1} \wedge \ldots \wedge dx^{i_k} \equiv \sum_{i_1<\ldots<i_k} T'_{j_1\ldots j_k}\, dx'^{j_1} \wedge \ldots \wedge dx'^{j_k},$$

where $dx^i = \dfrac{\partial x^i}{\partial x'^j}\, dx'^j$, and the components $T'_{j_1\ldots j_k}$ are obtained form $T_{i_1\ldots i_k}$ by the usual tensor law.

In a space of any dimension n, the differential form of degree 0 is simply the scalar function $T(z)$. By a zero-measure oriented region U we understand a tuple of points $\{p_\alpha\}$ with the assigned signs $\sigma(p_\alpha) = \pm 1$.

By definition, "the integral of a zero-form T over a zero-dimensional region U" is the quantity

$$\int_U T = \sum_{(\alpha)} T(p_\alpha)\, \sigma(p_\alpha).$$

In a two-dimensional space we can integrate covector fields along curves Γ and also skew-symmetric tensors of rank two over regions U on the plane.

In the three-dimensional case, we can integrate

a) covector fields (forms of degree 1) $T_\alpha\, dx^\alpha$ along curves,

b) tensor fields of type (0, 2) (forms of degree 2) over surfaces,

c) tensor fields of type (0, 3) (forms of degree 3 or an expression of the form $f(x)\,dx^1 \wedge dx^2 \wedge dx^3$, where $f(x)$ is a function) over regions. Forms of degree 2 in a three-dimensional space with coordinates x^1, x^2, x^3 are given by

$$f_{12}\,dx^1 \wedge dx^2 + f_{13}\,dx^1 \wedge dx^3 + f_{23}\,dx^2 \wedge dx^3,$$

where f_{ij} are the components of the skew-symmetric tensor $(f_{ij}) = \sum_{i<j} f_{ij}\,dx^i \wedge dx^j$. The tensor f_{ij} in Euclidean coordinates is often associated with the "vector" $T^1 = f_{23}$, $T^2 = -f_{13}$, $T^3 = f_{12}$. As a concluding remark, we shall explain the assignment of the symbols $dx^{i_1} \wedge \ldots \wedge dz^{i_q}$ to the tensor basis $e_{\alpha_1} \circ \ldots \circ e_{\alpha_p} \circ e^{\beta_1} \circ \ldots e^{\beta_s}$. We know that the symbols e^i correspond to dz^i. What does the symbol $dz^i \wedge dz^j$ correspond to? We have two basis tensors e^ie^j and e^je^i. The expression $dz^i \wedge dz^i = -dz^j \wedge dz^i$ is skew-symmetric with respect to (i, j). There exists a skew-symmetric expression $(e^ie^j - e^je^i)$ composed of basis tensors. It is only to this expression that we assign the symbol $dz^i \wedge dz^j$:

$$dz^i \wedge dz^j \leftrightarrow (e^ie^j - e^je^i).$$

We shall verify this. The skew-symmetric tensor can be written in two ways:

$$T = (T_{ij}) = \sum_{i<j} T_{ij}\,dz^i \wedge dz^j = \sum_{i,j} T_{ij}\,e^ie^j = T_{ij}e^iT^j \qquad \text{(all pairs } i, j\text{)}.$$

For example: $T_{12}\,dz^1 \wedge dz^2 = T_{12}e^1e^2 + T_{21}e^2e^1$, where $T_{12} = -T_{21}$, whence

$$T_{12}\,dz^1 \wedge dz^2 = T_{12}e^1e^2 - T_{12}e^2e^1 = T_{12}(e^1e^2 - e^2e^1).$$

This is valid for any pair (i, j), where $i \neq j$.

Thus, we obtain that $dz^i \wedge dz^j = e^ie^j - e^je^i$. Similarly, for tensors of any rank, we have

$$dz^{i_1} \wedge \ldots \wedge dz^{i_q} = \sum_{j_1,\ldots,j_q} (-1)^{\alpha(j_1,\ldots,j_q)} e^{j_1} \circ \ldots \circ e^{j_q},$$

where $(j_1, \ldots, j_q)$ is permutation of the set of indices $(i_1, \ldots, i_q)$ and $\alpha(j_1, \ldots, j_q)$ is the sign of the permutation. For example

$$dz^1 \wedge dz^2 \wedge dz^3 = e^1e^2e^3 + e^3e^1e^2 + e^2e^3e^1 - e^2e^1e^3 - e^3e^2e^1 - e^1e^3e^2.$$

We have introduced, above, the following symbolism for writing skew-symmetric tensors of type (0, k):

$$T = (T_{i_1 \dots i_k}) \leftrightarrow \sum_{i_1<\dots<i_k} T_{i_1 \dots i_k} dx^{i_1} \wedge \dots \wedge dx^{i_k},$$

where $dx^i \wedge dx^j = - dx^j \wedge dx^i$.

We have also introduced the operation of restriction of a skew-symmetric tensor (of a differential form) to a q-dimensional surface:

$$x^i = x^i(z^1, \dots, z^q),$$

$$T = \sum_{i_1<\dots<i_k} T_{i_1 \dots i_k} dx^{i_1} \wedge \dots \wedge dx^{i_k} \mapsto \tilde{T} =$$

$$= \sum_{i_1<\dots<i_k} T_{i_1 \dots i_k} (x(z))\, dz^{i_1} \wedge \dots \wedge dz^{i_k},$$

where $dx^i = \dfrac{\partial x^i}{\partial z^j} dz^j$ and $dz^j \wedge dz^p = - dz^p \wedge dz^j$. This operation was defined and needed, only for $q = k$ in the theory of integration where the tensor $\tilde{T}$ is the k-th rank tensor in a k-dimensional space with coordinates $z^1, \dots, z^k$ (however, the restriction operation itself is also meaningful for $q \neq k$).

DEFINITION 2. The integral (of the second kind) of a skew-symmetric rank-k tensor T of type (0, k) over any k-dimensional surface $x = x^i(z^1, \dots, z^k)$ or over a region U on this surface is a usual multiple integral in the k-dimensional space $(z^1, \dots, z^k)$ of the restriction $\tilde{T}$ of this tensor T to the surface with the coordinates $x^i = x^i(z^1, \dots, z^k)$.

The basic invariance properties of this integral are as follows:

1) the integral is independent of the coordinates on the surface: its value remains invariant under the change $z = z(z')$;

2) the integral is independent of the coordinates in space: its value is invariant under the change $x = x(x')$.

This is the integral of a tensor (of a differential form), given in the entire space, over a region on any surface which depends neither on the coordinates in the space nor on the coordinates on this surface.

In Euclidean space on a surface $x^i = x^i(z^1, \ldots, z^k)$, $i = 1, \ldots, n$, the Riemannian metric is defined as follows: if $(x^1, \ldots, x^n)$ are Euclidean coordinates, then $g_{ij}\, dz^i dz^j = \sum_{q=1}^{n} (dx^q)^2$, g_{ij} being equal to g_{ji} and $dz^i dz^j = dz^j dz^i$,

$$dx^q = \frac{\partial x^q}{\partial z^\alpha} dz^\alpha.$$

The volume element on the surface is given, as always, by

$$d\sigma = (|g|)^{1/2} dz^1 \wedge \ldots \wedge dz^n, \quad g = \det(g_{ij}).$$

Let an arbitrary function $f(z^1, \ldots, z^n)$ be given on a surface.

DEFINITION 3. The integral (of the first kind) of a function $f(z^1, \ldots, z^n)$ over a surface is the integral of the function over the element $d\sigma$ of volume on the surface:

$$\text{the integral of the first kind} = \int \underset{U}{\cdots} \int f(z^1, \ldots, z^n)\, (|g|)^{1/2} dz^1 \wedge \ldots \wedge dz^k,$$

(on the surface).

It is important to note that the integral of the second kind is not related to the Riemannian geometry in space or on the surface, whereas the integral of the first kind is related to it through the volume element $(|g|)^{1/2} dz^1 \wedge \ldots \wedge dz^n$, which is a skew-symmetric rank-k tensor (under changes with a positive Jacobian) defined only on the given surface by the Riemannian metric. The Riemannian metric itself is determined by the Euclidean metric in the entire space.

EXAMPLE 0. A trivial example of the "integral" of a zero-rank tensor (a scalar $f(x)$) over a surface of dimension 0 (over a point P) is, by definition, the value of the function $f(x)$ at this point P: "the integral" is equal to $f(P)$. We have made this trivial remark intentionally — it will be useful when we come to discuss the general Stokes formula.

EXAMPLE 1. We have already discussed the integral of a covector field or of a differential form of degree 1 $(T_\alpha) = T_\alpha dx^\alpha$ along a curve $x^i = x^i(t)$, $a \le t \le b$. The

integral along a curve (of the first kind) $= \int_a^b T_\alpha \frac{dx^\alpha}{dt} dt$.

EXAMPLE 2. The integral of a tensor field $(T_{ij}) = \sum_{i<j} T_{ij}\, dx^i \wedge dx^j$ (i.e. of a differential form of degree 2) over a surface $x^i = x^i(z^1, z^2)$, $i = 1, \dots, n$, is given by

$$\text{the integral over the surface} = \iint_U \sum_{i<j} T_{ij}\, dx^i \wedge dx^j,$$

where $T_{ij} = T_{ji}\,(x(z))$ and $dx^i = \frac{\partial x^i}{\partial z^j} dz^j$, $dz^i \wedge dz^j = -dz^j \wedge dz^i$.

In a three-dimensional space ($n = 3$) with Euclidean coordinates $x^1, x^2 x^3$, where $(dl)^2 = \sum (dx^i)^2$, these integrals are usually written as follows:

1. The integral of a covector field along a curve ("circulation") is given by

$$\oint_P^Q \langle T, \xi\rangle\, dt = \int_a^b T_\alpha \frac{dz^\alpha}{dt} dt = \oint_P^Q T\xi\, dt,$$

where $\xi = \dot{z}$, $T = T_\alpha = T^\alpha$ (relative to Euclidean coordinates the concepts of a vector and a covector coincide, and this is also the case under rotations),

$$P = (z^1(a),\ z^2(a),\ z^3(a)),$$

$$Q = (z^1(b),\ z^2(b),\ z^3(b)),\quad a \le t \le b,$$

$T\xi = \langle T, \xi\rangle$ is a scalar product.

2. The integral of a skew-symmetric tensor field of type (0, 2) (i.e. of the form of degree 2) over a surface (sometimes referred to as "flux") is given by

$$\iint_{U} T_{ij}\, dx^i \wedge dx^j = \iint_{U} T_{ij}(x(z)) \left(\frac{\partial x^i}{\partial z^q} dz^q\right) \wedge \left(\frac{\partial x^i}{\partial z^q} dz^q\right) =$$

(on the surface)

$$= \iint_{U} [\sum_{i<j} T_{ij} \left(\frac{\partial x^i}{z^1} \frac{\partial x^j}{\partial z^2} - \frac{\partial x^j}{\partial z^1} \frac{\partial x^i}{\partial z^2}\right)\}\, dz^1 \wedge dz^2$$

(on the surface)

Note that $\frac{\partial x^i}{z^1} \frac{\partial x^j}{\partial z^2} - \frac{\partial x^j}{\partial z^1} \frac{\partial x^i}{\partial z^2} = J^{ij}_{12}$ is the minor of a (3×2) Jacobian matrix $\frac{\partial x^i}{\partial z^q}$, $i = 1, 2, 3$, $q = 1, 2$. Therefore, we finally obtain

$$= \iint_{U} [\sum_{i<j} T_{ij}\; J^{ij}_{12}]\, dz^1 \wedge dx^2.$$

In a three-dimensional Euclidean space, with Euclidean coordinates x^1, x^2, x^3 on a surface $x^i = x^i(z^1, z^2, z^3)$, the vectors

$$\xi = (\xi^i) = \left(\frac{\partial x^i}{\partial z^1}\right) = \frac{\partial x^i}{\partial z^1} e_i,$$

$$\eta = (\eta^i) = \left(\frac{\partial x^i}{\partial z^2}\right) = \frac{\partial x^i}{\partial z^2} e_i,$$

form a basis for the tangent plane at each point of the given surface. In this Euclidean context the vector product $[\xi, \eta]$ of these vectors is normal to the surface.

The vector product is essentially the tensor $J^{ij} = (\xi^i \eta^j - \xi^j \eta^i)$, to which the "vector" is assigned

$$J^1 = J^{23},\; J^2 = -J^{13},\; J^3 = J^{12},\; J = (J^1, J^2, J^3).$$

It is obvious that $J^{ij} = J^{ij}_{12}$.

The vector $(J^i) = J$ is normal to the surface. Its length is equal to $(|g|)^{1/2}$ where $g = \det(g_{ij})$ (on the surface) and $g_{ij}\, dz^i dz^j = \sum_{i=1}^{3} (dx^i)^2$, $dx^i = \frac{\partial x^i}{\partial z^j} dz^j$ (see Part I).

Therefore, the integral over a region U on the surface $x^i = x^i(z^1, z^2)$ in a Euclidean space relative to Euclidean coordinates (x^1, x^2, x^3) reduces to the form

$$\iint_U T_{ij}\, dx^i \wedge dx^j = \iint_U \Big(\sum_{i<j} T_{ij}\, J^{ij}_{12}\Big) dz^1 \wedge dz^2 =$$

$$= \iint_U \langle T, [\xi, \eta]\rangle\, dz^1 \wedge dz^2 = \iint_U \langle T, n\rangle\, (|g|)^{1/2}\, dz^1 \wedge dz^2,$$

where n is the unit vector of the normal, $n = \dfrac{[\xi, \eta]}{|[\xi, \eta]|} = \dfrac{[J, \eta]}{(|g|)^{1/2}}$.

REMARK. In a four-dimensional space $n = 4$, the integrals of forms of degree 2 over surfaces ($k = 2$) cannot be reduced to operations on vectors only, even if the space is Euclidean.

For the three-dimensional case, we have proved

THEOREM 2. *In Euclidean 3-space the integral of a form of degree 2 over a surface coincides with the integral of the second kind*:

$$\iint_U T_{ij}\, dx^i \wedge dx^j = \iint_U \langle T, n\rangle\, (|g|)^{1/2}\, dz^1 \wedge dz^2,$$

(on the surface)

where n is the unit normal to the surface, T is the vector, in Euclidean coordinates (x^1, x^2, x^3) associated with the tensor (T^{ij}),

$$g_{ij}\, dz^i dz^j = \sum_{i=1}^{3} (dx^i)^2, \quad dx^i = \Big(\frac{\partial x^i}{\partial z^j} dz^j\Big), \quad \langle T, n\rangle$$

is the scalar product, the surface is given by

$$x^i = x^i(z^1, z^2), \quad i = 1, 2, 3.$$

This formula has been derived above.

Since, in the three-dimensional case in Euclidean coordinates (x^1, x^2, x^3), skew-symmetric tensors (forms) of type (0, 2) are associated with vectors and vectors are associated with covectors, we usually consider the integral of a vector field $T_\alpha = T^\alpha$:

a) along a curve $\oint_\Gamma T_\alpha \, dx^\alpha$,

b) over a surface $\iint_U \langle T, n\rangle (|g|)^{1/2} \, dz^1 \wedge dz^2$, where n is the unit normal.

Let us recall the definition from analysis.

A. If a curve Γ is closed (i.e. Γ has the form $z^i(t)$, where $x^i(a) = x^i(b)$, $i = 1, 2, 3$, $a \le t \le b$), then the integral of a covector field

$$\oint_\Gamma T_\alpha \xi^\alpha \, dt$$

is called *field circulation along the curve* Γ.

B. If a surface $U = \{f(x^1, x^2, x^3) = \text{const.}\}$ is closed in the sense that it is the boundary of the region $f(x^1, x^2, x^3) \le C$, which is bounded in space, then the integral $\iint_U T_{ij} \, dx^i \wedge dx^j$ is called the *total flux through the surface of the tensor field* $(T_{ij}) = -(T_{ji})$ or, in the Euclidean case, *the flux of the vector field* $T = (T^1, T^2, T^3)$ through this surface: $T^1 = T_{23}, T^2 = -T_{13}, T^3 = T_{12}$ provided that the coordinates x^1, x^2, x^3 are Euclidean.

It is possible that a surface as a whole cannot be given parametrically in the form $x^i = x^i(z^1, z^2)$ if it is given by the equation $F(x^1, x^2, x^3) = C$. However, this can be done in a neighbourhood of each non-singular point. The integral does not depend on the choice of coordinates on the surface. Therefore, in calculating the integral over the entire surface, we should divide this surface into pieces in such a way that each piece, separately, is represented parametrically; then we should take the sum of the integrals by pieces. A sphere, for example, can be divided into two pieces, namely, the upper and lower hemispheres. Now we shall turn to the operation of skew-symmetric gradient which we introduced above.

For functions $\nabla^s f_i = \dfrac{\partial f}{\partial x^i}$.

For covectors $(\nabla^s T)_{ij} = \dfrac{\partial T_i}{\partial x^j} - \dfrac{\partial T_j}{\partial x^i}$.

For any skew-symmetric tensors of type $(0, k)$

$$(\nabla^s T)_{i_1 \dots i_k} = \sum_{j=1} (-1)^j \frac{\partial T_{i_1 \dots \hat{i}_j \dots i_k}}{\partial x^{i_j}},$$

where $\hat{i}_j$ implies that this index is omitted.

The skew-symmetric gradient was not related to the metric, and therefore this operation is simpler than the covariant differentiation of any tensors (see Part I).

How do we write this operation in the symbolism of "differential forms"?

If $f(x)$ is a function, then its differential has the form $df = \dfrac{\partial f}{\partial x^\alpha} dx^\alpha$. In analysis (for functions) the gradient operation in the differential symbolism was denoted by the letter d. We shall follow this notation in the symbolism of differential forms of all degrees.

If we are given a form of degree 1 (a covector field) $T_\alpha dx^\alpha$, we can apply to it the operation d (by definition) by the rules

a) $dT_\alpha = \dfrac{\partial T_\alpha}{\partial x^j} dx^j$,

b) $d(dx^\alpha) \equiv 0$.

Applying these formulae and differentiating the form $T_\alpha dx^\alpha$, we obtain

$$d(T_\alpha dx^\alpha) = dT_\alpha \wedge dx^\alpha = \frac{\partial T_\alpha}{\partial x^j} dx^j \wedge dx^\alpha.$$

Next, we require the following:

c) a Leibniz type formula should hold in differentiating a product,

$$d(f(x)T) = df \wedge T + f dT,$$

where $f(x)$ is a function and

$$T = \sum_{i_1<\ldots<i_k} T_{i_1\ldots i_k}\, dx^{i_1} \wedge \ldots \wedge dx^{i_k}$$ is an arbitrary differential form.

Indeed, we have

$$d(T_\alpha\, dx^\alpha) = dT_\alpha \wedge dx^\alpha + T_\alpha\, d(dx^\alpha) =$$

$$= \left(\frac{\partial T_\alpha}{\partial x^\beta} dx^\beta\right) \wedge dx^\alpha = \sum_{\alpha<\beta} \left(\frac{\partial T_\alpha}{\partial x^\beta} - \frac{\partial T_\beta}{\partial x^\alpha}\right) dx^\alpha \wedge dx^\beta.$$

This is the skew-symmetric gradient, or the curl, as already introduced, represented as a differential form.

We now impose the last requirement:

d) for all k: $d(x^{i_1} \wedge \ldots \wedge dx^{i_k}) \equiv 0$.

Proceeding from the requirements a), b), c) and d) we can calculate dT, where T is a form of arbitrary degree.

EXAMPLE 1. Let $n = 3$, $k = 2$. How shall we calculate dT, where

$$T = (T_{ij}) = \sum_{i<j} T_{ij}\, dx^i \wedge dx^j\ ?$$

By definition

$$d\left(\sum_{i<j} T_{ij}\, dx^i \wedge dx^j\right) = \sum_{i<j} d\left(T_{ij}\, dx^i \wedge dx^j\right) =$$

$$= \sum_{i<j} dT_{ij} \wedge dx^i \wedge dx^j = \sum_{i<j} \left(\frac{\partial T_{ij}}{\partial x^\alpha} dx^\alpha \wedge dx^i \wedge dx^j\right) =$$

$$= \frac{\partial T_{12}}{\partial x^3} dx^3 \wedge dx^1 \wedge dx^2 + \frac{\partial T_{13}}{\partial x^2} dx^2 \wedge dx^1 \wedge dx^3 +$$

$$+ \frac{\partial T_{23}}{\partial x^1} dx^1 \wedge dx^2 \wedge dx^3 =$$

$$= \left(\frac{\partial T_{12}}{\partial x^3} - \frac{\partial T_{13}}{\partial x^2} + \frac{\partial T_{23}}{\partial x^1}\right) dx^1 \wedge dx^2 \wedge dx^3 = \left(\frac{\partial T^i}{\partial x^i}\right) dx^1 \wedge dx^2 \wedge dx^3.$$

Here $T^1 = T_{23}, T^2 = -T_{13}, T^3 = T_{12}$.

In the Euclidean case $\partial T^i/\partial x^i$ is called the *divergence of the vector field* (T^i). Instead of the term "curl" we sometimes say "vortex".

EXAMPLE 2. (a simpler one). On a plane (x^1, x^2) we have

$$d(T_\alpha\, dx^\alpha) = \left(\frac{\partial T_1}{\partial x^2} - \frac{\partial T_2}{\partial x^1}\right) dx^1 \wedge dx^2.$$

We recommend the reader to verify, directly, the important property of the operation d (which coincides with the operation of the skew-symmetric gradient of tensors of type $(0, k)$):

$$d(dT) \equiv 0 \text{ for all } k,$$

where

$$T = \sum_{i_1 < \ldots < i_k} T_{i_1 \ldots i_k}\, dx^{i_1} \wedge \ldots \wedge dx^{i_k}.$$

Thus, the operation d is our skew-symmetric gradient which is written using the differential symbolism.

We now introduce the most general concept of the integral of the second kind over k-dimensional surfaces in an n-dimensional space (in a manifold) proceeding from the requirements of integral invariance under coordinate changes both in space and on the surface (see above).

DEFINITION. The general non-linear k-form is a smooth fuction $\omega(x, \eta_1, \ldots, \eta_k)$, such that

a) $\omega(x, \eta_{i_1}, \ldots, \eta_{i_k}) = \omega(x, \eta_1 \wedge \ldots \wedge \eta_k)$,

b) $\omega(x, \lambda\eta_1 \wedge \ldots \wedge \eta_k) = \lambda\omega(x, \eta_1 \wedge \ldots \wedge \eta_k)$

for $\lambda \geq 0$.

Here x is a point in an n-dimensional manifold and $\eta_1, \ldots, \eta_k$ is a tuple of k tangent vectors at the point x.

A particular (linear) case of the non-linear k-forms are the differential k-forms, i.e. skew-symmetric tensors of rank k. The operation of restriction of the general non-linear k-form to an arbitrary sub-manifold is defined similarly to the restriction of the differential form.

Under the coordinate changes with positive Jacobian, the general n-forms in an n-dimensional space behave as ordinary differential n-forms. Thus, they are the ordinary n-forms on orientable manifolds.

The restriction of a general non-linear k-form to a k-dimensional orientable surface determines an ordinary differential k-form on this surface, which can be integrated over any region on this surface.

Hence, we have defined the integral of the general non-linear k-form over k-dimensional orientable surfaces. This integral is invariant under coordinate changes (preserving orientation) in space and on the surface.

EXAMPLE 1. Let $g_{ij}(x)$ be any Riemannian (pseudo-Rimannian) metric on a manifold. The metric gives rise to the element of k-dimensional area

$$\omega(x, \eta_1, \wedge \ldots \wedge \eta_k) = \left(|\det \gamma_{ij}|\right)^{1/2}$$

where $\gamma_{ij} = \langle \eta_i, \eta_j \rangle$ is the scalar product of vectors η_i and η_j in this Riemannian metric g_{ij}. For $k = 1$, we obtain the element of length, for $k = n$ we obtain the volume element. The integral of this quantity over any k-dimensional surface is its k-dimensional volume.

EXAMPLE 2. Let $k = 1$ and let the function $\omega(x, \eta)$ be strictly positive for any $\eta \neq 0$:

$$\omega(x, \lambda\eta) = \lambda\omega(x, \eta) \quad \text{for } \lambda > 0,$$

$$\omega(x, \eta) > 0 \qquad \text{for } \eta \neq 0.$$

In such a case we say that on a manifold we are given a Finsler metric. The Finsler length of any smooth curve $\gamma = x(t)$ is given by the formula $l_\Phi(\gamma) = \int_\gamma \omega(x, \dot{x})\, dt$.

Obviously, $l_\Phi(\gamma) > 0$, and the quantity $l_\Phi(\gamma)$ does not change under a monotonic change of parameter: $t = t(\tau)$, $d\tau/dt > 0$.

EXAMPLE 3. Let $k = 1$, as in Example 2, and let a Riemannian metric g_{ij} and a differential 1-form $A = A_i\, dx^i$ be given on a manifold. For any smooth curve $\gamma = x(t)$ we put

$$l_\Phi(\gamma) = \int_\gamma \left((g_{ij}\dot{x}^i\dot{x}^j)^{1/2} + \lambda A_\alpha \dot{x}^\alpha\right) dt,$$

where λ in constant number ("a charge").

EXERCISE. Prove that on a compact manifold this formula determines a Finsler metric for all sufficiently small λ.

Functionals of this type play an important role in the description of motion of charged particles in electro-magnetic fields.

Consider (locally) a manifold of dimension $n + 1$ with the following metric which does not depend on the coordinate x^{n+1}:

$$\hat{g}_{ab}(x) = \left(\frac{g_{ij}}{\lambda A_j}\Big|\frac{\lambda A_j}{1}\right).$$

Prove that after being projected onto the space $x^1, \dots, x^n$, the "horizontal" geodesics of the metric coincide with the extremals of the functional $l_\Phi(\gamma)$ from Example 3 (the Kaluza theorem). See Appendix 5 which contains elements of variational calculus.

Thus the class of objects which can, in an invariant manner, without a Riemannian metric, be integrated over surfaces is appreciably wider than the class of differential forms. However, no analogue of the operator d on general non-linear k-forms can be determined.

EXERCISE. We shall call a general non-linear k-form closed if its surface integral has an identically zero variation, i.e. does not change upon a small variation of the surface. Prove that any closed smooth general k-form is linear, i.e. is an ordinary closed differential form. Stokes type formulae which we discuss in the following section also have sense only for ordinary linear differential forms.

REMARK. There exists an important modern generalization of geometry, essential for quantum theory. This is the so-called "super-geometry" in which most natural analogues of differential forms, including closed ones, are already not tensors and are non-linear.

2.11 The General Stokes Formula and Examples

As noticed in Section 2.10, the definition of the integral of a form of degree k over a k-dimensional surface in an n-dimensional space does not necessarily require parametrization of this surface as a whole in the form $x^i = x^i(z^1, \dots, z^k)$. Since the integral over the sum of regions is equal to the sum of the integrals, we can sub-divide the surface into several pieces and then coordinatize each piece separately. After this, and having integrated over each piece, we should sum up the results and obtain the integral over the surface.

Another remark is that on a surface we can introduce coordinates $z^1, \dots, z^k$ (always) which have singular points (see Part I) on a set of smaller dimension, making no contribution to the integral of the form of degree k. Such coordinates are often employed in the theory of integration. For example, such are polar coordinates r, ϕ on a plane (the singular point is $r = 0$), cylindrical and spherical coordinates in a space:

r, ϕ, z are cylindrical coordinates; singular points are $r = 0$;

r, θ, ϕ are spherical coordinates; singular points are $r = 0$ and $\theta = 0, \pi$, r is arbitrary.

On a sphere there are coordinates (θ, ϕ) where singular points are $\theta = 0, \pi$. The sphere is the simplest surface on which we cannot, in principle, introduce coordinates (a unique system) without singular points. In all of these examples the set of singular points of coordinate systems was small, making no contribution to the integration, so we could ignore it.

In any case, we can determine the integral of a form of degree k (a tensor of type $(0, k)$) over a region on a surface of dimension k in an n-dimensional space.

From the course in analysis, the reader recalls the relation between the integral over a surface and the integral over the boundary of the surface. We mean Green's formula for $n = 2$, Gauss-Ostrogradskii formula for $n = 3$, $k = 3$ and Stokes' formula for $n = 3$, $k = 2$. We shall now treat these formulae from the point of view of the theory of skew-symmetric tensors (differential forms).

In view of the additive character of the integral, it suffices to know the basic definitions of pieces of surfaces. Suppose we are given a region U in k-dimensional space $z^1, \dots, z^k$ in the form of the inequality $f(z^1, \dots, z^k) \le C$, and suppose Γ is the boundary of this region specified by the equation $f(z^1, \dots, z^k) = C$.

Suppose that we are also given an embedding of this region together with the boundary into an n-dimensional space $(x^1, \dots, x^n)$,

$$x^i = x^i(z^1, \dots, z^k), \quad i = 1, 2, \dots, n.$$

We obtain the parametrically given surface, the region U on the surface, and the boundary Γ of this region which is a surface of dimension $(k-1)$.

What is the relation between the integral over the region U and the integral over the boundary Γ of this region of the various forms of degree k in an n-dimensional space $(x^1, \dots, x^n)$?

A particularly simple case is $k = 1$, where $x^i = x^i(f)$, $z^1 = t$ is a curve, U is a segment $a \le t \le b$, its boundary Γ is a pair of points $(t = a)$ and $(t = b)$, the point a being taken with the minus sign and the point b with the plus sign.

Specifically mentioned is the trivial case of the "integral" of a scalar (i.e. of a form of degree 0) over a zero-dimensional surface consisting, by definition, of several points with signs.

"The surface of dimension 0" is a formal sum of points $\sum \pm P_i$, where P_i are points of the space. The "integral" of a function $f(x)$ over "a surface of dimension 0" is, by definition, the sum of the values of the function at these points with corresponding signs:

$$\text{the integral} = \sum_i \pm f(P_i).$$

If in a space we are given a curve $x^i(t)$ and a sement of the curve U $(a \le t \le b)$ with boundary $\Gamma = Q - P$, then, as is known from analysis, the following formula holds

$$\int_\Gamma f = f(Q) - f(P) = \oint_U df = \int_a^b \frac{\partial f}{\partial x^\alpha} \frac{dx^\alpha}{dt} dt.$$

This is the simplest "Stokes' formula", familiar from the first course, relating the integral over the boundary to the integral over the region.

The multi-dimensional Stokes type formulae are, in a sense, its direct generlization and, moreover, can be formally reduced to it.

We shall now return to the general case of a region U in coordinates $z^1, \dots, z^k$ on a surface $x^i = x^i(z^1, \dots, z^k)$, $i = 1, \dots, n$, with boundary Γ given by the equation $f(z^1, \dots, z^k) = C$ (the region U is specified by the inequality $f(z) \le C$).

If in a space $x^1, \dots, x^n$ we are given a form of degree $(k-1)$, i.e. a skew-symmetric tensor of type $(0, k-1)$, written as

$$(T_{i_1 \dots i_{k-1}}) = \sum_{i_1 < \dots < i_{k-1}} (T_{i_1 \dots i_{k-1}})\, dx^{i_1} \wedge \dots \wedge dx^{i_{k-1}},$$

it can be integrated over the $(k-1)$-dimensional surface Γ, which is the boundary of the region on a surface $x^i(z^1, \ldots, z^k)$, $i = 1, \ldots, n$.

The general theorem holds.

THEOREM 1. *For any differential form*

$$\sum_{i_1<\ldots<i_{k-1}} T_{i_1 \ldots i_k} dx^{i_1} \wedge \ldots \wedge dx^{i_k} = T$$

with smooth coefficients $T_{i_1 \ldots i_k}$, any smooth surface $x^i(z^1, \ldots, z^k)$ and a bounded region U (on this surface) with a sooth boundary Γ consisting of one piece, there holds the formula $\pm \int_\Gamma T = \int_U dT$.

The trivial version of this formula for $k = 1$, $k-1 = 0$ has been given above. Here dT is a form of degree k (the skew-symmetric gradient of the tensor $(T_{i_1 \ldots i_{k-1}})$ or the differential of the form of degree $k-1$).

The various two- and three-dimensional cases of this formula are named after Green, Gauss-Ostrogradskii, and Stokes, and are usually given separate proofs in analysis courses. We shall now examine these cases. The reduction of the general formula to these cases is just the proof of the theorem for $n = 2$ and $n = 3$.

I. The planar case ($n = 2$). Here Γ is a closed curve on a plane $x^i = x^i(t)$, where $x^i(a) = x^i(b)$. Suppose this curve is bounding a region of the plane (Figure 47).

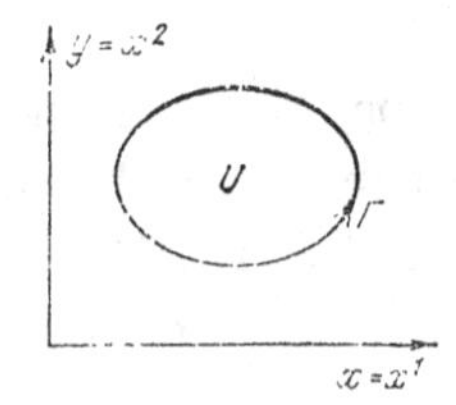

Figure 47

For any (co)vector field $T_\alpha dx^\alpha$ (form of degree 1) the integral around Γ is defined. If the form $T_\alpha dx^\alpha$ is defined and has no singularities inside the region U, then either

$$\int_\Gamma T_\alpha \frac{dx^\alpha}{dt}\, dt = \iint_U \left(\frac{\partial T_1}{\partial x^2} - \frac{\partial T_2}{\partial x^1}\right) dx^1 \wedge dx^2$$

or

$$\int_\Gamma (T_1\, dx + T_2\, dy) = \iint_U \left(\frac{\partial T_1}{\partial x^2} - \frac{\partial T_2}{\partial x^1}\right) dx \wedge dy\ .$$

This is Green's formula which is a particular case of the general Stokes formula.

EXAMPLE. Consider the supplement referring to the complex variable. Suppose $i^2 = -1$ (i is imaginary unity), $z = x + iy$, $dz = dx + idy$ and $f(z) = f(x, y) = u(x, y) + i\,v(x, y)$ is a pair of functions u and v, or one complex-valued function.

Consider the integral

$$\oint_\Gamma f(z)\, dz = \oint_\Gamma (u + iv)\,(dx + i\, dy) =$$

$$= \oint_\Gamma (u\ dx - v\, dy) + i \oint_\Gamma (v\, dx + u\, dy).$$

Applying Green's formula

$$\oint f(z)\, dz = \iint_U \left(\frac{\partial u}{\partial y} + \frac{\partial v}{\partial x}\right) dx \wedge dy + i \iint_U \left(\frac{\partial v}{\partial y} - \frac{\partial u}{\partial x}\right) dx \wedge dy.$$

we arrive at the conclusion that the identity $\oint_\Gamma (z)\, dz = 0$ holds provided that everywhere inside the region U the function $f(z)$ is smooth, and the identitites

$$\frac{\partial u}{\partial y} = -\frac{\partial v}{\partial x}, \qquad \frac{\partial v}{\partial y} = \frac{\partial u}{\partial x}$$

are satisfied. In this case $f(z)$ is called a complex analytic function.

Note that for the functions $f(z) = z^n$ for any integer n the following identities hold

$$\oint z^n\, dz = \begin{cases} 0 & \text{if } n \neq -1, \\ 2\pi i & \text{if } n = -1 \text{ and the contour of } \Gamma \text{ embraces } 0. \end{cases}$$

(Verify this by choosing Γ in the form of a circumference $|z| = 1$). This underlies the important residue formula.

II. The three-dimensional space with coordinates x^1, x^2, x^3. Here we distinguish between two cases: $k = 2$ and $k = 3$.

1. Suppose $k = 3$, U is a region and Γ is the boudary of this region. In this case

$$\iint_U \sum_{i<j} T_{ij}\, dx^i \wedge dx^j = \iiint_U \left(\frac{\partial T_{12}}{\partial x^3} + \frac{\partial T_{23}}{\partial x^1} - \frac{\partial T_{13}}{\partial x^2}\right) dx^1 \wedge dx^2 \wedge dx^3.$$

If the coordinates x^1, x^2, x^3 are Euclidean, if T is a vector where $T^i = T_{23}$, $T^2 = -T_{13}$, $T^3 = T_{12}$ and n is a unit vector of the normal k to the surface Γ then, according to the theorem of Section 2.10 and to the general Stokes formula, we obtain

$$\iint_U \sum_{i<j} T_{ij}\, dx^i \wedge dx^j = \iint_U \langle T, n\rangle\, (|g|)^{1/2}\, dz^1 \wedge dz^2,$$

where z^1, z^2 are coordinates on the surface, $d\sigma = (|g|)^{1/2}\, dz^1 \wedge dz^2$ is the element of area on this surface.

Next,

$$\frac{\partial T_{12}}{\partial x^3} - \frac{\partial T_{13}}{\partial x^1} + \frac{\partial T_{23}}{\partial x^1} = \frac{\partial T^i}{\partial x^i} = \operatorname{div} T.$$

We finally come to

$$\iint_\Gamma \langle T, n\rangle\, (|g|)^{1/2}\, dz^1 \wedge dz^2 = \iint_\Gamma \langle T, n\rangle\, d\sigma = \iiint_\Gamma (\operatorname{div} T)\, dx^1 \wedge dx^2 \wedge dx^3.$$

This is the Gauss-Ostrogradskii formula in Euclidean 3-space.

2. Suppose now $k = 2$, U is a region on a surface $x^i = x^i(z^1, z^2)$, $i = 1, 2, 3$, Γ is a (curve) boundary of this region. We have

$$\oint T_\alpha\, dx^\alpha = \iint_U [(\frac{\partial T_1}{\partial x^2} - \frac{\partial T_2}{\partial x^1})\, dx^1 \wedge dx^2 +$$

$$+ (\frac{\partial T_1}{\partial x^3} - \frac{\partial T_3}{\partial x^1})\, dx^1 \wedge dx^3 + (\frac{\partial T_2}{\partial x^3} - \frac{\partial T_3}{\partial x^2})\, dx^2 \wedge dx^3].$$

In the Euclidean case, where we need not distinguish between vectors and covectors, and where we can assign to a skew-symmetric vector (T_{ij}), a vector $T = (T^i)$, we obtain (the Stokes formula)

$$\oint T_\alpha\, dx^\alpha = \iint_U \langle \mathrm{rot}\, T, n\rangle\, (|g|)^{1/2}\, dz^1 \wedge dz^2,$$

where rot T is the vector assigned to the skew-symmetric tensor

$$T = (T_\alpha); \quad (\mathrm{rot}\, T)_{\alpha\beta} = \frac{\partial T_\alpha}{\partial x^\beta} - \frac{\partial T_\beta}{\partial x^\alpha}.$$

In all these cases the general Stokes formula has transformed into different integral formulae from the course in analysis, which means that it is proved for the three-dimensional space.

As a concluding remark we note that in the formulation of the general "Stokes Theorem" it is not necessary to assume that the boundary Γ consists of one piece. If the boundary Γ consists of several pieces, then the integrals over the different pieces will enter with either plus or minus sign which should be chosen appropriately. This has already been mentioned for the trivial case $k = 1$, where the boundary Γ of a segment of a curve consists of two points — one (terminal) with a plus sign and the other (initial) with a minus sign. It is relevant to note here that the choice of the sign in the expression $\iint_\Gamma \langle T, n\rangle\, (|g|)^{1/2}\, dz^1 \wedge dz^2$ (or, alternatively, the "orientation" of the boundary Γ) is determined by the direction of the unit normal n.

We shall now examine another important application of the general Stokes formula.

Let us consider a four-dimensional space $x^0 = ct, x^1, x^2, x^3$ with the metric

$$g_{ij} = \begin{pmatrix} -1 & & & 0 \\ & 1 & & \\ & & 1 & \\ 0 & & & 1 \end{pmatrix}, \text{ where } c \text{ is the speed of light and } t \text{ is the time.}$$

Suppose $F_{ik} = -F_{ki}$ is the tensor of an electro-magnetic field, $i, k = 0, 1, 2, 3$. Now we shall only consider the behaviour and the property of this tensor for a constant time $x^0 = ct$, where we can transform only spatial coordinates:

$$x^{i,} \quad i = 1, 2, 3:$$

$$x'^0 = x^0, \quad x'^i(x^1, x^2, x^3).$$

In this case, the tensor F_{ik} in a four-dimensional space is determined by the *covector of the electric field* $E_\alpha = F_{0\alpha}$, $\alpha = 1, 2, 3$, and by *the tensor of the magnetic field* $H_{\alpha\beta} = -H_{\beta\alpha}$, $\alpha, \beta = 1, 2, 3$. If the coordinates x^1, x^2, x^3 are Euclidean, then the magnetic field is determined by the *axial vector* of the magnetic field

$$H^1 = H_{23}, \quad H^2 = -H_{13}, \quad H^3 = H_{12}.$$

In the notation of differential forms, we have

$$d(F_{ij}\, dx^i \wedge dx^j) = 0 \quad \text{(the first pair of Maxwell's equations)}$$

or in the three-dimensional Euclidean form

$$\text{a)} \quad \operatorname{div} H = \frac{\partial H^i}{\partial x^i} = 0,$$

$$\text{b)} \quad \operatorname{rot} E + \frac{1}{c}\frac{\partial H}{\partial t} = 0.$$

From equation a) and from the Gauss-Ostrogradskii formula, we have (Γ is the boundary of the region U)

$$\iiint_U \operatorname{div} H\, dx^1 \wedge dx^2 \wedge dx^3 = \iint_U \langle H, n\rangle\, d\sigma = 0$$

("the magnetic field flux through a closed surface is always equal to zero").

From equation b) and from the Stokes formula we have

$$\iint_U \langle \operatorname{rot} E, n\rangle \, d\sigma = \oint_\Gamma E_\alpha \, dx^\alpha,$$

(on the surface)

$$\iint_U \left\langle \frac{1}{c}\frac{\partial H}{\partial t}, n \right\rangle d\sigma = \oint_\Gamma E_\alpha \, dx^\alpha,$$

(the boundary of the region on the surface)

("the time derivative of the magnetic field flux through a surface is equal to the electric field circulation along the boundary of the region on the surface").

The second pair of Maxwell's equations has the form

$$\sum_{k=1}^{3} \frac{\partial F_{ik}}{\partial x^k} - \frac{1}{c}\frac{\partial F_{0i}}{\partial t} = \frac{4\pi}{c} j_{(4)i}, \quad \text{where } j_{(4)} = (c\,\rho, j_1, j_2, j_3)$$

(the four-dimensional divergence of the tensor F_{ik} in the metric

$$G_{ij} = \begin{pmatrix} -1 & & & 0 \\ & 1 & & \\ & & 1 & \\ 0 & & & 1 \end{pmatrix}$$

is equal to the four-dimensional vector of the electric current, ρ is the charge density and $j = (j_1, j_2, j_3)$ is the usual vector of current density).

In the three-dimensional form this yields

a) $\operatorname{div} E = 4\pi\rho,$

b) $\operatorname{rot} H + \dfrac{1}{c}\dfrac{\partial E}{\partial t} = \dfrac{4\pi}{c} j.$

Equation a) together with the Gauss-Ostrogradskii formula imply

$$\iiint_U 4\pi\rho \, dx^1 \wedge dx^2 \wedge dx^3 = \iint_U \langle E, n\rangle \, d\sigma$$

(the electric field flux through the surface Γ is equal, with an accuracy to 4π, to the total charge in the region U).

Equation b) and the Stokes formulae imply

$$-\frac{1}{c}\iint_U \frac{\partial E}{\partial t} n\, d\sigma + \iint_U \frac{4\pi}{c} jn\, d\sigma = \oint_\Gamma H_\alpha\, dx^\alpha,$$

(on a surface) (on a surface) (the curve is a boundary)

("the total current through the surface ± the derivative of the electric field flux through this surface is equal to the magnetic field circulation around the boundary Γ of the surface ").

We can see that the geometric content of the first and the second pairs of the Maxwell equations is different here.

The first pair of the Maxwell equations is not related to the space metric, the second pair of equations cannot be written without a metric. The natural form of these equations is closely related to Euclidean coordinates x^1, x^2, x^3. Further on, in the course in the mechanics of a continuous medium and in many other places, the reader will encounter various applications of integral formulae from analysis, whose geometric and tensorial content we have already discussed.

We have shown above that the classical integral Green and Gauss-Ostrogradskii formulae are particular cases of the general Stokes formula. Now we are in a position to prove the general Stokes formula for the case where the inegration region is a k-dimensional cube.

A singular cube σ of a space $\mathbb{R}^n$ is defined as a smooth mapping $\sigma\colon I^k \to \mathbb{R}^n$, $n > k$, where I^k is a Cartesian cube of dimension k, i.e. $I^k = \{(x^1, \ldots, x^k \mid 0 \le x^\alpha \le 1\}$. The equations $x^\alpha = 0$ and $x^\alpha = 1$ specify two $(k-1)$-dimensional faces I_α^- and I_α^+. We denote the boundary of the cube I^k by σI^k, i.e. $\partial I^k = \bigcup_\alpha (I_\alpha^+ \cup I_\alpha^-)$. Let ϕ^{k-1} be a $(k-1)$-form in $\mathbb{R}^n$ and let $d\phi^{k-1}$ be its exterior differential. Next, let $\sigma\colon I^k \to \mathbb{R}^n$ be a singular cube.

THEOREM 2. *The following equality holds* $\displaystyle\int_{\sigma(\partial I^k)} \phi^{k-1} = \int_{\sigma(I^k)} d\phi^{k-1}$. *The orientation on the cube boundary ∂I^k is taken to be that induced by the standard orientation of the cube I^k by means of the exterior normal.*

By the integral $\int_{\sigma(\partial I^k)} \phi$ we mean the sum of integrals over all the faces of the cube.

Proof. Consider the form $\omega = \sigma^*(\phi)$, where $\sigma^*(\phi)$ denotes the form obtained through the change of variables in the form ϕ by means of the map σ. From the fact that the operation σ^* and d are permutable, we have $d\omega = d\sigma^*(\phi) = \sigma^*(d\phi)$, and, therefore, it suffices to prove our assertion in the form $\int_{\partial I^k} \omega = \int_{I^k} d\omega$. We have made use of the fact that, by definition of the form ω

$$\int_{\sigma(I^k)} d\phi = \int_{I^k} \sigma^*(d\phi) \quad \text{and} \quad \int_{\sigma(\partial I^k)} \phi = \int_{\partial I^k} \sigma^*(\phi).$$

Suppose that $\omega = a_\alpha(x^1, \ldots, x^k)\, dx^1 \wedge \ldots \wedge d\hat{x}^\alpha \wedge \ldots \wedge dx^k$, where $a_\alpha(x^1, \ldots, x^k)$ are smooth functions and the differential dx^α is omitted. We are led to

$$d\omega = \sum_\alpha \frac{\partial a_\alpha}{\partial x^\alpha} dx^\alpha \wedge dx^1 \wedge \ldots \wedge \widehat{dx^\alpha} \wedge \ldots \wedge dx^k =$$

$$= \sum_\alpha (-1)^{\alpha-1} \frac{\partial a_\alpha}{\partial x^\alpha} d^k x,$$

where $d^k x = dx^1 \wedge \ldots \wedge dx^k$. For the sake of simplification in the following, we assume that the functions $a_\alpha(x^1, \ldots, x^k)$ are represented in the form of products $a_\alpha(x^1, \ldots, x^k) = \prod_{q=1}^{k} b^q_\alpha(x^q)$. Here the functions b^q_α are assumed to be smooth functions of one variable x^q. Recall that in the course in analysis there exists a theorem to the effect that any smooth function can be uniformly approximated by linear combinations of products of smooth functions of one variable. We shall not, of course, prove this here.

We shall calculate, in an explicit form, the expression $\int_{I^k} d\omega$;

$$\int_{I^k} d\omega = \int_{I^k} \Big(\sum_\alpha (-1)^{\alpha-1} \frac{\partial a_\alpha}{\partial x^\alpha}\Big)\, d^k x =$$

$$= \int_{I^k} \Big(\sum_\alpha (-1)^{\alpha-1} \frac{\partial}{\partial x^\alpha} \Big(\prod_{q=1}^{k} b_\alpha^q(x^q)\Big)\Big)\, d^k x =$$

$$= \int_{I^k} \Big(\sum_\alpha (-1)^{\alpha-1}\, b_\alpha^1(x^1) \ldots \frac{\partial b_\alpha^\alpha(x^\alpha)}{\partial x^\alpha} \ldots b_\alpha^k(x^k)\Big)\, d^k x =$$

$$= \sum_\alpha \int_{(x^1)} \ldots \int_{(\hat{x}^\alpha)} \ldots \int_{(x^k)} (-1)^{\alpha-1} b_\alpha(x^1) \ldots$$

$$\ldots b_\alpha^{\alpha-1}(x^{\alpha-1})\, b_\alpha^{\alpha+1}(x^{\alpha+1}) \ldots b_\alpha^k(x^k) \times$$

$$\times \Big[(-1)^{\alpha-1} \int \frac{\partial b_\alpha^\alpha(x^\alpha)}{\partial x^\alpha} dx^\alpha\Big]\, dx^1 \wedge \ldots \wedge \widehat{dx^\alpha} \wedge \ldots \wedge dx^k =$$

$$= \sum_\alpha \int_{(x^1)} \ldots \int_{(\hat{x}^\alpha)} \ldots \int_{(x^k)} \big(b_\alpha^1(x^1) \ldots \hat{b}_\alpha^\alpha(x^\alpha) \ldots b_\alpha^k(x^k) \times$$

$$\times \big[b_\alpha^\alpha(1) - b_\alpha^\alpha(0)\big]\big)\, dx^1 \wedge \ldots \wedge \widehat{dx^\alpha} \wedge \ldots \wedge dx^k =$$

$$= \sum_\alpha \int_{(x^1)} \ldots \int_{(\hat{x}^\alpha)} \ldots \int_{(x^k)} \big(b_\alpha^1(x^1) \ldots b_\alpha^\alpha(1) \ldots b_\alpha^k(x^k) -$$

$$- b_\alpha^1(x^1) \ldots b_\alpha^\alpha(0) \ldots b_\alpha^k(x^k)\big)\, dx^1 \wedge \ldots \wedge \widehat{dx^\alpha} \wedge \ldots \wedge dx^k =$$

$$= \sum_\alpha \int_{(x^1)} \dots \int_{(\hat{x}^\alpha)} \dots \int_{(x^k)} (a_\alpha(x^1, \dots, x^k)|_{x^\alpha=1} -$$

$$- a_\alpha(x^1, \dots, x^k)|_{x^\alpha=0} \; dx^1 \wedge \dots \wedge \widehat{dx^\alpha} \wedge \dots \wedge dx^k = \int_{\partial I^k} \omega.$$

With this we have completed the proof of the theorem. As is seen from the computation, this theorem is a simple consequence of the one-dimensional Newton-Leibniz formula.

PART III

BASIC ELEMENTS OF TOPOLOGY

3.1 Examples of Differential Forms

We shall first examine several especially interesting examples of differential forms.

EXAMPLE 1. A skew-symmetric scalar product of vectors in an even ($2n$)-dimensional space with coordinates $x^1, \dots, x^{2n}$ is given by the 2-form

$$\Omega = g_{\alpha\beta}\, dx^\alpha \wedge dx^\beta, \quad g_{\alpha\beta} = -g_{\beta\alpha}.$$

Non-degeneracy implies that the matrix $g_{\alpha\beta}$ has the inverse $g^{\alpha\beta}$, where

$$g^{\alpha\beta} g_{\beta\gamma} = \delta^\alpha_\gamma.$$

THEOREM 1. *A skew-symmetric scalar product is non-degenerate if and only if the $2n$-form*

$$\Omega^n = \underbrace{\Omega \wedge \dots \wedge \Omega}_{n\text{-times}} = n!\, f(x^1, \dots, x^{2n})\, dx^1 \wedge \dots \wedge dx^{2n}$$

is non-zero, i.e. $f(x^1, \dots, x^{2n}) \neq 0$ *and* $f = (g)^{1/2}$.

Proof. We wish to verify directly that $f^2 = \det(g_{\alpha\beta}) = g$. (The expression $f = (\det g_{\alpha\beta})^{1/2}$ itself for skew-symmetric matrices is called 'Pfaffian'.) Indeed, by the definition of multiplication we have

$$\Omega^n = \wedge \dots \wedge \Omega = (g_{\alpha\beta}\, dx^\alpha \wedge dx^\beta) \wedge \dots \wedge (g_{\alpha\beta} dx^\alpha \wedge dx^\beta).$$

By virtue of the fact that our expression does not depend on the derivatives, we can, without loss of generality, assume the matrix $g_{\alpha\beta}$ to be constant. Both the forms Ω^n and $(g)^{1/2} d^n x = d\sigma$ are well defined (in an invariant manner) under the changes. We shall verify the equality $\Omega^n = n!(g)^{1/2} d^n x$ in a special coordinate system. Namely, we shall find coordinates $(x^1, y^1, \dots, x^n, y^n)$, where $\Omega = \sum dx^i \wedge dy^i$. For a constant skew-symmetric matrix $g_{\alpha\beta}$ such coordinates can be chosen using a linear change. In this coordinate system the theorem is trivial: for example, for $n = 2$ we have

$$2\, dx^1 \wedge dy^1 \wedge dx^2 \wedge dy^2 = (dx^1 \wedge dy^1 + dx^2 \wedge dy^2)^2,$$

because $(dx^i \wedge dy^i)^2 = 0$. The verification of the theorem is similar for all $n > 2$, and the result follows.

Especially important are *closed* 2-forms determining scalar products where $d\Omega \equiv 0$. For example, in a $2n$-dimensional space $x^1, \ldots x^{2n}$ we can choose coordinates $q^1, \ldots, q^n, p^1, \ldots, p^n$, such that the 2-form Ω is given by

$$\Omega = \sum_{\alpha=1}^{n} dp^\alpha \wedge dq^\alpha.$$

Such a scalar product is called *Hamiltonian* (or *symplectic*) and a space with such a scalar product is called the *phase space.*

EXAMPLE 2. Given a *complex* space with coordinates $z^1, \ldots, z^n, \bar{z}^1, \ldots, \bar{z}^n$ where $z^j = x^j + iy^j$, $\bar{z}^\alpha = x^\alpha - iy^\alpha$, all the differential forms can be written as

$$T = \sum T^{(p,q)}, \quad p + q = k,$$

where the summands

$$T^{(p,q)} = T_{i_1 \ldots i_p j_1 \ldots j_q}\, dz^{i_1} \wedge \ldots \wedge dz^{i_p} \quad d\bar{z}^{j_1} \wedge \ldots \wedge d\bar{z}^{j_q}$$

are called 'forms of type (p, q)'.

For example, we are given a form $\Omega = T_{\alpha\beta}\, dz^\alpha \wedge d\bar{z}^\beta$, where $T_{\alpha\beta} = -\bar{T}_{\beta\alpha}$.

Then the matrix $(i\bar{T}_{\alpha\beta})$ has the form $i\bar{T}_{\alpha\beta} = \tilde{T}_{\alpha\beta}$ and is the matrix of the Hermitian quadratic form $\sum \tilde{T}_{\alpha\beta}\, dz^\alpha\, d\bar{z}^\beta$. Thus, in the complex case the Hermitian metric is given by a form of type (1, 1). Of particular importance here is the case of *Kählerian* metrics, where $d\Omega \equiv 0$.

EXAMPLE 3. Let $n = 1$ and let the coordinates of a one-dimensional complex space have the form $z, \bar{z}$. Let there be given a 1-form

$$\omega = f(z, \bar{z})\, dz.$$

Obviously, the condition $d\omega = 0$ is equivalent to the condition $\partial f/\partial \bar{z} \equiv 0$ or to the complex analyticity condition. From the general Stokes formula (in this case, from Green's formula) we have

$$\underset{\text{along path 1}}{\int_P^Q f(z, \bar{z})\, dz} \;\equiv\; \underset{\text{along path 2}}{\int_P^Q f(z, \bar{z})\, dz},$$

provided that in the region between paths 1 and 2 the $\partial f/\partial \bar{z}$ is everywhere identically zero (Figure 48).

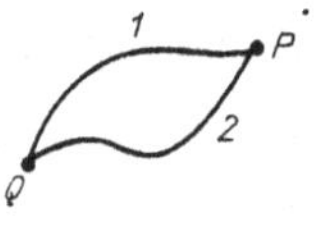

Figure 48.

Or, the integral along a closed path is equal to zero provided that inside the path the function $f(z, \bar{z})$ is analytic, i.e. $\partial f/\partial \bar{z} \equiv 0$. It can be directly verified that for the powers $f(z) = (z - a)^n$, where n are integers, the integrals are given by

$$\oint f(z)\, dz = \begin{cases} 2\pi i, & n = -1, \\ 0, & n \neq -1, \end{cases}$$

where $z = a + \rho e^{i\phi}$ (Figure 49). Indeed, let us consider a differential 1-form $\omega = f(z)\, dz$ and a contour γ embracing several singular points $a_1, \ldots, a_N$ of the function $f(z)$. We may consider another contour γ', sketched in Figure 50, which embraces the same singular points.

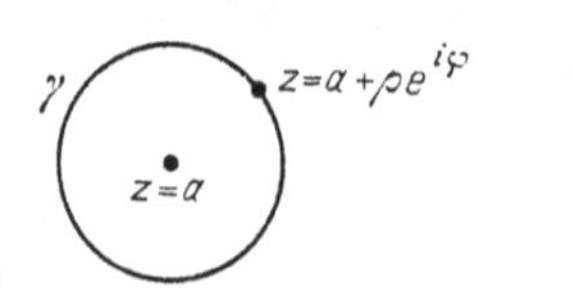

Figure 49.

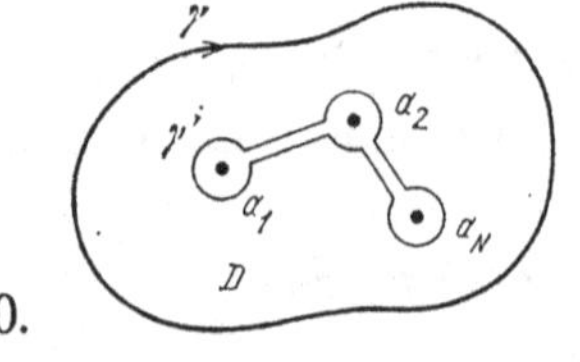

Figure 50.

Clearly the Stokes formula implies the equality $\int_\gamma f(z)\,dz = \int_{\gamma'} f(z)\,dz$. The latter integral falls into the sum of N summands each of which is calculated along the circumference of small radius, which embraces one singular point. Each of these integrals is calculated as follows

$$\int \left(\sum c_k z^k\right) dz = \sum_k c_k \int_0^{2\pi} \left(re^{i\phi}\right)^k d\left(re^{i\phi}\right) =$$

$$= \sum_k c_k r^{k+1} \int_0^{2\pi} e^{i(k+1)\phi}\, d\phi = 2\pi i c_{-1}.$$

Therefore, for uniformly convergent series $f(z) = \sum_{-\infty}^{\infty} (z-a)^n c_n$ the formula holds (the contour γ embraces the point a and lies inside the region of uniform convergence of the series)

$$\int_\gamma f(z)\,dz = 2\pi i c_{-1}; \quad \int_\gamma (z-a)^{-k}\,dz = 2\pi i c_{k-1}.$$

For the analytic function $f(z)$, these formulae allow us to determine the coefficients of its Taylor series (if all powers $n \ge 0$) or of its Laurent series (if $-\infty < n < \infty$) through the integrals.

Let $f\colon M^p \to N^q$ be a smooth map of one smooth manifold into another and let $T = \{T_{i_1 i_2 \ldots i_n}\}$ be a tensor field of rank n on the manifold N^q. Then we can define a new tensor field f^*T on the manifold M^p. Let us introduce local coordinates $x^1, \ldots, x^p$ on the manifold M^p and local coordinates $y^1, \ldots, y^q$ on the manifold N^q Then $f^*T = \{S_{j_1 \ldots j_n}(x)\}$, where

$$S_{j_1 \ldots j_n}(x) = \frac{\partial y^{i_1}(x)}{\partial x^{j_1}} \cdots \frac{\partial y^{i_n}(x)}{\partial x^{j_n}} T_{i_1 \ldots i_n}(y(x)).$$

Here $y = f(x)$, i.e. $y^i = f^i(x^1, \ldots, x^p)$, $1 \le i \le q$.

Let, on the manifold N^q, there be given a differential form

$$T = \sum_{i_1 < \ldots < i_n} dy^{i_1} \wedge \ldots \wedge dy^{i_n}.$$

Then the differential form f^*T on the manifold M^p is defined to be

$$j*(T) = \sum_{j_1<\ldots<j_n} \frac{\partial y^{i_1}}{\partial x^{[j_1}} \cdots \frac{\partial y^{i_n}}{\partial x^{j_n]}} T_{i_1 \ldots i_n} dx^{j_1} \wedge \ldots \wedge dx^{j_n}.$$

EXAMPLE 4. Suppose we are given a hyper-surface M^{n-1} in an n-dimensional Euclidean space

$$F(x^1, \ldots, x^n) = 0, \quad \nabla F \neq 0,$$

or, locally, $x^\alpha = x^\alpha (y^1, \ldots, y^{n-1})$. Here $x^1, \ldots, x^n$ are Euclidean coordinates. The "curvature form" is defined to be

$$K\, d\sigma = K\, (|g|)^{1/2}\, dy^1 \wedge \ldots \wedge dy^{n-1},$$

where K is curvature (for $n = 2$ this is the curvature of the curve in a plane, while for $n = 3$ this is the Guaussian curvature).

Consider a sphere S^{n-1} given by the equation $\sum_{\alpha=1}^{n} (x^\alpha)^2 = 1$. We denote by Ω_{n-1} the $(n-1)$-dimensional volume element on the sphere, invariant under rotation, which for $n = 2, 3$, has the form

$$n = 2: \ \Omega_{n-1} = d\phi,$$

$$n = 3: \ \Omega_{n-1} = |\sin \theta|\, d\theta\, d\phi.$$

We define the Gauss spherical map of the manifold M^{n-1} to the sphere S^{n-1}: consider at a point P of the manifold M^{n-1}, the unit normal n_p to the surface and transport this vector n_p to the origin of coordinates (Figure 51). The map $P \to n_p$ defines the Gauss map

$$\phi: M^{n-1} \to S^{n-1}$$

(the point P is set to the tip of the vector n_p after the vector n_p has been transported so that its tail is at the origin of coordonates). The Gaussian curve of a hyper-surface is defined to be the ratio det Q/det G, where Q and G are the matrices of the first and second quadratic forms, respectively.

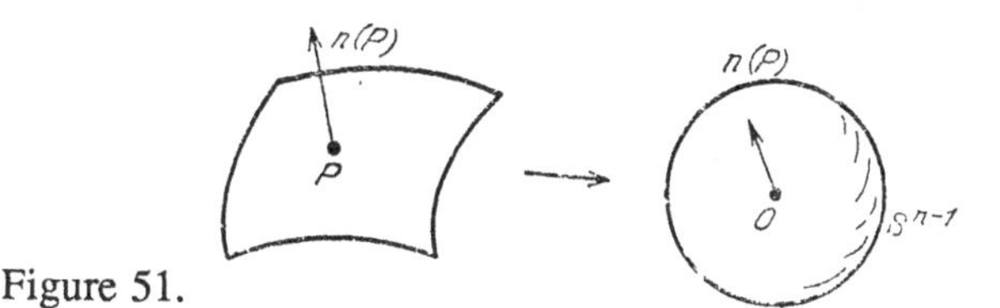

Figure 51.

THEOREM 2. *The following formula holds*

$$K\,d\sigma = \phi^*\Omega; \quad \begin{cases} K\,d\sigma = f^*(d\phi), & n=2, \\ K\,(|g)^{1/2}\,dy^1 \wedge dy^2 = f^*(d\,\Omega), & n=3, \end{cases}$$

where $d\sigma = (|g)^{1/2}\,dy^1 \wedge \ldots \wedge dy^{n-1}$ *is the element of an* $(n-1)$*-dimensional volume in local coordinates* $y^1, \ldots, y^{n-1}$ *on the surface.*

Proof. The proof is similar for all $n \geq 2$. We give the proof for $n = 3$ only. We choose Euclidean coordinates in $\mathbb{R}^3$, where the axis $x^3 = z$ is orthogonal to the surface at a point P, and $x = x^1$ and $y = x^2$ are tangent to the surface. Then $y^1 = x$, $y^2 = y$ and in the neighbourhood of the point P the surface is specified by the equation $z = f(x, y)$, where $df|_P = 0$. In this case, we have

$$K = \det \begin{vmatrix} f_{xx} & f_{xy} \\ f_{yx} & f_{yy} \end{vmatrix}; \quad g_{ij} = \delta_{ij}$$

at the point $P(f_x = f_y = 0)$. On the sphere $S^{n-1} = S^2$, we choose the same coordinates at the point $\phi(P) = Q$, where $\bar{z} \perp S^2$. The form Ω at the point P is given by $\Omega|_P = d\bar{x} \wedge d\bar{y}$. In the neighbourhood of the point Q the sphere is given by the equation $\bar{z} = \left(1 - \bar{x}^2 - \bar{y}^2\right)^{1/2}$, where $\bar{x} = 0$, $\bar{y} = 0$ at the point Q, and the metric of the sphere at this point has the form $g_{ij} = \delta_{ij}$.

The coordinates of the normal vector at the point P' near the point P are

$$n_{P'} = (f_x, f_y, -1)\left(\frac{1}{(1 + f_x^2 + f_y^2)^{1/2}}\right)$$

$(f_x = f_y = 0$ at the point $P)$.

Therefore, in a neighbourhood of the point P the Gauss map is written as

$$\bar{x} = \frac{f_x}{(1+f_x^2+f_y^2)^{1/2}}, \qquad \bar{y} = \frac{f_y}{(1+f_x^2+f_y^2)^{1/2}},$$

(here $P'$with coordinates x, y goes over to $Q'$with coordinates $\bar{x}, \bar{y}$).

By the definition of $\phi^*(\Omega_{n-1})$ we have

$$\phi^*(\Omega_{n-1})\,|_P = \left(\frac{\partial\bar{x}}{\partial x}\frac{\partial\bar{y}}{\partial y} - \frac{\partial\bar{x}}{\partial y}\frac{\partial\bar{y}}{\partial x}\right)\Big|_P\, dx\, dy = J\, dx \wedge dy,$$

where J is the Jacobian of the map ϕ at the point P. Obviously, since $f_x = f_y = 0$ at the point P, it follows that

$$J = f_{xx}f_{yy} - f_{xy}f_{yx} = K$$

(the Gaussian curvature). Since $|g|$ at the point P is equal to unity, we have, finally, that in the chosen system of coordinates the following formula holds (at the point P)

$$K\,(|g|)^{1/2} dx \wedge dy = \phi^*(\Omega_{n-1}), \quad n = 3$$

which implies the theorem for $n = 3$. For all other n the proof is similar.

REMARK. When $n = 2$, we have $\Omega_{n-1} = d\phi$, and for a curve $x^1 = x^2(y), x^2 = x^2(y)$ we can see that $K\,(|g|)^{1/2}\,dy$ becomes $K\,dl$, where dl is the element of length (the natural parameter).

3.2 The Degree of Mapping. Homotopy

In the preceding section, we gave several examples of differential forms. In particular we proved two theorems.

1. For hyper-surfaces M^{n-1} in Euclidean space $\mathbb{R}^n$ the following formula holds

$$K\,(g)^{1/2}\,dy^1 \wedge \ldots \wedge dy^n = f^*(\Omega),$$

where $f\colon M^{n-1} \to S^{n-1}$ is the Gauss map. Ω is the volume element on the sphere (for $n = 3$, $\Omega = |\sin\theta|\,d\phi\,d\theta$) and K is the Gaussian curvature (for $n = 2$ this is the curvature of the curve M^1 and $\Omega = d\,\phi$ on the circumference S^1).

2. For skew-symmetric scalar products $g_{\alpha\beta} = -g_{\beta\alpha}$: det $|g_{\alpha\beta}| \neq 0$, the following formula holds

$$\Omega \wedge \ldots \wedge \Omega = \pm n! \quad (g)^{1/2}\,dx^1 \wedge \ldots \wedge dx^{2n};$$

$$\alpha, \beta = 1, 2, \ldots, n, \ldots, 2n; \quad \Omega = g_{\alpha\beta}\,dx^\alpha \wedge dx^\beta.$$

In connection with Theorem 1 of Section 3.1 and with the fact that $f^*\Omega = K\,(g)^{1/2}\,du\,dv$ for surfaces $M^2 \subset \mathbb{R}^3$ we have mentioned the specifics of the important case where the surface M^2 is closed.

DEFINITION 1. A manifold M^n is called *closed* if it is compact (i.e. any infinite sequence of points has a limiting point) and has no boundary.

For example, a sphere S^n, a torus T^n, projective spaces $\mathbb{R}P^n$, $\mathbb{C}P^n$ of the group SO_n, U_n, surfaces with k handles in a three-dimensional space etc. (Figure 52).

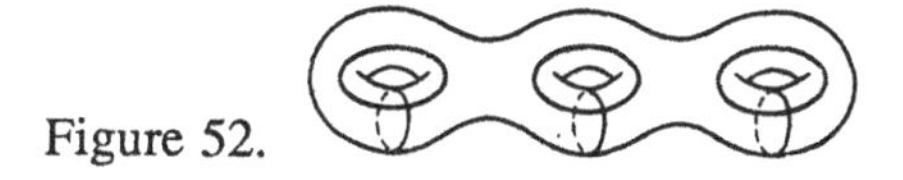
Figure 52.

Recall that a manifold M^n is thought of as *oriented* if it is sub-divided into regions of action of local coordinates, $M^n = \bigcup_\alpha U_\alpha$; $x^1_\alpha, \ldots, x^n_\alpha$, where in the intersections of the regions $U_\alpha \cap U_\beta$ functions $x^q_\beta(x^1_\alpha, \ldots, x^n_\alpha)$, $q = 1, \ldots, n$, are such that the Jacobian $J > 0$, where $J = \det(dx^q_\beta/dx^p_\alpha)$.

Suppose there exist two manifolds M_1^m and M_2^n (for example, both are spheres S^n). Suppose we are given a smooth map

$$f: M_1^m \to M_2^n .$$

DEFINITION 2. A point $P \in M_1^m$ is called a *regular point for the map f* if the Jacobian matrix J of the map at the point P has rank m.

DEFINITION 3. A point $P' \in M_2^n$ is called a *regular* value if *all* points $P \in f^{-1}(P')$ of the complete pre-image are regular.

The following important lemma holds.

LEMMA 1. (Sard's lemma). *If the mapping f is smooth, then almost all points $Q \in M_2^n$ are regular values. The words 'almost all' should be understood in the sense of measure: this means that arbitrarily close to each point $Q \in M_2^n$ there exist regular values.*

We do not give the proof of Sard's lemma here but refer the reader to the book [1].

EXAMPLES.

1. If $m < n$, then only points $Q \in M_2^n$ are regular values, where the complete pre-image $f^{-1}(Q)$ is empty (i.e. there is not a single point P such that $f(P) = Q$).

2. If $m = n$, then the complete pre-image of a regular value $f^{-1}(Q) = P_1 \cup \ldots \cup P_N$ consists of a certain number of points P_α. At each point P_α there is a sign: sign (P_α), where

$$\operatorname{sgn} P_\alpha = \operatorname{sgn}\left(\det \frac{\partial x^p}{\partial y^q}\right),$$

Here x^p are local coordinates at the point Q and y^q are local coordinates at the point P.

We shall list several essential properties of regular values.

THEOREM 1. *If* $f: M^m \to M^n$ *is a smooth map and if* $Q \in M^n$ *is a regular value, then the complete pre-image* $f^{-1}(Q) \in M^m$ *is a smooth manifold of dimension* $m - n$. *Furthermore, at any point* $P \in f^{-1}(Q)$ *the differential of the map f (the linear map of tangent spaces* $\hat{J}: \mathbb{R}^q \to \mathbb{R}^n$ *given by the Jacobian matrix of the map f) has rank n.*

Proof. Suppose $x^1, \dots, x^n$ are local coordinates in a neighbourhood of the point Q on the manifold M^n and $y^1, \dots, y^m$ are local coordinates in a neighbourhood of the point $P \in f^{-1}(Q)$ on the manifold M^m. In the region served by the coordinates $y^1, \dots, y^m$, the map f is specified by the formulae

$$x^\alpha = x^\alpha(y^1, \dots, y^m),$$

$$\alpha = 1, \dots, n,$$

provided that the image of the point $y^1, \dots, y^m$ comes under the action of the coordinates $x^1, \dots, x^n$ (this is, of course, the case within a certain neighbourhood of the point P).

In a neighbourhood of the point P the complete pre-image is given by the equation $x^\alpha(y^1, \dots, y^m) = x_0^\alpha$ ($x_0^1, \dots, x_0^n$ are coordinates of the point Q), and by the condition that Q is a regular value and P is regular point we have

$$\mathrm{rk}\left(\frac{\partial x^\alpha}{\partial y^\beta}\right) = n,$$

where J is the matrix $\left(\frac{\partial x^\alpha}{\partial y^\beta}\right)$ at points from $f^{-1}(Q)$. Therefore, the vectors

$$\left(\frac{\partial x^\alpha}{\partial y^1}, \dots, \frac{\partial x^\alpha}{\partial y^m}\right) = \eta^{(\alpha)}; \quad \alpha = 1, \dots, n,$$

are linearly independent at points from $f^{-1}(Q)$. Such a set of equations $x^\alpha(y^1, \dots, y^m) = 0$; ($\alpha = 1, \dots, n$) determines in a non-degenerate way a manifold of dimension $M - n$ (by the implicit function theorem). Hence, in a neighbourhood of any of its points $P \in f^{-1}(Q)$ a complete pre-image is given in a non-degenerate way and is a manifold. With this, we have completed the proof.

COROLLARY 1. *If $m = n$ and if a manifold M_1^n is compact (where $f: M_1^n \to M_2^n$), then the complete pre-image of a regular value $Q \in M_2^n$ consists of a finite number of points P_j ($j = 1, \dots, N$); when the point Q is moved little, $Q \to Q'$, the new value $Q' \in M_2^n$ is also regular, its pre- image also being moved little in the manifold M_1^n.*

COROLLARY 2. *If $m = n$ and if both manifolds M_1^n, M_2^n are oriented and M_1^n is compact, then at each point of the complete pre-image $P \in f^{-1}(Q)$ the sign is well defined:*

$$\mathrm{sgn}_\beta (P) = \mathrm{sgn} \det \left(\frac{\partial x^\alpha}{\partial y^\beta}\right)_P .$$

DEFINITION 4. The *degree of mapping of oriented manifolds* $M_1^n \xrightarrow{f} M_2^n$ at a regular value $Q \in M_2^n$, where the complete pre-image $f^{-1}(Q)$ consists of a finite number of points P_α, is the sum

$$\deg_Q f = \sum_{P_\alpha \in f^{-1}(Q)} \mathrm{sgn}\,(P_\alpha).$$

REMARK. For non-compact manifolds M_q^1 the class of *proper* maps $M_1^q \xrightarrow{f} M_2^n$, such that the complete pre-image of a compact set N (in particular, a single point $Q = N$) is compact itself, is defined; $f^{-1}(N)$ is compact provided that N is compact.

We recall that compact is a set of points, such that any infinite sequence P_α of its points $P_\alpha \in N$, has in this set, a limiting point $P \in N$.

For proper maps of oriented manifolds $M_1^n \xrightarrow{f} M_2^n$ (possibly non-compact, e.g. $M_1^n = \mathbb{R}^n$) the degree $\deg_Q (f)$ is defined, where $Q \in M_2{}^n$ is a regular value (i.e. a regular point in the image)

$$\deg_Q f = \sum_{P_\alpha \in f^{-1}(Q)} \mathrm{sgn}_{P_\alpha} (f).$$

For example, suppose we are given a map of a straight line $\mathbb{R}^1 \xrightarrow{f} \mathbb{R}^1$ where $y = f(x)$.

Suppose $y \to \pm\infty$ in the case when $x \to \pm\infty$, then this map is proper (for instance, if $y = f(x)$ is a polynomial).

EXERCISE. For proper maps of a straight line $\mathbb{R}^1 \to \mathbb{R}^1$ the degree of a map can be equal to zero or unity.

EXAMPLE 1. Suppose we are given a map of circumference into circumference

$$f\colon\ S^1(x) \to S^1(y).$$

This map is described by the function $y = f(x)$, where the numbers x, as well as $x + 2\pi n$ and $y = 2\pi m$ for integer n and m, define identical points of both circumferences.

The function $y = f(x)$ satisfies the condition

$$f(x + 2\pi) = f(x) + 2k\pi,$$

where k is an integer since the points x and $x + 2\pi$ coincide; therefore, the points $y_1 = f(x)$ and $y_2 = f(x + 2\pi)$ must also coincide. The number k is constant since the map is continuous. A simple equality holds:

$$k = \deg(f)$$

(in this case the degree of a mapping is frequently referred to as the *rotational number*). We can verify that the rotational number

$$k = \frac{1}{2\pi}\,(f(x + 2\pi) - f(x)) = \frac{1}{2\pi}\int_0^{2\pi} \frac{df}{dx}\,dx$$

coincides with the degree of mapping at any regular point Q on the circumference. The simplest maps of degree k are linear:

$$f(x) = kx$$

(the points x and $f(x)$ lie on the circumference).

Obviously, $f(x + 2\pi) = k(x + 2\pi) = kx + 2k\pi$ and the number k must be an integer.

EXAMPLE 2. Suppose we are given a two-dimensional sphere as a projective complex space $\mathbb{C}P^1$ coordinatized by complex projective coordinates $(u, v) \cong \lambda(u, v)$; $\lambda \neq 0$,

$$w = u/v = 1/z, \quad \text{where } z = 1/w.$$

Let in the z-plane there exist a polynomial of degree n

$$z^1 = f_n(z) = a_0 z^n + \dots + a_n.$$

This polynomial describes a map $\mathbb{C}P^1 \xrightarrow{f} \mathbb{C}P^1$. The equation $z^1 = A = f_n(z)$ for almost all A has n roots.

Thus, the complete pre-image $f^{-1}(A)$ has the form $(z_1, \dots, z_n)$, and all the Jacobians have positive signs (prove this!). Therefore, we have $\deg f = n$. We are now in a position to formulate the important theorem.

THEOREM 2. *The degree of the map $M_1^n \xrightarrow{f} S^n$ of any closed, oriented manifold onto a sphere S^n does not depend on the choice of a regular value $Q \in S^n$. Furthermore, the degree remains unchanged under smooth homotopies (deformations) of the map f.*

We shall give an important definition: the *homotopy* (or *deformation*) of any map f: $X \to Y$, $y = f(x)$ is a continuous map $y = F(x, t)$ of a cylinder with base X of length 1, $0 \le t \le 1$,

$$X \times I(0, 1) \to Y; \quad y = F(x, t),$$

where $F(x, 0) = f$; all the maps $f_t(x) = F(x, t)$ are called *homotopic* to the initial map f, $0 \le t \le 1, f_0 = f$. We shall consider only maps of smooth manifolds and assume the maps $f(x)$, $F(x, t)$ to be smooth (smooth homotopy).

We shall construct the proof as indicated in the following scheme.

We shall use Sard's lemma and its corollaries (Theorem 1). To begin with we prove that at a given point Q (of the image) which is regular for two homotopic maps

$$f\colon M_1^n \to S^n; \quad g\colon M_1^n \to S^n,$$

the degrees are the same: $\deg_Q(f) = \deg_Q(g)$.

The steps of the proof are as follows.

Step 1. If M_1 is closed and if $Q \in S^n$ is a regular value, then all the points sufficiently close to Q are regular values and have the same degree of the map f.

Step 2. If $F(x, t)$ is a smooth homotopy between maps f and g, such that $F(x, 0) = f(x)$, $F(x, 1) = g(x)$, then there exists a point Q^1 arbitrarily close to the point Q, which is a regular value for all the homotopy of the map F.

Step 3. If we are given a point Q^1 regular for the whole map $F(x, t)$, then the complete pre-image $F^{-1}(Q^1)$ is a one-dimensional manifold which has boundary only when $t = 0$ and $t = 1$ (Figure 53).

Step 4. Under the conditions formulated in Lemma 3, the degrees $\deg_{Q^1} f$ and $\deg_{Q^1} g$ are exactly equal to each other at the point Q^1 which is regular for all the homotopy $F(x, t)$.

Step 5. Since using a smooth homotopy we can (by rotating the sphere S^n) carry any value Q^1 of the sphere into any other value Q^2, from the invariance of the degree under homotopy at *a given point* Q^1 *there follows independence of the degree of mapping of the choice of a regular value.*

Steps 1 to 5 imply Theorem 2. The motivation of Step 1 can be found in Theorem 1 and its Corollaries 1 and 2.

Step 2 is immediate from Sard's lemma.

Step 3 follows from the definition of a regular value (Theorem 1).

Step 4 is crucial. Let us view the schematic picture (Figure 54) showing the complete pre-image $F^{-1}(Q^1)$.

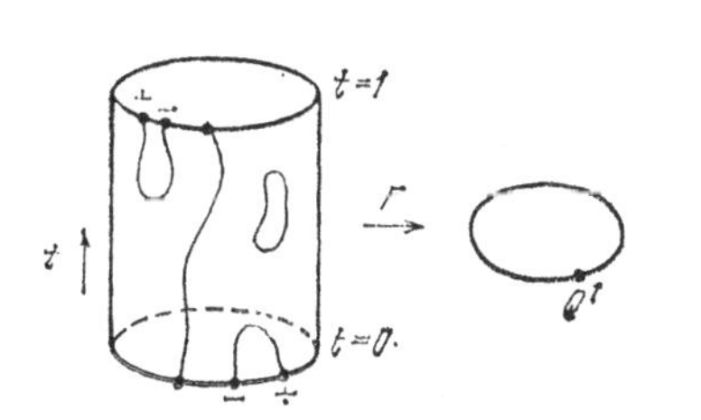

Figure 53.

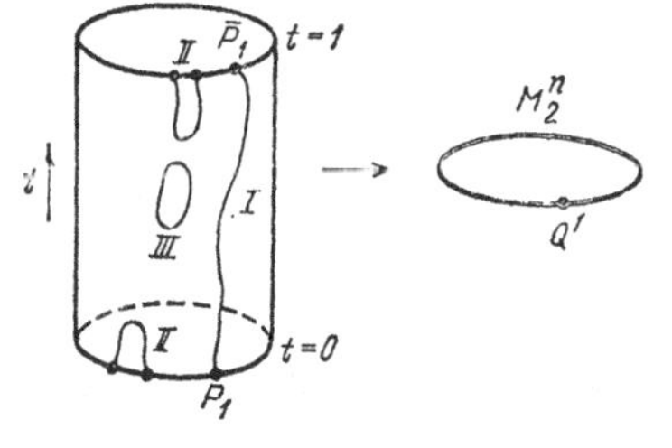

Figure 54.

This pre-image is a one-dimensional compact manifold which has boundary only when $t = 0, 1$ because the value $Q^1 \in M_2^n$ is regular. This pre-image falls into the following pieces:

1) a smooth line going from $t = 0$ to $t = 1$ (type I);

2) a smooth line with both ends lying either at $t = 0$ or at $t = 1$ (type II);

3) a closed line without ends (type III).

Obviously, for type I the ends are points of the complete pre-images, $P_1 \in f^{-1}(Q^1)$ and $\bar{P}_1 \in g^{-1}(Q^1)$. These points are of the same sign. For type II the ends of the segment are pairs of points

$$(P_2, P_3) \in f^{-1}(Q^1) \quad \text{or} \quad (\overline{P_2}, \overline{P_3}) \in g^{-1}(Q^1)$$

which have opposite signs

$$(\mathrm{sgn}_{P_2} f = -(\mathrm{sgn}_{P_3} f).$$

The pre-image $F^{-1}(Q^{-1})$ of type III does not intersect $t = 0$ and $t = 1$. Hence, the sums are equal

$$\deg_{Q^1} f = \deg_{Q^1} g = \sum_{f^{-1}(Q^1)} \mathrm{sgn}_{P_1}(f) = \sum_{g^{-1}(Q^1)} \mathrm{sgn}_{\overline{P}_1}(g)$$

which implies the assertion of Step 4.

Step 5. Suppose $A(t)$ is a rotation of the sphere S^n, where $A(0) = 1$ and $A(1)Q_1 = Q_2$, Q_1 and Q_2 being two regular values of the map f: $M_1^n \to S^n$. Obviously, the homotopy (deformation) of the map f is defined to be

$$F(x, t) = A(t) f(x); \quad 0 \le t \le 1,$$

where $F(x, 0) = f(x)$. Under the map $f_1 = F(x, 1) = A(1) f(x)$ we have

a) $f_1^{-1}(Q_2) = f_0^{-1}(Q_1) \equiv f^{-1}(Q_1) \quad (f_0 \equiv f)$,

b) therefore, $\deg_{Q_2} f_1 = \deg_{Q_1} f_0$,

c) by virtue of Theorem 2 we have $\deg_{Q_1} f_1 = \deg_{Q_1} f_0$, whence

$$\deg_{Q_1} f_1 = \deg_{Q_1} f_0 = \deg_{Q_2} f_1.$$

Thus, the degree does not depend on the choice of a regular point, as required.

COROLLARY 3. *If* $z^1 = f_n(z)$ *is a polynomial of degree* n*, then for all* $z^1 = A$ *the equation* $f_n(z) = A$ *is soluble (Gauss theorem).*

Proof. Suppose, by contradiction, the complete pre-image is empty. Then the value z^1 is regular, and the degree of the map $\mathbb{C}P^1 \xrightarrow{f_n} \mathbb{C}P^1$ specified by the polynomial $z^1 = f_n(z)$ must be equal to zero. There exists, however, at least a single point A, such that the pre-image is non-empty and the roots are aliquant. By virtue of the complexity of the map, all the signs are positive, and therefore $\deg f_n \neq 0$. This is a contradiction.

REMARKS on Theorem 2.

1. Theorem 2 on the degree of a mapping can also be applied to smooth maps between non-compact manifolds $M_1^n \to M_2^n$ provided that both the map f itself and all the homotopy $F(x, t)$ $(0 \le t \le 1)$ are proper maps (in the case where the maps are not proper, Theorem 2 is invalid). Indeed, let f: $\mathbb{R}^1 \to \mathbb{R}^1$ be the map of a straight line, where $y = \arctan(x)$; we can see that the complete pre-image $f^{-1}(y_0)$ is empty when $|y_0| \ge \pi/2$; on the contrary, the complete pre-image $f^{-1}(y_0)$ consists of one point if $|y_0| < \pi/2$; this map is, however, not proper since the pre-image of the closed segment $f^{-1}([-\pi/2; +\pi/2])$ is non-compact, it is the whole straight line: $(-\infty, +\infty)$.

2. If the manifold M_1^n is closed (and, therefore, compact), and the manifold M_2^n is non-compact (e.g. M_2^n is a Euclidean space $\mathbb{R}^n$), then the degree of any map f: $M_1^n \to M_2^n$ is equal to zero (prove this!).

3. If one of the manifolds M_1^n, M_2^n is non-orientable (e.g. a projective plane $\mathbb{R}P^2$), then the degree can be defined only as the residue modulo 2 since the signs of the points from the complete pre-image cannot be well defined.

4. A very important case of the degree of a mapping is its adaptation to manifolds with boundary.

DEFINITION 5. *A manifold N^n with boundary ∂N* is a region in a closed manifold M^n described by the equation

$$f(x) \le 0,$$

where $f(x)$ is a smooth function on the manifold M^n, such that its gradient ∇f is non-zero wherever $f(x) = 0$.

The boundary of the manifold N is a hyper-surface $f(x) = 0$; the boundary is denoted by ∂N^n. Suppose we are given manifolds with boundaries N_1^n and N_2^n and their smooth map f: $N_1^n \to N_2^n$ such that the image of the boundary $f(\partial N_1^n)$ always goes onto the boundary N_2^n. Then for any point $Q \in N_2^n$ which does not belong to the boundary, the complete pre-image $f_n^{-1}(Q) \in N_1^n$ consists of interior points (which do not belong to the boundary) of N_1^n. The degree of the map is the sum

$$\deg_Q f = \sum_{P_\alpha \in f^{-1}(Q)} \operatorname{sign}_{P_\alpha}(f)$$

where both the manifolds are oriented.

We shall formulate the following theorem without proof.

THEOREM 3. *If the homotopy $F(x, t)$ of the maps $f_t = F(x, t)$: $N_1^n \to N_2^n$ is such that the image of the boundary $F(\partial N_1^n; t)$ for all $t \leq 1$ goes onto the boundary ∂N_2^n, then the degree of the mapping f_t: $N_1^n \to N_2^n$ remains unchanged under homotopy and does not depend on the choice of a regular value. Moreover, if the map f: $N_1^n \to N_2^n$ is one-to-one on the boundary, then* $|\deg f| = 1$.

The proof of Theorem 3 is identical to the proof of Theorem 2 except for the last point concerning boundaries. This point follows from the lemma below.

LEMMA 2. *The degree of a smooth one-to-one mapping of closed manifolds is equal to* ± 1; *the degree of mapping of boundaries is equal to the degree of mapping of the interior points of manifolds with boundary.*

We shall not give the proofs of Lemma 2 and Theorem 3.

3.3 Applications of the Degree of a Mapping

Suppose there exists a smooth map of closed manifolds $M_1^n \xrightarrow{f} M_2^n$ and suppose Ω is a differential form of degree n on M_2^n. In terms of local coordinates $(y^1, \dots, y^n)$ this form is given by $\Omega = g(y)\, dy^1 \wedge \dots \wedge dy^n$.

THEOREM 1. The following formula holds

$$\int_{M_1^n} f^*(\Omega) = (\deg f) \int_{M_2^n} \Omega.$$

Proof. By Sard's lemma, almost all (in the sense of measure) values $Q \in M_2^n$ are regular. Consider a regular value Q_1 and its small regular neighbourhood U relative to coordinates $(y^1, \dots, y^n)$. The complete pre-image $f^{-1}(U) = V_1 \cup \dots \cup V_N$ consists of regular points if the neighbourhood U is small. In neighbourhoods V_α on the manifold M_1^n there lie points P_α of the complete pre-image of the point Q:

$$f^{-1}(Q) = P_1 \cup \dots \cup P_N;$$

$$f^{-1}(U) = V_1 \cup \dots \cup V_N .$$

Let us denote the local coordinates in the neighbourhoods V_α by $(x_\alpha^1; \dots ; x_\alpha^n)$; $\alpha = 1, \dots , N$, assuming that these coordinates x_α are chosen corresponding to the orientation of the manifold M_1^n and the coordinates $(y^1, \dots , y^n)$ in the region U corresponding to the orientation of the manifold M_2^n. In the regions V_α the map f is given by

$$y^j = f^j_{(\alpha)} (x_\alpha^1; \dots ; x_\alpha^n); \quad j = 1, 2, \dots , n$$

and the degree has the form

$$\deg_Q (f) = \sum_{\alpha=1}^{N} \operatorname{sgn} \deg \left(\frac{\partial y^q}{\partial x^p_{(\alpha)}}\right)_{P_\alpha} .$$

For each region V_α with coordinates $x_{\alpha,}$ by the theorem on the change of variables in the integral, the following formula holds

$$\int_{V_\alpha} f^*(\Omega) = \left[\operatorname{sgn} \det \left(\frac{\partial y^q}{\partial x^p_{(\alpha)}}\right)_{V_\alpha}\right] \int_U \Omega,$$

where locally we have $\Omega = g(y)\, dy^1 \wedge \ldots \wedge dy^n$. Summing this equality over all $\alpha = 1, \ldots, N$, we obtain either

$$\int f^*(\Omega) = \left(\sum_{\alpha=1}^{N} \operatorname{sgn} \det \left(\frac{\partial y^N}{\partial x^p_\alpha}\right)\right)_{V_\alpha} \cdot \int \Omega$$

or

$$\int_{f^{-1}(U)} f^*(\Omega) = (\deg f) \cdot \int_U \Omega.$$

If, from the manifold M_2^n, we discard the sub-set of all irregular values $N \subset M_2^n$ of measure zero and partition the remainder $M_2^n \backslash N$ into a union of regular points $\bigcup_k U_k = M_2^n \backslash N$, we shall come to

$$\int_{M^n} f^*(\Omega) = \int_{\cup f^{-1}(U)} f^*(\Omega) = (\deg f) \cdot \int_{M_2^n \backslash N} \Omega$$

since $M_2^n \backslash N = \bigcup_k U_k$, and the theorem follows. (Note that the form $f^*(\Omega)$ vanishes at all irregular points of M_1^n, and these points can be discarded from M_2^n without changing the integral).

COROLLARY 1. *If $M^2 \subset \mathbb{R}^3$ is a closed surface in a three-dimensional Euclidean space and if K is the Gaussian curvature, then the following formula holds*

$$\int_{M^2} K\, d\sigma = 4\pi \times (\text{an integer}), \quad d\sigma = (g)^{1/2}\, dy^1 \wedge dy^2$$

where the integer is equal to the degree of the Gauss spherical map

$$M^2 \xrightarrow{f} S^2.$$

Proof. If θ, ϕ are spherical coordinates, then we know that $K\, d\sigma = f^2(\Omega)$, where $\Omega = |\sin \theta|\, d\theta \cdot d\phi$. Next we note that

$$\int_{S^2}(\Omega) = \int_0^{2\pi}\left(\int_0^{\pi}|\sin\theta|\,d\theta\right)d\phi = 2\pi\int_0^{\pi}|\sin\theta|\,d\phi = 4\pi.$$

By the theorem just proved we have

$$\int_{M^2} K\,d\sigma = \int_{M^2} f^*(\Omega) = (\deg f)\cdot\int_{S^2}\Omega = (4\pi)\deg f$$

which complétes the proof of the corollary.

For curves in the plane $M^1 \subset \mathbb{R}^2$ we also have the spherical Gauss map $f: M^1 \to S^1$. If a curve is closed $x^\alpha = x^\alpha(t)$; $\alpha = 1, 2$; $0 \le t \le 2\pi$, we obtain (the curve is assumed to be regular, i.e. $|d\bar{x}/dt| \ne 0$) $f^*(d\phi) = k\,dl$, $t = 2\pi$, where k is curvature,

$$\int_M f^*(d\phi) = \int_{t=0}^{t=2\pi} k\,dl =$$

$$= \text{(the rotational number of the normal)}\cdot\int d\phi = 2\pi n.$$

In this case the degree $(\deg f)$ is equal to the rotational number of the normal along the curve.

DEFINITION 1. We shall call a curve $x^\alpha(t)$ on a plane $\mathbb{R}^2$ *typical* (*in general position*) if:

a) all the points of self-intersection are double, and

b) the tangent vectors at these points are not parallel (the curve is thought of as regular: $|dx/dt| \ne 0$).

EXERCISE. Prove that $l + 1$, where l is the number of self-intersection points of a typical flat curve, is equal to the rotational number of the normal (modulo 2). We can see that for curves in a plane the degree of a Gauss map depends on the position of the curve on the plane, and this degree remains unchanged under regular homotopies of the curve (i.e. under such homotopies that the curve is regular at all instants of time). If the tangent vector of the curve becomes degenerate under deformation, then the rotational number $\deg f$ can change as shown in Figure 55. Thus it is easy to investigate the degree of a Gauss map for curves in a plane. It is much more complicated to calculate this degree for surfaces in a three-dimensional space $M^2 \subset \mathbb{R}^3$.

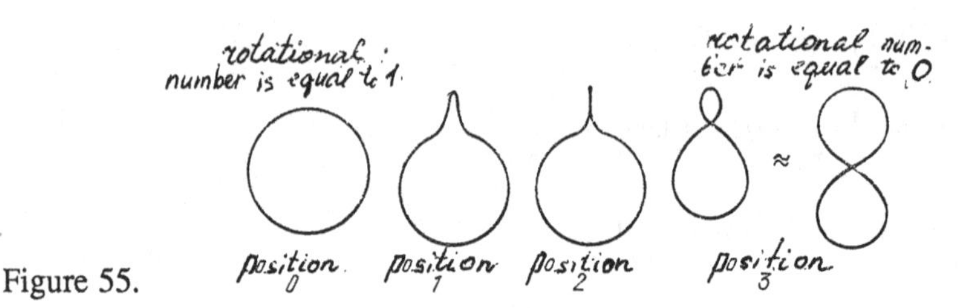

Figure 55.

On the basis of Corollary 1 above, we obtain $\int_{M^2} K\, d\sigma = 4\pi$ (an integer) $= 4\pi$ (deg f).

We may assume at once that this integer does not depend on the embedding of the closed surface in $\mathbb{R}^3$. The point is that $K = R/2$, where R is the scalar curvature of the Riemannian metric on the surface itself, and therefore $\int_{M^2} K\, d\sigma$ is an inner invariant coinciding with $1/2 \int R\, d\sigma$, where $d\sigma$ is the Riemannian element of volume (area).

REMARK. We have so far considered the expression $\int R\, d\sigma$ only for metrics of the surface M^2 induced by the embedding in three-dimensional Euclidean space $\mathbb{R}^3$. As a matter of fact, in the two-dimensional case this expression remains unchanged upon variations of the metric, for any metrics: let M be an n-dimensional manifold and let $g_{ij}^{(\alpha)}$ be a metric smoothly depending on the parameter α, such that, outside a compact region, g_{ij} does not depend on α (for example, the entire manifold is compact and closed). The general formula holds

$$\frac{d}{d\alpha}\left[\int_{M^n} R\, d\sigma\right] = \int_{M^n} (R_{ij} - 1/2\, R\, g_{ij}) \frac{dg^{ij}}{d\alpha} d\sigma \tag{1}$$

(see reference [27]). In particular, for $n = 2$ we have $R_{ij} = 1/2\, R\, g_{ij}$ (always), and therefore $(d/d\alpha)\,(\int R\, d\sigma) \equiv 0$. Besides, any two Riemannian metrics can be joined via deformation with parameter α. In particular, a torus T^2 has a Euclidean metric where $R = 0$. Therefore, for any metric $\int_{T^2} R\, d\sigma = 0$. For a sphere we have

$1/2 \int_{S^2} R\, d\sigma = 4\pi$ in the standard metric of a unit sphere. We, therefore, always have $1/2 \int_{S^2} R\, d\sigma = 4\pi$; for an oriented surface M_g^2 with g handles we have

$$1/2 \int_{M_g^2} R\, d\sigma = 4\pi \cdot (1 - g) \quad \text{(whatever the metric).}$$

We shall not make use of formula (1) and for a surface in $\mathbb{R}^3$ shall investigate the degree of the Gauss map using another method. The idea of this method is as follows. We consider a Gauss map $f\colon M^2 \to S^2$ for the surface $M^2 \subset \mathbb{R}^3$. Suppose that a pair of opposite points of the sphere $(n, -n)$ are both regular values. We also consider the "height" function on the surface $g(P)$, $P \in M^2$, whose value at the point P is equal to the orthogonal projection of the surface onto the straight line going through the origin of coordinates in the direction of n. The critical (stationary) points of the function $g(P)$ are points P_i, where $(\nabla_g)(P_i) = 0$. The stationary points P_i are, obviously, such that the vector n (or $-n$) is orthogonal to the surface. Thus, we have: the set of stationary points P_i of the function $g(P)$ is $f^{-1}(n) \cup f^{-1}(-n)$ (which is the union of two complete pre-images). It can be readily shown that the stationary point P_i is non-degenerate (i.e. $\det (\partial^2 g/\partial y^\alpha\, \partial y^\beta)_{P_i} \neq 0$) if and only if the point is regular for the Gauss map $f\colon M^2 \to S^2$.

Let us ascribe to the maxima and minima of the function the multiplicity, namely, a numeral 1, and to the saddles a numeral -1. On the surface, the sum of stationary points with multiplicities appears to be independent of the choice of the function and is equal to a doubled degree of the Gauss map.

We shall investigate these questions in the sections which follow.

3.4 Vector Fields

In this section we are primarily concerned with the presentation of the simplest concepts associated with vector fields in the plane and space. The reader is already familiar with one example of the vector field in Euclidean space — this is grad $f(x)$, where f is a smooth function on the space $\mathbb{R}^n$. It should be noted that grad f is, in fact, not a vector, but a covector field and can be interpreted as a vector field only if the space is endowed with a *Riemannian metric*; a more thorough treatment was given in Part II. Recall that we are already acquainted with the concept of a derivative of the function $f(x)$ in the direction a: df/da. We also proved the formula: $df/da = (a, \operatorname{grad} f)$ (a scalar product).

Suppose we are given a smooth function $f(x)$ on $\mathbb{R}^n$; consider its level hyper-surfaces (if $n = 3$, we shall simply speak of level surfaces and if $n = 2$, we shall speak of level lines), i.e. a tuple of all points $x \in \mathbb{R}^n$, for which $f(x) = c$, where c is a fixed constant. The level hyper-surface is descibed by $n - 1$ parameters (since one constraint, namely, the equation $f(x) = c = \text{const.}$ is imposed upon n parameters in $\mathbb{R}^n$), and, therefore, the dimension of the hyper-surface $\{f = c\}$ is equal to $n - 1$.

DEFINITION 1. A point $x_0 \in \{f = c\}$ is called *non-singular* if $\operatorname{grad} f(x_0) \neq 0$; it is called *singular* if $\operatorname{grad} f(x_0) = 0$.

EXAMPLE. Suppose on $\mathbb{R}^2$ we are given a function $z = x^2 - y^2$; consider the level line $\{z = 0\} = \{x^2 - Y^2 = 0\}$. The level line consists of two straight lines $x = \pm y$ intersecting at the origin of the coordinates. The point O is the only singular point of this level line.

Let $x_0 \in \{f = c\}$ be a non-singular point. The vector a applied to the point x_0 is called *tangent* to the hyper-surface $\{f = c\}$ if there exists a smooth curve $\gamma(t)$, the whole of which belongs to the hyper-surface $\{f = c\}$, such that $\gamma(0) = x_0$ and $d\gamma(t)/dt\,|_{t=0} = a$.

PROPOSITION 1. *Let $f(x)$ be a smooth function on $\mathbb{R}^n$ and let $x_0 \in \{f = c\}$ be a non-singular point. Then the vector grad $f(x_0)$ is orthogonal to the hyper-surface $\{f = c\}$ at the point x_0, i.e. grad $f(x_0)$ is orthogonal to any vector a tangent to the hyper-surface $\{f = c\}$.*

Proof. Since $(a, \operatorname{grad} f) = df/da$, it suffices to calculate the derivative of f with respect to the direction a, i.e. $\frac{df(\gamma(t))}{da}\Big|_{t=0}$. However, $f(\gamma(t)) = \text{const.} = c$ since the whole of $\gamma(t)$ lies on the hyper-surface $\{f = c\}$, i.e. $df/da = 0$, as required.

If $x_0 \in \{f = c\}$ is a singular point, then $\operatorname{grad} f(x_0) = 0$, and therefore, we can formally take $\operatorname{grad} f(x)$ to be orthogonal to the hyper-surface $\{f = c\}$at any of its points.

Since the direction and the magnitude of the vector $\operatorname{grad} f$ indicate the direction and the rate of the function increase, we have proved that the function f always increases along the normal to an arbitrary level hyper-surface $\{f = c\}$. Now we are in a position to formulate the general concept of a vector field $v(x)$ in a space $\mathbb{R}^n$.

DEFINITION 2. We say that in a certain region $G \in \mathbb{R}^n$ a *vector field* $v(x)$ is defined if at each point $x \in G$ we are given a vector $v(x) = (v^1(x), \dots, v^n(x))$ with coordinates $v^1(x), \dots, v^n(x)$ which are functions of the point $x \in G$. The vector field $v(x)$ is called *continuous* (respectively, *smooth*) if the functions $v^i(x)$, $1 \leq i \leq n$, are continuous (respectively, smooth) in the region G. The point x is called a *non-singular point of the vector field* $v(x)$ if this field $v(x)$ is continuous (smooth) at a certain neighbourhood of the point x and if, as well, $v(x) \neq 0$. Otherwise, the point x is called a *singular point of the vector field* $v(x)$. A singular point x_0 of the field $v(x)$, such that $v(x_0) = 0$, is called the *zero of the field* $v(x)$ or the *equilibrium position.*

Vector fields having discontinuity points and essentially singular points play an important role in physics and mechanics (for instance, in hydromechanics). We shall become acquainted with such fields later.

DEFINITION 3. Let $v(x)$ be a smooth vector field. A trajectory (a curve) $\gamma(t)$ is called the *integral trajectory* of this vector field if $\dot{\gamma}(t) = v(\gamma(t))$, i.e. if tangent vectors to the curve $\gamma(t)$ are vectors of the field v.

Recall that by the term "curve $\gamma(t)$" we always mean a curve with parametrization and not simply a geometric image of a curve γ.

Let us examine the simplest examples of integral trajectories. Suppose $f(x)$ is a function on a plane and $v(x) = \operatorname{grad} f(x)$:

a) $f(x) = (x^1)^2 + (x^2)^2$; $\operatorname{grad} f(x) = (2x^1, 2x^2)$ (Figure 56). Here the integral trajectories form a bundle of rays;

b) $f(x) = -(x^1)^2 + (x^2)^2$; $\operatorname{grad} f(x) = (-2x^1, 2x^2)$ (Figure 57). Here the integral trajectories are hyperbolas;

c) $f(x) = -(x^1)^2 - (x^2)^2$; $\operatorname{grad} f(x) = (-2x^1, -2x^2)$ (Figure 58).

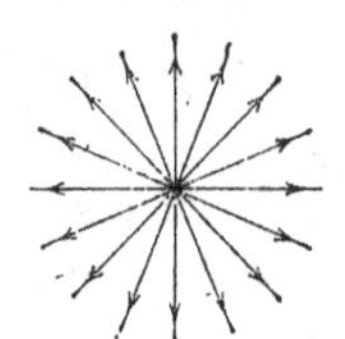

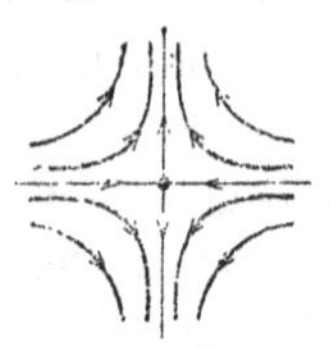

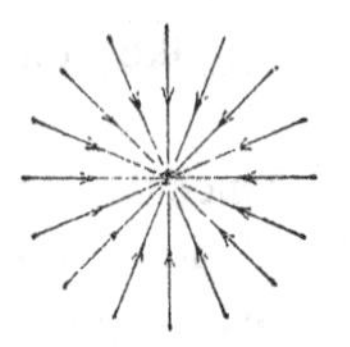

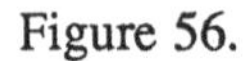

Figure 56. Figure 57. Figure 58.

All the fields a), b), c) have, at the origin of coordinaters, a singular point — the zero of the field $v(x)$. The function $f(x)$ in example a) has a minimum at the point 0; in example c) it has a maximum at the point 0; in example b) the point 0 is the saddle of the function $f(x)$.

The vector field $v(x)$ is often interpreted as a *flow of liquid* through a space or a region of space. Then it is assumed that an arrow is attached to each particle of this liquid, which is the velocity vector of the particle. Singular points of the vector field are singular points of the *flow* of liquid. Sometimes, the flow of a liquid through a space is called, for brevity, simply the flow. So, for instance, in example a) the singular point 0 of the flow $v(x)$ is the *source*, while in example c) the point 0 is the *discharge*.

Integral trajectories of the field $v(x)$ are occasionally referred to as *lines of flow* of a liquid whose motion is described by this velocity field.

Of course, it is not each real (physical) flow of liquid that generates a vector field in the above-mentioned sense. The point is that the coordinates of our velocity field vectors do not depend on time (they depend only on the point in space); in other words, the flows of liquid corresponding to such fields are *stationary flows*. Time-dependent flows are called *non-stationary flows*.

Let the field $v(x)$ be a *gradient*, i.e. $v(x) = \operatorname{grad} f$. Consider an arbitrary integral trajectory $\gamma(t)$ of this field and consider a function $h(t) = f(\gamma(t))$. Then $h(t)$ is a strictly monotonically increasing function of all of those t for which $\gamma(t)$ is a non-singular point of the field $v(x)$. Indeed,

$$\frac{dh(t)}{dt} = \frac{d(f(\gamma(t)))}{dt} = \frac{df}{dv} = (v, \operatorname{grad} f) = |\operatorname{grad} f|^2 > 0$$

(at a non-singular point). Thus the function $f(x)$ increases monotonically along each integral trajectory of the field $v(x) = \operatorname{grad} f(x)$.

CLAIM 1. *Let* $v(x) = \text{grad} f(x)$. *Then among integral trajectories of the vector field* $v(x)$ *there is not a single closed integral trajectory.*

Proof. If such a trajectory $\gamma(t)$ existed, then moving along γ from the point $\gamma(t_0)$ in the direction of the field $v(x)$, we would go back to the initial point $\gamma(t_0)$ within a finite time, which contradicts the strictly monotonic increase of the function $f(x)$ along $\gamma(t)$, and the result follows.

For example, the vector field $v(x) = v(x^1, x^2) = (-x^2, x^1)$ cannot be the gradient field for the function $f(x)$ as long as the integral trajectories of this field are *closed* (Figure 59).

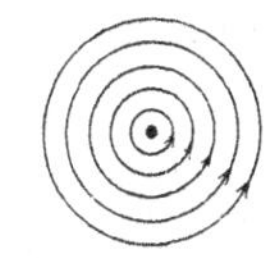

Figure 59.

Suppose we are given a smooth field $v(x)$. The following practical question arises: How shall we find the explicit form of the integral trajectories (lines of flow) of this field? The definition of an integral trajectory leads us to a system of differential equations. Let $\gamma(t) = (x^1(t), \dots, x^n(t))$, where $x^i(t)$; $1 \le i \le n$ are the unknown functions to be defined. Since $d\gamma(t)/dt = v(\gamma(t))$, we arrive at the system

$$\frac{dx^1(t)}{dt} = v^1(x^1(t), \dots, x^n(t)),$$

$$\dots\dots\dots\dots\dots$$

$$\frac{dx^n(t)}{dt} = v^n(x^1(t), \dots x^n(t)).$$

The solutions of this system of equations are, obviously, just the integral trajectories of the field $v(x)$. We should note that the right-hand sides of this system do not contain the parameter (time) in an explicit form. Hence to each vector field there corresponds a *system of differential equations*. The inverse is also valid.

Suppose that we are given a system of differential equations and assume that the right-hand side of this system does not involve time in an explicit way. Such systems of differential equations are called *autonomous systems*:

$$\frac{dx^1}{dt} = f_1(x^1, \dots, x^n),$$

$$\dots\dots\dots\dots$$

$$\frac{dx^n}{dt} = f_n(x^1, \dots, x^n).$$

Then we can construct a vector field $v(x)$, setting $v^i(x) = f_i(x) = f_i(x^1, \dots, x^n)$, where $x^1, \dots, x^n$ vary within a certain region $G \subset \mathbb{R}^n$. Thus, we have established a relationship between vector fields $v(x)$ and autonomous systems of differential equations. Note that the singular points of the vector field $v(x)$ are exactly the singular points of the system of differential equations, and vice versa. We shall use this simple relationship for the geometric description of an important concept,namely, the integral of a system of equations.

By definition, the integral of a system of equations is a function $f(x^1, \dots, x^n)$ constant on all the trajectories, i.e. on all the solutions of this system. Using this, we shall construct a vector field $v(x)$ in $\mathbb{R}^n$; then the solutions of the system will be integral trajectories of the field $v(x)$. Let us consider an arbitrary hyper-surface $\{f = c\}$ (where c is fixed, but arbitrary). From the definition of the integral of a system it is immediate that if an integral trajectory has at least one common point with $\{f = c\}$, then the whole of the trajectory lies on the hyper-surface $\{f = c\}$, i.e. the vector field $v(x)$ is tangent to $\{f = c\}$ at each point of this hyper-surface (Figure 60).

Figure 60.

Consequently, each vector $v(x)$ is tangent exactly in one hyper-surface $\{f = c\}$ through the point x. This allows us to lower the initial system of equations from n to $n - 1$, restricting it (i.e. restricting the vector field $v(x)$) to the hyper-surface $\{f = c\}$. Recall that the dimension of $\{f = c\}$ is equal to $n - 1$. This simple geometric procedure of restricting the field $v(x)$ precisely corresponds to the well-known assertion that the definition of the first integral of a system makes it possible to lower its order *by unity*. If the second, the third etc., integrals of the system are known, geometrically this means that we can continue restricting the vector field $v(x)$ to level surfaces of increasingly small dimensions.

If it so happens that we have lowered the dimension of the level surface to unity (i.e. we have obtained a one-dimensional curve), this means that we have managed to completely integrate the system, i.e. to find its solutions.

Since a stationary flow of liquid may be not only through a Euclidean space, but also through a surface (for example, a liquid may spread about a two-dimensional sphere), we can study vector fields and the corresponding differential equations on surfaces in Euclidean space. Differential equations are of great importance, for instance, on a two-dimensional torus (the boundary of a roll), but we shall not go into detail. The differential equations on the torus will be discussed in celestial mechanics.

As a concluding remark, we shall note an interesting fact that any stationary flow of liquid on a two-dimensional sphere must necessarily have at least one singular point (such as, for example, a source or a discharge). This distinguishes the sphere from other two-dimensional surfaces. For example, on the torus there exist vector fields *without singular points* (Figure 61).

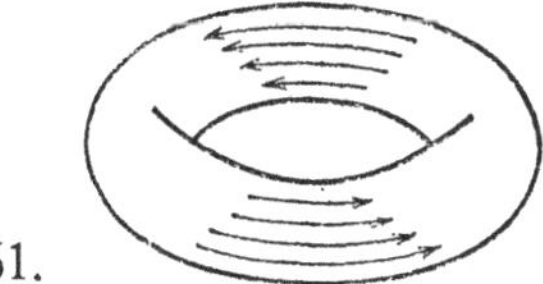

Figure 61.

It should be noted that a vector field without singular points obviously exists on the circumference, which is a one-dimensional sphere. On the three-dimensional sphere there also exist vector fields without singularities. This fact is generally valid for any sphere of odd dimension.

It turns out that the presence or absence of singular points is connected with the global properties of surfaces, with the so-called *topological properties*.

We now proceed to a more detailed study of a certain special class of vector fields on the plane. To begin with we shall make a statement which will be of special importance throughout this section.

An essential part of the theory of two-dimensional Riemannian geometry of surfaces in some exact sense comes out as the geometry of functions of one complex variable.

We shall demonstrate this general principle on an example of vector fields existing, for instance, in hydromechanics.

We shall emphasize once again that the hydromechanical interpretation of vector fields has a deep *physical meaning*, and therefore we shall not divert from using the hydromechanical terminology in our further presentation.

Suppose through a plane with Cartesian coordinates x, y there is a flow of liquid$v = (P, Q)$, where $P(x, y)$ and $Q(x, y)$ are smooth functions on the plane. Suppose also that the flow v is stationary and the liquid is incompressible; let its density ρ be equal to unity (i.e. ρ is constant). Let D be a region on the (x, y)-plane, and let the boundary of the region D be a piecewise smooth curve. We shall denote the mass of liquid escaping from the region D per unit time by $m_1(D)$ and the mass of liquid emerging in the region D per unit time by $m_2(D)$. Suppose $\Delta m(D) = m_1(D) - m_2(D)$ is the change of the mass in the region D (since the liquid is incompressible, $\Delta m(D) = 0$, but for the present we shall not use this fact because the final formula for $\Delta m(D)$ which we are now going to derive is also satisfied in the case of a compressible liquid). Consider an infinitesimal rectangle Π with sides Δx and Δy parallel to the coordinate axes. Then for $\Delta m(\Pi)$ we are led to the following picture (Figure 62).

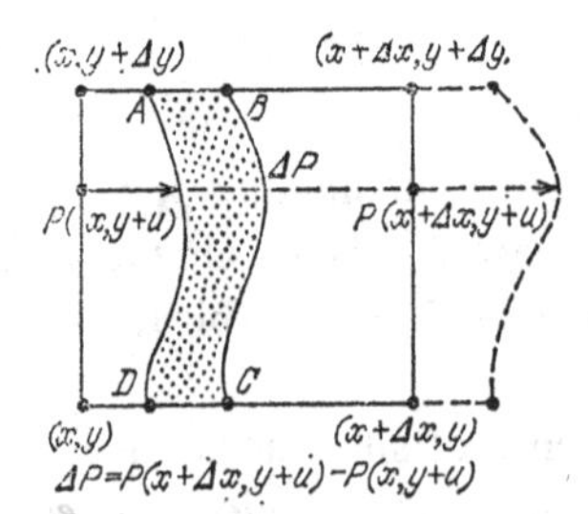

Figure 62.

Since the flow v can be expanded into the sum of two flows $v = (P, Q) = (P, Q) + (0, Q)$, it suffices to be able to calculate the change of the mass for each of these two flows. In Figure 62, the region ABCD shows the variation of mass of the flow (P, Q) (recall that the density is constant).

By virtue of the mean-value theorem known from analysis, we derive

$$\Delta m\,(\Pi) \;=\; \Big(\frac{\partial P(s)}{\partial x} + \frac{\partial Q(t)}{\partial y}\Big)\,\Delta x\,\Delta y,$$

where s and t are points situated somewhere inside our rectangle Π. Approximating the region D by rectangles Π, we finally come to

$$\Delta m\,(\Pi) \;=\; \iint\limits_D \Big(\frac{\partial P}{\partial x} + \frac{\partial Q}{\partial y}\Big) \;=\; \iint\limits_D \operatorname{div}(v)\,dx\,dy,$$

where $\operatorname{div}(v) = \partial P/\partial x + \partial Q/\partial y$. The result obtained can be reformulated as follows.

Consider the region D and the integral trajectories $\gamma(t)$ of the flow v. Let $x \in D$ and $x = \gamma(0)$; then we examine the point $x(t) = \gamma(t)$. If we now fix some value of t, we shall oabtain a set of points $\{x(t)\}$, where $x(0) \in D$. The points $\{x(t)\}$ form a region $D(t)$ which is the image of the region D under translation by t along all the integral trajectories. Let $S(D(t))$ be the area of the region $D(t)$. Then we have, in fact, proved that

$$\frac{d}{dt}\,[S(D(t))]_{t=0} \;=\; \iint\limits_D \operatorname{div}(v)\,dx\,dy.$$

Since, in our case, the liquid is incompressible, it follows that $\Delta m(D) = 0$ for any region D, i.e. $\operatorname{div}(v) = 0$.

We shall introduce another important class of flows v. Recall that if c is an arbitrary piecewise smooth closed contour on a plane, then the circulation of the flow v along the contour c is the integral $\int_C P\,dx + Q\,dy$. The flow v is called *vortex-free* if its circulation along any closed contour is equal to zero.

Suppose the flow v is vortex-free; then if D is a region bounded by an arbitrary closed contour C, we obtain by the Stokes formula

$$0 = \int_C P\,dx + Q\,dy \;=\; \iint\limits_D \Big(\frac{\partial P}{\partial y} - \frac{\partial Q}{\partial x}\Big)\,dx\,dy$$

and by virtue of the arbitrariness of the contour C we have $\partial P/\partial y = \partial Q/\partial x$. This equality is the *necessary and sufficient condition for the flow to be vortex-free.*

PROPOSITION 2. *Let v be a vortex-free flow. Then the flow v is potential (possesses a potential), i.e. there exists a function $a(x, y)$ such that $\operatorname{grad} a(x, y) = v$.*

In particular, the form $P\,dx + Q\,dy$ is the total differential of this function $a(x, y)$. The function $a(x, y)$ is uniquely defined with an accuracy to an additive constant.

A flux v such that $v = \text{grad}\,(a)$ will also be called a *gradient flow.*

Proof. We shall integrate the following system of differential equations in partial derivatives: $P = \partial a/\partial x$; $Q = \partial a/\partial y$ under the condition that $\partial P/\partial y = \partial Q/\partial x$. Integrating the first equation over x, we obtain $a(x, y) = \int_0^x P(x, y)\,dx + g(y)$. Differentiation with respect to y yields

$$\frac{\partial a(x, y)}{\partial y} = \int_0^x \frac{\partial P(x, y)}{\partial y}\,dx + \frac{dg(y)}{dy},$$

whence either

$$Q(x, y) = \int_0^x \frac{\partial Q(x, y)}{\partial x}\,dx + \frac{dg(y)}{dy},$$

or

$$Q(x, y) = Q(x, y) - Q(0, y) + \frac{dg(y)}{dy}.$$

From this we find

$$g'(y) = Q(0, y), \qquad g(y) = \int_0^y Q(0, y)\,dy + c,$$

where $c = \text{const}$. Finally, we are led to

$$a(x, y) = \int_0^x P(x, y)\,dx + \int_0^y Q(0, y)\,dy + c.$$

If we started our integration with the equation $Q = \partial a/\partial y$, we would obtain

$$a(x, y) = \int_0^y Q(x, y)\,dy + \int_0^x P(x, 0)\,dx + c'.$$

The function $a(x, y)$ is called the *potential of a flow* and is uniquely defined to an accuracy of a constant. This function can be described geometrically.

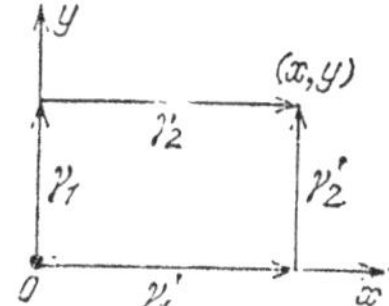

Figure 63.

We shall consider two piecewise smooth paths: $\gamma = \gamma_1 \cup \gamma_2$ and $\gamma' = \gamma_1' \cup \gamma_2'$ (Figure 63). It is clear that

$$a(x, y) = \int_0^x P(x, y)\, dx + \int_0^y Q(0, y)\, dy = \int_\gamma (P\, dx + Q\, dy);$$

$$a(x, y) = \int_0^y Q(x, y)\, dy + \int_0^x P(x, 0)\, dy = \int_{\gamma'} (P\, dx + Q\, dy),$$

that is, the value of $a(x, y)$ can be obtained via integration of the differential form $\omega = P\, dx + Q\, dy$ either along the path γ or along the path γ' which both lead us form the point $(0, 0)$ to the point (x, y). We can make the general statement.

PROPOSITION 3. *Suppose the flow v is vortex-free. Then the flow is potential, and the potential* $a(x, y)$ *can be represented as follows:*

$$a(x, y) = \int_\gamma \omega = \int_\gamma (P\, dx + Q\, dy),$$

where γ *is an arbitrary piecewise smooth path leading from the point* $(0, 0)$ *to the point* (x, y). *In particular* $\int_\gamma \omega$ *is independent of the choice of the path* γ.

Proof. To begin with we prove that the integral $\int_\gamma P\, dx + Q\, dy$ does not depend on the choice of path (under the condition that the initial and the final points are fixed). Indeed, let γ' be any other path from $(0, 0)$ to (x, y); examine

$$\alpha = \int_\gamma \omega - \int_{\gamma'} \omega = \int_{\gamma \cup (-\gamma')} \omega\, ;$$

here by $(-\gamma')$ we denoted the path γ' oriented in the backward direction (Figure 64).

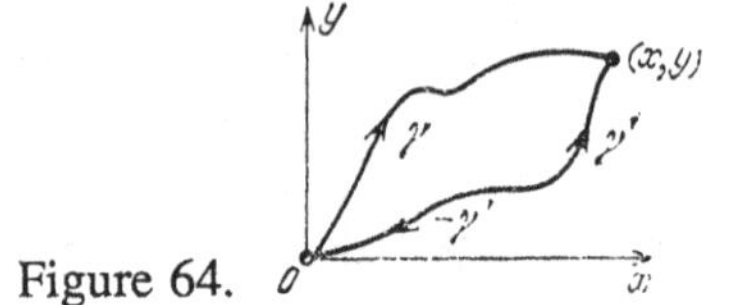

Figure 64.

Then $\int_{\gamma\cup(-\gamma')} (P\,dx + Q\,dy) = \int_C (P\,dx + Q\,dy)$ since $C = \gamma \cup (-\gamma')$ is a closed contour and the flow is vortex-free. So, $\alpha = 0$ and $\int_\gamma \omega = \int_{\gamma'} \omega$. Since we have proved the indpendence of $\int_\gamma \omega$ of the path γ, it follows that to find the numerical value of the integral $\int_\gamma \omega$ we can, say, take one of the paths depicted in Figure 64, which will give us the equality $\int_\gamma \omega = a(x, y)$. This completes the proof.

The change of the initial point of the integration path, obviously, changes the potential $a(x, y)$ by and *additive constant.*

Now let the flow v be both vortex-free and incompressible (which is exactly a *flow of incompressible liquid*). Then the coordinates P, Q of this flow satisfy the following equations

$$\frac{\partial P}{\partial x} = -\frac{\partial Q}{\partial y}; \quad P = \frac{\partial a}{\partial x}; \quad Q = \frac{\partial a}{\partial y};$$

whence we find $\dfrac{\partial^2 a}{\partial x^2} + \dfrac{\partial^2 a}{\partial y^2} = 0.$

DEFINITION 4. The linear differential operator Δ of order two, $\Delta = \partial^2/\partial x^2 + \partial^2/\partial y^2$, is called the *Laplace operator*. The function $f(x\ y)$ satisfying the equation $\Delta f = 0$ is called *harmonic*.

Thus, we have proved that the *potential of a vortex-free and incompressible flow is a harmonic function* on a plane (x, y). The potential $a(x, y)$ is conventionally considered in pair with another potential, $b(x, y)$, which is called a conjugate potential or the potential of a conjugate flow. To define this potential, we shall consider the following system of differential equations: $\frac{\partial b}{\partial x} = -Q$; $\frac{\partial b}{\partial y} = P$. The function $b(x, y)$ is the solution of this system (if this solution does exist) and is called a *conjugate potential*. We shall now prove the existence of a solution and its uniqueness to an accuracy of an arbitrary additive constant. We introduce a new notation: $\tilde{P} = -Q$; $\tilde{Q} = P$. Then we have $\partial b/\partial x = \tilde{P}$; $\partial b/\partial y = \tilde{Q}$ under the condition that $\frac{\partial \tilde{Q}}{\partial y} = -\frac{\partial \tilde{P}}{\partial x}$; $\frac{\partial \tilde{Q}}{\partial x} = \frac{\partial \tilde{P}}{\partial y}$. This system of equations and conditions we recognize as the one that we have just integrated to find the potential $a(x, y)$. Consequently, the potential $b(x, y)$ exists and plays the role of the potential $a(x, y)$ for the flow $(\tilde{P}, \tilde{Q}) = (-Q, P)$. We should note that the flow $\tilde{v} = (\tilde{P}, \tilde{Q})$ is called *conjugate* to the flow (P, Q). Obviously, the inverse is also valid: the potential $a(x, y)$ is conjugate to the potential $b(x, y)$, that is, a potential doubly conjugated to $a(x, y)$ coincides with the latter. Note that the flows v and $\tilde{v}$ are orthogonal: $(v, \tilde{v}) = -PQ + QP = 0$.

We shall now take an important step in the study of the geometry of our flows.

Consider a plane (x, y) as a plane of one complex variable $z = x + iy$ and consider the following complex-valued function: $f(x, y) = a(x, y) + ib(x, y)$, where a and b are the potential and the conjugate potential of an incompressible flow $v = (P, Q)$. In the sequel we shall write, for simplicity, g_y, g_x instead of $\partial g/\partial y$, $\partial g/\partial x$, respectively.

Since $a_x = P$; $a_y = Q$; $b_x = -Q$; $b_y = P$, it follows that $a_x = b_y$; $a_y = -b_x$. Such functions $f(x, y)$ - $a + ib$ are called *complex analytic functions* and the equations for the functions $a(x, y)$ and $b(x, y)$ are called the *Cauchy-Riemann equations* (*conditions*). The functions $a(x, y)$ and $b(x, y)$ are respectively called the *real and imaginary parts of the function f* and are customarily denoted by $a = \mathrm{Re}\,(f)$ and $b = \mathrm{Im}\,(f)$. We shall recall some properties of the complex analytic function.

Suppose $z = x + iy$; $\bar{z} = x - iy$; then $x = 1/2(z + \bar{z})$, $y = -i/2(z - \bar{z})$, and therefore *any* function $g(x, y) = u + iv$ can be written in the form $g(x, y) = \tilde{g}(z, \bar{z})$.

By the rule of differentiation of a composite function

$$\frac{\partial}{\partial z} = \frac{\partial x}{\partial z}\frac{\partial}{\partial x} + \frac{\partial y}{\partial z}\frac{\partial}{\partial y} = \frac{1}{2}\left(\frac{\partial}{\partial x} - i\frac{\partial}{\partial y}\right),$$

similarly we have $\frac{\partial}{\partial \bar{z}} = \frac{1}{2}\left(\frac{\partial}{\partial x} + i\frac{\partial}{\partial y}\right)$. From among all the functions $g(z, \bar{z})$ we shall identify those which depend only on z (i.e. do not depend on $\bar{z}$). Analytically this property can be written as $\frac{\partial g(z, \bar{z})}{z} \equiv 0$. Only these functions are called complex analytic; they can be expanded only in power series of the variable z (this is one of the possible definitions). Since $\partial g/\partial \bar{z} = 0$, it follows that $g_x + ig_y = 0$, i.e. $u_x + iv_x + i(u_y + iv_y) = 0$ and we arrive at the conclusion that the condition $\partial g/\partial \bar{z} = 0$ is exactly equivalent to the Cauchy-Riemann equations: $u_x = v_y$; $u_y = -v_x$.

Thus, we have proved the following statement.

THEOREM 1. *Any vortex-free incompressible flow* $v = (P, Q)$ *can be represented in the form* $v =$ grad $(a(x, y))$, *and the conjugate flow* $\tilde{v} = (\tilde{P}, \tilde{Q})$ *in the form* $\tilde{v} =$ grad $b(x, y)$, *where the function* $f(x, y) = a(x, y) + ib(x, y)$ *is complex analytic and is uniquely defined up to an arbitrary additive constant. The inverse is also valid: if* $f(z)$ *is an arbitrary complex analytic function, then the flows* $v =$ grad Re $f(z)$ *and* $v =$ grad Im $f(x)$ *are vortex-free and incompressible and are, in addition, mutually conjugate.*

The integral trajectories of the flows v and $\tilde{v}$ are orthogonal to one another at each point. The function $f = a + ib$ is called the *complex potential of the flow*. Let $f(z) = a + ib$ be an analytic function. How shall we go about finding the zeros of the flows v and $\tilde{v}$? From the Cauchy-Riemann equations we obtain

$$f'_z(z) = 1/2\,(f_x - if_y) = a_x - ia_y = b_y + ib_x.$$

This implies the assertion: the points at which the derivative f_z vanishes coincide with the zeros of the flow v (or, which is the same, with the zeros of the flow $\tilde{v}$). Hence, to find the zeros of the flows v, $\tilde{v}$, it suffices to solve the equation $f_z(z) = 0$.

The flows v, $\tilde{v}$ may have singular points other than zeros (points of discontinuity), which are not of course roots of the equation $f_z(z) = 0$.

We ask a practical question: if $v =$ grad Re $f(z)$, where $f(z)$ is the kown analytic function, then in practice how shall we find the integral trajectories of this vector field? It turns out to be unnecessary to solve, in an explicit form, the corresponding system of differential equations.

PROPOSITION 4. *Let* $f = a + ib$ *be a complex analytic function,* $v =$ grad (a); $\tilde{v} =$ grad (b). *Then the function* $b(x, y)$ *is a first integral for the vector field* v, *and the function* $a(x, y)$ *is a first integral for the vector field* $\tilde{v}$, *i.e. the integral trajectories of the flow* v *are exactly all the level lines of the function* b, *and the integral trajectories of the flow* $\tilde{v}$ *are exactly all the level lines of the functions* a.

Proof. It suffices to calculate the following derivatives: $da/d\tilde{v}$ and db/dv. For example, $db/dv = (v, \text{grad } b) = a_x b_x + a_y b_y = b_y b_x - b_x b_y = 0$. Similarly, $da/d\tilde{v} = 0$, i.e. the functions a and b are constant on corresponding integral trajectories. This completes the proof.

We shall give some examples. Let $f(z) = z^k$, $k \geq 2$; $f' = kz^{k-1}$, $f'(z) = 0$ only at the point 0; $f = r^k (\cos k\phi + i \sin k\phi)$; i.e. $a = r^k \cos k\phi$; $b = r^k \sin k\phi$. Figure 65 shows the integral trajectories of a flow grad (a) (for $k = 4$). The origin of coordinates is a singular point which can be obtained through merging of several singular points of higher order.

Let $f = z^{-k}$; $k = 1$, $f = r^k(\cos k\phi - i \sin k\phi)$. Figure 66 shows the integral trajectories of a flow grad (a) (for $k = 4$).

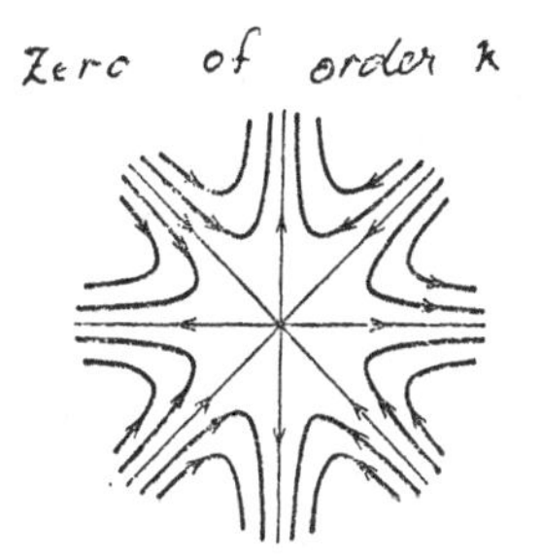

Figure 65.

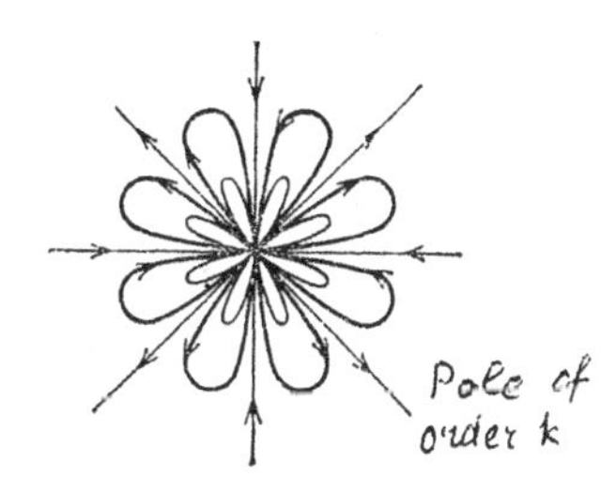

Figure 66.

Let $f(z) = \ln z$. Figure 67 shows the integral trajectories of flows v and $\tilde{v}$.

An example of a more composite function is depicted in Figure 68. (The Zhukovsky function $f(z) = z + 1/z$.)

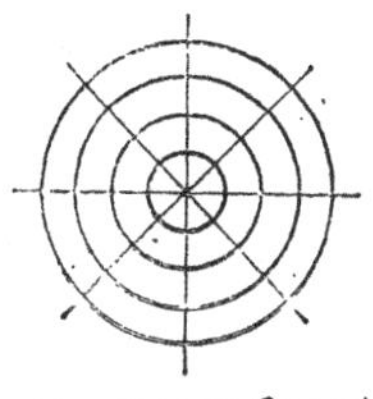

Figure 67. Logarithmic singularity

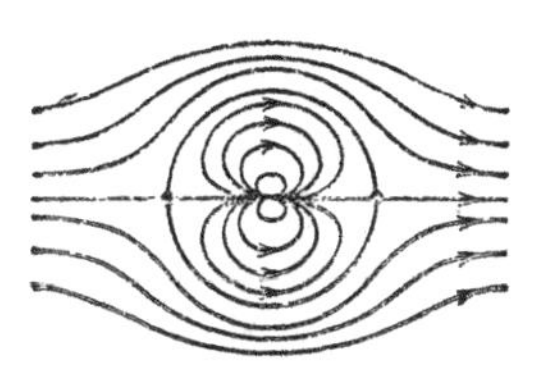

Figure 68.

We shall give another example: $f(z) = (\ln(z + \alpha) - \ln z)/\alpha$ (Figure 69).

On this example we may demonstrate how singularities merge. Let $\alpha \to 0$. Then clearly $f(z) = (\ln z)'_z = 1/z$, and geometrically it is also obvious that the *field of the dipole* becomes the field of the flow corresponding to the first-order pole (Figure 70).

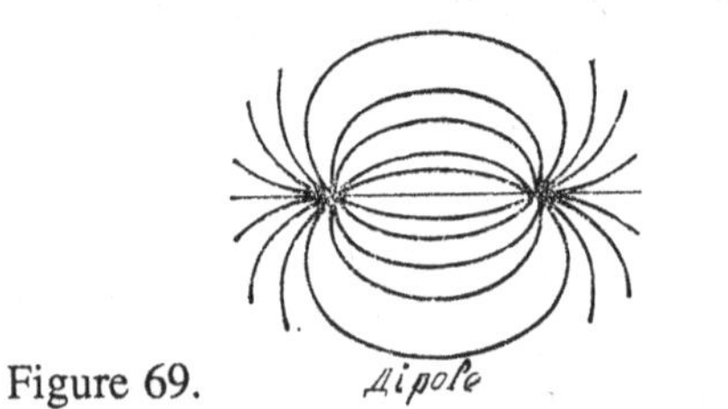

Figure 69.

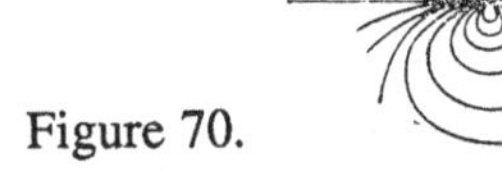

Figure 70.

We have considered vector fields on the plane; all these fields can, however, be mapped onto a two-dimensional sphere S^2. Recall that a stereographic projection (see Part I) establishes a one-to-one correspondence between all points of a two-dimensional sphere and points of an extended complex plane (an extended complex plane is a plane of one complex variable endowed with an infinitely remote point). It is sometimes more convenient to view vector fields on a sphere rather than on a plane.

In conclusion, we note that a constant flow v on a plane (i.e. $v = \text{grad}\,(z)$) has the only singular point at infinity; more precisely, infinity for this flow is a first-order pole. A particularly illustrative example may be the sphere S^2 (Figure 71). The flow for $f = z + 1/z$ on the sphere is given in Figure 72.

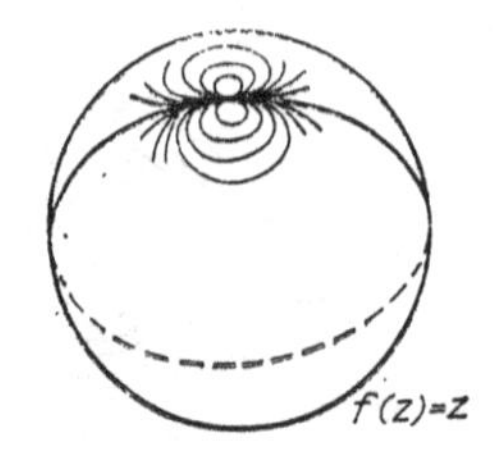

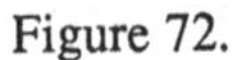

Figure 71.

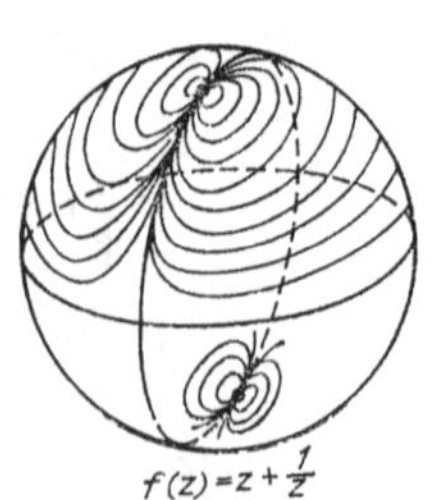

Figure 72.

As an exercise we recommend constructing, on a plane, a qualitative picture of the behaviour of integral trajectories of the fields grad Re $f(z)$ and grad Im $f(z)$ for the following functions $f(z)$:

$$(\alpha + i\,\beta)\ln\left(\frac{az+b}{cz+d}\right);$$

$$1/z + \ln z\,;$$

$$z^k + \ln z.$$

3.5 Functions on Manifolds and Vector Fields

Suppose we are given a smooth function on a manifold M^n.

DEFINITION 1. A point P is called *critical* or *stationary* (*extreme*) for a function f on a manifold if $(df)_P = 0$ or, in terms of local coordinates $(x^1, \dots, x^n)$, we have $(\partial f/\partial x^1, \dots, \partial f/\partial x^n) = 0$.

DEFINITION 2. The critical point P of a function f is called *non-degenerate* if (in terms of local coordinates $(x^1, \dots, x^n)$) the matrix $(\partial^2 f/\partial x^\alpha \partial x^\beta)_P$ is non-degenerate ($\det \neq 0$).

DEFINITION 3. A function f on a manifold M^n is called *typical* (or in *general position*, or a *Morse function*) if *all* its critical points are non-degenerate.

We shall describe an important class of functions. If a manifold M^n is smoothly embedded in Euclidean space $\mathbb{R}^N$ and if a straight line $\xi_n(t)$ goes in n- or $-n$-direction through the origin of coordinates in $\mathbb{R}^n$, then we define the function $g_n(P)$ (the "height" or "coordinate" function) whose value at the points of the manifold is equal to the orthogonal projection of the points of the manifold onto the straight line $\xi_n(t)$.

We shall enlist the properties of the height (or coordinate) function.

1. Such functions are in correspondence with pairs of diametrically opposite points of the sphere $S^{n-1}(n, -n)$ or, equivalently, with the points of the projective space (this is obvious).
2. A point $P \in M^n$ is a stationary point of a height function g_n if the vector n (or $-n$) is orthogonal to the manifold M^n at the point P (this is obvious).
3. We should discover conditions under which a critical point P of a height function g_n is non-degenerate.

LEMMA 1. *For hyper-surfaces $M^n \subset \mathbb{R}^{n+1}$ the Gauss map $M^n \to S^n \to \mathbb{R}P^n$ is defined; the point P is a non-degenerate critical point of a height function $g_{(n,-n)}$ ($= g_n$) if it is a regular point of the Gauss map F: $M^n \to \mathbb{R}P^n$, where the vector n is orthogonal to the surface $M^n \subset \mathbb{R}^{n+1}$ at the point P.*

Proof. We take the vector n to be the x^{n+1}-axis and the vectors $x^1, \dots, x^n$ to be tangent to the manifold M^n. In a neighbourhood of the point P the manifold is given by the equation $x^{n+1} = \phi(x^1, \dots, x^n)$ and $(d\phi)_P = 0$. In the region near the point P

the coordinates $x^1, \ldots, x^n$ serve as local coordinates on the surface; the "height" g_n in that neighbourhood is the function $\phi(x^1, \ldots, x^n) = x^{n+1}$ of the point on the surface. Similar coordinates $\bar{x}^1, \ldots, \bar{x}^n$ are chosen on the sphere in a neighbourhood of the point n (or $-n$). Repeating the calculations, as in the proof of the theorem saying that $K\,d\sigma = f^*\,\Omega$, we obtain that in coordinates $x^1, \ldots, x^n, \bar{x}^1, \ldots, \bar{x}^n$ at the point P, the matrix

$$\left(\frac{\partial^2 g_n}{\partial x^\alpha \partial x^\beta}\right)_P = \left(\frac{\partial^2 \phi}{\partial x^\alpha \partial x^\beta}\right)_P = \left(\frac{\partial \bar{x}^\alpha}{\partial x^\beta}\right)_P$$

and

$$\det\left(\frac{\partial^2 g_n}{\partial x^\alpha \partial x^\beta}\right)_P = K.$$

Therefore, the condition of regularity of the Gauss map at the point P, namely $\left(\frac{\partial \bar{x}^\alpha}{\partial x^\beta}\right) \neq 0$, is equivalent to $\det\left(\frac{\partial^2 g_n}{\partial x^\alpha \partial x^\beta}\right) \neq 0$. This completes the proof of the lemma.

An embedding $M^n \subset \mathbb{R}^N$, where $N > n + 1$, defines a "normal manifold" whose points are pairs (P, n_P), where $P \in M^n$, $n_P \perp M^n$ at the point P. This normal manifold is denoted by $N(M^n)$ and has dimension $N - 1$. The Gauss map is defined to be

$$N(M^n) \to S^{N-1} \to \mathbb{R}P^{N-1},$$

$$(P, n_P) \to n_P \to (n, -n).$$

The lemma is also valid in this case: if a point (P, n_P) and a point P are regular, then the "height" g_{n_P} has P as a non-deenerate critical point. The proof is identical to that of the above lemma. Carry it out for a circumference S^1 in space $\mathbb{R}^3$.

The lemma implies

THEOREM 1. *A height function $g_{(n,-n)}$ on a hyper-surface $M^n \subset \mathbb{R}^{n+1}$ is typical (i.e. all of its critical points are non-degenerate) if and only if the Gauss map $F: M^n \to \mathbb{R}P^n$ has $(n, -n) = Q$ as a regular value. Almost all height functions are typical.*

Proof. The lemma implies that the point $P \in F^{-1}(n, -n)$ is non-degenerate for the "height" $g_{(n,-n)}$ if and only if P is a regular point. Therefore, the theorem follows from the lemma, from the definition of a regular value (all pre-images are regular) and from Sard's lemma (almost all the values $a \in \mathbb{R}P^n$ are regular). This completes the proof.

Thus, the set of critical points of a height function g_n is a union of two pre-images under the Gauss map $F: M^n \to \mathbb{R}P^n$; this union coincides with

$$F^{-1}(n) \cup F^{-1}(-n).$$

DEFINITION 4. A non-degenerate critical point of a function is called a *point of type* $(k, n-k)$ if at this point a second differential $(d^2g)_P$ is a quadratic form with k positive and $(n-k)$ negative squares in the canonical form (obviously, the sign of the determinant $\det(\partial^2 g/\partial x^\alpha\, \partial x^\beta)_P$ is equal to $(-1)^{n-k}$).

The following theorem holds

THEOREM 2. *Suppose we are given an oriented hyper-surface $M^n \subset \mathbb{R}^{n+1}$ and a typical height function g_n; if f is a Gauss map, then the "sign of a point" P for the map f (i.e. the detree of f at the point P) coincides with the sign of the determinant* $\operatorname{sgn}\det(\partial^2 g_n/\partial x^\alpha\, \partial x^\beta)_P = (-1)^{n-k}$), *where the point P has type $(k, n-k)$. If n is even, the functions g_n and $g_{-n} = g_n$ are of the same sign. For even $n = 2l$ the degree of the Gauss map is calculated by the formula*

$$2 \deg f = \sum_P \left(\operatorname{sgn}\det\left(\frac{\partial^2 g_n}{\partial x^\alpha \partial x^\beta}\right)\right) = \sum_P (-1)^{n-k}$$

where P is a critical point, and the summation is over all the critical points of the "height" g_n.

Proof. The sign of the point $P \in f^{-1}(Q)$ is defined as the sign of the Jacobian of the map f in local coordinates near points P and Q. In terms of the chosen coordinates on the hyper-surface, where $\bar{x}^{n+1} = g(x^1, \dots, x^n)$ and $(dg)_P = 0$ and on the sphere $\bar{x}^{n+1} = \left((\bar{x}^1)^2 + \dots + (\bar{x}^n)^2\right)^{1/2}$, where $(d\bar{x}^{n+1})_Q = 0$ we had

$$\left(\frac{\partial \bar{x}^\alpha}{\partial x^\beta}\right)_P = \left(\frac{\partial^2 g}{\partial x^\alpha \partial x^\beta}\right)_P.$$

Therefore, the sign of the point P for the map f coincides with the sign of the determinant $\det(\partial^2 g/\partial x^\alpha \partial x^\beta)_P = (-1)^{n-k}$. If n is even, we have $(-1)^{n-k} = (-1)^k$. Under the change $g \mapsto -g$ the numbers k and $n-k$ in the type of a critical point change places

$$(k, n-k) \mapsto (n-k, k).$$

From this we have

$$\deg f = \sum_{P \in j^{-1}(n)} \operatorname{sgn} \det \left(\frac{\partial^2 g_n}{\partial x^\alpha \partial x^\beta}\right)_P = \sum_{P \in j^{-1}(n)} (-1)^{n-k}.$$

Similarly, at the point $-n$ we have (since $g_{-n} = -g_n$)

$$\deg f = \sum_{P \in j^{-1}(n)} \operatorname{sgn} \det \left(\frac{\partial^2 g_n}{\partial x^\alpha \partial x^\beta}\right)_P = \sum_P (-1)^{k-n}.$$

For even n we obtain

$$2 \deg f = \sum_{P \in j^{-1}(n) \cup j^{-1}(-n)} \operatorname{sgn} \det \left(\frac{\partial^2 g_n}{\partial x^\alpha \partial x^\beta}\right)_P = \sum_P (-1)^k.$$

This completes the proof of the thoerem.

Examples are given in Figures 73, 74 and 75.

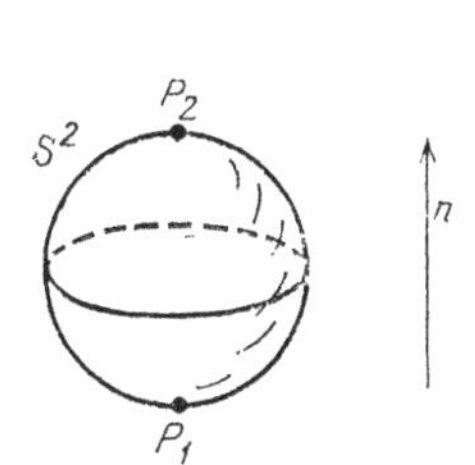

Figure 73.

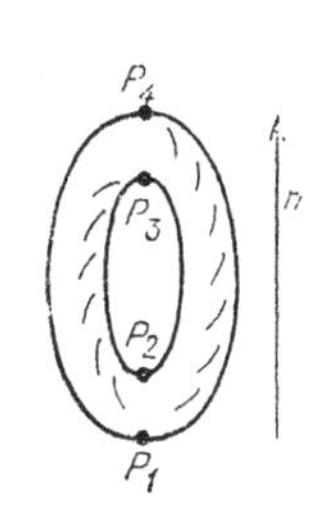

Figure 74.

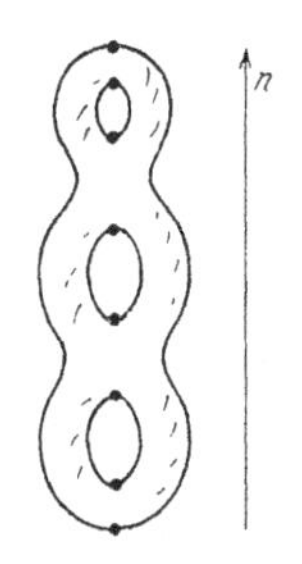

Figure 75.

1) The sphere (the boundary of a convex figure — Figure 73):

$$2\deg f = 2 = 1_{P_1} + 1_{P_2};$$

2) The torus (see Figure 74):

$$2\deg f = 1_{P_1} - 1_{P_2} - 1_{P_3} + 1_{P_4} = 0;$$

3) The pretzel with g handles (see Figure 75):

$$2\deg f = 2 - 2g, \quad \deg f = 1 - g.$$

We have illustrated such embeddings (positions) of a surface with g handles for which the formula $\deg f = 1 - g$ is satisfied. But the formula for the degree of the Gauss map is not yet proved for all embeddings of a surface. For example, a torus can be knotted (Figure 76). A torus embedded in such a manner cannot be deformed regularly into a usually embedded one.

Figure 76.

On any manifold M^n with a Riemannian metric (g_{ij}) a smooth function g defines the vector field ∇_g-gradient by the formula (in terms of local coordinates)

$$(\nabla_g)^\alpha = g^{\alpha\beta} \frac{\partial g}{\partial x^\beta}.$$

The critical points of the function g are such that $(\nabla_g)_P = 0$. Each point is ascribed the sign $(-1)^{n-k}$, where $n - k$ is the number of negative squares of the quadratic form $(d^2 g)_P$. In calculating the degree of the Gauss map (for even n) we considered the expression

$$\sum_P (-1)^{n-k},$$

where $(k, n - k)$ is the type of the critical point P of the height function g. Now we shall examine an arbitrary vector field (ξ^α) on the manifold M^n.

DEFINITION 5. A point P is called *singular* for a vector field (ξ^α) if $\xi^\alpha = 0$, $\alpha = 1, \dots, n$.

DEFINITION 6. A singular point P of a vector field is called *non-degenerate* if $\det\left(\frac{\partial \xi^\alpha}{\partial x^\beta}\right) \neq 0$, where $(x^1, \dots, x^n)$ are local coordinates and $(\xi^1, \dots, \xi^n)$ are components of the vector field in terms of these coordinates.

DEFINITION 7. The sign $\det(\partial \xi^\alpha/\partial x^\beta)_P = \pm 1$ is called the *index* of a non-degenerate singular point P of the vector field (ξ^α).

A simple lemma holds.

LEMMA 2. *Let a Riemannian metric be positive and a vector field (ξ^α) be the gradient $\xi^\alpha = g^{\alpha\beta}\, \partial g/\partial x^\beta$ of a function g. If P is a critical point, then the following equality holds*

$$\operatorname{sgn}\det\left(\frac{\partial \xi^\alpha}{\partial x^\beta}\right)_P = \operatorname{sgn}\det\left(\frac{\partial^2 g}{\partial x^\alpha \partial x^\beta}\right)_P = (-1)^{n-k},$$

where $(k, n-k)$ is the type of the critical point P.

Proof. If the metric is Euclidean, $g_{ij} = \delta_{ij}$, then $\xi^\alpha = \partial g/\partial x^\alpha$, and the lemma is obvious. If the metric g_{ij} is not Euclidean but $(g_{\alpha\beta})_P = \delta_{\alpha\beta}$ and $(\partial g_{\alpha\beta}/\partial x^j)_p = 0$, then the lemma is also obvious. On the surface $M^2 \subset \mathbb{R}^3$ we could always choose coordinates $x^1, \dots, x^n$ in a neighbourhood of the point P, such that $(f_{\alpha\beta})_P = 0$, $(\partial g_{\alpha\beta}/\partial x^j)_P = 0$. This implies the assertion for surfaces $M^2 \subset \mathbb{R}^3$ (similarly, for hyper-surfaces $M^n \subset \mathbb{R}^{n+1}$). In the general case, for any (positive) metric $g_{\alpha\beta}$ in coordinates $x^1, \dots, x^n$ we can consider the deformation

$$g_{\alpha\beta}(t) = (1-t)g_{\alpha\beta} + t\,\delta_{\alpha\beta}, \quad 0 \geq t \leq 1.$$

For all $0 \leq t \leq 1$ this metric $g_{\alpha\beta}(t)$ is positive and

$$g_{\alpha\beta}(0) = g_{\alpha\beta}, \quad g_{\alpha\beta}(1) = g_{\alpha\beta}.$$

The sign $\operatorname{sgn}\det(\partial \xi^\alpha/\partial x^\beta)$ remains unchanged for all $0 \leq t \leq 1$ since for all those t we have $\det(\partial \xi^\alpha/\partial x^\beta) \neq 0$. This completes the proof.

DEFINITION 8. A vector field on a manifold M^n is called "*typical*" (in general position) if all of its singular points are non-degenerate.

Let $\xi^\alpha(x^1, \dots, x^n)$, $\alpha = 1, \dots, n$ be a vector field in a region of Euclidean space with coordinates $x^1, \dots, x^n$. Let η^α be a constant vector (independent of $x^1, \dots, x^n$). The following theorem holds.

THEOREM 3. *Almost all of the vector fields* $(\xi + \eta)^\alpha = \eta^\alpha + \xi^\alpha(x^1, \dots, x^n)$, *where* $\eta^\alpha = \text{const.}$, *are typical.*

Proof. The components $\xi^\alpha(x^1, \dots, x^n)$ define the map of the region U:

$$\phi_\xi \colon U \to \mathbb{R}^n, \text{ where } \phi_\xi^\alpha(x^1, \dots, x^n) = \xi^\alpha(x^1, \dots, x^n).$$

A simple lemma holds.

LEMMA 3. *A point* $P \subset U$ *is regular for a map* ϕ *if and only if*

$$\det \left(\frac{\partial \xi^\alpha}{\partial x^\beta}\right)_P \neq 0.$$

The proof is obvious by definition.

Next, the value O (the origin) is regular in $\mathbb{R}^n$ if and only if *all* points P_0 of the pre-image $\phi^{-1}(O)$ are regular. This means that the value is regular if and only if the vector field ξ^α is typical. Sard's lemma tells us that almsot all values $Q \in \mathbb{R}^n$ are regular. Suppose η is a vector going from the origin O into a regular value $Q \in \mathbb{R}^n$. Then the vector field $\xi + \eta$ is such that $\phi_{\xi+n}(O) = \phi_\xi^{-1}(O)$. Since $\eta = \text{const.}$, we obtain that for almost all η the field $\xi + \eta$ is typical, as required.

Suppose now that we are given an *arbitrary* vector field (ξ^α) in a region U of Euclidean space $\mathbb{R}^n$. Suppose $V \subset U$ is a sub-region in which the vector field ξ^α does not have singular points $\xi = 0$.

The "spherical" map $V \xrightarrow{f_\xi} S$ is defined by the formula

$$f_\xi(x^1, \dots, x^n) = \frac{\xi}{|\xi|}.$$

This formula has sense in the region V, where $\xi \neq 0$. On the sphere S^{n-1} we have defined the form Ω of degree $n - 1$ (the volume element) and also defined the expression $\Omega_\xi = f^*(\Omega)$ (in the region V). The form Ω_ξ has degree $n-1$ in the region V of n-dimensional space.

If P is an isolated singular point of a field ξ in a region U, then in a small neighbourhood of the point P everywhere $\xi \neq 0$ except at the point P. Consider a sphere

$$S_P^{n-1} = \{ \sum_{\alpha=1}^{n} (x^\alpha - x_0^\alpha)^2 = \varepsilon \}$$

where ε is a small number > 0 and $x_0^1, \ldots, x_0^n$ are coordinates of the point P. Everywhere on the sphere we have $\xi \neq 0$ and in the interior of the sphere P is the only singular point. On the sphere S_P^{n-1} the "spherical" map is defined to be

$$f_\xi^P : S_P^{n-1} \to S^{n-1}.$$

DEFINITION 9. The degree f_ξ^P is called the *index* (the *rotational number*) of a vector field at a singular point P.

If $a(n) = \int_{S^{n-1}} \Omega \neq 0$, then (by the theorem on degree) the following formula holds

$$\deg f_\xi^P = \frac{1}{a(n)} \int f_\xi^* (\Omega).$$

For $n = 2$ we have $n - 1 = 1$ and $\Omega = d\phi$. For $n = 3$ we have $n - 1 = 2$ and $\Omega = |\sin \theta| d\theta\, d\phi$. Let *all* singular points of the vector field ξ in a region U of space $\mathbb{R}^n$ be isolated (for example, non-degenerate). Let $M^{n-1} \subset U \subset \mathbb{R}^n$ be a closed hyper-surface on which there is not a single singular point.

The following theorem holds:

THEOREM 4. *The integral* $\frac{1}{a(n)} \int_{M^{n-1}} f_\xi^* (\Omega)$ *is equal to the sum of the indices of all singular points of the field* ξ *which lie in the interior of the surface* M^{n-1}, *where*

$$a(n) = \int_{S^{n-1}} \Omega.$$

Proof. The form Ω on the sphere s^{n-1} has degree $n-1$ and, therefore, is closed: $d\Omega = 0$. Consequently, the form Ω is also closed in the region $V \subset \mathbb{R}^n$, where $\xi \neq 0$

since $f_\xi^*(d\Omega) = d(f_\xi^*(\Omega)) = 0$. Consider the region between the spheres V surrounding each singular point P_i in the interior of M^{n-1} and the M^{n-1} itself. The boundary of the region $\bar{V}$ is $M^{n-1} \cup S_{P_1}^{n-1} \cup \ldots \cup S_{P_N}^{n-1}$, where N is the number of singular points in the interior of M^{n-1}. By the Stokes formula we have

$$0 = \int_{\bar{V}} d(f_\xi^* \Omega) = \int_{\partial \bar{V}} f_\xi^* \Omega =$$

$$= -\int_{M^{n-1}} f_\xi^* \Omega + \int_{S_{P_1}^{n-1}} f_\xi^* \Omega + \ldots + \int_{S_{P_N}^{n-1}} f_\xi^* \Omega =$$

$$= -\int_{M^{n-1}} f_\xi^* \Omega + \sum_{P_\alpha} \deg f_\xi^{P_\alpha}$$

and the theorem follows.

On the plane $\mathbb{R}^2$ the theorem becomes more illustrative since the index of a singular point is the rotational number of the vector field in going round this singular point.

For a vector field ξ^α at a non-degenerate singular point P the matrix $a_\beta^\alpha = (\partial \xi^\alpha / \partial x^\beta)_P$ and its eigenvalues $\lambda_1, \ldots, \lambda_n$ are defined. Let Re $\lambda_i \neq 0$, $i = 1, \ldots, n$. We have k eigenvalues λ_i, where Re $\lambda_i > 0$ and $n - k$ eigenvalues, where Re $\lambda_i < 0$. The type of the singular point is $(k, n-k)$. Reduction to the canonical form yields

$$\operatorname{sgn} \det \left(\frac{\partial \xi^\alpha}{\partial x^\beta}\right)_P = (-1)^{n-k}.$$

We have the following theorem.

THEOREM 5. *The index of a field* ξ *at a singular* (*non-degenerate*, Re $\lambda_i \neq 0$) *point* P *is equal to* $(-1)^{n-k}$.

Proof. In a neighbourhood of the point P we have $\xi^\alpha = \alpha_\beta^\alpha x^\beta + \Delta^\alpha(x)$, where Δ^α have a higher order of smallness, $a_\beta^\alpha = (\partial \xi^\alpha / \partial x^\beta)_P$.

Consider a field $\xi^\alpha(t) = a_\beta^\alpha x^\beta + \Delta^\alpha(1-t)$, $0 \le t \le 1$, in a small neighbourhood of the point P non-degenerate for the reason that $\det(a_\beta^\alpha) \neq 0$. For $t = 1$ we obtain

$$\xi^{\alpha}(1) = \alpha^{\alpha}_{\beta} x^{\beta}, \text{ where } a^{\alpha}_{\beta} = \text{const.}$$

Under deformation $0 \le t \le 1$ the index of the point P remains unchanged. The field $\xi^{\alpha}(1)$ is linear. Obviously, the degree of a singular point is the degree of the linear map (on spheres, since under the map $x \xrightarrow{\phi} a(x)$ a ray is transformed into a ray):

$$\Psi(x)^{\alpha} = a^{\alpha}_{\beta} x^{\beta}.$$

Obviously, this degree is det a^{α}_{β}, and the result follows.

EXAMPLE. On the plane ($n = 2$):

1) a knot, a focus, a centre have index equal to 1,
2) a saddle has index equal to -1.

Next, let on a plane $\mathbb{R}^2$ there exist a curve such that:

a) either the field is everywhere tangent to the curve (a periodic solution),
b) or the field is nowhere tangent to the curve (a contactless cycle).

We arrive at the following corollary.

Inside such a curve there necessarily exists a singular point. Rotation of the field along such a curve is equal to 1, and the theorem given above tells us that the rotational number of the field around a curve is equal to the sum of the indices of all interior singular points. Note that the index of a non-singular point is equal to zero (verify it).

3.6 Singular Points of Vector Fields. The Fundamental Group

For an arbitrary closed, oriented manifold M^n and for a vector field ξ, on this manifold, with isolated singularities it turns out that the sum of the indices of these singular points $\sum_P \mathrm{ind}_P \xi$ is *independent of the vector field*; this sum is called the *Euler characteristic* of the manifold. This fact can be explained quite simply: two vector fields ξ and η with isolated singularities appear to be homotopic, and throughout the homotopy process the vector field has isolated singularities. But to prove this fact is not very easy. We shall prove this assertion only in the simplest cases.

THEOREM 1. 1) *If we are given a two-dimensional disc D^2 and if on this disc D^2 a vector field ξ is defined which is not equal to zero on the boundary Γ and is such that:*

a) *either the field ξ or the boundary Γ of the disc D^2 is evberywhere tangent to the boundary* (Γ *is an integral trajectory*),

b) *or the field ξ on Γ is nowhere tangent to Γ* (*the boundary is a "cycle without contact*"), *then the following equality holds*

$$1 = \sum_{P \text{inside } D^2} \mathrm{ind}_{P_2} \xi$$

and, in particular, in the interior there exists at least one singular point.

2) *If S^2 is a two-dimensionl sphere and ξ is a vector field on it, then the following formula holds*

$$2 = \sum_P \mathrm{ind}_P \xi.$$

Proof. Item 1 was, in fact, proved in the preceding section. The point is that under the conditions of the theorem, the field rotation along Γ is equal to 1. Let us prove item 2 concerning the sphere S^2. Suppose Q is a non-singular point and D_0^2 is a small disc with a centre at this point. Suppose D_1^2 is a complementary disc $S^2 = D_0^2 \cup D_1^2$. In terms of local coordinates (y_1, y_2) the vector field ξ in the interior of the disc D_0^2 is approximately constant, and therefore in these coordinates (y) it is homotopic to a constant one. Hence, on the contour Γ the field ξ is homotopic to such a field which has constant components in terms of the coordinates (y) in the interior of the disc D_0^2. In coordinates (x) in the interior of the disc D_1^2 the field ξ

does not have constant components: taking D_0^2 as the lower hemisphere of the sphere S^2 and D_1^2 as the upper one with standard coordinates, we see that the field ξ on the contour Γ in the coordinates (x) of the disc D_1^2 is homotopic (Figure 77) to an absolutely standard half-field η (on Γ) which has constant components under change to the coordinates (y). What is the rotation of the field η along Γ in the coordinates (x) of the disc D_1^2? This rotation is the sum of indices of all singular points of the field ξ, which lie in the interior of the disc D_1^2; this sum does not depend on the field ξ. The vector field (the gradient of the height function), where the sum of indices of the singular points is equal to 2 is directly specified (see Figure 77), Ind P_1, = Ind P_2 = 1, i.e. $\sum_{P_\alpha} \mathrm{Ind}_{P_\alpha} \xi$. This implies the theorem.

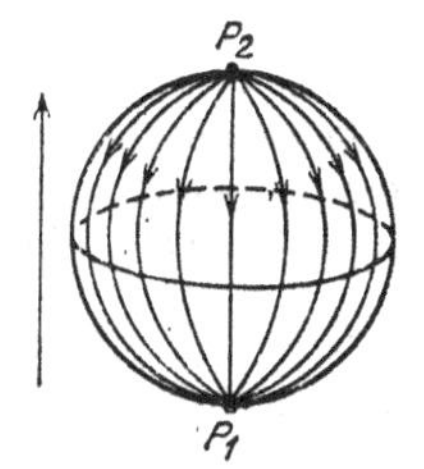

Figure 77.

REMARK. The proof of the theorem works for spheres of all dimensions n but $\sum_{P_\alpha} \mathrm{Ind}_{P_\alpha} \xi = 1 + (-1)^n$. (Prove it!)

How can we verify , in practice, for particular fields that along the boundary Γ of the disc D^2 the vector field ξ is transverse to Γ? Suppose we are given a (Lyapunov) function $F(x^1, x^2)$, such that a) Γ is the level surface $F(x^1, x^2) = c$ and b) the function F is such that $\nabla_\xi F < 0$ for $F = c$ (on the contour Γ). Then the field Γ is transverse to the boundary Γ (verify it). If, in addition, in the interior of the disc there exists *exactly one* singular point P of index 1 and if this point "repulses" the integral trajectories of the field ξ (a repulsing focus or a knot), then somewhere there exists a periodic trajectory (the Poincaré-Bendixson theorem).

For example, such conditions are satisfied by a vector field given by the equations $\dot{x} = v$, $\dot{v} = f(x, v)$, where $f(x, v) = -\sin x + kv$. Find the contour Γ and the Lyapounov function F, where $\nabla_\xi F < 0$ on Γ.

How are two-dimensional, compact, connected, smooth manifolds without boundary organized? It turns out that their description is quite simple. We shall describe two infinite series of manifolds. The direct product of a segment by a circumference will be called a handle. A handle is homeomorphic to a cylinder. The edge of a handle consists of two non-intersecting circumferences. Discard from a two-dimensional sphere two non-intersecting closed discs. Then glue a handle to the boundary of the manifold obtained by identifying each of its boundary circumferences with one of the circumferences, which are the edge of the sphere with two holes. We obtain a two-dimensional manifold without boundary. We shall call this operation the glueing of a handle. It is clear that in an analogous manner we can define glueing of several handles to a sphere.

The second series of manifolds is obtained as follows. Consider a Möbius strip (band) and its boundary circumference. Discard a disc from a sphere and glue the hole with a Möbius strip (i.e. identify the boundary of the Möbius strip with the boundary of the hole in the sphere). We shall call this operation the glueing of the Möbius strip. Naturally, we can define the glueing of several Möbius strips in a similar way.

The classification theorem for 2-manifolds. *Any smooth, compact, connected, two-dimensional manifold without boundary is homeomorphic either to a sphere with a certain number of handles or to a sphere to which several Möbius strips are glued.*

The two series of manifolds described above may be regarded as smooth manifolds. Then we appear to be in a position to substitute differomorphism for homeomorphism in the classification theorem.

We shall not prove this thoerem here.

We shall emphasize an essential property of smooth two-dimensional manifolds. To begin with we shall give an important definition. We say that a compact two-dimensional manifold without boundary admits a finite *triangulation* if on this manifold there exist a finite number of points (called vertices of triangulation) joined, in some order, by a finite number of smooth curve segments on the surface. Given this, it is required that each curve segment should join two distinct vertices of triangulation and that it should not pass through any other vertices. It is also required that the set of all these curve segments partition the manifold into a finite number of closed triangles with vertices from the set of vertices of triangulation. The sides of the triangles are called the edges of triangulation. Finally, it is required that any two triangles of our partition should either not intersect, or intersect in one common vertex or intersect on a single common side, i.e. on a common edge of triangulation.

It can be proved that an arbitrary two-dimensional smooth, compact, connected manifold without boundary allows a finite triangulation.

We fix such a triangulation (it is not uniquely defined) on a manifold M^2. The number of vertices of triangulation we denote by a_0, the number of edges of triangulation by a_1, and the number of triangles in triangulation by a_2. We obtain the following equality: $\chi(M) = a_0 - a_1 + a_2$.

THEOREM. *The number $\chi(M)$ does not depend on the choice of finite triangulation of the manifold. This number coincides with the Euler characteristic of the manifold.*

Occasionally, the Euler characteristic is introduced, by definition, as the expression $a_0 - a_1 + a_2$.

We now proceed to an important concept of the fundamental group of a manifold defined via classes of homotopic closed paths with their tails (and tips) at a fixed point $P \in M^n$.

We shall now introduce simple concepts.

The path is a continuous (or even piecewise smooth) map of a segment I ($a \le t \le b$) into a manifold γ: $I \to M^n$ (or $\gamma(t)$ are points of the manifold M^n)

The cyclic path is a map of a circumference into a manifold γ: $S^1 \to M^n$ (the initial and terminal points coincide, but are not fixed).

The closed path is a map of a segment γ: $I \to M^n$, where $\gamma(a) = \gamma(b) = P \in M^n$, a and b are the end-points of the segment (the initial and terminal points are fixed and coincide).

NOTATION. $\Omega_{PQ}(M^n)$ is the totality of paths from a point P to a point Q along the manifold M^n; $\Omega_{PP} = \Omega_P$ (closed paths beginning at P).

DEFINITION 1. a) Two paths $\gamma_1(t)$ and $\gamma_2(t)$ of Ω_{PQ} are called *homotopic* if they are homotopic as a map of a segment, such that in the homotopy process the beginning and the end are *stagnant* (are homotopic in the interior of Ω_{PQ}). This type of homotopy will be denoted by $\gamma_1 \sim \gamma_2$.

b) Homotopy of cyclic paths is an arbitrary continuous homotopy $\gamma(t, \tau)$, where $\gamma_1 = \gamma(t, 1)$ and $\gamma_2 = \gamma(t, 2)$, $1 \le \tau \le 2$ (this is homotopy of two maps of a circumference, where for any τ $\gamma(t, \tau)$: $S^1 \to M^n$).

Homotopy of cyclic paths will also be denoted by $\gamma_1 \sim \gamma_2$.

LEMMA 1. *If two paths* $\gamma_1(t)$ and $\gamma_2(t)$, *where* $0 \le t \le 1, 0 \le \tau \le 1$, *are such that* $\gamma_2(\tau)$ and $\gamma_1(t(\tau))$, *then these paths are homotopic.*

Proof. Let us consider a function $t(\tau)$ and let us deform it to the function $t \equiv \tau$. Then the lemma will be proved. Let us examine the graphs of the functions $t(\tau)$ and $t = \tau$. Figure 78 makes obvious the deformation of the graph $t(\tau)$ and $t = \tau$ illustrated by the arrows. We can see that the end-points do not move, and the lemma follows.

For the proof of the lemma it is of no importance that $dt/d\tau > 0$, as is seen in Figure 78. The change need not be necessarily one-to-one.

According to the lemma we shall not distinguish (up to homotopy) between paths which differ only by the introduction of a parameter. Moreover, we shall choose (or change) the parameter or even the range of its variation as convenient, namely, make parallel transport and extend without distinguishing between cooresponding paths. Obviously, with such operations we can always reduce the range of parameter variation to a segment of 0 to 1.

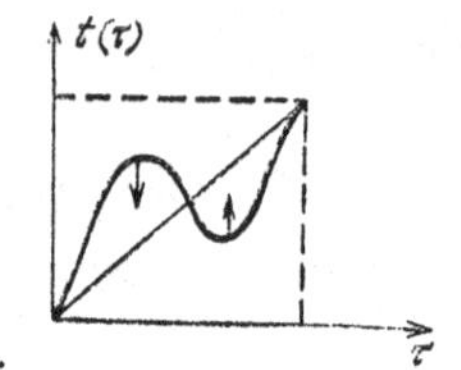

Figure 78.

Algebraic operations on paths.

1. The product of paths. Given two paths $\gamma_1 \in \Omega_{PQ}$ and $\gamma_2 \in \Omega_{QR}$, their "product" is defined to be

$$\gamma_1 \circ \gamma_2 \in \Omega_{PR},$$

where we first go along the path γ_1 (t ranges from 0 to 1) and then along the path γ_2 (t ranges from 1 to 2). A simple lemma holds.

LEMMA 2. *Given three paths* $\gamma_1 \in \Omega_{PQ}$, $\gamma_2 \in \Omega_{QR}$ and $\gamma_3 \in \Omega_{RS}$, *the product is associative* (*up to homotopy*)

$$(\gamma_1 \circ \gamma_2) \circ \gamma_3 \sim \gamma_1 \circ (\gamma_2 \circ \gamma_3).$$

Proof. We shall choose, making use of Lemma 1, the parameter between 0 and 1 (for γ_1), between 1 and 2 (for γ_2) and between 2 and 3 (for γ_3). Then Lemma 2 is obvious.

2. The inverse way. For the path $\gamma(t) \in \Omega_{PQ}$, $0 \leq t \leq 1$, the inverse path $\gamma^{-1} \in \Omega_{QP}$ is defined; this is exactly the same path but in the opposite direction. $\gamma^{-1}(t) = \gamma(2-t)$, where $1 \leq t \leq 2$. Then we have a simple lemma

LEMMA 3. *Both the products* $\gamma^{-1} \circ \gamma \in \Omega_{PP}$ *and* $\gamma \circ \gamma^{-1} \in \Omega_{QQ}$ *are homotopic to the constant paths* $\gamma^{-1} \circ \gamma \sim 1$, $\gamma \circ \gamma^{-1} \sim 1$, *where* 1 *is a path* j *such that* $j(t) = P$ *or* $j(t) = Q$.

The proof is almost obvious (see Figure 79).

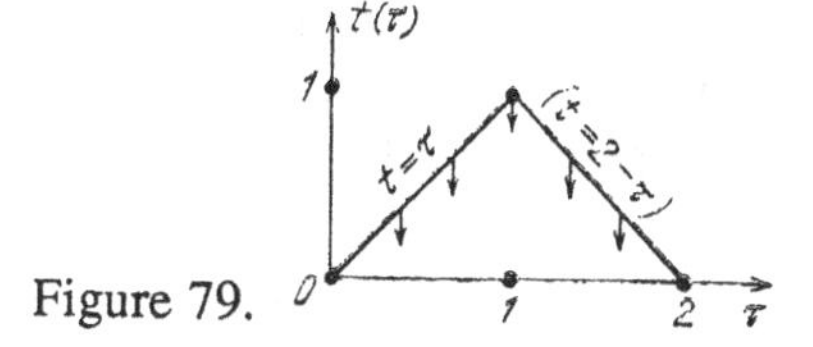

Figure 79.

It is carried out, in fact, along the path γ itself. Since $\gamma^{-1}(t) = \gamma(2-t)$, we have for the path $\gamma \circ \gamma^{-1} = x$

$$\gamma \circ \gamma^{-1} = x; \quad x(\tau) = \begin{cases} \gamma(t), & 0 \le t \le 1, \quad t = \tau, \\ \gamma(2-t), & 0 \le t \le 1, \quad 2 - t = \tau. \end{cases}$$

We shall consider the graph for the parameter $t(\tau)$ (Figure 79).

The deformation of this graph to $t = 0$ is indicated by arrows in Figure 79. Since $t = 0$, Lemma 3 is, in effect, similar to Lemma 1, but here $t(0) = 0$ and $t(2) = 0$, i.e. the whole graph is deformed to $t \equiv 0$, which implies the lemma.

Let us now consider the closed paths $\Omega_P = \Omega_{PP}$ with their starting and terminal points at the point P. From Lemmas 2 and 3 there follows

THEOREM. 2. *The classes of homotopic closed paths on an arbitrary manifold with the starting and terminal points lying at a point* P *form a group (possibly, non-commutative). We shall always denote this group by* $\pi_1(M^n, P)$. *This group is called the fundamental group.*

EXAMPLES.

1. For the Euclidean space $\mathbb{R}^n$, the disc D^n, the group π_1 is trivial (identity). The proof is obvious since the whole space $\mathbb{R}^n$, D^n (and, therefore, any closed path in this space) can be deformed to a single point.

2. For the sphere S^n for $n > 1$ the group π_1 is the identity group.

Proof. An arbitrary path γ: $I \rightarrow S^n$ is homotopic to a smooth path $\tilde{\gamma}$: $I \rightarrow S^n$, which is close to it. For all t the functions $\tilde{\gamma}(t)$ and $\gamma(t)$ are close to one another (continuous functions are approximated by smooth ones). By Sard's lemma, a smooth path leaves at least one point $Q \in S^n$ free. The image of the path lies, therefore, in $\mathbb{R}^n = S^n \backslash 0$. In a Euclidean space, any closed path is homotopic to a constant (see item 1). Hence, $\pi_1(S^n, P) = 1$.

3. For a circumference, $\pi_1(S^1, P)$ is a cyclic (infinite group). Prove it.

3a. If we discard a point or a disc from a plane, then the remaining region U has an infinite cyclic fundamental group, as a circumference. The proof consists of shrinking this region $U = \mathbb{R}^2 \backslash Q$ along itself to the circumference $S^1 \subset U$ (a detour around Q).

4. If from the plane $\mathbb{R}^2$ we discard a finite number of points $U_n = \mathbb{R}^2 \backslash (Q_1 \cup \ldots \cup Q_n)$, we shall obtain a region U_n. What is the group $\pi_1(Un, P)$? Let $n = 2$ and let two points Q_1 and Q_2 be discarded from $\mathbb{R}^2$ (Figure 80). Verify directly (by drawing) that the path $aba^{-1}b^{-1} = \gamma$ is not homotopic to 1 (to a constant path) although the integrals of the analytic function $\oint f(z)\, dz \equiv 0$ are always zero (if singular points for $f(z)$ are only Q_1 and Q_2) and also that the rotation of any vector field ξ with singularities at points Q_1 and Q_2 only along the path γ is zero as well. It turns out that the fundamental group $\pi_1(U_2, P)$ is a free group with two generatrices a and b. All the elements of the indicated form are distinct:

$$a^{\alpha_1} b^{\alpha_2} \ldots a^{\alpha_k} b^{\alpha_{k+1}}$$

for any k and any integer $\alpha_1, \ldots, \alpha_{k+1}$ (perhaps $\alpha_1 = 0$ or $\alpha_{k+1} = 0$, but the remaining $\alpha_j \neq 0$).

Similarly, $\pi_1(U_n, P)$ is a free group with n generatrices. Generally, the fundamental group of any region on a plane *always* appears to be a free group with a certain (possibly infinite) number of generatrices.

We shall pay attention to an interesting circumstance: for a region $U_2 = \mathbb{R}^2 \backslash (Q_1 \cup Q_2)$ the paths $aba^{-1} = \gamma_1$ and $b = \gamma_2$ are different elements of the group $\pi_1(U_2, P)$ (Figure 81).

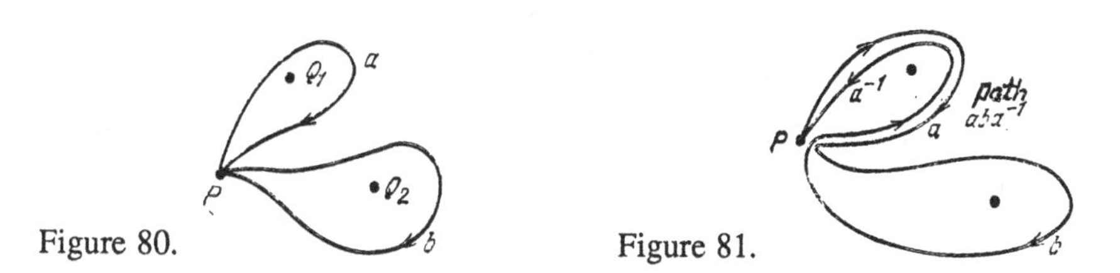

Figure 80. Figure 81.

However, these paths are homotopic. How can we account for this fact? The point is that under deformation of the path aba^{-1} to the path b the starting point is not motionless, it moves along the path a detouring the point Q_1. Such a deformation is not allowed by the definition of the group π_1. Let the path aba^{-1} be homotopic to the path b only as a cyclic path (its starting and terminal points are not marked). We have

THEOREM 3. *The classes of homotopic cyclic paths on a connected manifold M^n (where any two points can be joined by a path) are in one-to-one correspondence with conjugation classes in the group $\pi_1(M^n, P)$.*

Proof. Two elements a and b from the group $\pi_1(M^n, P)$ are called, as usual, *conjugate* if there exists an element $x \in \pi_1(M^n, P)$, such that $b = xax^{-1}$. Let us prove this theorem. By virtue of connectedness of the manifild, any cyclic path γ can be deformed to a path through the point P (Figure 82). Now the point P will be thought of as the starting and terminal point of the path γ.

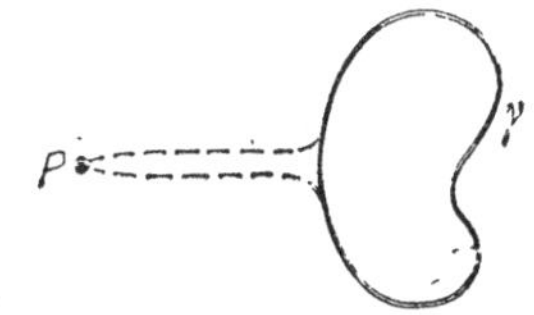

Figure 82.

Our task is the following: there exist two paths with the starting and terminal points at the point P, or two elements $a, b \in \pi_1(M^n, P)$. These paths are homotopic as cyclic paths (i.e. in the deformation process $0 \le \tau \le 1$ the starting and terminal points traverse the path $\mathcal{H}(\tau)$, where $\mathcal{H}(0) = P$ and $\mathcal{H}(1) = P$). How are the elements a and b related in the group $\pi_1(M^n, P)$? The path $\mathcal{H}(\tau)$ is the motion of the starting point P under deformation of cyclic paths. But the path $\mathcal{H}(\tau)$ is closed and determines the element $x \in \pi_1(M^n, P)$ since $\mathcal{H}(0) = \mathcal{H}(1) = P$. Let us verify the equality $b \sim xax^{-1}$ in the group $\pi_1(M^n, P)$ or $a \sim xbx^{-1}$. The transition from a to xax^{-1} is equivalent to the deformation of the starting point P of the path a along the path x^{-1} (Figure 83). This implies the theorem.

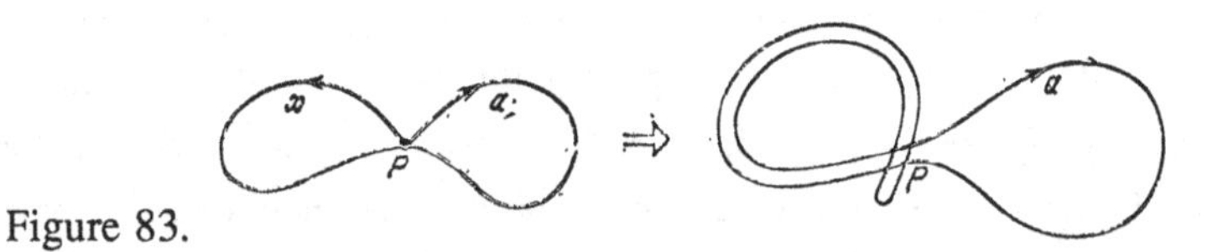

Figure 83.

The following, almost obvious, theorem will be useful.

THEOREM 4. *For any pair of points P, Q on a connected manifold M^n the set of classes of homotopic paths with the beginning at the point Q and the end at the point P is in one-to-one correspondence with the elements of the group $\pi_1(M^n, P)$ (or $\pi_1(M^n, Q)$).*

Proof. Let us choose a path $\gamma_0(t)$ leading from the point P to the point Q. Let $\gamma_1(t)$ be any other path from P to Q. Then $\gamma_0 \circ \varphi_1^{-1}$ determines the closed path $\gamma(t)$ from P to P, i.e. the element of the group $\pi_1(M^n, P)$. If the path γ_0 is fixed, then there is a correspondence (determined by the path γ_0) $\gamma \leftrightarrow \gamma_0\, \gamma_1^{-1}$, γ_1 (from P to Q), γ (from P to P). Going over to homotopy classes, we arrive at the statement of the theorem in a trivial way.

3.7 The Fundamental Group and Coverings.

Our aim in this section is to learn to calculate the fundamental group of some simple manifolds and to give examples of its application. To begin with we shall give an important definition of "covering": suppose we are given a smooth map of two manifolds of the same dimension

$$f\colon M_1^n \to M^n,$$

such that at *all* points $P \in M_1^n$ the rank of the Jacobian matrix is equal to n. Moreover, suppose for any point Q from M_2^n the complete pre-image $f^{-1}(Q) = P_1 \cup \ldots \cup P_N \cup \ldots$ consists of the same number of points continuously dependent on the point Q if it moves along the manifold M_2^n. We shall require, in fact, that any point $Q \in M_2^n$ should have a neighbourhood U, where $Q \in U$, such that the pre-image $f^{-1}(U) = U_1 \cup \ldots \cup U_N \cup \ldots$ of this neighbourhood consists of a union of pairwise non-intersecting regions U_α in M_1^n and on each region U_α the map $f\colon U_\alpha \to U$ is a smooth, with non-zero Jacobian, one-to-one map $U_\alpha \cong U$. In this case the map f is called *covering*. For coverings above a connected manifold M_2^n it is obvious that the number of pre-images of distinct points is equal (draw a path from a point Q_1 to another point Q along the manifold M_2^n; each of the pre-images of the point Q will continuously move along this point, follwowing it and imitating its motion). Coverings, in which the number of pre-images of a point is equal to k are called k-sheeted.

EXAMPLES.

1. The trivial k-sheeted covering. Here $M_1^n = M_2^n \cup \ldots \cup M_2^n$ (k-sheeted), and the projection $f\colon M_1^n \to M_2^n$ *is one-to-one on each piece* $M_2^n \subset M_1^n$. To eliminate trivial coverings, we shall further on require that the manifold M_1^n be connected (i.e. any two points can be joined by a continuous path).

2. $f\colon \mathbb{R}^1_{(x)} \to S^1_{(\phi)}$, where $f(x) = e^{2\pi i x}$. Since $e^{2\pi i n} = 1$, where n is an integer, we can readily see that this is a non-trivial infinite-sheeted covering.

3. $f\colon \underset{|z|=1}{S^1} \to \underset{|x|=1}{S^1}$, $f\colon z \to z^n$; the covering here is n-sheeted.

4. $f\colon S^m \to \mathbb{R}P^m$; if n is a unit vector in $\mathbb{R}^{m+1}$, then the pair $(n, -n)$ defines one point from $\mathbb{R}P^m$. This covering is 2-sheeted.

5. $f\colon \mathbb{R}^2 \to T^2$, where the points of the torus T^2 are represented by equivalence classes of the points $\{(x, y)\}$ of the Euclidean plane, and $(x + n, y + m)$

define one and the same point of the torus provided that m and n are integers. The map F: $\mathbb{R}^2 \to T^2$ is obvious since to the point $(x\ y)$ there corresponds its equivalence class.

6. f: $\mathbb{R}^2 \to K^2$, where K^2 is a Klein bottle. The points K^2 are equivalence classes of the points of the plane $\mathbb{R}^2$. Namely: let us take two transformations T_1 and T_2 of the plane

$$T_1(x, y) = (x + 1, y), \quad T_2(x, y) = (1 - x, y + 1).$$

The equivalence class of a point consists of all points which can be obtained from the given point through transformations $T_1, T_2, T_1^{-1}, T_2^{-1}$ and their repeated super-positions.

7. Let a Riemannian surface M^2 in a two-dimensional complex space $\mathbb{C}^2(z^1, z^2)$, where $z^1 = z$, $z^2 = w$, be given by the equation

$$P(z, w) = z^n + P_1(w)\, z^{n-1} + \dots + P_n(w) = 0,$$

$P_1, \dots, P_n$ being polynomials.

The manifold M_2^n is a region U on the w-plane, where the equation does not have roots that are multiples of z. The region U is $\mathbb{R}^2$ with branching points $Q_1, \dots, Q_n$ punctured out: $U = \mathbb{R}^2 \backslash (Q_1 \cup \dots \cup Q_N)$. The branching points are obtained form the solution of the two equations (so that the roots of the first equation be multiples of z):

$$P(z, w) = 0, \quad \frac{\partial P}{\partial z}(z, w) = 0.$$

The manifold M^2 (the Riemannian surface) is projected onto the w-plane $\mathbb{R}^2$:

$$F\colon M^2 \to \mathbb{R}^2.$$

The manifold M_1^2 is $M^2 \backslash (f^{-1}(Q_1) \cup \dots \cup f^{-1}(Q_N))$, which means that the pre-images of all the branching points are removed.

The map

$$F\colon M_1^2 \to M_2^2$$

is an n-sheeted covering, i.e. for any non-singular $w \in M_2^n = \mathbb{R}^2 \backslash (Q_1 \cup \dots \cup Q_N)$ the equation $z^n + P_1(w)\, z^{n-1} + \dots + P_N(w) = 0$ has n roots which are multiples of z. How many branching points are there? If the polynomials $P_i(w)$ have the degree i in the variable w (the total degree of $P_i(w)\, z^{n-i}$ is equal to n), then the number of branching points in general position is equal to $n(n-1)/2$, and above each of these branching points there merges, roughly speaking, one pair of n roots (altogether $n(n-1)/2$ pairs).

8. Let M_1^n and M_2^n be connected, closed manifolds and let $f\colon M_1^n \to M_2^n$ be a map such that the rank of the Jacobian matrix be *always* non-zero. Then this is a finite-sheeted covering map (prove it).

9. Let Γ be a discrete group of transformations of the manifold M_1^n. This means that to each element $g \in \Gamma$ there corresponds a non-trivial transformation $g\colon M_1^n \to M_1^n$ with the property ("discreteness") that for any point $P \in M_1^n$ and for any transformation $g \neq 1$ the distance between the points P and $g(P)$ is not less than a certain number $\varepsilon(P)$. In other words, there exists a neighbourhood of the point P, such that the points $g(P)$ *all* lie *outside* this neighbourhood (for all $g \neq 1$).

The covering map $f\colon M_1^n \to M_2^n$ is defined as follows. Points of the manifold M_2^n are, by definition, the equivalence classes of points from M_1^n or, equivalently, orbits of the group Γ. In addition to the point P, the equivalence class involves all points of the form $g(P)$ for all $g \in \Gamma$, where Γ is a discrete transformation group. The map f assigns to a point its equivalence class. The number of covering sheets is equal here to the number of elements of the group Γ.

Examples 2, 3, 4, 5, 6 are all of this type; in these examples $M_1^n = \mathbb{R}^1, S^1, S^n, \mathbb{R}^2, \mathbb{R}^2$. The group Γ in example 2 is infinitely cyclic, in example 3 it is cyclic of order n, in example 4 — cyclic of order 2, in example 5 — the direct sum of two infinitely cyclic groups (a lattice on a plane), and in example 6 the group is non-commutative; it is generated by transformations T_1 and T_2 related as $T_1 \circ T_2 \circ T_1 = T_2$. In example 7, for Riemannian surfaces of the form $z^2 + P_n(w) = 0$ (the roots of the polynomial $P_n(w)$ are aliquant) the group Γ is cyclic of order n. For the other Riemannian surfaces the covering maps from example 7 do no, generally, refer to this class. To make the cause of this clear, we should define the so-called "monodromy group" of the covering.

We shall now define the monodromy group of the covering

$$f\colon M_1^n \to M_2^n.$$

Let Q be a point in M_2^n and let $\gamma(t)$ be an arbitrary closed path beginning and ending at the point Q. Let $\gamma(t)$ define the element $\gamma \in \pi_1(M_2^n, Q)$ and let $P_1 P_2, P_3, \ldots$ be all points from the pre-image $f^{-1}(Q)$ in the manifold M_1^n. If the point Q moves along the path $\gamma(t)$, $0 \leq t \leq 1$, then by the definition of covering each point moves "above it".

Namely, suppose that $Q_t = \gamma(t)$ and suppose that $P_\alpha(t)$ is a point in M_1^n such that $P_\alpha(0) = P_\alpha$ and $f(P_\alpha(t)) = Q(t)$ for all $0 \le t \le 1$. The point $P_\alpha(t)$ is uniquely defined by the initial point P_α and by the path $\gamma(t)$ (see the definition of covering). But when in the manifold M_2^n the path $\gamma(t)$ has become closed at $t = 1$ and the point Q has returned to the initial position, the covering point $P_\alpha(t)$ at $t = 1$ may fail to coincide with the initial point $P_\alpha(0) = P_\alpha$ (Figure 84).

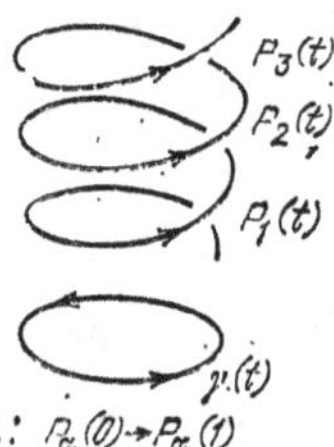

Figure 84.

What does the position of the point $P_\alpha(1)$ depend on? Obviously it depends on the initial point $P_\alpha(0)$ and on the homotopy class of the path $\gamma(t)$, i.e. on the element $\gamma \in \pi_1(M_2^n, Q)$. A monodromy transformation arises: M_γ: $P_\alpha(0) \to P_\alpha(1)$ (along the path $\gamma(t)$). Obviously, M_γ is a permutation of points from the complete pre-image $f^{-1}(Q) = P_1 \cup P_2 \cup \dots$. The properties are obvious (here $1, \gamma, \gamma_1, \gamma_2 \in \pi_1(M_2^n, Q)$, M_1 is a unit permutation):

$$M_{\gamma^{-1}} = (M_\gamma)^{-1}, \quad (M_{\gamma_1\gamma_2}) = M_{\gamma_1} \cdot M_{\gamma_2}.$$

There arises homomorphism of the fundamental group into a group of permutations ("monodromy"). We have already dealt with monodromy transformations for the simplest Riemannian surfaces $z = (P_n)^{1/2}(w)$.

EXERCISE. For more complicated Riemannian surfaces

$$P_n(z, w) = z^n + P_1(w)\, z^{n-1} + \dots + P_n(w) = 0$$

in the "typical case" (when there exists $n(n-1)$ distinct branching points and the degree of the polynomial $P_i(w)\, z^{n-2}$ is equal to n for all i) the monodromy group of the covering above the plane without branching points is the total permutation group of all n pre-images of the point.

We now proceed to a calculation of the fundamental group of simplest manifolds. We know already that the Euclidean space $\mathbb{R}^n$ for all $n \geq 1$ and the sphere S^2 for $n \geq 2$ are *simply-connected* (i.e. $\pi_1(\mathbb{R}^n) = 1$ and $\pi_1(S^n) = 1 (n > 1)$).

A manifold is called *simply-connected* if $\pi_1(M^n, Q) = 1$ for any point Q.

THEOREM 1. *If on a simply-connected manifold M_1^n there acts a discrete group of transformations Γ and if a manifold M_2^n is defined as the totaility of the equivalence classes of the points of the initial manifold M_1^n with respect to the group Γ, then the equality $\pi_1(M_2^n, Q) = \Gamma$ holds at any point $Q \in M_2^n$* (the map $f: M_1^n \to M_2^n$ *is the covering from example* 9).

Proof. Let us take any point P on the manifold M_1^n. Its equivalence class $\{g(P)\}$ for all $g \in \Gamma$ defines a point Q of the manifold M_2^n. How shall we descibe the closed paths on M_2^n? It is convenient to represent them as paths beginning at the point P and terminating at any point $g(P)$ of the same equivalence class for $g \in \Gamma$.

Because of the simple-connectedness of the manifold M_1^n, the homotopy classes of such paths on M_1^n with fixed end-points are completely defined by the initial and terminal points. Therefore, there exist precisly the same number of homotopy classes of closed paths from $\pi_1(M_2^n, Q)$ as there are elements of the group Γ. Obviously, the multiplication law in the group Γ and in $\pi_1(M_2^n, Q)$ coincides as well. This completes the proof of the theorem.

We can thus calculate the fundamental group in all the examples of manifolds which we may produce in the form of a torus, a Klein bottle, a projective plane, a circumference (see the examples of coverings). All surfaces and regions on the plane can be represented in the same manner, but this is somewhat more complicated.

Suppose there exists a closed differential form of rank 1 on a manifold M^n

$$\omega = \sum_{\alpha} f_{\alpha}(x)\, dx^{\alpha}$$

(in terms of local coordinates $x^1, \ldots, x^n$; $0 \leq \alpha \leq n$), $d\omega = 0$.
How shall we calculate the integral $\int_{\gamma} \omega$, where γ is a closed path? Let γ determine the element of the group $\pi_1(M^n, Q)$. We shall denote by $H_1(M^n)$ the factor group $\pi_1(M^n, Q)$ with respect to the commutation relation $ab = ba$ ("a commutated group").

A simple claim holds

CLAIM 1. *If the element γ has a finite order in the factor group $H_1(M^n)$, then the integral of any closd form along the path γ vanishes.*

Proof. If ω is a closed form and if a, b are closed paths beginning at the point Q, then we have

$$\int_{a \circ b} \omega = \int_{b \circ a} \omega.$$

Since the form ω is closed ($d\omega = 0$), the integral remains unaltered under homotopy (by the Stokes formula). The integral is therefore well defined on the factor group of the group π_1 with respect to the commutation relation. If, for the path γ, the path $\gamma \circ \ldots \circ \gamma$ becomes 1 in the group π_1 or in the factor group $H_1(M^n)$, then $\int_{\gamma^n} \omega = n \int_{\gamma} \omega = 0$. Therefore, $\int_{\gamma} \omega = 0$, and the claim follows.

This implies that in the calculation of the integrals of closed forms, of importance is only the factor group $H_1(M^n)$ with respect to all finite-order elements. Non-commutativity of the group π_1 does not play a role here.

EXAMPLES.

1. For a region on a plane $U_m = \mathbb{R}^2 \backslash (Q_1 \cup \ldots \cup Q_n)$ the group $\pi_1(U_m, P)$ is free and the group $H_1(U_m)$ is the direct sum of m infinite cyclic groups.

2. For a projective space $\mathbb{R}P^n$ the group $\pi_1(\mathbb{R}P^n, Q)$ is of second order. Consequently, the factor group $H_1(\mathbb{R}P^n)$/(the finite-order elements) is trivial (identity) and the integral $\int_{\gamma} \omega$ vanishes provided that $d\omega$ is equal to zero and the path is closed.

3. For a Klein bottle the group $\pi_1(K^2, P)$ is generated by two elements T_1 and T_2 linked by the relation $T_1 \circ T_2 \circ T_1 = T_2$. Therefore, with additive notation of the group operation, in the group $H_1(K^2)$ we obtain $2[T_1] + [T_2] = [T_2]$ or $2[T_1] = 0$. The factor group $H_1(K^2)$/(the finite-order elements) have a single generatrix $[T^2]$. Hence the integral $\int_{\gamma} \omega$ of the form ω (where $d\omega = 0$) along any closed path is an integer multiple of $\int_{T_1} \omega$.

APPENDIX 1

THE SIMPLEST GROUPS OF TRANSFORMATIONS OF EUCLIDEAN AND NON-EUCLIDEAN SPACES

We shall consider two-dimensional Riemannian manifolds. The conformal class of metrics $g_{ij} = g\delta_{ij}$ in the two-dimensional case is invariant under confomal (complex analytic) changes of coordinates. On any complex surface such a metric can be defined.

If a surface is given in a space (z^1, z^2) of two complex variables by a complex analytic (e.g. a polynomial) equation

$$P(z^1, z^2) = 0$$

and if the space (z^1, z^2) is endowed with a Euclidean metric

$$dl^2 = |dz^1|^2 + |dz^2|^2 = \sum_{\alpha=1}^{4} (dx^\alpha)^2,$$

where

$$x^1 = x^1 + ix^2, \quad z^2 = x^3 + ix^4,$$

then on the surface there arises a conformal metric in natural conformal coordinates. Namely, suppose $z^2 = f(z^1)$ is a local solution of the equation if (z^1) is an analytic function,

$$\frac{dz^2}{d\bar{z}^1} \equiv 0,$$

where the metric on the surface has the form

$$dz^1 d\bar{z}^1 + dz^2 d\bar{z}^2 = |dz^1|^2 + \left|\frac{dz^2}{dz^1}\right|^2 |dz^1|^2 =$$

$$= \left(1 + \left|\frac{dz^2}{dz^1}\right|^2\right) dz^1 d\bar{z}^1 .$$

Similarly, if a parametric surface is given in the form $z^1 = z^1(\omega)$, $z^2 = z^2(\omega)$, where $z^1(\omega)$, $z^2(\omega)$ are complex analytic functions, then the metric on the surface is

$$|dz^1|^2 + |dz^2|^2 = \left(\left|\frac{dz^1}{d\omega}\right|^2 + \left|\frac{dz^2}{d\omega}\right|^2\right) d\omega\, d\bar{\omega}$$

(the reader should verify it!).

We shall now make an essential assertion which we formulate as a theorem.

THEOREM 1. 1) *If we are given a conformal metric* $g\, dz\, d\bar{z}$, *then the expression*

$$K = -\frac{1}{2g}\frac{\partial^2}{\partial z\, \partial \bar{z}}(\ln g) = -\frac{1}{2g}\left(\frac{\partial^2}{\partial x^2} + \frac{\partial^2}{\partial y^2}\right)\ln g$$

remains unchanged under conformal coordinate changes $z = z(\omega)$: *if* $g(z)\, dz\, d\bar{z} = \bar{g}(\omega)\, d\omega\, d\bar{\omega}$, *then*

$$-\frac{1}{2g}\frac{\partial^2}{\partial z\, \partial \bar{z}}(\ln g) = -\frac{1}{2\bar{g}}\frac{\partial^2}{\partial \omega\, \partial \bar{\omega}}(\ln \bar{g}).$$

2) *If a surface is defined in a three-dimensional real Euclidean space* $x^\alpha = x^\alpha(y^1, y^2)$, $\alpha = 1, 2, 3$, *and coordinates* $y^1 = u, y^2 = v$ *are conformal* (*i.e.* $g_{ij} = g(y^1, y^2)\, \delta_{ij}$), *then the expression*

$$-\frac{1}{2g}\frac{\partial^2}{\partial z\, \partial \bar{z}}\ln g$$

coincides with the Gaussian curvature K of the surface ,

$$z = u + iv, \quad \frac{\partial}{\partial z} = \frac{1}{2}\left(\frac{\partial}{\partial u} - i\frac{\partial}{\partial v}\right),$$

$$\frac{\partial}{\partial \bar{z}} = \frac{1}{2}\left(\frac{\partial}{\partial u} + i\frac{\partial}{\partial v}\right).$$

Verify this assertion by direct calculation!

Thus, for conformal metrics the Gaussian curvature is a rather simple function of the metric on the surface itself.

We shall analyze several examples.

1. The Lobachevskian plane. In a z-plane we are given a region $y > 0$, $z = x + iy$;

$$dl^2 = \frac{1}{y^2}(dx^2 + dy^2) = \frac{-1}{|(z - \bar{z})|^2}\, dz\, d\bar{z} = -\frac{dz\, d\bar{z}}{|z - \bar{z}|^2}.$$

It can be checked that

$$-\frac{1}{2g}\frac{\partial^2}{\partial z\, \partial \bar{z}}(\ln g) \equiv -1.$$

In a unit circle $|z| < 1$ the metric has the form

$$dl^2 = \frac{dz\, d\bar{z}}{(1 - |z|^2)^2}.$$

2. The sphere. Conformal coordinates on the sphere are introduced proceeding from the fact that the sphere is precisely the same manifold as $\mathbb{C}P^1$. A finite region is served by a coordinate z, and at infinity there exists a coordinate ω, where $\omega + 1/z$ in the region $z + 0$, $\omega + 0$. The point $\omega = 0$ is $z = \infty$ (infinity). In a finite z-region, the metric is

$$dl^2 = \frac{dz\, d\bar{z}}{(1 + |z|^2)^2}, \quad K = +1.$$

The total group of linear fractional complex (projective) transformations of the manifold $\mathbb{C}P^1$ has the form

$$z \to \frac{az + b}{cz + d},$$

where the matrices $\begin{pmatrix} a & b \\ c & d \end{pmatrix} = A$ and $\lambda A = \begin{pmatrix} \lambda a & \lambda b \\ \lambda c & \lambda d \end{pmatrix}$ for $\lambda \neq 0$ define one and the same transformation; if

$$z = \frac{x^1}{x^0} \text{ and } \begin{cases} x^0 \to ax^1 + bx^0, \\ x^1 \to cx^1 + dx^0, \end{cases}$$

then $z \to \dfrac{az + b}{cz + d}$, which means that the transformation is projective.

Since $\mathcal{A}$ and $\lambda\mathcal{A}$ yield one and the same transformation, it follows that an appropriate choice of λ may lead to det $\mathcal{A} = 1$. From this we can see that the linear fractional group is isomorphic to the factor group $SL(2, \mathbb{C})/\pm 1$ since $\lambda = \pm 1$ yield one and the same linear fractional transformation. The group $SL(2, \mathbb{C})$ consists of two sub-groups

1) $SL(2, \mathbb{R}) \subset SL(2, \mathbb{C})$

(the motions of the upper Lobachevskian half-plane $y > 0$),

2) the sub-group $SU_2 \subset SL(2, \mathbb{C})$, where

$$\begin{pmatrix} a & b \\ c & d \end{pmatrix} = \begin{pmatrix} a & b \\ -\bar{b} & \bar{a} \end{pmatrix}, \quad |a|^2 + |b|^2 = 1.$$

This is the group of motions of the metric of the sphere

$$SU_2/\pm 1 \subset SL(2, \mathbb{C})/\pm 1,$$

and the group $SU_2/\pm 1$ coincides with SO_3.

The most symmetric two-dimensional metrics are the Euclidean metric, the metric of the sphere and the metric of the Lobachevskian plane. We have already pointed out the parallelism in the study of the geometry of the sphere and the geometry of the Lobachevskian plane. Now we shall investigate this parallelism in more detail.

Consider a sphere $S^2 \subset \mathbb{R}^3$ of radius R with centre at the origin. Let (r, θ, ϕ) be spherical coordinates in $\mathbb{R}^3$; then, as is known, the Euclidean metric $dx^2 + dy^2 + dz^2$ takes, in terms of these coordinates, the following form: $ds^2 = dr^2 + r^2\, d\theta^2 + r^2 \sin^2\theta\, d\phi$ (verify it!). From this we obtain $ds^2(S^2) = R^2(d\theta^2 + \sin^2\theta\, d\phi^2)$. Here $0 \leq \phi \leq 2\pi$; $0 \leq \theta \leq \pi$ (Figure 85).

The distance between two points A and B measured along a circumference of radius π is equal to zero, i.e. the whole boundary of the circle is glued together into one point, which yields a two-dimensional sphere.

In a small neighbourhood of the point O we have $\sin\theta \sim \theta$, i.e. $ds^2\,(S)^2$ becomes the Euclidean metric: $d\theta^2 + \theta^2\, d\phi^2$.

Consider a stereographic projection of a sphere onto a plane (Figure 86) (depicted in this figure is a plane cross-section of the sphere).

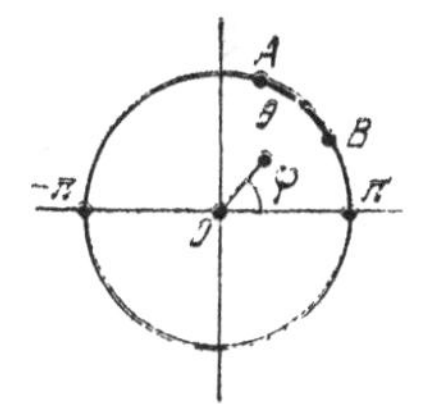

Figure 85.

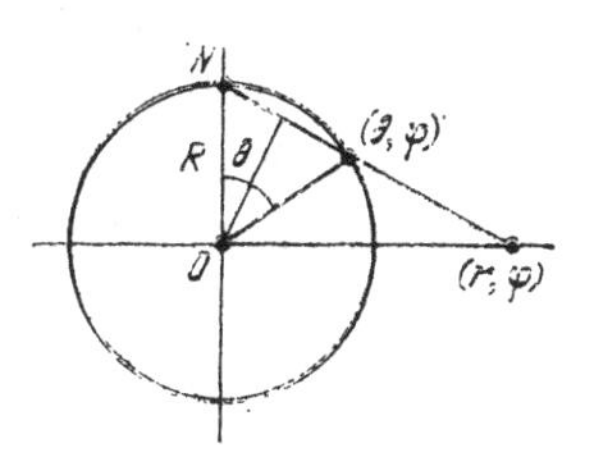

Figure 86.

Here (θ, ϕ) are coordinates on the sphere and (r, ϕ) are polar coordinates on the plane. From Figure 86 it follows that $\phi = \phi$; $r = R \operatorname{ctg}(\theta/2)$. Using these formulae for transition, we can rewrite the metric $ds^2(S^2)$ in terms of the coordinates (r, ϕ):

$$ds^2(S^2) = \frac{4R^2 \cdot (dr^2 + r\, d\phi^2)}{(R^2 + r^2)^2}$$

(verify it!). Clearly,

$$ds^2(S^2) = \frac{4R^2}{(R^2 + r^2)^2} \cdot ds^2(\mathbb{R}^2).$$

Now we shall proceed to Lobachevsky geometry. Consider a pseudo-Euclidean space $\mathbb{R}^3_1$ and a pseudo-sphere of imaginary radius iR. Then the stereographic projection of a hyperboloid of two sheets onto a (y, z)-plane is given by the formulae (see Part I):

$$x = -R \cdot \frac{(\bar{u}, \bar{u}) + R^2}{(\bar{u}, \bar{u}) - R^2}; \quad y = \frac{-2R^2 u^1}{(\bar{u}, \bar{u}) - R^2}; \quad z = \frac{-2R^2 u^2}{(\bar{u}, \bar{u}) - R^2},$$

where $(\bar{u}, \bar{u}) = (u^1)^2 + (u^2)^2$, u^1, u^2 being coordinates in a ring of radius R on the (y, z)-plane.

Consider the restriction of an indefinite metric $ds^2(\mathbb{R}^3_1 = -dx^2 + dy^2 + dz^2$ to a pseudo-sphere of radius iR; direct calculationtions show that this induced metric $ds^2(L_2)$ has, in terms of the coordinates (u^1, u^2), the form

$$ds^2(L_2) = \frac{4R^2((du^1)^2 + (du^2)^2)}{((u^1)^2 + (u^2)^2 - R^2)^2}$$

(verify it!).

Introducing in a ring or radius R polar coordinates ρ, ϕ (i.e. $\rho^2 = (u^1)^2 + (u^2)^2$; $\tan\phi = u^2/u^1$), we obtain $ds^2(L_2) = 4R^2(d\rho^2 + \rho^2\, d\phi^2)/(R^2 - \rho^2)^2$.

We have earlier dealt with the affine definition of Lobachevsky geometry, and now we have obtained its metric definition: the Lobachebvskian metric is a metric induced from a hyperboloid (i.e. from a pseudo-sphere of purely imaginary radius) in a pseudo-Euclidean space $\mathbb{R}^3_1$. Clearly, this metric can be given by

$$ds^2(L_2) = 4R^2 \; ds^2(\mathbb{R}^2)/(R^2 - \rho^2)^2.$$

Comparing this notation with the corresponding notation of the metric of a sphere we can see that the only difference between them is the sign before ρ^2 and r^2.

In what follows we shall assume, for simplicity, that $R = 1$. Next, by analogy with the sphere S^2, we shall examine the following transition formulae: $\phi = \phi$; $\rho = \mathrm{cth}\,(\chi/2)$. Representing the metric $ds^2(L_2)$ in terms of coordinates (χ, ϕ), we come to (verify it!) $ds^2(L_2) = d\chi^2 + \sinh^2\chi\, d\phi^2$.

The distinction of this metric from that of the sphere is that the function sin is replaced by the function sh. We shall tabulate all these forms of the metrics of the sphere and Lobachevskian metrics as shown in Figure 87.

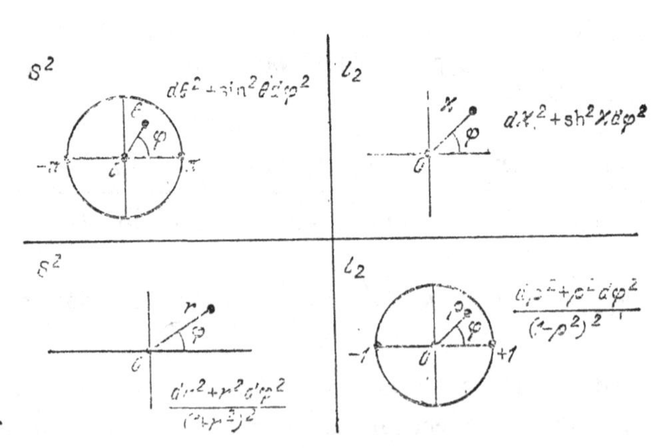

Figure 87.

We shall note an interesting fact: although the metric $ds^2(\mathbb{R}^3_1) = -dx^2 + dy^2 + dz^2$ is indefinite, its restriction to a pseudo-sphere of radius iR is a positive definite metric. The geometric interpretation of this fact is presented in Figure 88.

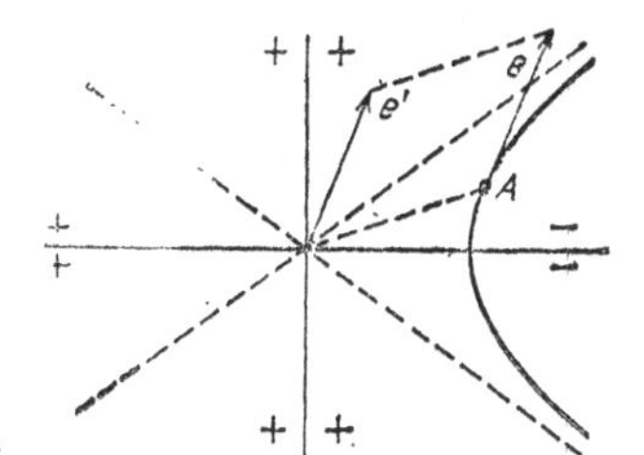

Figure 88.

Indeed, it suffices to verify that the scalar squarc of an vector e tangent to a hyperboloid is positive. For such vectors e, vectors e^1 parallel to them face a "positive" region of $\mathbb{R}^3_1$, which completes the proof.

On a hyperboloid we can also introduce some other coordinates, for example, (y, z), i.e. we can project a pseudo-sphere onto a (y, z)-plane parallel to the x-axis (precisely such coordinates are often considered in the special theory of relativity in the space $\mathbb{R}^4_1$). Direct calculation yields

$$ds^2(L_2) = \frac{(1 + z^2)\,dy^2 - 2yz\,dy\,dz + (1 + y^2)\,dz^2}{1 + y^2 + z^2}$$

(verify it!).

The positive definiteness of this form is already not so obvious as in the case of the metric in the Poincaré model, but it can be readily established through calculation of the determinant of this form (verify it!).

Now we shall examine the Poincaré model and employ it for writing the Lobachevskian metric in the complex form. Suppose $d\rho^2 + \rho^2\,d\phi^2 = dz\,d\bar{z} = dx^2 + dy^2$; $\rho = |z|^2$, i.e. we are led to

$$ds^2(L_2) = 4\,dz\,d\bar{z} \,/\, (1 - |z|^2)^2.$$

While a pair of points is tending to the boundary of a circle, the distance between these points tends to infinity; the boundary of the circle is sometimes called an *absolute*; recall that the points of this circumference do not belong to the set of points of the Lobachevsky geometry.

Let us consider another form of notation of the Lobachevskian metric. It is a well-known fact from the theory of functions of one complex variable that there exists a linear fractional transformation of a complex plane into itself, which maps the upper half-plane into a unit ring. (The linear fractional transformation is a map of the form $w = f(z) = \dfrac{cz + b}{cx + d}$. In the case $ad - bc = 0$, the transformation f maps the whole plane into one point. To eliminate this trivial case, we customarily assume that $ad - bc \neq 0$). One of such transformations is $z = \dfrac{1 + iw}{1 - iw}$; $z = g(w)$, depicted in Figure 89.

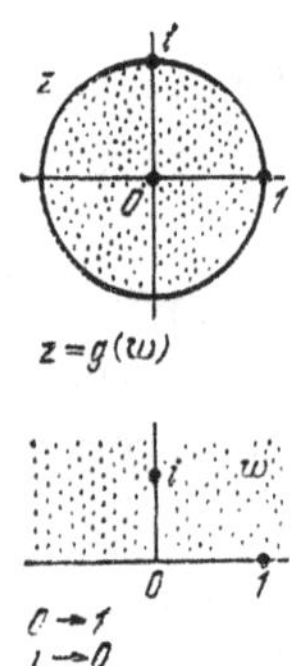

Figure 89.

Thus, we have introduced new coordinates w on the ring. If we write the metric $ds^2(L_2)$ in terms of the coordinates w, we see that the direct calculation yields

$$ds^2(L_2) = \frac{(dx^1) + (dy^1)^2}{(y^1)^2}, \quad \text{where } w = x^1 + iy^1.$$

Now we shall proceed to the *groups of motions* of the metrics of the sphere and those of the Lobachevskian plane. Recall that the group is a set of elements G on which two operations are defined: $(x, y) \to x \cdot y, x \to x^{-1}$ (where $xy \in G$) with the properties

1) $(xy)z = x(yz)$;
2) there exists an $e \in G$ such that $ex = xe = x$;
3) $xx^{-1} = x^{-1}x = e$.

A very essential role in geometry is played by the so-called topological groups (and a more special class of them — the Lie groups). A group G is called a *topological group* if the set G on which the group operation is defined is a topological space and both operations — multiplication and taking the inverse element — are continuous on this space (continuity is understood here as that of the mapping). So, for example, the totality of all transformations preserving some Riemannian metric is a topological group.

As has already been proved, the set of all possible motions of a sphere S^2 coincides with the set of all orthogonal (3×3) matrices $A^{-1} = A^{\mathrm{T}}$. This group is denoted by $O(3)$ and is called the *complete orthogonal group*. Being a topological space, this group consists of two connected components (two pieces): one component consists of those matrices A for which $\det A = +1$, the other consists of those matrices for which $\det A = -1$. Tose matrices for which $\det A = +1$, form a sub-group which is denoted by $SO(3)$. The second connected component is not a group (since $(-1) \cdot (-1) = 1$). The elements of the sub-group $SO(3)$ are occasionally called *proper rotations*, while the elements of the other component are called *improper rotations*.

REMARK. We have seen that all posssible transformations of the sphere defined by orthogonal matrices are motions, and therefore the group $O(3)$ is *contained* in the group of all motions of the sphere S^2. But we cannot, as yet, prove that the group $O(3)$ does actually *coincide* with the group of all motions. This coincidence takes place, and *any* metric-preserving transformation on S^2 is a linear and orthogonal transformation in $\mathbb{R}^3$; a rigorous proof of this fact requires, however, an application of the concept of a geodesic line.

To make the notation shorter, we shall introduce the following: we shall denote by $G(\mathbb{R}^n_s)$ the group of motions of the pseudo-Euclidean space $\mathbb{R}^n_s$ under which the point O (the origin) remains motionless. It is clear that this group coincides exactly with the group of all motions of the pseudo-sphere $S^{n-1}_s \subset \mathbb{R}^n_s$ (with the centre at 0). In particular, for $s = 0$ we have

$$G(\mathbb{R}^3_0) = G(\mathbb{R}^3) = O(3), \quad G(\mathbb{R}^2_0) = O(2).$$

Before proceeding to the group of motions of the metric $ds^2(L_2)$, we shall go back and examine the group of motions of the Euclidean plane $\mathbb{R}^2$ and that of the pseudo-Euclidean plane $\mathbb{R}^2_1$, which keep the point O motionless, i.e. the groups $G(\mathbb{R}^2_0)$ and $G(\mathbb{R}^2_1)$. We have calculated both of these groups before. So, $G(\mathbb{R}^2) = G(\mathbb{R}^2_0)$ and consists of *two* connected components. The group of hyperbolic rotations $G(\mathbb{R}^2_1)$ consists of *four* connected components. Figure 90 illustrates four transformations: g_1, g_2, g_3, g_4 which preserve the pseudo-sphere S^1_1 (of dimension one) and belong to four distinct connected components of the group $G(\mathbb{R}^2_1)$.

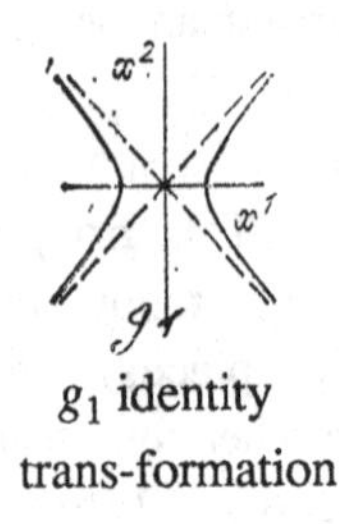

g_1 identity trans-formation

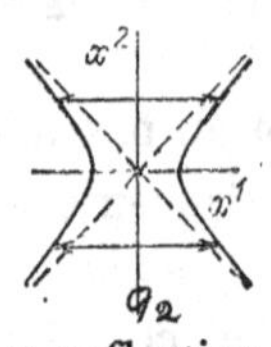

g_2 reflection in the x^2-axis

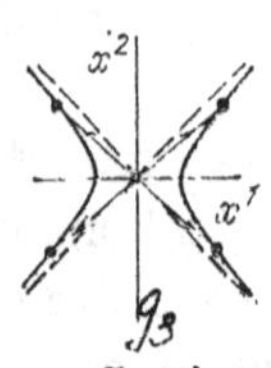

g_3 reflection at the point O

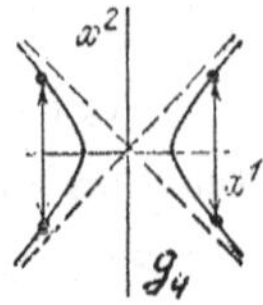

g_4 reflection in the x^2-axis

Figure 90.

Now we shall consider the Lobachevskian plane L_2. We shall calculate the group of motions of L_2. Examine the upper half-plane and $ds^2(L_2) = (dx^2 + dy^2)/y^2$; $y > 0$. Let $w = f(z) = \dfrac{az+b}{cz+a}$ be an arbitrary linear fractional transformation of the upper half plane into itself. Then it can readily be shown (verify it!) that the map f preserves the condition $y > 0$ if and only if a, b, c, d are real numbers and $ad - bc > 0$.

We claim that any such transformation f is a motion of the Lobachevskian metric.

Indeed,

$$ds = \frac{dx^2 + dy^2}{y^2} = \frac{-4\,dz\,d\bar{z}}{(z-\bar{z})^2}; \quad dw = \frac{ad-bc}{(cz+d)^2}\cdot dz;$$

whence

$$dz = \frac{(cz+d)^2}{ad-bc}dw; \quad d\bar{z} = \frac{(c\bar{z}+d)^2}{ad-ba}d\bar{w};$$

(a, b, c, d are real!). Substituting these formulae into $ds^2 = \dfrac{-4\,dz\,d\bar{z}}{(z-\bar{z})^2}$ and carrying out all the calculations, we shall come to $ds^2 = \dfrac{-4\,dw\,d\bar{w}}{(w-\bar{w})^2}$ (verify it!), that is, the transformation f preserves the metric $ds^2(L_2)$. Thus, the group of motions $ds^2(L_2)$ of course contains the sub-group of transformations $w = f(z) = \dfrac{az+b}{cz+a}$, where $a, b, c,$

d are real and $ad - bc > 0$. We denote the set of all such transformations by D_1.

Note that all these transformations are conformal (angle-preserving) transformations. The transformations D_1, however, do not at all exhaust the set of motions of the Lobachevskian metric. Indeed, let us consider the transformation g_0: $z \to -\bar{z}$ which is obviously a self-image of the upper half plane (reflection from the y-axis) and which, furthermore, preserves the Lobachevskian metric:

$$ds^2 = -\frac{4\,dz\,d\bar{z}}{(z-\bar{z})^2} \equiv \frac{-4\,dz\,d\bar{z}}{(\bar{\bar{z}} - z)^2}.$$

At the same time it is obvious that the tansformation g_0: $z \to -\bar{z}$ cannot be represented in the form $g_0(z) = (az + b)/(cz + d)$ (this transformation changes orientation of the angles).

All this means that we must examine all possible transformations $g(z)$ of the form $w = g(z) = -(d\bar{z} + \beta)/(\gamma\bar{z} + \delta)$, where $\alpha, \beta, \gamma, \delta$ are real and $\alpha\delta - \beta\gamma \geq 0$. Clearly we can write $w = (\alpha\bar{z} + \beta)/(\gamma\bar{z} + \delta)$, where $\alpha, \beta, \gamma, \delta$ are real and $\alpha\delta - \beta\gamma < 0$.

We shall denote the set of all such transformations by D_2. Note that the sets of transformations D_1 and D_2 are homeomorphic as topological spaces. Since any transformaion $g \in D_2$ has the form of the composition $g = g_0 f$, where $f \in D_1$, and since g_0 and f are motions, it follows that g is also a motion.

Two sets of transformations $D_1$and D_2 have an empty intersection as long as $\dfrac{az+b}{cz+d} \not\equiv \dfrac{\alpha\bar{z}+\beta}{\gamma\bar{z}+\delta}$. Their union $D = D_1 \cup D_2 = \{f\} \cup \{g\}$, obviously, forms a group in which D_1 is a sub-group and D_2 is not a sub-group. The group D is already the complete group of motions of the Lobachevskian plane. In the same way as the group of motions of the two-dimensional sphere $O(3)$, the group D consists of two connected components.

Let $f(z) = az + b)/(cz + d)$ be an arbitrary transformation from D_1. Since $\dfrac{az+b}{cz+d} \equiv \dfrac{\lambda az + \lambda b}{\lambda cz + \lambda d}$, we may assume that $ad - bc = 1$. Using similar arguments, we may assume that if $g(z) = (\alpha\bar{z} + \beta)/(\gamma\bar{z} + \delta)$ is an arbitrary transformation from D_2, then $\alpha\delta - \beta\gamma + -1$.

Let us consider the set of all real matrices of order 2 with determinant ± 1; these matrices, obviously, form a group which we denote by $L(2)$:

$$L(2):\ L(2) = \left\{ \begin{pmatrix} a & b \\ c & d \end{pmatrix} : ad - bc = \pm 1 \right\}.$$

The group $L(2)$ is disconnected: it consists of two components : $L(2) = L_1(2) \cup L_2(2)$, where

$$L_1(2) = \left\{ \begin{pmatrix} a & b \\ c & d \end{pmatrix} : ad - bc = +1 \right\};$$

$$L_2(2) = \left\{ \begin{pmatrix} \alpha & \beta \\ \gamma & \delta \end{pmatrix} : \alpha\delta - \beta\gamma = -1 \right\}.$$

The sub-group $L_1(2)$ is customarily denoted by $SL(2, \mathbb{R})$. Let us construct a map ϕ: $L(2) \to D$ by the following rule: if

$$A = \begin{pmatrix} a & b \\ c & d \end{pmatrix} \in L_2(2),$$

then $\phi(A) = f; f \in D_1$ and $f(z) = (az + b)/\,(cz + d)$; $ad - bc = +1$. If

$$B = \begin{pmatrix} \alpha & \beta \\ \gamma & \delta \end{pmatrix} \in L_2(2), \quad \text{then } \phi(B) = g,\ g \in D_2.$$

The map ϕ is an epimorphism (i.e. the image of ϕ covers the whole group D), but not one-to-one. Clearly, $\phi(A) = \phi(-A)$ and $\phi(B) = \phi(-B)$. However, if in the group $L(2)$ we identify matrices of the form C and $-C$, i.e. consider the factor group $L(2)/(\pm E)$, where $E = \begin{pmatrix} 1 & 0 \\ 0 & 1 \end{pmatrix}$, then the map ϕ': $L(2)/(\pm E) \to D$ will already be one-to-one. Furthermore, the map ϕ' establishes the algebraic isomorphism of these groups. For this it suffices to verify that the identity $\phi'(C'_1 \cdot C'_2) = \phi'(C'_1) \cdot \phi'(C'_2)$ is satisfied. This can be established through a direct calculation (verify it!). Thus, we have proved that the group of all motions of the Lobachevskian plane is isomorphic to the group $L(2)/(\pm E)$. In particular, the group D_1 is isomorphic to the group $SL(2, \mathbb{R})/(\pm E)$.

Now let us turn to the group $G(\mathbb{R}^1_3)$ (for the definition of this group, see above). This group, of course, contains the group D as a sub-group which, however, does not exhaust it.

The inclusion: $D \subset G(\mathbb{R}_1^2)$ follows from the fact that the Lobachevskian metric is an induced metric on a hyperboloid in $\mathbb{R}_1^3$ (pseudo-sphere of imaginary radius), and therefore any of its motions is the element of the group $G(\mathbb{R}_1^3)$. The group $G(\mathbb{R}_1^3)$ (the same as the group $G(\mathbb{R}_1^2)$) consists of four connected components (Figure 91).

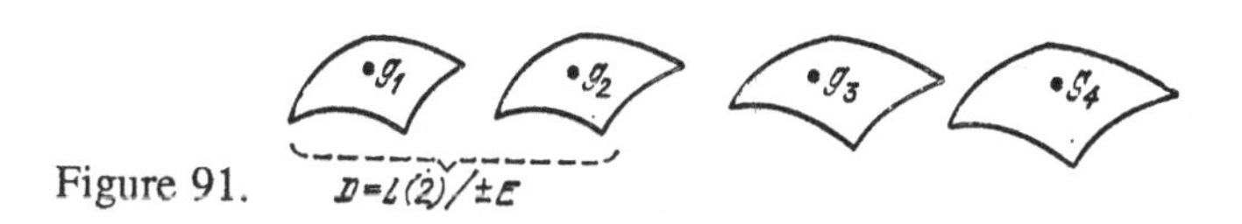

Figure 91.

Figure 92 presents four transformations: g_1, g_2, g_3, g_4 which preserve the pseudo-sphere S_2^1 ($\supset L_2$) and belong to distinct connected components of the group $G(\mathbb{R}_1^3)$.

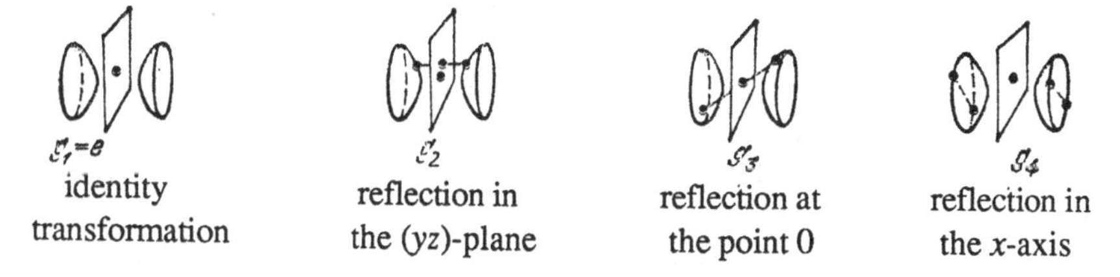

Figure 92.

The Lobachevsky geometry is realized separately on each of the sheets of the hyperboloid; we can say that a pseudo-sphere of imaginary radius is a union of two copies of the Lobachevskian plane.

In the space $\mathbb{R}_1^4$ of the special theory of relativity, a a pseudo-sphere of imaginary radius is also a three-dimensional hyperboloid of two sheets. The Riemannian metric induced on the latter by the envelope indefinite metric $ds^2\,(\mathbb{R}_1^4) = -\,c^2\,dt^2 + dx^2 + dy^2 + dz^2$ is positive definite (verify it!) and is called the metric of the three-dimensional Lobachevskian space.

We recommend the reader to study, repeating the arguments analogous to those used above, the geometry and the metric arising on a pseudo-sphere of *real radius*, which is (in the Euclidean model) a hyperboloid of one sheet.

APPENDIX 2

SOME ELEMENTS OF MODERN CONCEPTS OF THE GEOMETRY OF THE REAL WORLD

A.1 Introduction. Basic concepts

The principal types of physical forces determining the geometry of the surrounding macroscopic phenomena are gravitational and electro-magnetic forces. At the present time we know about the existence of four types of fundamental interactions, namely, nuclear ("strong"), electro-magnetic, "weak" and gravitational. Strong as they are, nuclear forces are rather short range, with the characteristic effective range of about 10^{-13} cm, called also the nuclear size. It is only electro-magnetic forces that eventually create (however, not without the aid of quantum theory) the surrounding matter, i.e. fasten together particles so that they form solids, liquids and gases, determine Mendeleyev's law etc. We shall not dwell on weak interactions — they are (for the present!) less noticeable. Gravitational forces keep us on the surface of the Earth, form the solar system, fasten galaxies together ($\sim 10^{20}$ cm) and are, possibly, responsible for the whole evolutionary process in the Universe ($\sim 10^{28}$ cm). It is relevant to note here that nuclear and weak interactions can be treated solely in the framework of quantum theory and have no classical equivalent. So, the basic types of fundamental forces which come under consideration without involving quantum theory and proceeding exclusively from the ideology of geometry and classical mechanics, are electro-magnetic and gravitational forces.

From the contemporary viewpoint, which basically took shape during the first two decades of the 20th century, space, time and gravity (together) form the space-time manifold M^4. The decisive role in the development of this theory was played by the papers by Einstein. An important contribution was also due to Lorentz, Poincaré, Minkowski and Hilbert. Points of the manifold M^4 are "events" which occurred in a certain place in space at a certain instant of time. It would, of course, be more precise to say that the assignment to each event of a point in a certain four-dimensional space-time manifold M^4 is a very convenient tool for the systematization of a large number of events.

Recall that a manifold M^4 is defined by an "atlas" $M^4 = \bigcup_q X_q$ consisting of "charts" X_q with local coordinates x^a_q, $a = 0, 1, 2, 3$. Each chart X_q is a region in a space R^4 with coordinates (x^a_q). In the common region of action of two coordinate systems, i.e. in the region $X_q \cap X_s$, all the expressions of coordinates in terms of one another

$$x^a_q(x^0_s, x^1_s, x^2_s, x^3_s)$$

are smooth functions for all pairs (q, s), where the intersections are non-empty and have a non-zero Jacobian

$$\left\| \frac{\partial x^a_q}{\partial x^b_s} \right\| \neq 0.$$

We shall use the term "Cartesian space" for the case where M^4 is described by one chart X with coordinates (x^a) running through all the real values.

According to modern concepts dating back to the famous paper by Einstein and Grossman (1913), the gravitational field is an *indefinite metric* on a space M^4, which has at each point the signature $(+---)$. This means that in each chart X_q with coordinates (x^a_q) there are given tensor fields $g^{(q)}_{ab} = g^{(q)}_{ba}$ $(a, b = 0, 1, 2, 3)$ such that in the region $X_q \cap X_s$ they are mapped to each other

$$\frac{\partial x^a_q}{\partial x^{a'}_s} g^{(q)}_{ab}(x_q(x_s)) \frac{\partial x^b_q}{\partial x^{b'}_s} = g^{(s)}_{a'b'}(x_s),$$

$x_q = x_q(x_s)$ in the region $X_q \cap X_s$.

It is assumed that at each point x_q of the region X_q the quadratic form $g_{ab}(x_q)\, \xi^a \xi^b$ is brought through a linear change to a diagonal form of the type $\eta_0^2 - \sum_{\alpha=1}^{3} \eta_\alpha^2$ (i.e. has the signature $(+---)$. Given this, $\det g_{ab} \neq 0$. As a rule we shall, in fact, work only in one coordinate system and shall not write the index q.

A *Minkowski space* (or a pseudo-Euclidean space) $M^4 = \mathbb{R}^{3,1}$ can by definition be given in the form of one Cartesian space $\mathbb{R}^4$ with the so-called Euclidean coordinates (x^0, x^1, x^2, x^3) and endowed with the metric

$$g_{ab} = \begin{pmatrix} 1 & & & 0 \\ & -1 & & \\ & & -1 & \\ 0 & & & -1 \end{pmatrix}, \quad dl^2 = (dx^0)^2 - \sum_{\alpha=1}^{3} (dx^\alpha)^2.$$

In the case of Minkowski space we say that the gravitational field is trivial or zero. The group of motions of the Minkowski space is called the Poincaré group.

The electro-magnetic field is determined by the covector field ("vector-potential") A_a (or by the 1-form $A = A_a\,dx^a$) on the space of events $A_a(x)$ which in the local system of coordinates x^q is given by the components $A_a(x)$: under the coordinate change $x(y)$ the components are transformed by the formula

$$\frac{\partial x^a}{\partial y^{a'}} A_a(x(y)) = A_{a'}(y), \quad A_a\,dx^a = A_{a'}\,dy^{a'}.$$

By definition, this covector field (or the differential 1-form) is given with an accuracy to a gauge (gradient) transformation:

$$A_a\,dx^a = A \approx A + d\phi = A'_a\,dx^a,$$

$$A_a \approx A_a + \frac{\partial\phi}{\partial x^a} = A'_a, \tag{1}$$

where $\phi(x)$ is an arbitrary scalar function.

The electro-magnetic field *tensor* is the expression (the skew-symmetric tensor F_{ab} or the differential 2-form)

$$F_{ab} = \frac{\partial A_a}{\partial x^b} - \frac{\partial A_b}{\partial x^a} = -F_{ba},$$

$$F_{ab}\,dx^a \wedge dx^b = F = dA, \tag{2}$$

independent of the choice of the field A_a up to the gauge transformation (1).

In the Minkowski space $M^4 = R^{3,1}$ with psuedo-Euclidean coordinates (x^a), where the metric has the form

$$g_{ab}\,dx^a\,dx^b = (dx^0)^2 - \sum_{\alpha=1}^{3} (dx^\alpha)^2,$$

the tensor F_{ab} is thought of as having the electric part $E_\alpha = F_{0\alpha}$ and the magnetic part $H_{\alpha\beta} = F_{\alpha\beta}$, $\alpha, \beta = 1, 2, 3$; the three-dimensional skew-symmetric tensor $H_{\alpha\beta}$ is formally associated with the so-called axial vector of the magnetic field

$$H^1 = H_{23}, \quad H^2 = -H_{13}, \quad H^3 = H_{12}.$$

In the three-dimensional formalism and notation we construct the 1-form of the electric field $E_\alpha \, dx^\alpha$ and the 2-form of the magnetic field $H = H_{\alpha\beta} \, dx^\alpha \wedge dx^\beta$. In the Minkowski space the quantity x^0/c is called the "world time", where c is the speed of light in a vacuum ($c \approx 3.10^{10}$ cm/s).

In any manifold M^4 with an indefinite metric $g_{ab}(x)$ of signature $(+---)$ and at any non-singular point x, where $\det g_{ab} \neq 0$, there exist three types of vectors:

a) time-like vectors,

b) space-like vectors,

c) isotropic (light-like) vectors.

We shall further denote $\xi^a \eta^b g_{ab}$ by $\langle \xi, \eta \rangle$, which is the scalar product of vectors ξ and η, at a point x, generated by the metric $g_{ab}(x)$. The quantity $\eta^a \eta^b g_{ab}$ will often be denoted by $\|\eta\|^2$. So, we have three types of vectors:

$$\text{a) } \|\eta\|^2 > 0, \quad \text{b) } \|\eta\|^2 < 0, \quad \text{c) } \|\eta\|^2 = 0.$$

Correspondingly, the line $x^a(\tau)$ in M^4 is called:

a) *time-like* in the case $\|dx/dt\|^2 > 0$ (everywhere);

b) *space-like* in the case $\|dx/d\tau\|^2 < 0$ (everywhere), and

c) *isotropic* (light-like) in the case $\|dx/d\tau\|^2 = 0$ (everywhere).

The evolution of any point particle during its lifetime is represented by a line in the space of all events M^4 (the "world line" of a particle). The following fundamental idea is hypothesized:

a) the world line of a particle of any mass $m > 0$ (a massive particle) is always time-like;

b) the world line of a massless particle $(m = 0)$ is isotropic.

By virtue of this hypothesis, for massive particles the "length" of any curve $x^a(\tau)$ is positive definite

$$0 < l = \int \left\| \frac{dx}{d\tau} \right\| d\tau = \int dl = \int \left(g_{ab}(x) \frac{dx^a}{d\tau} \frac{dx^b}{d\tau} \right)^{1/2} d\tau. \tag{3}$$

The following hypothesis (postulate) is assumed: the lifetime of a massive object along the world line $x^a(\tau)$ is proportional to the length of this line (this lifetime is referred to as "proper time"):

$$T_{prop} = 1/c \int dl = l/c. \tag{4}$$

EXAMPLE. Let

$$g_{ab}\, dx^a\, dx^b = (dx^0)^2 - \sum_{\alpha=1}^{3} (dx^\alpha)^2.$$

Consider two curves (a motionless point at zero and the motion along the world line 2) depicted in Figure 93.

Figure 93.

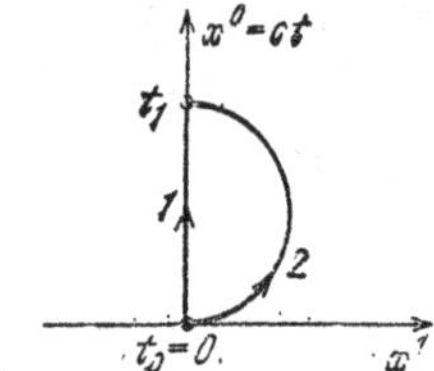

We can show that the following always holds

$$l_2 = \int_2 dl,$$

$$l_1 = \int_1 dl = c(t_1 - t_0) = ct_1 > l_2$$

(verify this!). Thus, the lifetime of a moving object is less than that of a motionless one (with the same beginning and end).

It should be noted that for any world line (which does not concinde with 1) the proper time is always less than the difference of the time coordinate at the initial or final moments: $ct_{\text{final}} - ct_{\text{initial}} > cT_{\text{proper}}$. Any world line of a massive particle (i.e. a time-like line) can be parametrized by the element of length (i.e. by the proper time), by the natural parameter, as is usually said in geometry:

$$\tau = l/c, \quad d\tau = dl/c. \tag{5}$$

The 4-dimensional velocity vector of a particle is always referred (by definition) precisely to the natural parametrization

$$u^a = \frac{dx^a}{d\tau}, \quad \langle u, u \rangle = \| u \|^2 = c^2. \tag{6}$$

In the Minkowski space with standard pseudo-Euclidean coordinates x^a, $dl^2 = (dx^0)^2 - \sum_\alpha (dx^\alpha)^2$ there exists another natural parametrization of world lines, the one using the world time — the coordinates $x^0/c = t$:

$$x^\alpha = x^\alpha(t), \quad x^0 = ct.$$

A 3-dimensional velocity vector arises $v^\alpha = dx^\alpha / dt$ which is related to the 4-velocity vector by the formula

$$1/c\, v^\alpha = u^\alpha / u^0, \quad \alpha = 1, 2, 3. \tag{7}$$

The element of length takes the form

$$\frac{dl}{c} = \left(1 - \frac{v^2}{c^2}\right)^{1/2} dt, \quad v^2 = \sum_{\alpha=1}^{3} (v^\alpha)^2. \tag{8}$$

Since $(u^0)^2 - \Sigma(u^\alpha)^2 = c^2$, for the 4-velocity vector, we obtain the formulae

$$u^0 = \frac{c}{\left(1 - v^2/c^2\right)^{1/2}}, \quad u^\alpha = \frac{v^\alpha}{\left(1 - v^2/c^2\right)^{1/2}}. \tag{9}$$

The laws of relativistic dynamics, i.e. of the motion of a massive particle in external electro-magnetic and gravitational fields, are defined proceeding from the extreme *action* principle

$$S(\gamma) = \int_{x_0}^{x_1} (\alpha\, dl + \beta A_a\, dx^a)$$

along the world time-like line.

Given this,

$\alpha = -mc$, where m is the particle mass,

$\beta = e/c$, where e is the particle charge. In accordance with the general rules of variational calculus for an action of the form

$$S(\gamma) = \int_\gamma L(x, w)\, d\tau,$$

where $w^a = dx^a/d\tau$ is the formal velocity, the 4-momentum is defined to be

$$p_a = \frac{\partial L}{\partial w^a} = \frac{-mcw^\alpha g_{ab}}{\left((w^0)^2 - \sum (w^\alpha)^2\right)^{1/2}} + \frac{e}{c} A_a .$$

Now, if $A_a \equiv 0$, then choosing $d\tau = dl/c$ we obtain

$$p^a = -mu^a,$$

$$p^a = g^{ab} p_b,$$

where u^a are the components of the 4-velocity vector with respect to the natural parameter $\tau = l/c$. In the Minkowski space $\mathbb{R}^{3,1}$ we obtain

$$-p_0 = \frac{mc}{\left(1 - \frac{v^2}{c^2}\right)^{1/2}}, \quad p_\alpha = \frac{mv^\alpha}{\left(1 - \frac{v^2}{c^2}\right)^{1/2}}. \tag{10}$$

Note that in the three-dimensional formalism, which is convenient for comparison with non-relativistic mechanics, we must choose the quantity $t = x^0/c$ as the parameter along the world line. We obtain the Lagrangian and the action of three-dimensional curves:

$$S(\gamma) = \int \left(-mc^2 \left(1 - v^2/c^2\right)^{1/2} + \frac{e}{c} A_\alpha v^\alpha + eA_0 \right) dt. \tag{11}$$

Suppose that $A_\alpha = 0$. We have the standard concepts of energy and 3-momentum for (11):

$$\varepsilon = v^\alpha \frac{\partial L}{\partial v^\alpha} - L = \frac{mc^2}{\left(1 - v^2/c^2\right)^{1/2}} + eA_0, \tag{12}$$

$$p_\alpha = \frac{mv^2}{\left(1 - v^2/c^2\right)^{1/2}} + \frac{e}{c} A_a . \tag{13}$$

This definition agrees with the preceding one, where

$$\varepsilon = -cp_0. \tag{14}$$

The energy is, therefore, proportional to the component p_0. We wish to emphasize an essential consequence: *in the absence of electro-magnetic field,* $A_a = 0$, *the following equality holds*

$$\langle p, p\rangle = g^{ab} p_a p_b = m^2 c^2 = m^2 \langle u, u\rangle, \tag{15}$$

where $g^{ab} g_{bc} = \delta^a_c$.

The surface (15) is called the "mass shell". Thus, the 4-momentum of free particles lies on the mass shell. The geometry of the mass shell, generated by the restriction to the mass shell of the Minkowski metric, is Lobachevsky geometry (see Section 1.4). Thus, Lobachevsky geometry is the geometry on the set of states of a free particle in the momentum space. In this context Lobachevsky geometry is a fundamental part of modern concepts of the real world.

A.2 Conservation laws. The Lorentz group

In the framework of variational calculus (or analytic mechanics) the laws of conservation are naturally associated with invariance of a system under simplest space-time transformation groups: time invariance and invariance under space translations and rotations. Mathematically, this can be explained as follows. Let there be given an arbitrary "Lagrangian" of the form $L(x, x^{\cdot})$ independent of the "time" τ and an "action" functional S on curves $x^q(\tau)$

$$S(\gamma) = \int L\left(x, \frac{dx}{d\tau}\right) d\tau,$$

$$x = x^1, \dots, x^m, \ v^q = dx^q/d\tau. \tag{1}$$

The conservation laws have the form:

1) the quantity $\varepsilon = v^q\,(\partial L/\partial v^q) - L$ (the "formal energy") is conserved along the trajectory, i.e. by virtue of the Euler-Lagrange equation

$$\frac{d\varepsilon}{d\tau} = 0, \quad \frac{dp^q}{d\tau} = \frac{\partial L}{\partial x^q}, \quad p_q = \frac{\partial L}{\partial v^q}$$

called the "formal momentum";

2) if the Lagrangian L does not depend on one of the coordinates $\partial L/\partial x^k = 0$, then $dp^k/d\tau = 0$.

More generally, if the Lagrangian L does not change along the vector field $Y^q(x)$

$$Y^q(x)\frac{\partial L}{\partial x^q}+\frac{\partial Y^i}{\partial x^j}v^j\frac{\partial L}{\partial v^i} = 0, \tag{2}$$

then (by virtue of the Euler-Lagrange equations) the conservation law holds

$$\frac{dp_Y}{d\tau} = 0, \quad p_Y = Y^q(x)\,p_q.$$

Here p_Y is the component of the momentum along the field Y. This form of the momemtum conservation law is equivalent to the previous one inasmuch as in a neighbourhood of any non-singular point x_0 of the vector field $Y(x_0) \neq 0$ there exists a coordinate system $z^1, \dots, z^m$ in which the field Y has the form $Y = (1, 0, 0, \dots, \dots, 0)$. Therefore, equation (2) reduces to the form $\partial L/\partial z^1 = 0$.

In classical mechanics we consider a system of n particles in a three-dimensional space (i.e. $m = 3n$). We have the total Lagrangian and the action

$$L = L(x_1^\alpha, \dots, x_n^\alpha, v_1^\alpha, \dots, v_n^a), \quad \alpha = 1, 2, 3.$$

The energy conservation law reduces to conservation of the quantity

$$\varepsilon_{\text{total}} = \sum_{q=1}^{n}\sum_{\alpha=1}^{3} v_q^\alpha \frac{\partial L}{\partial v_q^\alpha} - L.$$

The law of conservation of momentum and angular momentum arises from the requirement that the Lagrangian L be invariant under all *general* translations and rotations in $\mathbb{R}^3$, where the mutual position of particles remains unaltered:

$$L(Ax_1, \dots, A_{x_n}, A_* v_1, \dots, A_* v_n) = L(x_1, \dots, x_n, v_1, \dots, v_n). \tag{3}$$

Here A is any motion of the entire space $\mathbb{R}^3$, A_* is an induced transformation of variables v_i.

EXAMPLE 1. A is a translation along some axis in $\mathbb{R}^3$ (for example, x^1). For L we have

$$\sum_q \frac{\partial L}{\partial x_q^1} = 0 \leftrightarrow Y^{q\alpha} \frac{\partial L}{\partial x_q^\alpha} = 0, \quad \frac{\partial Y^{q\alpha}}{\partial x_p^i} = 0,$$

$$Y^{q\alpha} = (1, 0, 0, 1, 0, 0, \dots , 1, 0, 0).$$

Condition (3) gives rise to the law of conservation of total momentum

$$P_{\text{total}, \alpha} \sum_{q=1}^{n} p_{q,\alpha}, \quad p_{q,\alpha} = \frac{\partial L}{\partial v_q^\alpha}.$$

Here three-dimensional vectors $p_q = (p_{q\alpha})$ are called, by definition, *momenta* of each particle. If

$$L = \sum_q \frac{1}{2} m_q v_q^2 - V(x_1, \dots , x_n),$$

then the condition of translation invariance of the Lagrangian L has the form

$$V = V(x_1 - x_2, \dots , x_1 - x_n). \tag{4}$$

Naturally we have $V = \sum_{p<q} V(x_p - x_q)$. The total momentum is given by

$$P_{\text{total}} = \sum_q m_q v_q = \sum_q p_q.$$

In an electro-magnetic field (or in a co-moving coordinate system) the Lagrangians take the form

$$L = \sum_q \frac{1}{2} m_q v_q^2 - \sum_{p<q} V(x_p - x_q) + e \sum_q (A_\alpha v_q^\alpha / c + A_0(x_q)). \tag{5}$$

The energy and the total momenta are expressed as

$$\varepsilon = \sum_q \frac{1}{2} m_q v_q^2 + V + e \sum_q A_0(x_q),$$

$$p = \sum_q p_q = \sum_q m_q v_q + \frac{e}{c} A(x_q).$$

In the presence of an *external field* A_a the system (3) loses its translation invariance, and the total momentum is not conserved.

EXAMPLE 2. Let A be the matrix determining rotation (e.g. in a plane $(x^1, x^2) = (x, y)$). In this case, the vector field Y corresponding to the one-parameter group of rotations around the axis $z = x^3$, has the form

$$Y = (-x_1^2, x_1^1, 0, -x_2^2, x_2^1, 0, \dots, -x_n^2, x_n^1, 0).$$

We shall introduce in $\mathbb{R}^3$ a cylindrical system of coordinates $(z, \rho, \phi) = (y^1, y^2, y^3)$

$$x = x^1 = \rho \cos \phi,$$

$$y = x^2 = \rho \sin \phi, \quad z = x^3. \tag{6}$$

The vector field Y in coordinates $(y_1^1, y_1^2, y_1^3, \dots, y_n^1, y_n^2, y^3{}_n)$ becomes

$$Y = (0, 0, 1, \dots, 0, 0, 1).$$

The corresponding law of conservation of the component of momentum will be given by

$$p_Y = \sum_{q=1}^{n} p_{q,\phi} = \sum_q \frac{\partial L}{\partial \dot{y}_q^3} = \sum_q \frac{\partial L}{\partial \dot{\phi}_q}.$$

Because of (6), an elementary calculation gives

$$L(y, \dot{y}) = L(x, \dot{x}) =$$

$$= L(\rho_1 \cos \phi, \ \rho_1 \sin \phi, \ z_1, \dots; \ \rho_1 \cos \phi_1 -$$

$$- \rho_1 \sin \phi_1 \dot{\phi}_1, \ \dot{\rho}_1 \sin \phi + \rho_1 \cos \phi_1 \ \dot{\phi}_1, \dot{z}_1, \dots), \tag{7}$$

$$\frac{\partial L}{\partial \dot{\phi}^1} = \frac{\partial L}{\partial \dot{x}_1^1} x_1^2 + \frac{\partial L}{\partial \dot{x}_1^2} x_1^1 = x_1 p_{y_1} - y_1 p_{x_1}.$$

The quantity

$$\frac{\partial L}{\partial \dot{\phi}^q} = x_q p_{y_q} - y_q p_{x_q} = p_{\phi_q}$$

is called the z-component of the angular momentum M_q^3 of a q-th particle. The total vector of the angular momentum M_q of the q-th particle results from the change of z

by x or y. From formula (7) we obtain the law of conservation of the total angular momentum of a system of particles

$$M = \sum_{q=1}^{n} M_q = \sum_q [x_q \times p_q],$$

where x_q and p_q are respectively the radius-vector and momentum of the q-th particle.

It would be more correct (and this also refers to the case of the magnetic field) to assume the angular momentum to be the skew-symmetric tensor of rank two

$$M_q^{\alpha\beta} = x_q^{\alpha} p_q^{\beta} - x_q^{\beta} p_q^{\alpha}. \tag{8}$$

As is well known, in mechanics there exists a requirement of system invariance under Galilean transformations (the Lagrangian L of a system may change by a full derivative; given this, the "action" S remains unaffected to within boundary terms):

$$x = x' + wt, \quad w = \text{const.}, \quad t' = t \tag{9}$$

and for a system of n particles in $\mathbb{R}^3$ we shall have

$$L'(x', \dot{x}', t) = L'(x' + wt, \dot{x}' + w) + \frac{df(x)}{dt} = L(x, \dot{x}), \tag{10}$$

where $\dfrac{df}{dt} = \dfrac{df}{dx^i} \dot{x}^i$, $f(x)$ is a certain function.

For the total momentum this implies

$$p_i' = \frac{\partial L'}{\partial \dot{x}'^i} = \frac{\partial L}{\partial \dot{x}^i} - \frac{\partial f(x)}{\partial x^i}; \quad p' = p - \nabla f.$$

Thus, momentum may change only by a total gradient (note that in classical mechanics, Lagrangian is generally defined with an accuracy to a full derivative. momentum is thus defined within the gradient).
For the energy we obtain

$$\varepsilon' = v' \frac{\partial L'}{\partial v'} - L' = (v - w) \frac{\partial L}{\partial v} - L = \varepsilon - pw. \tag{11}$$

For the important case

$$L = \sum \frac{1}{2} m_q v_q^2 - V(x)$$

under Galilean transformation we shall obtain by virtue of (11)

$$p' - p - Mw, \quad M = \sum m_q \quad (p = P_{\text{total}})$$

$$\varepsilon' = \varepsilon - pw + \frac{Mw^2}{2},$$

$$\varepsilon = \varepsilon' + pw - \frac{Mw^2}{2}. \tag{12}$$

Now turning to relativistic mechanics, we should consider the whole system in a four-dimensional space (x^a), $a = 0, 1, 2, 3$. We shall deal with the laws of conservation *only* in a Minkowski space endowed by the metric $(dl)^2 = (dx^0)^2 - \sum_\alpha (dx^\alpha)^2$. The Lagrangian $L(x_1^a, \ldots, x_n^a, \dot{x}_1^a, \ldots, \dot{x}_n^a)$ will be referred to a certain parameter $\dot{x} = dx_q / d\tau$.

In this case we should require invariance of the system under the general motion A of the Minkowski space $\mathbb{R}^{3,1}$. The group of all motions (generated by linear transformations around the origin O and translations) is called the Poincaré group. The sub-group of linear transformations is denoted by $O(3, 1)$ and is referred to as the Lorentz group (the group of pseudo-orthogonal transformations). The connected component of unity in the Lorentz group is denoted by $SO(3, 1)$. As distinct fom ordinary groups $O(n)$, i.e. from the orthogonal transformations of Euclidean space $\mathbb{R}^n$, the Lorentz group $O(n, 1)$ consists of four connected components.

THEOREM. *There exists a continuous homomorphism* ϕ *of the group* $O(n, 1)$ *into the group* $\mathbb{Z}_2 \times \mathbb{Z}_2$ *(involving four elements) defined as*

$$\phi(A) = (\operatorname{sgn} \det A, \ \operatorname{sgn} \langle e_0, Ae_0 \rangle).$$

In particular, if $\phi(A) = (1, 1)$, *then we have* $A \in SO(1, 1)$, *i.e.* A *looks like*

$$A = \begin{pmatrix} \operatorname{ch} \psi & \operatorname{sh} \psi \\ \operatorname{sh} \psi & \operatorname{ch} \psi \end{pmatrix}.$$

This is a connected group. The connectedness of the group $SO(n, 1)$ for $n > 1$ is proved in a somewhat more complicated way (we leave it as an exercise to prove it for $n = 2, 3$).

Transformations belonging to the group $SO(1, 1)$ are called "elementary Lorentz transformations". The whole group $O(n, 1)$ has the form

$$O(n, 1) = SO(n, 1)\, P^s\, T^q,$$

where $s = 0, 1,\ q = 0, 1.$

$$P(x^a) = \begin{cases} x^a, & a = 0, \\ -x^a, & a = 1, 2, \dots, n \end{cases}$$

("spatial reflection"),

$$T(x^a) = \begin{cases} -x^a, & a = 0, \\ x^a, & a = 1, 2, \dots, n \end{cases}$$

("time reflection").

The transformations $A \in SO(1, 1)$ are written in the form

$$A = \begin{pmatrix} \dfrac{1}{\left(1 - v^2/c^2\right)^{1/2}} & \dfrac{v/c}{\left(1 - v^2/c^2\right)^{1/2}} \\ \dfrac{v/c}{\left(1 - v^2/c^2\right)^{1/2}} & \dfrac{1}{\left(1 - v^2/c^2\right)^{1/2}} \end{pmatrix},$$

where v is the three-dimensional velocity and c is the speed of light in a vacuum. For $v/c \ll 1$, we obtain the Galilean transformations

$$v/c \to 0,\quad x^0 = ct,\quad x'^0 = ct',$$

$$A : \begin{cases} x = \dfrac{x'}{\left(1 - v^2/c^2\right)^{1/2}} + \dfrac{v t'}{\left(1 - v^2/c^2\right)^{1/2}} \mapsto x \approx x' + vt', \\ ct = \dfrac{x' v/c}{\left(1 - v^2/c^2\right)^{1/2}} + \dfrac{c t'}{\left(1 - v^2/c^2\right)^{1/2}} \mapsto t \approx t'. \end{cases} \tag{13}$$

We now proceed to the laws of transformation of a system of n parrticles in a space $\mathbb{R}^{3,1}$, invariant under all the motions

$$L(Ax_1, \dots, Ax_n, A_* x^{\cdot}_1, \dots, A_* x^{\cdot}_n) = L.$$

1. If A are translations, we are led as before to the law of conservation of the 4-momentum of the system

$$P_{\text{total}} = \sum_q p_q,$$

where $\varepsilon = -cp_0$.

2. If A are rotations in a three-dimensional space, then we obtain, as before, the law of conservation of the total three-dimensional angular momentum

$$M = \sum_q [x_q \times p_q], \quad M^{\alpha\beta} = x_q^\alpha p_q^\beta - x_q^\beta p_q^\alpha.$$

3. If A are the various elementary Lorentz transformations in planes $(0, \alpha)$, where $\alpha = 1, 2, 3$, then we arrive at

$$M^{0\alpha} = \sum_q x_q^0 p_q^\alpha - x_q^\alpha p_q^0, \tag{14}$$

where $p_q^a = g^{ab} p_{ab}$. For each particle we have $x_q^0 = ct$, $p_q^0 = -\varepsilon_q/c$, x_q^α are coordinates in $\mathbb{R}^3$.

Thus we are led to

$$\frac{1}{c} M^{0\alpha} = \sum_q (tp_q^\alpha - \frac{1}{c^2} \varepsilon_q x_q^\alpha) = \text{const.}$$

Applying to the other laws of conservation:

$$\sum_\varepsilon \varepsilon_q = \varepsilon = \text{const.}, \quad \sum_q p_q^\alpha = \text{const.},$$

we finally have

$$tp^\alpha_{\text{total}} + \text{const} = \frac{\sum \varepsilon_q x_q^\alpha}{c^2}.$$

This is a uniform motion of the "relativistic centre of mass":

$$\bar{x}^{\alpha} = \left(\sum_q \varepsilon_q x_q^{\alpha}\right) / \sum_q \varepsilon_q, \quad \dot{\bar{x}}^{\alpha} = \text{const.}$$

A. 3 Free particles. Mass shell. Velocity addition. The simplest scattering processes

The energy and momentum of free massive particles in Minkowski space $\mathbb{R}^{3,1}$, as mentioned in A.1, are of the form

$$p_0 = \frac{-mc}{\left(1 - v^2/c^2\right)^{1/2}}, \quad p_{\alpha} = \frac{mv^{\alpha}}{\left(1 - v^2/c^2\right)^{1/2}}, \tag{1}$$

where $-\varepsilon = cp_0, p_{\alpha} = \partial L/\partial v^{\alpha}$, and v^{α} are components of the three-dimensional velocity of the particle. The action is given by (in three-dimensional formalism)

$$\begin{aligned} S &= \int L(v)\, dt = -mc \int dl, \\ L(v) &= -mc^2 \left(1 - v^2/c^2\right)^{1/2}, \quad x^0 = ct. \end{aligned} \tag{2}$$

The quantities (p_0, p_{α}) together form the 4-covector. The corresponding 4-vector has the form

$$p^{\alpha} = g^{\alpha b} p_b, \quad p^0 = p_0, \quad p^{\alpha} = -p_{\alpha} \tag{3}$$

on the "mass shell"

$$(p^0)^2 - \sum_{\alpha} (p^{\alpha})^2 = m^2 c^2 \tag{4}$$

or

$$-p^0 = \frac{\varepsilon}{c} = \left(m^2 c^2 + \sum_{\alpha} (p^{\alpha})^2\right)^{1/2} \quad (\text{let } \varepsilon > 0).$$

The restriction of the Minkowski metric g_{ab} to the mass shell (4) is a fixed-sign metric with coordinates p^{α} or v^{α} related by formula (1).

The metric has the form

$$-g_{\alpha\beta}\, dp^{\alpha}\, dp^{\beta} = (dp^0)^2 - \sum_{\alpha=1}^{3} (dp^{\alpha})^2, \tag{5}$$

where (p^0, p^1, p^2, p^3) is defined by formula (4); a simple computation (independent of the dimension of the space) shows that the metric (5) has the volume element

$$\left(\det g_{\alpha\beta}\right)^{1/2} dp^1 \wedge \ldots \wedge dp^n = \frac{mc}{p^0}\, dp^1 \wedge \ldots \wedge dp^n. \tag{6}$$

EXERCISE 1. Prove formula (6). 2. Find the coordinates where we obtain the so-called Poincaré model of Lobachevsky geometry in a unit ball (see Section 1.4), for the metric on the mass shell.

The group of motions of Lobachevskain space coincides with the group $O^{+}(3, 1)$ which is the orthochronic sub-group of the Lorentz group $O(3, 1)$, where the upper portion of the light cone is sent to itself. Indeed, any motion from the group $O^{+}(3, 1)$ preserves the Minkowski metric, the mass shell (4) and the condition $\varepsilon > 0$. It can be shown that Lobachevskian space has no other motions.

Let us consider the so-called "velocity addition law". The accurate statement of the problem is this: let a particle move at a velocity $v = (v^1, v^2, v^3)$ relative to a pseudo-Euclidean coordinate system $K = (x, y, z, t)$ in a space endowed with a metric

$$dl^2 = (dx^0)^2 - \sum (dx^{\alpha})^2,$$

$$x^0 = ct, \quad x^1 = x, \; x^2 = y, \quad x^3 = z.$$

Here $v^2 \leq c^2$.

Now examine the elementary Lorentz transformation, say, in an (x, t)-plane (determined by the velocity $\bar{v}$, where $\bar{v}^2 < c^2$) to a new coordinate system (t^1, x^1). What will the velocity $\bar{\bar{v}}$ of the particle be in the new coordinate system? By definition, $\bar{\bar{v}}$ is the "sum" of the velocities v and $\bar{v}$. Obviously, v and $\bar{v}$ are non-symmetric here. Calculate the quantity $\bar{\bar{v}}(v, \bar{v})$. By definition

$$dx = \frac{1}{\left(1-\bar{v}^2/c^2\right)^{1/2}}\,(dx^1 + \bar{v}\,dt^1),$$

$$dt = \frac{1}{\left(1-\bar{v}^2/c^2\right)^{1/2}}\left(\frac{\bar{v}}{c^2}\,dx^1 + d\,t^1\right),$$

$$dy = dy^1, \quad dz = dz^1. \tag{7}$$

Given this, $dx/dt = v^1$, $dy/dt = v^2$, $dz/dl = v^3$. In view of the fact that $v^{=i} = dx^j/dt^1$, we can readily derive the formulae of velocity addition

$$\bar{\bar{v}} = \bar{\bar{v}}(v, \bar{v}).$$

For parallel velocities (i.e. $v^2 = v^3 = 0$) the result is

$$v \,\|\, \bar{v}, \quad \bar{\bar{v}} = \frac{v + \bar{v}}{1 + \dfrac{v\bar{v}}{c^2}}. \tag{8}$$

Formula (8) implies that if the particle velocity $|v| = c$, then in the new coordinate system the velocity $|v|$ is also equal to c in the absolute value.

This conclusion remains valid for non-parallel velocities inasmuch as Lorentz transformations preserve the light cone.

Thus, the results of the special theory of relativity are consistent with the well-known Michelson-Morley experiment on the invariance of the speed of light in a vacuum under the change to a uniformly moving coordinate system created by material bodies (i.e. moving with a velocity $\bar{v} < c$).

We have already said that as $\bar{v}/c \to 0$, the Lorentz transformation (7) becomes a Galilean transformation

$$x = x^1 + \bar{v}\,t, \quad t = t^1$$

and the velocity addition law becomes the ordinary sum of vectors. It should be emphasized that the velocity addition law in the theory of relativity is non-commutative and even non-associative for non-parallel velocities.

Now let us see what effect the Lorentz transformation has upon the energy momentum 4-vector:

$$(p^0, p^\alpha) = (p_0, -p_\alpha) = -\left(\frac{mc}{(1-v^2/c^2)^{1/2}}, \frac{mv^\alpha}{(1-v^2/c^2)^{1/2}}\right),$$

$$p^1 = \frac{1}{(1-\bar{v}^2/c^2)^{1/2}}\,(p'^1 + \frac{\bar{v}}{c}\,p'^0),$$

$$p^0 = \frac{1}{(1-\bar{v}^2/c^2)^{1/2}}\,(\frac{\bar{v}}{c}\,p'^1 + p'^0). \tag{9}$$

For low velocities we have

$$-p^0 = \frac{\varepsilon}{c} = mc + \frac{mv^2}{2c} + O(1/c^2),$$

$$-p^2 = \frac{mv^\alpha}{(1-\bar{v}^2/c^2)^{1/2}}, \quad \bar{v} = (\bar{v}^1, 0, 0), \quad v = (v^1, v^2, v^3). \tag{9'}$$

From formulae (9) and (9') we obtain

$$p \approx p' + m\bar{v} + O(1/c),$$

$$\varepsilon \approx \varepsilon' + p'\bar{v} + \frac{m\bar{v}^2}{2} + O(1/c). \tag{10}$$

Thus, from the Lorentz transformations for 4-momentum we have derived the law of energy-momentum transformation under Galilean transformations.

Now we shall view the simplest processes of relativistic particle scattering. It is assumed that before the beginning of the process ($t \to -\infty$) we had a set M of free particles $x_1, \ldots, x_M$ with masses $m_1, \ldots, m_M$, 4-momenta $p_1, \ldots, p_M$ and angular momenta $M_1, \ldots, M_M$ relative to the origin. After the scattering process ($t \to +\infty$) we obtain N free particles with masses $m'_1, \ldots, m'_N$, momenta $p'_1, \ldots, p'_N$ and angular 4-momenta $M'_1, \ldots, M'_N$. Whatever the process we must have the law of conservation of total 4-momentum and angular 4-momentum ("general geometric conservation laws"):

$$\sum_{j=1}^{M} M_j = \sum_{k=1}^{N} M'_k ,$$

$$\sum_{j=1}^{M} P_j = \sum_{k=1}^{N} P'_k . \tag{11}$$

All the assumptions on the character of the process must be invariant under the Lorentz (and Poincaré) group.

EXAMPLE 1. Spontaneous decay of one in-flight particle, of mass M, into two particles of masses m_1, m_2. Let the velocity of the primary particle be $(v, 0, 0)$ and let the 4-momentum of this particle have the form

$$-p^0 = \frac{Mc}{\left(1 - v^2/c^2\right)^{1/2}}, \quad -p^1 = \frac{Mv}{\left(1 - v^2/c^2\right)^{1/2}}, \quad p^2 = p^3 = 0.$$

To consider the decay process, it is convenient to employ the following step-by-step procedure.

Step 1. We change to a moving coordinate system (C-system), in which the first particle is at rest, its 4-momentum being $-p' = (Mc, 0, 0, 0)$.

Step 2. Consider the process in the C-system. For 4-momenta we have the relation

$$p' = p''_1 + p''_2 = (Mc, 0, 0, 0),$$

where p''_i are momenta of the decaying particles, in the form

$$-p''_1 = \left(\frac{m_1 c}{\left(1 - v_1^2/c^2\right)^{1/2}}, \frac{m_1 v_1^\alpha}{\left(1 - v_1^2/c^2\right)^{1/2}}\right) = (-p_1''^0, -p_1''^\alpha),$$

$$-p''_2 = \left(\frac{m_2 c}{\left(1 - v_2^2/c^2\right)^{1/2}}, \frac{m_2 v_2^\alpha}{\left(1 - v_2^2/c^2\right)^{1/2}}\right) = (-p_2''^0, -p_2''^{0\alpha}).$$

Obviously, $Mc = m_1c \left|(1 - v_1^2/c^2)^{1/2} + m_2c \left|(1 - v_2^2/c^2)^{1/2}\right.\right.$. Therefore, $m_1 + m_2 < M$ provided $v_1 \neq 0$ or $v_2 \neq 0$. Next, we obtain

$$p_1''^{(\alpha)} + p_2''^{\alpha} = 0 \text{ (see Figure 94).}$$

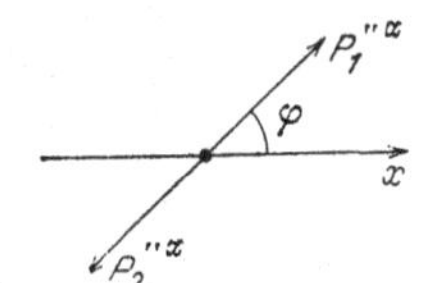

Figure 94.

Thus, the whole process is characterized by the angle ϕ (which is the slope angle of the momenta $p_j''^{\alpha}$ of decaying particles to the x-axis in the C-system) and by the absolute value of the vectors

$$\sum_{\alpha} |p_1''^{(\alpha)}|^2 = \sum_{\alpha} |p_2''^{(\alpha)}|^2, \quad \alpha = 1, 2, 3.$$

Step 3. We return to the original coordinate system and assume that, for example, the distribution of decaying particles does not depend on the angle ϕ (for the calculations we refer the reader to the book [29]).

EXAMPLE 2. Elastic scatterings of a pair of particles with masses m_1 and m_2 (i.e. the particles themselves are assumed to remain unaffected). It is, in fact, assumed that one particle of mass $m_1 = M$ is at rest ($v_1 = 0$) while the other is incident on it with the velocity $v_2 = (v_2, 0, 0)$ and mass $m_2 = m$.

Step 1. We go over to the C-system, where the resultant 3-momentum is equal to zero:

$$-(p_1' + p_2') = (\varepsilon'/c, 0, 0, 0).$$

Before our transition to the C-system we had

$$-p_1 = \left(\frac{mc}{\left(1 - v_2^2/c^2\right)^{1/2}}, \frac{mv_2}{\left(1 - v_1^2/c^2\right)^{1/2}} \right),$$

$$-p_2 = (Mc, 0, 0, 0).$$

Step 2. Consider the process in the C-system. We have

$$p'^{\alpha}_2 + p'^{\alpha}_1 = 0, \quad \alpha = 1, 2, 3,$$

$$-p'_1 = \left(\frac{mc}{\left(1 - v'^2_1/c^2\right)^{1/2}}, \frac{mv'_2}{\left(1 - v'^2_1/c^2\right)^{1/2}}, 0, 0 \right)$$

$$-p'_2 = \left(\frac{Mc}{\left(1 - v'^2_2/c^2\right)^{1/2}}, \frac{Mv'_2}{\left(1 - v'^2_2/c^2\right)^{1/2}}, 0, 0 \right).$$

By virtue of the law of conservation of 3-momentum, after the process we obtain the new 4-momenta p''_1, p''_2 and the equality

$$p''^{\alpha}_1 + p''^{\alpha}_2 = 0.$$

Applying now to the law of conservation of energy, we obtain

$$\sum_{\alpha=1}^{3} (p'^{\alpha}_1)^2 = \sum_{\alpha=1}^{3} (p''^{\alpha}_1)^2,$$

$$\sum_{\alpha=1}^{3} (p'^{\alpha}_2)^2 = \sum_{\alpha=1}^{3} (p''^{\alpha}_2)^2.$$

Thus, the elastic process in the C-system is characterized only by the angle ϕ — the rotation of a pair of 3-vectors $(p'^{\alpha}_1) \rightarrow (p''^{\alpha}_1)$, $p'^{\alpha}_2 = -p''^{\alpha}_1$.

Step 3. As before, we return to the original ("laboratory") system in which all physical conclusions are drawn (see ref. [29]).

An elastic process in the C-system is considered exactly in the same manner as in classical mechanics. The final difference is *only* due to the distinction between the Lorentz and the Galilean transformations in the change to the C-system.

A. 4 An electro-magnetic field

As mentioned above (see A.1), an electro-magnetic field (in a vacuum) is defined by the so-called vector-potential $A_a(x)$, more precisely, by the class of equivalent vector-potentials

$$A_a \sim A_a + \partial_a \psi,$$

where ψ is an arbitrary single-valued smooth function.

Directly observable are the strengths, i.e. a skew-symmetric tensor of the form

$$F_{ab} = \partial_a A_b - \partial_b A_a.$$

In pseudo-Euclidean coordinates (x^a) in a space $\mathbb{R}^{3,1}$ we introduce the electric (E) and magnetic (H) field strengths:

$$E_\alpha = F_{0\alpha}, \; H_{\alpha\beta} = F_{\alpha\beta}, \; \alpha, \beta = 1, 2, 3,$$

$$H_{12} \to H^3, \; H_{13} \to -H^2, \; H_{23} \to H^1.$$

To each particle, there corresponds, beside mass, an electric charge e (either positive or negative) with the result that particles in the field are affected by the force, both in classical and relativistic mechanics (v is a three-dimensional velocity, the notation is three-dimensional):

$$\dot{p}_\alpha = eE_\alpha + e\left[\frac{v}{c} \times H\right]_\alpha = eE_\alpha + \frac{ev^\beta}{c} H_{\alpha\beta},$$

where p is 3-momentum of particle before the field is on:

$$p_\alpha = mv^\alpha \qquad \text{in classical mechanics}$$

$$p_\alpha = \frac{mv^\alpha}{\left(1 - v^2/c^2\right)^{1/2}} \qquad \text{in relativistic mechanics.}$$

For the Lagrangian we have

$$S(\gamma) = \int_\gamma \left(L_{\text{free}} + \frac{e}{c} A_\alpha \dot{x}^\alpha + \frac{e}{c} A_0 \dot{x}^0\right) dt,$$

$$A_0 = \phi, \; x^0 = ct.$$

For time-independent fields

$$E = \nabla\phi, \quad H = \operatorname{rot} A, \quad A = (A_1 A_2, A_3), \quad \phi = A_0.$$

A "gauge" is a choice of one or another representative in a class of equivalent vector-potentials:

$$A \sim A + \nabla\psi.$$

The most popular gauges are the following:
a) Lorentz (relativistically invariant) gauge

$$\partial_a A_a = 0 \quad (a = 0, 1, 2, 3),$$

b) Coulomb gauge

$$\partial_a A_\alpha = 0 \quad (\alpha = 1, 2, 3),$$

c) Hamiltonian gauge

$$A_0 = 0, \quad E_\alpha = \mathring{A}_\alpha, \quad H = \operatorname{rot} A\,.$$

Now we shall give particularly simple exercises from particle mechanics in the fields E and H.

EXERCISE 1. The motion of a relativistic particle in a constant (independent of x, y, z, t) electric field $E = (E, 0, 0)$. Here $H = 0$.

For the classical case this is the motion in a constant field with constant acceleration. Find the relativistic analogue of this motion — the "particle acceleration" by an electric field up to relativistic velocities.

EXERCISE 2. The motion of a classical (and a relativistic) particle in a constant magnetic field $H = (0, 0, H)$; express the (Larmor) radius of the orbit (in an (x, y)-plane) in terms of the energy and the magnetic field.

We leave the consideration of this to the reader (see ref. [29]).

The following elementary statements hold.

1. On switching on a magnetic field, the energy-momentum vector of a particle is shifted by the vector-potential (see A.1):

$$p_a \to p_a + \frac{e}{c} A_a = p'_a, \quad cp_0 = -\varepsilon,$$

$$\varepsilon \to \varepsilon - e\phi.$$

2. In particular, in the absence of an electric field the energy remains unchanged provided that the magnetic field is constant in time: here we should take $A_0 = \phi = 0, \mathring{A}_\alpha = 0, \alpha = 1, 2, 3.$

An electric and a magnetic field together form a skew-symmetric tensor $F_{ab}(x)$ which, as pointed out above, is dependent on a point in space. It is natural to ask the following questions which require geometric solution:

a) What functions of the tensor F_{ab} at a given point x_0 are invariant under Lorentz transformations?

b) To what simplest form can the tensor be brought by a Lorentz transformation?

The general algebraic rules tell us that to seek the invariants of the bilinear form F_{ab} in a space endowed with a metric g_{ab}, we should construct the equation

$$P(\lambda) = \det (F_{ab} - \lambda g_{ab}) = 0.$$

The coefficients of the polynomial $P(\lambda)$ are just these unknown invariants (symmetric functions of $\lambda_1, \lambda_2, \lambda_3, \lambda_4$ if $\det g_{ab} = \pm 1$). In view of the skew symmetry of the matrix F_{ab}, the polynomial $P(\lambda)$ acquires the form

$$P(\lambda) = \lambda^4 + a\lambda^2 + b,$$

where

$$a = \text{const} \cdot (E^2 - H^2), \quad b = \text{const} \cdot (EH)^2$$

(verify this!) and g_{ab} is the Minkowski metric. The invariance of the quantities $E^2 - H^2$ and $(EH)^2$ is beyond doubt also for the reason that

$$F_{ab} F^{ab} = F_{ab} F_{cd} g^{ac} g^{bd} = \langle F, F \rangle = \text{const} (E^2 - H^2),$$

$$F_{ab} F_{cd} \varepsilon^{abcd} = (EH) (\text{const}).$$

The absence of invariants other than eigenvalues λ_i — or the coefficients of the characteristic polynomial — is fully analogous to the corresponding theorem in an ordinary Euclidean space $\mathbb{R}^4$ for the group $O(4)$.

However, for the Lorentz group , due to the indefiniteness of the metric g_{ab}, the reduction of the matrix F_{ab} to the classical block-diagonal form consisting of 2×2 matrices appears to be generally impossible; the amount of possible cases and types of eigenvalues is mauch larger. Depending on invariants, we single out the following cases.

1. $E^2 - H^2$ is arbitrary, $EH \neq 0$. A Lorentz transformation may result (at a given point x_0) in $E \| H$.

2. $EH = 0$, $E^2 - H^2 \neq 0$: a) $E^2 > H^2$, b) $E^2 < H^2$. Here we can always come to the form $H = 0$ in case a) and to the form $E = 0$ in case b).

These cases are therefore referred to as "purely electric" or "purely magnetic".

3. $E^2 = H^2$, $EH = 0$. Here all the eigenvalues λ_i are equal to zero: $\lambda_i = 0$, $P(\lambda) = \lambda^4$.

Lorentz transformations do not change the property that $|E| = |H|$ and $E \perp H$. This case corresponds to electro-magnetic waves propogating in one direction (say , along x), i.e. to vector-potentials of the form $A_0 = 0$, $A_\alpha\,(x - ct)$, but we shall not discuss Maxwell's equations here (see ref. [29]). The only thing worth noticing is the dimension of the fields and charges which is determined from the following requirements:

1) eE has the dimension of force,
2) the dimensions of E and H coincide,
3) the quantity $(E^2 + H^2)/8\pi$ has the dimension of energy density.

The total energy of the field itself (in the absence of charges) is given by the formula

$$E = \int (E^2 + H^2)/8\pi \, d^3x$$

whence

$$[e] = m^{1/2}\, T^{-1}\, L^{3/2},$$

$$[E] = [H] = m^{1/2}\, L^{-1/2}\, T^{-1}.$$

Here [ ] implies the dimension of the quantity in brackets — the product of the scales of mass m, length L and time T raised to corresponding powers. We see that the dimensions of the fields appear to be fractional.

The simplest fields are:

1) $E = \text{const}$, $H = \text{const}$, which are constant fields; here E and H are any three-dimensional vectors;

2) $A_0 = e/|x|$, $|x| = \sum_\alpha (x^\alpha)^2$, $A_\alpha = 0$, $\alpha = 1, 2, 3$ (the Coulomb potential of a point charge);

3) $A = \mathrm{Re}\left(A^{(0)} \exp(\pm i(\omega t - k_\alpha x^\alpha))\right)$, $\omega = c|k|$, $A^{(0)} = \text{const.}$, $A_0{}^{(0)} = 0$, $\sum_\alpha A_\alpha{}^{(0)} k_\alpha = 0$, i.e. the 3-vectors (k_α) and $(A_\alpha{}^{(0)})$ are orthogonal. These are running waves with "polarization" $A^{(0)}$.

A. 5 The simplest information on gravitational fields.

Recall that by the definition given in A.1, the relativistic gravitational field is simply an indefinite metric g_{ab} in the space of events, or equivalently, in a four-dimensional manifold M^4. The first question that naturally arises with such a definition of a gravitational field is the question of consistency with the classical Newton theory of gravitation. In a non-relativistic theory, gravitation is determined by a scalar field (potential) $\phi(x, y, z, t)$. The potential itself satisfies the (Poisson) equation

$$\Delta\phi = 4\pi\rho, \quad \Delta = \sum_{\alpha=1}^{3} \frac{\partial^2}{(\partial x^\alpha)^2}, \quad x^1 = x, \; x^2 = y, \; x^3 = z,$$

where ρ is mass density. The motion of a massive particle in a field is specified by the (Newton) equation

$$\ddot{x}^\alpha = -\partial_\alpha \phi. \tag{1}$$

It is noteworthy that the mass m drops out of these equations, and it is only important that $m \neq 0$. This property unites the gravitational forces with the so-called "geometric forces" which are due to a lame choice of the coordinate system. For example, let a particle move freely (without forces); let us change to a coordinate system moving with acceleration $a(t)$. Then the equation of motion in a moving coordinate system assumes the form

$$\ddot{x} = -a(t),$$

where the particle mass does not enter. Due to this, the relation between gravitation and geometry was hypothesized long ago, but this idea assumed its ultimate shape only in the course of the creation of the general theory of relativity, i.e. after the appearance of the special theory of relativity where the concept of a "four-dimensional indefinite metric" had appeared for the first time.

Zero-mass particles in non-relativistic theory did not interact with gravitational fields.

In the general theory of relativity, the motion of a test particle which has no retroaction upon the gravitational field $g_{ab}(x)$, is assumed to proceed along time-like geodesic lines of the metric g_{ab}. The motion of light particles (photons) proceeds along light-like, or isotropic, geodesics. The particle is assumed to be unaffected by any force other than gravitational. To what procedure can we apply to compare the Newton equation (1) with the equation of time-like geodesic in a certain metric?

We shall introduce the class of "weak" gravitational fields g_{ab} in the form

$$g_{ab} = \overset{(0)}{g}_{ab} + \frac{1}{c^2}\eta_{ab} + O\left(\frac{1}{c^3}\right), \tag{2}$$

where formally $1/c \to 0$ (c is the speed of light in a vacuum, regarded here as a formal large parameter). The metric $g^{(0)}_{ab}$ coincides with the Minkowski metric, the quantities $\eta_{ab}(x)$ and any of their derivatives with respect to x^α, t should be finite. the remainder $O(1/c^3)$ should be of the order of $1/c^3$ together with the derivatives with respect to the coordinates x^α, t. The equation of geodesics is determined by the so-called "Christoffel symbols" Γ^k_{ij}:

$$\ddot{x}^k + \Gamma^k_{ij}\dot{x}^i\dot{x}^j = 0 \tag{3}$$

where

$$\Gamma^k_{ij} = \frac{1}{2}g^{kl}\left(\frac{\partial g_{il}}{\partial x^j} + \frac{\partial g_{jl}}{\partial x^i} - \frac{\partial g_{ij}}{f x^l}\right).$$

CLAIM. *For the weak gravitational field* (2) *equation* (3) *of "slow" time-like geodesics has the form of Newton's equations* mod $O(1/c)$:

$$\ddot{x}^\alpha = -\partial_\alpha\phi + O(1/c),$$

where

$$g_{00} = 1 - \frac{2\phi}{c^2} + O\left(\frac{1}{c^3}\right), \quad |v| = \left|\frac{dx^\alpha}{dt}\right| \ll c .$$

Proof. Christoffel's formula expresses the symbols Γ^a_{bc} in terms of first derivatives of the metric $g_{ab} = g^{(0)}_{ab} + 1/c^2\,\eta_{ab} + O(1/c^3)$, where $g^{(0)}_{ab}$ are constants and all the derivatives $\partial\eta_{ab}/\partial x^\alpha$, $\partial\eta_{ab}/\partial t$ have zero order in $1/c$. The parameter of running through a geodesic is natural: equation (3) just involves a natural parameter

$$c\,d\tau = \left(g_{ab}\frac{dx^a}{dt}\frac{dx^b}{dt}\right)^{1/2} dt, \quad t = x^0/c.$$

From this and equation (2) we have

$$d\tau = \left(1 - v^2/c^2 + O(1/c^2)\right)^{1/2} dt = dt(1 + O(1/c^2)).$$

So, instead of the parameter τ we can use the absolute time $t = x^0/c$. Next, we have

$$\ddot{x}^\alpha = \Gamma^\alpha_{ab}\dot{x}^a\dot{x}^b.$$

Taking into account the formula for Christoffel symbols, we obtain the only term which has zero order with respect to $1/c$:

$$\Gamma^\alpha_{00}\dot{x}^0\dot{x}^0 = (c^2 + O(c))\,\Gamma^\alpha_{0\,0},$$

$$\Gamma^\alpha_{00} = \frac{1}{2}g^{\alpha\alpha}\left(-\frac{\partial g^{00}}{\partial x^\alpha}\right) + O(1/c^3) = -\frac{1}{2c^2}\frac{\partial\phi}{\partial x^\alpha} + O(1/c^3)$$

(there is no sum over α!). This, obviously, implies the assertion.

APPENDIX 3

CRYSTALLOGRAPHIC GROUPS

Our prime concern in this appendix is the role of tensors in solid state theory; in particular, we shall acquaint ourselves with the physical tensors of crystals and with crystallographic groups.

We shall consider an ideal crystal, i.e. a crystalline structure which occupies an entire three-dimensional Euclidean space or a Euclidian plane. A real crystal, of course, has boundaries, but by way of introduction of periodicity conditions on the boundary we shall bring our investigation to an analysis of "infinite" crystals. The crystal is regarded as consisting of a few types of atoms fixed rigidly in space (or in the plane) and distributed throughout space (or the plane) in a regular way. We shall suppose, as is convenient in such cases, that a crystal contains a sub-set of atoms defined by the following vectors: $\bar{a} = n_1\bar{\alpha}_1 + n_2\bar{\alpha}_2 + n_3\bar{\alpha}_3$, where n_1, n_2, n_3 are arbitrary integers and $\bar{\alpha}_1, \bar{\alpha}_2, \bar{\alpha}_3$ are linearly independent.

Because of the important role played by planar symmetry in nature, we shall also dwell on planar crystalline structures. For a deeper insight into planar symmetry and its manifestations in organic and inorganic nature, we refer the reader to the remarkable lecture by Herman Weyl published in the book Symmetry (Princeton, 1953).

DEFINITION 1. The *lattice R of a crystal* is the set of all atoms of the crystal. In other words, we may primarily assume the lattice to be an arbitray set of points either in $\mathbb{R}^3$ or in $\mathbb{R}^2$.

In so far as the excessive generality of this definition of latttices does not allow us to make any concrete statement concerning them, we shall appreciably narrow down the class of lattices to be analyzed here and restrict our consideration in the sequel to the so-called translation-invariant lattices which we are going to define a little later. Precisely this type of lattice corresponds to real physical crystals.

A crystal lattice always contains, as a sub-set, the set of all points (atoms) with position vectors of the form $\alpha = n_1\alpha_1 + n_2\alpha_2 + n_3\alpha_3$ (or in the planar case, $\alpha = n_1\alpha_1 + n_2\alpha_2$). Here n_1, n_2, n_3 are arbitrary integers. (Figure 95).

DEFINITION 2. The vectors $\alpha_1, \alpha_2, \alpha_3$ are called the *vectors of basic translations*. The vectors $\alpha_1, \alpha_2, \alpha_3$ are occasionally called the *primitive vectors of the lattice*, and they are always assumed to be linearly independent.

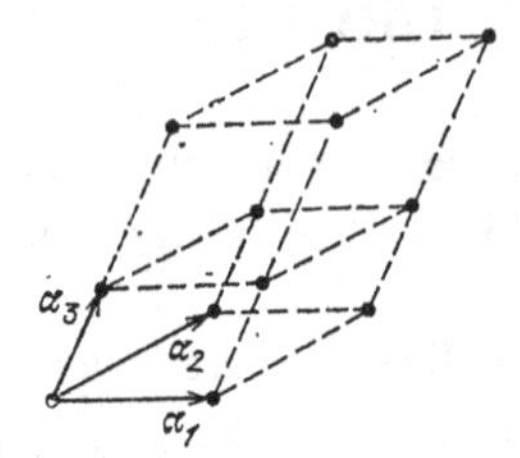

Figure 95.

Translation along the vector α is parallel transport in the direction of the vector α over a distance equal to the length of this given vector $|\alpha|$.

REQUIREMENT. The lattice R of a crystal is customarily assumed to *be sent to itself* under basic translations along $\alpha_1, \alpha_2, \alpha_3$ and their integer linear combinations; i.e. we require that the crystalline structure *remain invariant* under all translations generated by the vectors $\alpha_1, \alpha_2, \alpha_3$. This is one of the basic properties of real ("infinite") crystals.

We shall denote translations along $\alpha_1, \alpha_2, \alpha_3$ respectively by τ_1, τ_2, τ_3. Then any translation can be written in the form

$$T = n_1 \tau_1 + n_2 \tau_2 + n_3 \tau_3.$$

DEFINITION 3. The lattice R is called *translation-invariant* if it is sent to itself under an arbitrary translation of the form $T = n_1 \tau_1 + n_2 \tau_2 + n_3 \tau_3$.

Translation-invariant planar lattices are defined analogously.

So, we restrict ourselves to considering only translation-invariant lattices (on the plane or in space). Of course, the boundaries of a real crystal will be shifted, but we shall be concerned only with the interior of the crystal (this is just the reason why we have introduced an ideal infinite crystal into our consideration).

An important REMARK. Suppose that we are given a certain lattice R. We shall always assume the vectors $\alpha_1, \alpha_2, \alpha_3$ (or α_1, α_2 in the planar case) to be the *smallest* vectors, translations along which preserve the crystal (i.e. the crystal slides along itself).

DEFINITION 4. A parallelepiped with vectors $\alpha_1, \alpha_2, \alpha_3$ as sides is called a *primitive* (unit) *cell of the crystal lattice*.

Figure 96 illustrates the simplest two-dimensional lattice.

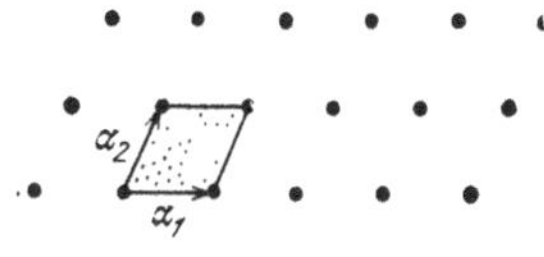

Figure 96.

Clearly, by virtue of translation invariance, the whole crystal consists of a union of primitive cells which underwent translation.

The simplest lattice depicted in Figure 96 is characterized by the property that each atom (i.e. each point of the lattice) is obtained by applying translation $T = n_1 \tau_1 + n_2 \tau_2 + n_3 \tau_3$ to any one of them. This fact can be formulated as follows: the *set of all translations is transitive on the lattice*. This is, however, far from being valid for all lattices. In particular, this may not be the case for the following reason: a crystal lattice is in general composed of several types of atoms, and so it is natural to require that under translations atoms of one type be again sent to atoms of the same type, and not to those points which are already occupied by atoms of a differen type. To put it differently, the set of translations may well be not transitive on the lattice. Such a lattice is shown in Figure 97. Here atoms of type A cannot be translated to atoms of type B.

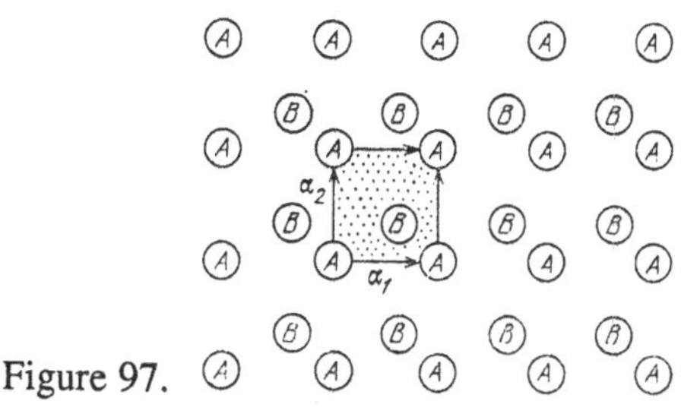

Figure 97.

Therefore, in order to specify a crystal lattice completely, it is not enough to give the set of translations. On the other hand it is clear that since the whole lattice is a union of primitive cells, it follows that to describe the lattice *completely*, we should determine, in addition to the set of translations, the position of atoms in one particular primitive cell.

DEFINITION 5. A lattice R (in two or three dimensions) whose atoms all are positioned at points of the form $n_1 \alpha_1 + n_2 \alpha_2 + n_3 \alpha_3$ (or $n_1\alpha_1 + n_2\alpha_2$ in the planar case), where n_1, n_2, n_3 are arbitrary integers, is called a *Bravais lattice.*

We say that the *set of all translations is transitive on the Bravais lattice.* Different Bravais lattices will differ only in the shapes of their primitive cells. Any two Bravais lattices can be transformed into one another by means of a suitable affine transformation.

DEFINITION 6. Let $X_1, X_2, .., X_N$ be all atoms positioned inside a primitive cell. Then the vectors $X_1, X_2, .., X_N$ (all going from the origin of coordinates of the vertex of the primitive cell to the points $X_1, X_2, .., X_N$) together form a *basis of the lattice*, as shown in Figure 98.

CLAIM 1. *A lattice is completely determined by the set of vectors of the basic translations together with the basis for the lattice.*

The proof of the claim is obvious from our defintions of basis, of translations and from the property of translation invariance of the lattice.

We shall establish the agreement between the properties of an ideal infinite crystal and a real crystal with boundary, as say the real perfect three-dimensional crystal depicted in Figure 99.

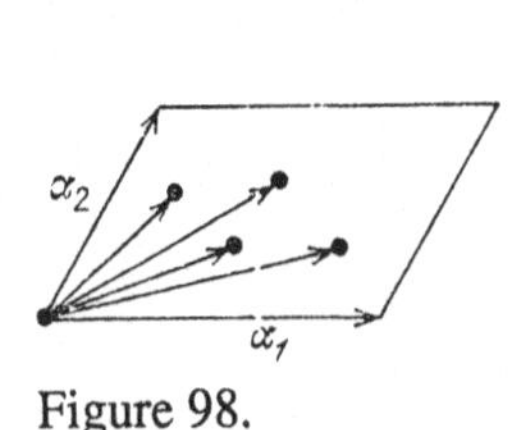

Figure 98.

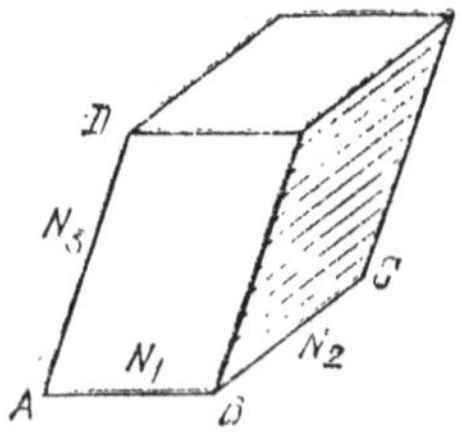

Figure 99.

Here N_1, N_2 and N_3 count the numbers of primitive cells stacked along the corresponding edges of our parallelepiped (crystal), i.e.

$$AB = N_1\alpha_1; \quad BC = N_2\alpha_2; \quad AD = N_3\alpha_3.$$

We shall apply translations along vectors multiple of α_i. The whole procedure will look like this: we translate the crystal along α_1, slice off the layer protruding beyond the right boundary of the crystal and glue it to the left boundary which now has moved into the parallelepiped along the vector α_1. The use of such a formal model appears to be justified: most of the physical results remain unchanged under this procedure. Clearly, this point of view is exactly equivalent to the consideration of an infinite ideal crystal, (a one-dimensional crystal is merely a linear chain of atoms in which the distance between neighbouring atoms is equal to one and the same number). The introduction of these conditions of periodicity on the boundaries of a crystal can be made illustrative, i.e. they can be explained geometrically. Since the boundary of a crystal consists of two atoms numbered 1 and N, any translation of the crystal becomes a rotation of the circle. We could, of course, believe that a three-dimensional crystal is also glued to become a three-dimensional ring in a four-dimensional space, but this idea does not seem natural.

With each cell there is associated, in a natural way, the concept of a symmetric cell (not to be confused with a primitive cell!). A symmetric cell has an atom as its centre.

DEFINITION 7. In a lattice R with the atom fixed, the *symmetric cell* is the set of points of space (or the plane in the planar case) situated closer to the fixed atom than to any other atom of the lacttice. The symmetric cell is sometimes also called the *Wigner-Seitz cell.*

Figure 100 illustrates a *hexagonal* two-dimensional lattice on which a primitive cell and a symmetric cell are indicated.

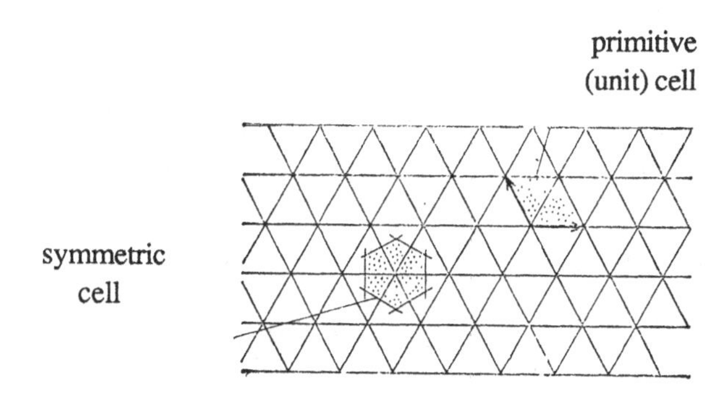

Figure 100.

The boundaries of a (two-dimensional) symmetric cell are perpendicular bisectors of the edges of the lattice, joining the fixed atom to all its nearest neighbours.

We now turn to transformations preserving a lattice (i.e. mapping a lattice onto itself). Consider the group of motions of a space (or a plane), i.e. the set of all linear transformations preserving the quadratic form $ds^2 = (dx^1)^2 + (dx^2)^2 + (dx^3)^2$ (correspondingly, $ds^2 = (dx^1)^2 + (dx^2)^2$). We shall denote this group by G_3 (by G_2 in the planar case). Any element g of the group G_3 (or of G_2) can generally be expressed in exactly one way, as a composition of two transformations, of which one is a parallel transport T and the other is a rotation α (proper or improper, i.e. with determinant $+1$ or -1), that is, $g = T\alpha$.

The set of parallel transports $\{T\}$ forms a sub-group T_3 (T_2) in G_3 (respectively, G_2) which is a normal divisor in G_3 (or in G_2).

From among all transformations of the group G_3 (G_2) we shall single out those mapping a certain fixed lattice R into itself.

DEFINITION 8. The set of transformations (motions) of the group G_3 (or G_2) realizing a self-map of a lattice R are called the *space-group* of this lattice, which we shall denote by $G_3(R)$ (or respectively $G_2(R)$).

It is clear that the set $G_3(R)$ (or $G_2(R)$) is the group in the usual algebraic sense.

We shall formulate all our further definitions only for the three-dimensional case, since for the planar case the corresponding properties hold similarly.

The group $G_3(R)$ contains a sub-group $T_3(R)$ which is the group of parallel transports (translations).

DEFINITION 9. The *translation group* of a crystal (i.e. of a lattice R) is the sub-group $T_3(R)$ of the group $G_3(R)$, consisting of all possible translations T (recall that any translation of our lattice R has the form $T = n_1\tau_1 + n_2\tau_2 + n_3\tau_3$, where τ_i $(1 \leq i \leq 3)$ are primitive translations generated by the basis vectors $\alpha_1, \alpha_2, \alpha_3$).

All *lattices* R under investigation are *translation-invariant*.

It can easily be shown that the sub-group $T_3(R)$ is normal in the group $G_3(R)$ (this fact will, however, be of no use in the sequel). Indeed, we shall show that if $g \in G_3(R)$ and $t \in T_3(R)$, then $gtg^{-1} \in T_3(R)$ for any g and t. In other words, we have to make sure that the transformation gtg^{-1} is again a translation. However, this property of the transformation gtg^{-1} is obvious.

We shall assume from now on that a certain point in space, say, the "origin of coordinates" is fixed; for example, this may be the vertex of a primitive cell from which the vectors of primitive translations emanate. All possible rotations of a lattice (with determinant ± 1) will be considered relative to this point O. Clearly, as the point O we can take any arbitrary point in space. In particular, we assume all rotations to preserve the point O.

It is well known that any transformation $g \in G_3$ admits a unique representatiuon in the form $g = T\alpha$ (see above), where T is parallel transport and α is rotation with det $(\alpha) = \pm 1$. Since $G_3(R) \subset G_3$, it follows that any element $g \in G_3(R)$ admits a representation of the form $g = T\alpha$ (note that generally speaking $T\alpha \neq \alpha T$). We should also emphasize an important fact: the transformations T and α need not necessarily belong to the group $G_3(R)$; in particular, T may not be included as an element of the translation group of a crystal. (Figure 101).

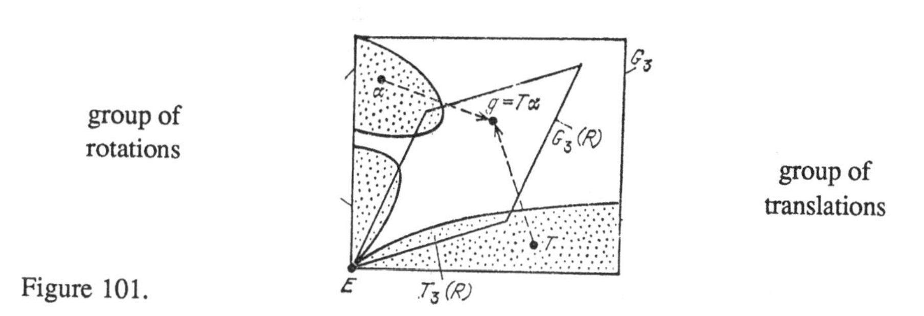

Figure 101.

Figure 101 shows the group of all rotations in $\mathbb{R}^3$ (i.e. the group of orthogonal matrices $O(3)$) in the form of two pieces, which agrees with the fact that the group $O(3)$ (the group of all three-dimensional orthogonal matrices) consists of two connected components (as a topological space), namely, one (the sub-group $SO(3)$) composed of matrices α with determinant + 1 and the other composed of matrices α with determinant − 1. The group of all parallel transports in $\mathbb{R}^3$ is connected (it is described by three parameters). At the same time, the group $G_3(R)$ is discrete, i.e. when we consider it as a topological space we see that it consists of a set of isolated points, each point being a transformation of the lattice R. To put it differently, discreteness of a group means that in the group there exists no transformations arbitrarily close to the identity transformation of this group (except, of course, the identity transformation itself).

Thus, any element $g \in G_3(R)$ has the form $g = T\alpha$. (The transformations T and α are reconstructed uniquely from the transformation g, i.e. from the equality $g = T\alpha = T'\alpha'$ it is immediate that $T = T'$, $\alpha = \alpha'$. Indeed, if $T\alpha = T'\alpha'$, then $(T')^{-1} = \alpha'\alpha^{-1}$, where $(T')^{-1}T$ is a translation and $\alpha'\alpha^{-1}$ is a rotation. A rotation may appear to be a translation only in the case when it is identical, which implies $T = T'$ and $\alpha = \alpha'$.) The group G_3 does not, however, fall into the direct product of its sub-groups: $O(3)$ and T_3 since, generally speaking, $T\alpha \neq \alpha T$.

DEFINITION 10. The set of all transformations $\alpha \in O(3)$ such that for a certain T (parallel transport) the transformation $g = T\alpha$ belongs to the group $G_3(R)$ is called the *point group of a crystal.*

In other words, α belongs to the point group of a crystal if and only if there exists a parallel transport T (not necessarily belonging to $T_3(R)$) with the property that the composition $T\alpha$ is the element of the space group of the crystal. We shall denote the point group by $S_3(R)$. This group is often referred to as the *symmetry group of a crystal* (*of a lattice*) and its elements as *symmetry operations.* Recall that the point O — the centre of rotations — need not necessarily be a point of a given lattice R. However, we have not yet proved that the set $S_3(R)$ forms a group.

CLAIM 2. *The set $S_3(R)$ is a group in the usual algebraic sense.*

Proof. Let $\alpha_1, \alpha_2 \in S_3(R)$. It should be proved that $\alpha_s = \alpha_1 \cdot \alpha_2$ also belongs to $S_3(R)$. By the definition of $S_3(R)$, there exist T_1 and T_2 such that $T_1\alpha_1 \in G_3(R)$ and $T_2\alpha_2 \in G_3(R)$; since $G_3(R)$ is a group, it follows that the transformation $(T_1\alpha_1)(T_2\alpha_2) \in G_1(R)$. Suppose that the translations T_1, T_2 are determined by vectors x_1, x_2 and the rotations α_1, α_2 are determined by matrices A_1, A_2, respectively. Then if r is the radious-vector in a space R^3, the following equality holds $(T_1\alpha_1)r = A_1r + x_1$; $(T_2\alpha_2)r = A_2r + x_2$ (rotation first and then parallel transport). From this we have $(T_1\alpha_1)(T_2\alpha_2)r = A_1A_2r + (A_1x_2 + x_1)$. Thus, the rotation $\alpha_3 = \alpha_1\alpha_2$ determined by the matrix A_1A_2 enters in the transformation $g = T_3\alpha_3 \in G_3(R)$, where the translation T_3 is given by the vector $x_3 = A_1x_2 + x_1$, which implies that $\alpha_1\alpha_2 \in S_3(R)$.

Next, let $\alpha \in S_3(R)$. We should prove that $\alpha^{-1} \in S_3(R)$, where α^{-1} stands, as usual, for the inverse transformation. Since $\alpha \in S_3(R)$, there exists a T such that $g = T\alpha \in G_3(R)$, i.e. $g(r) = (T\alpha)(r) = Ar + x$. From this we come to the conclusion that since $G_3(R)$ is a group, it follows that $g^{-1} \in G_3(R)$, i.e. $(g^{-1})(r) = (T\alpha)^{-1}(r) = A^{-1}r - A^{-1}(x) = Br + y$, where $B = A^{-1}$; $y = -A^{-1}(x)$. Finally we are led to $\alpha^{-1} \in S_3(R)$.

The associativeness of multiplication in $S_3(R)$ and the existence of a unit element can be verified directly, and our assertion follows.

We now give an example of a planar (two-dimensional) crystal lattice R for which there exists an element $g \in G_3(R)$ whose expression as a composition of the form $g = T\alpha$ has the property that $T \notin G_3(R)$ and $\alpha \notin G_3(R)$. The lattice is depicted in Figure 102.

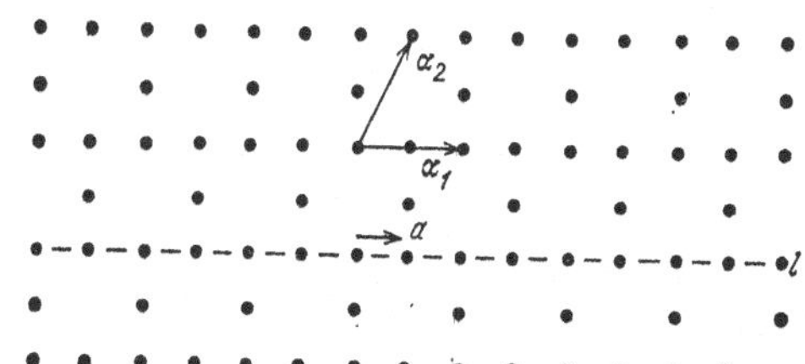

Figure 102.

The reflection $\alpha \in O(2)$ in the straight line l, obviously, does not preserve the lattice R. Next, the parallel transport along the vector a (note that the translation T generated by the vector a is not a primitive translation of the lattice) does not preserve the lattice R either, that is, neither of the transformations α and T belongs to the group $G_2(R)$, but the transformation $g = T\alpha$, obviously, maps the lattice R into itself. This operation (transformation) $g = T\alpha$ is called the *glide-reflection symmetry, and the lattice R possesses the glide-reflection symmetry*. It should be emphasized once again that the elements (transformations) of the point group of a crystal (of a lattice) do not, generally, map the crystal (the lattice) to itself. This group is of great importance in the theory of crystalline structure, and it is not for nothing that it is alternatively referred to as the *symmetry group of a lattice* since along with the "genuine" symmetries of the lattice it includes also those transformations which send the lattice into itself only after a translation is applied. Clearly, the lattice depicted in Figure 102 generates a three-dimensional lattice possessing glide-reflection symmetry. There exist crystals which possess, in addition to this type of symmetry, also *screw*, (or *axial screw*) symmetry. This symmetry is a composition of rotation $\alpha \in O(s)$ and translation T of the lattice R along the axis of this rotation. We recommend to the reader that he construct an example of a three-dimensional lattice possessing an axial screw symmetry.

Another group of transformations of a lattice R is sometimes considered.

DEFINITION 11. *The stationary group* $H_3(R)$ *of a lattice* is the sub-group $G_3(R)$ consisting of all lattice-preserving transformations which leave the origin O motionless. (Recall that the point O is fixed.)

It is clear that $H_3(R) = G_3(R) \cap O(3)$ since any transformation of a lattice which leaves the point O at its original place is an orthogonal transformation, i.e. a rotation around the point O, whose determinant may be either + 1 or – 1. Note that the group $H_3(R)$ is not, generally speaking, a factor group of the group $G_3(R)$ with respect to the sub-group $T_3(R)$: $H_3(R)/T_3(R)$. See, for instance, Figure 103.

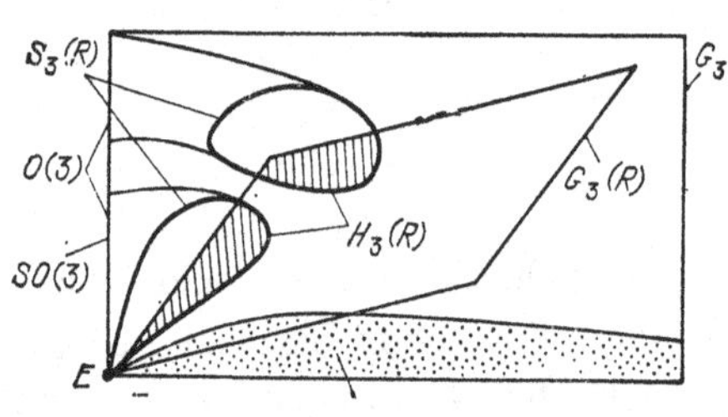

Figure 103. T_3 – parallel transports

REMARK. The group $H_3(R)$ does not generally coincide with the symmetry group (point group) $S_3(R)$.

CLAIM 3. *The groups* $H_3(R)$, $S_3(R)$ *and* $G_3(R)$ *regarded as sub-groups in the group* G_3 *are linked by the following relation*

$$H_3(R) = S_3(R) \cap G_3(R) \quad (H_3(R) \subset O(3)).$$

Proof. We shall first prove that $H_3(R) \subset S_3(R) \cap G_3(R)$. Let $\alpha \in H_3(R)$. Then α is, in particular, a rotation (and belongs to $G_3(R)$), and therefore we may put $g = T\alpha$, where $T = E$ is an identical transformation (translation along a zero vector), i.e. $g = \alpha = E\alpha$; whence by definition of $S_3(R)$ we have $\alpha \in S_3(R)$.

Inversely, we shall prove that $H_3(R) \supset S_3(R) \cap G_3(R)$. Suppose $\alpha \in S_3(R)$ and $\alpha \in G_3(R)$. This means that α preserves the lattice R and, besides, is a rotation, i.e. $\alpha \in O(3) \cap G_3(R)$ and leaves the point O motionless, that is, $\alpha \in H_3(R)$. The fact that α enters a certain decomposition $g = T\alpha$ is inessential for us now, which completes the proof.

If a lattice R is Bravais, then the groups $S_3(R)$ and $H_3(R)$ coincide. This statement follows from the definition of a Bravais lattice.

Figure 103 illustrates all the subgroups which we have introduced into our consideration and their interaction with one another.

The stationary group $H_3(R)$ may be said to consist of "genuine" symmetries of the lattice, and the group $S_3(R)$ of glide symmetries.

It is not every sub-group in the orthogonal group $O(3)$ that can be the point group (i.e. the symmetry group) of the lattice. Translation invariance of the lattice imposes very rigid limitations on the groups $S_3(R)$, $G_3(R)$ and $H_3(R)$. We shall denote by $H_3(R)_0$ the sub-group in $H_3(R)$, consisting of proper rotations only, i.e. $H_3(R)_0 = SO(3) \cap G_3(R)$; each rotation of $H_3(R)_0$ has determinant $+1$ and does not move the point O. As the point O we take an arbitrary atom of the lattice.

THEOREM. 1. *Let R be a translation-invariant lattice and let $H_3(R)$ be the stationary group of the lattice. Then the group $H_3(R)_0$ consists of a finite number of transformations each of which is a rotation of the point O through an angle ϕ multiple either of $\pi/3$ or $\pi/2$.*

Proof. Suppose $\alpha \in H_3(R)_0$ is a proper rotation. Then, as is well known from algebra, the rotation α is a rotation through a certain angle ϕ around a certain motionless axis l passing through the motionless point O.

Suppose Π is a plane orthogonal to l, through the point O. Each lattice R contains a sub-lattice R_1 consisting of points determined by vectors $\alpha = n_2\alpha_1 + n_2\alpha_2 + n_3\alpha_3$ (see the definition of a lattice). We shall project all (atoms) lattices R_1 parallel to the straight line l onto the plane Π and consider all those projections of the points which are the nearest to the point O. We fix one of such points A_1 (we find a few such points) as shown in Figure 104.

Since the lattice R_1 is symmetric about the point O (this is a consequence of translation invariance; the whole lattice R may appear not to be symmetric about the point O), it follows that along with each point $B_1 \in R_1$, the point B'_1 opposite to it also belongs to R_1. Under the rotation around l through the angle ϕ the point B_1 will be sent to a point of the lattice R_1 (since the lattice R_1 is preserved by the transformation α), i.e. the projection OA_1 will be sent to the projection OA_2 with the angle ϕ between them. Since the vectors OB_1 and OB_2 belong to the lattice, their difference — the vector B_1B_2 — also belongs to the lattice R_1 (we mean the vector parallel to the vector B_1B_2 and starting from the point O). From this it follows that the length of the vector A_1A_2 is not less than the length of the vector OA_1 ($|OA_1| = |OA_2|$) since the points A_1 and A_2 are separated from the point O by the smallest possible distance.

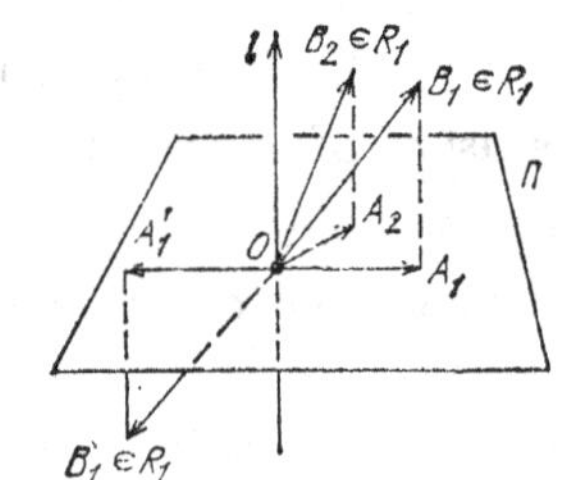

Figure 104.

So, in the triangle OA_1A_2 the side A_1A_2 is not smaller than $|OA_1| = |OA_2|$, that is, the angle ϕ is not smaller than $\pi/3$. Applying the transformation α successively, we obtain in the plane Π a rectilinear polygon with vertices $A_1, A_2, A_3, \ldots, A_m$ (where $A_{m+1} = A_1$), and since $\phi \geq \pi/3$, we have $1 \leq m \leq 6$. However, by virtue of the symmetry of the lattice R_1, the polygon $A_1, \ldots, A_m$ is also symmetric about the point O. This means that m can assume only values 2, 4, 6. Hence ϕ may be equal to $k\pi$, $k\pi/2$, $k\pi/3$, which implies the theorem.

We now turn to a planar lattice R and the group $H_2(R)_0$. The theorem proved above allows us to describe completely the set of groups $H_2(R)_0$ for arbitrary planar lattices R; in other words we shall now give the list of five groups G_1, G_2, G_3, G_4, G_6 such that any group $H_2(R)_0$ coincides with one of these five groups.

THEOREM 2. (Classification theorem in two dimensions.) *Let C_n (where $n = 1, 2, 3, 4, 6$) denote a group of n elements of the form*

$$\begin{pmatrix} \cos\frac{2\pi k}{n} & \sin\frac{2\pi k}{n} \\ -\sin\frac{2\pi k}{n} & \cos\frac{2\pi k}{n} \end{pmatrix}$$

(where $0 \le k \le n-1$), i.e. the group C_n consists of rotations through an angle $\frac{2\pi k}{n}$ about the point O. Then for any planar lattice R^n its group $H_2(R)$ coincides with one of these groups C_n ($n = 1, 2, 3, 4, 6$).

The result is immediate from the theorem proved above.

Thus, planar lattices R may possess symmetry only under five groups $H_2(R)_0$. (It should be noted that the group C_1 consists only of a single identity transformation.) The corresponding classification in the three-dimensional case gives the list of 32 possible symmetry groups. This classification is too sophisticated to be presented in this book.

We shall now again return to tensors. We shall consider, for example, such a macroscopic property of crystals as *electrical conductivity* which describes the relationship between the electric field vector and the current density vector j. This relationship is specified by the relation $j_k = \sigma^s_k \varepsilon_s$, where $j = \{j_n\}$, $\varepsilon = \{\varepsilon_s\}$, and $\{\sigma^s_k\}$ is the electrical conductivity tensor of the medium. In the case where the medium is isotropic, $\sigma^s_k = \sigma\delta^s_k$, where σ is a scalar, i.e. in this simplest case the electrical conductivity is given by a scalar σ. In the general case $\{\sigma^s_k\}$ is a tensor. We shall consider the electrical conductivity tensor of a crystal described by the cubic lattice in $\mathbb{R}^3$, i.e. a cubic crystal (Figure 105).

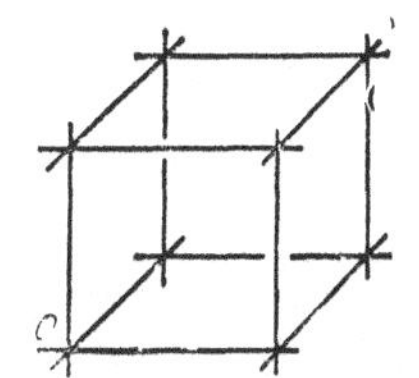

Figure 105.

Suppose we are dealing with an ideal crystal whose lattice R fills the entire space. Clearly, the group $S_3(R)$ of symmetries of this crystal contains, in particular the following three transformations:

$$\alpha_1 = \begin{pmatrix} 0 & 1 & 0 \\ -1 & 0 & 0 \\ 0 & 0 & 1 \end{pmatrix}; \qquad \alpha_2 = \begin{pmatrix} 0 & 0 & 1 \\ 0 & 1 & 0 \\ -1 & 0 & 0 \end{pmatrix}; \qquad \alpha_3 = \begin{pmatrix} 1 & 0 & 0 \\ 0 & 0 & 1 \\ 0 & -1 & 0 \end{pmatrix},$$

i.e. α_1 is a rotation through $\pi/2$ about the z-axis, α_2 a rotation through $\pi/2$ about the y-axis and α_3 a rotation through $\pi/2$ about the x-axis. Since the lattice R goes into itself, it follows that theses three symmetry operations preserve the tnesor $\{\sigma^s_k\}$. We shall write this. We shall denote by A the matrix $\{\sigma^s_k\}$. Then $A^1_i = \alpha_i A \alpha_i^{-1} = A$ for any i, $1 \le i \le 3$. Calculating the matrix A^1_1, we obtain

$$A^1_1 = \begin{pmatrix} \sigma^2_2 & -\sigma^2_1 & \sigma^2_3 \\ -\sigma^1_2 & \sigma^1_1 & -\sigma^1_3 \\ \sigma^3_2 & \sigma^3_1 & \sigma^3_3 \end{pmatrix} = \begin{pmatrix} \sigma^1_1 & \sigma^1_2 & \sigma^1_3 \\ \sigma^2_1 & \sigma^2_2 & \sigma^2_3 \\ \sigma^3_1 & \sigma^3_2 & \sigma^3_3 \end{pmatrix} = A .$$

From this it follows that $\sigma^1_1 = \sigma^2_2$. Next, calculations of the matrices A^1_2 and A^1_3 yield $\sigma^1_1 = \sigma^2_2 = \sigma^3_3$.

The group $S_3(R)$ contains three more transformations:

$$\beta_1 = \begin{pmatrix} -1 & 0 & 0 \\ 0 & -1 & 0 \\ 0 & 0 & 1 \end{pmatrix}; \quad \beta_2 = \begin{pmatrix} 1 & 0 & 0 \\ 0 & -1 & 0 \\ 0 & 0 & -1 \end{pmatrix}; \quad \beta_3 = \begin{pmatrix} -1 & 0 & 0 \\ 0 & 1 & 0 \\ 0 & 0 & -1 \end{pmatrix},$$

i.e. β_1 is a rotation through π about the z-axis, β_2 a rotation through π about the x-axis and β_3 a rotation through π about the y-axis. The lattice goes again into itself, which yields the relations $\tilde{A}_i = \beta_i A \beta_i^{-1}$, $1 \le i \le 3$; calculating the matrix $\tilde{A}_1$ we obtain

$$\tilde{A}_1 = \begin{pmatrix} \sigma^2_2 & \sigma^2_1 & -\sigma^2_3 \\ \sigma^1_2 & \sigma^1_1 & -\sigma^1_3 \\ -\sigma^3_2 & -\sigma^3_1 & -\sigma^3_3 \end{pmatrix} = \begin{pmatrix} \sigma^1_1 & \sigma^1_2 & \sigma^1_3 \\ \sigma^2_1 & \sigma^2_2 & \sigma^2_3 \\ \sigma^3_1 & \sigma^3_2 & \sigma^3_3 \end{pmatrix} = A ;$$

whence $\sigma_3^1 = \sigma_3^2 = \sigma_1^3 = \sigma_2^3 = 0$. Calculations of the matrices $\tilde{A}_2$ and $\tilde{A}_3$ similarly give us $\sigma_j^i = 0$ for $i \neq j$, i.e. we are finally led to

$$A = \{\sigma_k^s\} = \sigma \begin{pmatrix} 1 & 0 & 0 \\ 0 & 1 & 0 \\ 0 & 0 & 1 \end{pmatrix},$$

that is,

$$\sigma_k^s = \sigma \delta_k^s,$$

where σ is a scalar. We have thus proved the important statement that the *electrical conductivity of a cubic crystal is isotropic, i.e independent of direction, as is the electrical conductivity of any isotropic medium.* This result is not physically obvious since it would be natural to expect that the conductivity of a cubic crystal in directions parallel to the edges differs from that in the diagonal direction. We have thus demonstrated the important (although rather elementary) application of the symmetry group $S_3(R)$ by lowering drastically the number of independent components of the tensor $\{\sigma_k^s\}$.

We shall now proceed to three-dimensional lattices. The problem of classification and complete description of all types of the groups $H_3(R)$ and $G_3(R)$ is much more complex than in the planar case treated above. We shall therefore not carry out this classification in full detail but shall restrict our consideration to answering a simpler question: What is the structure of finite groups of proper rotations in the three-dimensional case? Since the stationary group for an arbitrary translation-invariant three-dimensional lattice is discrete (and, therefore, finite), it follows that by compiling a complete list of all finite sub-groups of the group $SO(3)$, we shall thus estimate "from above" the list of groups $H_3(R)_0$ and $S_3(R)_0$ for three-dimensional lattices.

We shall begin by presenting a list of finite groups of rotations of a three-dimensional space. To this end we first consider some straight line l through the point O and assume Π to be a plane orthogonal to the straigth line l and also passing through the origin O. We consider in the plane Π the action of the group C_n (the cyclic sub-group of rotations in the plane Π about the point O through the angle $2\pi/n$). Clearly, this group becomes the rotation group of the entire three-dimensional space about the axis l. We denote this group also by C_n. Here $n = 1, 2, 3, \ldots$, the group C_1 consisting of the single identity transformation. In addition to the group C_n, there is another group D_n, acting on the plane. The reflection of the plane Π

relative to a certain straight line q, lying in the plane Π, in a three-dimensional space can be realized using rotation about this straight line q through an angle π. Thus, these improper rotations of the plane can be complemented to become proper rotations of a three-dimensional space. We shall denote this newly appearing group by D'_n. The group D'_n will consist of the following transformations: all the transformations from the subgroup C_n and besides, the rotations by the angle π of the whole three-dimensional space about n axes lying in the plane Π and making group angles equal to $\frac{2\cdot\pi}{2n}=\frac{\pi}{n}$ with one another. It should be emphasized that the D'_1, as well as the group C_2, consists only of two elements: of identity transformation and half turn about the only straight line in the plane Π, and therefore these two groups are isomorphic. Therefore, if we wish to make the list of different (non-isomorphic) groups, we should eliminate the group D'_1. We thus arrive at the following list: C_n, $n = 1, 2, 3, \ldots$, ; D'_n, $n = 2, 3, 4, \ldots$.

Along with these two infinite series of discrete groups, in a three-dimensional space there also exist a few more exotic transformation goups. Indeed, we shall consider five regular polyhedra in a three-dimensional space, namely, cube, octahedron, icosahedron, tetrahedron and dodecahedron. With each of these we can associate a finite group of proper rotations sending a particular polyhedron on to itself. In doing so, we obtain five more finite groups, some of which however coincide. Only three of them are actually distinct, namely the groups of tetrahedron, cube and dodecahedron. We shall consider this point in more detail. We shall inscribe a sphere into a cube so that the sphere be tangent to the cube faces, and into this sphere describe an octahedron whose vertices are tangent to the sphere at those points where the sphere is tangent to the cube. It is clear that any rotation sending the cube into itself will also leave invariant the octahedron, and conversely, for which reason the symmetries of the cube and octahedron coincide. In exactly the same way we establish the coincidence of the symmetry groups (i.e. proper rotation) of dodecahedron and icosahedron. We shall denote by T, W, P the groups of proper rotations respectively of the tetrahedron, cube (and octahedron), dodecahedron (and icosahedron). We leave it to the reader to verify that the orders of these three groups are, respectively, 12, 24, 60. If we consider the complete groups of polyhedron rotations (i.e. those including also improper rotations), then the obtained groups $\tilde{T}, \tilde{W}, \tilde{P}$ will naturally have orders 24, 48, 120. It turns out that the rotation groups presented above fully exhaust the list of proper discrete groups of rotations of three-dimensional space.

THEOREM 3. *The exhaustive list of finite groups of proper rotations in three-dimensional space has the form* C_n $(n = 1, 2, 3, \dots)$, D'_n $(n = 2, 3, 4, \dots)$. *Here* C_n *is the group (cyclic) consisting of repeated applications of rotation about some axis* l *through an angle* α *equal to* $2\pi/n$, *where* n *is an integer;* D'_n *is the group of the same rotations, together with the reflections relative to* n *axes lying in the plane perpendicular to* l *and making an angle* $\alpha/2$ *with one another;* T, W, P *are the transformation groups preserving respectively the regular tetrahedron, cube (or octahedron) and dodecahedron (or icosahedron).*

APPENDIX 4

HOMOLOGY GROUPS AND METHODS OF THEIR CALCULATION

One of the most important geometric invariants of a manifold is the homology groups which we shall define below. We have already defined the one-dimensional homology group $H_1(M^n)$ as the factor group of the fundamental group over the first commutant.

Several ways to define the homology groups exist.

To begin with, consider closed differential forms of degree k on the manifold M^n (i.e. locally we have

$$\omega = \sum_{i_1<\ldots<i_k} a_{i_1 \ldots i_k} dx^{i_1} \wedge \ldots \wedge dx^{i_k}, \quad d\omega = 0).$$

A closed differential form is called *exact* (or *cohomologic to zero*) if $\omega = d\omega'$, where ω' is the form of degree $k-1$.

DEFINITION 1. The *cohomology group* $H^k(M^n; \mathbb{R})$ is the quotient group of all closed forms of degree k with respect to the sub-group of exact forms, or alternatively the group of equivalence classes of closed forms up to exact forms ($\omega_1 \sim \omega_2$ if $\omega_1 - \omega_2 = d\omega'$).

The simplest fact concerning calculation of the cohomology groups is the following statement.

CLAIM 1. *For any manifold M^n the group $H^0(M^n; \mathbb{R})$ is a linear space of dimension q equal to the number of connected pieces (components) of the manifold.*

Proof. The form of degree 0 is a scalar function $f(x)$ on the manifold. If this form is closed, then $df(x) = 0$. This means that the function $f(x)$ is locally constant, i.e. constant on each connected piece of the manifold. Closed forms of degree 0 are merely sets of q constants, where q is the number of pieces. The statement follows since there are no exact forms here.

If there exists a smooth map of manifolds, $f\colon M_1 \to M_2$, then there is defined the (well-known to us) map of the forms $\omega \mapsto f^*(d\omega)$, such that $d(f^*\omega) = f^*(d\omega)$. Therefore the map of homology groups is defined to be

$$f^*: H^k(M_2; \mathbb{R}) \to H^k(M_1; \mathbb{R}),$$

since the equivalence classes go into one another (under the map f^*, closed forms remain closed and exact forms remain exact).

The following theorem holds:

THEOREM 1. *Given two smooth homotopic maps*

$$f_1: M_1 \to M_2 \quad \text{and} \quad f_2: M_1 \to M_2,$$

the maps f_1^* and f_2^* of the cohomology group coincide:

$$f_1^* = f_2^*: H^k(M_2, \mathbb{R}) \to H^k(M_1; \mathbb{R}).$$

Proof. Suppose we are given a smooth homotopy $F: M_1 \times I \to M_2$, where I is a segment, $1 \leq t \leq 2$, and $F(x, 1) = f_1(x)$, $F(x, 2) = f_2(x)$. Any form Ω of degree k on $M_1 \times I$ is given by

$$\Omega = \omega_1 + \omega_2 \wedge dt, \quad \Omega|_{t=t_0} = \omega_1(t_0),$$

where ω_1 is the form of degree k containing no dt among differentials, and ω_2 is the form of degree $k-1$ containing no dt among differentials (local coordinates in $M_1 \times I$ are always chosen to be $(x^1, \ldots, x^n, t)$, where $(x^1, \ldots, x^n)$ are local coordinates on M^1). Let ω be any form of degree k on the manifold M_2. Then $F^*(\omega) = \Omega = \omega_1 + \omega_2 \wedge dt$, where locally we have

$$\omega_2 = \sum_{i_1 < \ldots < i_{k-1}} a_{i_1 \ldots i_{k-1}} dx^{i_1} \wedge \ldots \wedge dx^{i_{k-1}},$$

$$\omega_1 = \sum_{j_1 < \ldots < j_k} b_{j_1 \ldots j_k} dx^{j_1} \wedge \ldots \wedge dx^{j_k}.$$

We shall define the form $D\,\Omega$ of degree $k-1$ by the following formula (locally):

$$D\,\Omega = \sum_{i_1 < \ldots < i_{k-1}} \left(\int_1^2 a_{i_1 \ldots i_k}(x, t)\, dt \right) dx^{i_1} \wedge \ldots \wedge dx^{i_{k-1}} = \int_1^2 \omega_2\, dt.$$

Obviously, $D\omega$ is the form of degree $k-1$ on the manifold $M_1 \times I$. An important lemma follows:

LEMMA. 1. *The formula*

$$d(D(F^*(\omega))) \pm D(d(F^*(\omega))) = f_2^*(\omega) - f_1^*(\omega)$$

is valid, or if $\Omega = f^*(\omega)$, *where* ω *is the form on* M_2 and $F|_{t=2} = f_2$, $F|_{t=1} = f_1$, *then the formula* $dD(\Omega) \pm D(d\Omega) = \Omega|_{t=2} - \Omega|_{t=1}$ *is valid.*

Proof. Let us calculate $d(D(\Omega))$ for any form Ω on $M_1 \times I$, where $\Omega = \omega_1 + \omega_2 \wedge dt$. By the definition of d we locally have

$$dD\Omega = \sum_{i_1<\ldots<i_k} \sum_j \Big(\int_1^2 \frac{\partial a_{i_1 \ldots i_{k-1}}}{\partial x^j}\, dt\Big)\, dx^j \wedge dx^{i_1} \wedge \ldots \wedge dx^{i_{k-1}},$$

$$Dd\Omega = D(d\omega_1) + D(d\omega_2 \wedge dt) =$$

$$= D\Big(\sum_{j_1<\ldots<j_k} \sum_q \frac{\partial b_{j_1 \ldots j_k}}{\partial x^q}\, dx^q \wedge dx^{j_1} \wedge \ldots \wedge dx^{j_k} +$$

$$+ \sum_{j_1<\ldots<j_n} \frac{\partial b_{j_1 \ldots j_k}}{\partial t}\, dx \wedge dx^{j_1} \wedge \ldots \wedge dx^{j_k} \neq$$

$$\neq D\Big(\sum_{i_1<\ldots<i_k} \sum_p \frac{\partial a_{i_1 \ldots i_{k-1}}}{\partial x^p}\, dx^p \wedge dx^{i_1} \wedge \ldots \wedge dx^{i_k}\Big).$$

From this we can readily see

$$dD\Omega + (-1)^{k+1} Dd\Omega =$$

$$= \pm \sum_{j_1<\ldots<j_n} \big(b_{j_1 \ldots j_k}(x, 2) + b_{j_1 \ldots j_k}(x, 1)\big)\, dx^{j_1} \wedge \ldots \wedge dx^{j_k}.$$

Since locally

$$\Omega|_{t=t_0} = (\omega_1 + \omega_2 \wedge dt)\,|_{t=t_0},$$

we are led to

$$dD\Omega + (-1)^{k+1} Dd\Omega = \Omega|_{t=2} - \Omega|_{t=1},$$

which implies the first statement of our lemma. If now $\Omega = f^*(\omega)$, then $\Omega|_{t=t_0} = f^*_{t_0}(\omega)$ where $F(x, t_0) = f_{t_0}: M_1 \to M_2$. In particular for $t_0 = 1, 2$ the result follows, which completes proof of the lemma.

Continuation of Proof of Theorem 1. Suppose we are given a closed form ω on M_2 (i.e. $d\omega = 0$). Then the form

$$d(f_2^*(\omega) - f_1^*(\omega)) = d(D(F^*(\omega)) \pm Dd(F^*(\omega)).$$

However, $dF^*(\omega) = F^*(d\omega) = 0$. Therefore, we have $f_2^*(\omega) - f_1^*(\omega) = dDF^*(\omega)$, i.e. the difference of the forms is exact (or, the forms are homotopic). This just means, by definition, that the homomorphisms

$$f_1^*:\ H^k(M_2; \mathbb{R}) \to H^k(M_1; \mathbb{R}),$$

$$f_2^*::\ H^k(M_2; \mathbb{R}) \to H^k(M_1; \mathbb{R})$$

coincide on equivalence (cohomology) classes. This concludes the proof of the theorem.

DEFINITION 2. Manifolds M_1, M_2 are called *homotopically equivalent* if there exist (smooth) maps $f: M_1 \to M_2$, $g: M_2 \to M_1$, such that both the superpositions $f \circ g: M_2 \to M_2$ and $g \circ f: M_1 \to M_1$ are homotopic to the identity maps

$$M_1 \to M_1(x \mapsto x), \quad M_2 \to M_2(y \mapsto y).$$

For example:

1. Euclidean spaces $\mathbb{R}^n$ (or discs $D^n = \sum_{\alpha=1}^{n} (x^\alpha)^2 \leq R^2\}$) are homotopically equivalent to a point. The proof consists in the fact that $\mathbb{R}^n$ (or D^n) are deformed along themselves to a point. The precise meaning of this is that the identity map $1: \mathbb{R}^n \to \mathbb{R}^n$, where $x \mapsto x$, is homotopic to a constant $\mathbb{R}^n \to Q$, where Q is a point. This fact is obvious.

2. A space without a point $\mathbb{R}^n \setminus Q$ (or a ring between spheres of radii r_1 and r_2) contracts to a sphere S^{n-1}, and therefore $\mathbb{R}^n \setminus Q$ is homotopically equivalent to the sphere S^{n-1}. For $n = 2$ the region $\mathbb{R}^2 \setminus Q$ is homotopically equivalent to a circle. Note that $\mathbb{R}^2 \setminus (Q_1 \cup Q_2)$ contracts along itself to a figure-of-eight (Figure 106.) The figure-of-eight is not a manifold, but for it we can define the cohomology groups — they are the same, by definition, as those for the region $\mathbb{R}^2 \setminus Q_1 \cup Q_2)$ by virtue of Theorem 1.

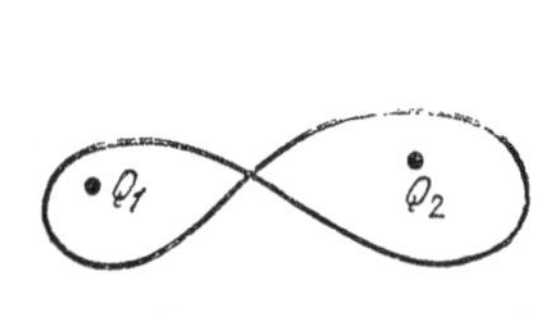

Figure 106.

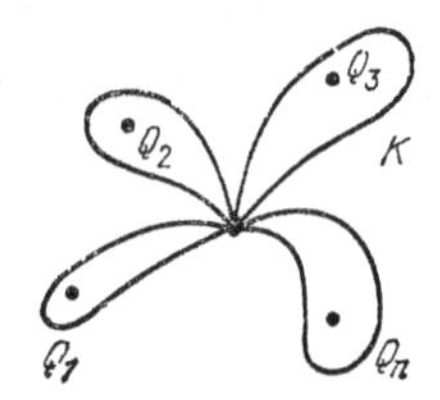

Figure 107.

Generally it is possible to define these groups for all bodies K for which there exists a manifold $M^n \supset K$ contracting to this particular body by assuming Theorem 1 (and Corollary 1 below) as the basic property of cohomologies:

$$H^k(M^n; \mathbb{R}) = H^k(K; \mathbb{R}).$$

For example, for a region on a plane $U_n = \mathbb{R}^2 \setminus (Q_1 \cup \ldots \cup Qn)$ as a body K we may take a bouquet of circles, as shown in Figure 107.

COROLLARY 1. *Homotopically equivalent manifolds* $M_1 \underset{g}{\overset{f}{\rightleftarrows}} M_2$ *have identical (co)homology groups.*

Proof. Consider the maps $f^*: H^k(M_2) \to H^k(M_1)$ and $g^*: H^k(M_1) \to H^k(M_2)$. Since the maps $f \circ g$ and $g \circ f$ are homotopic to the identity maps, the homomorphisms $(f \circ g)^* = g^* \circ f^*$ and $(g \circ f)^* = f^* \circ g^*$ are exactly identity homomorphisms of the cohomology groups, according to Theorem 1:

$$1 = g^* \circ f^*: H^k(M_2) \to H^k(M_2),$$

$$1 = f^* \circ g^*: H^k(M_1) \to H^k(M_1).$$

This, obviously, implies that the homomorphisms f^* and g^* themselves are mutually inverse isomorphisms, i.e. $f^* = (g^*)^{-1}$, and the result follows.

Exactly the same sort of argument proves that the *fundamental groups* π_1 *of homotopically equivalent manifolds coincide.*

COROLLARY 2. *The cohomology groups of a Euclidean space* $\mathbb{R}^n$ *or of a disc* D^n *are the same as those of a point, i.e.* $H^k(\mathbb{R}^n)$ *is a trivial group for* $k > 0$, $H^0(\mathbb{R}^n) = \mathbb{R}$ *(a one-dimensional linear space).*

This fact is occasionally referred to as the "Poincaré lemma": *locally, in a region around any point on a manifold* M^n, *any closed form* $d\omega = 0$ *is exact* $\omega = d\omega'$. Indeed, if we choose a disc D^n in local coordinates with centre at point Q, $\{\sum_{\alpha=1}^{n} (x^\alpha - x_0^\alpha)^2 \le \varepsilon\}$, we can apply to the disc Corollary 2 of Theorem 1 which tells us that $H^k(D^n) = 0$, $k > 0$.

For $k = 1$ the Poincaré lemma is familiar to the reader from the course in analysis since for 1-forms $\omega = f_k\, dx^k$ we have $\omega = dF$, where $F(P) = \int_Q^P dx^k$ on the path from a point Q to a point P in the disc D^n.

COROLLARY 3. *The cohomology groups of a Euclidean plane without a point* $\mathbb{R}^2 \backslash Q$ *(or of a ring) are the same as those of a circle* S^1 *and are given by*

$$H^k(S^1) = H^k(\mathbb{R}^2 \backslash Q) = 0, \ k > 1,$$

$$H^1(S^1) = H^1(\mathbb{R}^2 \backslash Q) = \mathbb{R}, \ k = 1,$$

$$H^0(S^1) = H^0(\mathbb{R}^2 \backslash Q) = \mathbb{R}, \ k = 0.$$

To prove Corollary 3, we shall calculate the groups $H^k(S^1)$. Obviously, they are trivial (we say equal to zero) if $k > 1$. Next, $H^0(S^1) = \mathbb{R}$ since the circle is connected (see Claim 1). To calculate the group $H^1(S^1)$, we introduce a coordinate ϕ, where ϕ and $\phi + 2\pi n$ represent one point for integer n. The form of degree 1 is specified by the equality $a(\phi)\, d\phi = \omega$, where $a(\phi)$ is a periodic function $a(\phi + 2\pi) = a(\phi)$. We always have $d\omega = 0$ since the dimension of the circle is equal to 1. When is the form $a(\phi)\, d\phi$ exact?

This means that $a(\phi)\, d\phi = dF$, where $F(\phi)$ is a periodic function. Obviously, $F(\phi) = \int_0^\phi a(\psi)\, d\psi + \text{const}$. Hence, the function $F(\phi)$ is periodic if and only if the condition $\int_0^{2\pi} a(\psi)\, d\psi = 0$ or $\int_{S^1} \omega = 0$ is satisfied.

Thus, the form of degree 1, $\omega = a(\phi)\, d\phi$, on a circle is exact if an only if there holds the condition $\int_{S^1} \omega = 0$. This implies that the two forms $\omega_1 = a(\phi)\, d\phi$ and $\omega_2 = b(\phi)\, d\phi$ define one and the same cohomology class if and only if $\int_{S^1} \omega_1 = \int_{S^1} \omega_2$. Therefore, we obtain $H^1(S^1; \mathbb{R}) = \mathbb{R}$, which completes the proof of Corollary 3.

COROLLARY 4. *The cohomology group $H^n(M^n, \mathbb{R})$ of an oriented closed (say, connected) Riemannian manifold M^n is nontrivial.*

Proof. Consider a volume element Ω, where (locally) we have $\Omega = |g|^{1/2}\, dx^1 \wedge \ldots \wedge dx^n$. If the set of regions of local coordinates is chosen in accordance with orientation (i.e. all the transformation Jacobians are positive), then Ω is the differential form of degree n and we have $\int_{M^n} \Omega > 0$ (this is the volume of the manifold M^n). Obviously, $d\Omega = 0$ since the degree of the form Ω is n. If we had $\Omega = d\omega$, then by the Stokes formula we would obtain $\int_{\partial M^n} \omega = \int_{M^n} \Omega = \int_{M^n} d\omega = 0$ (since M^n is closed and has no boundary). This is a contradiction which implies our corollary.

REMARK. If a closed manifold M^n is non-orientable (e.g. $\mathbb{R}P^2$), then the group $H^n(M^n, \mathbb{R})$ is trivial, but we shall not prove it here. In particular the volume element $d\sigma = |g|^{1/2}\, d^n x$ does not behave as a differential form under transformations with negative Jacobian.

We shall clarify the geometrical meaning of homology groups. If M^n is an arbitrary manifold and ω is a closed form of degree k, then its "integrals over cycles" are defined. The exact meaning of our statement is this. Suppose M^k is a closed orientable k-dimensional manifold. By the "cycle" in the manifold M^n we understand a smooth map

$$f: M^k \to M^n.$$

DEFINITION 3. *The period of the form* ω *with respect to the cycle* (M^k, f) *is the integral* $\int_{M^k} f^*(\omega)$.

Suppose now that N^{k+1} is an arbitrary oriented manifold with boundary $\Gamma = M^k$. The boundary is a closed oriented manifold (consisting, perhaps, of several pieces). By the "film" we shall understand the map $F: N^{k+1} \to M^n$ which coincides with f on M^k.

We shall make a simple assertion.

THEOREM 2. a) *For any cycle* (M^k, f) *the period of any exact form* $\omega = d\omega'$ *is equal to zero.*
b) *If the cycle* (M^k, f) *is the boundary of the film* (N^{k+1}, P), *where* M^k *is the boundary of* N^{k+1} *and* $F|_{M^k} = f$, *then the period of any closed form with respect to such a cycle* (M^k, f) *is equal to zero.*

Proof. a) If $\omega = d\omega'$, then by the Stokes formula we have

$$\int_{M^k} f^*(\omega) = \int_{M^k} f^*(d\omega') = \int_{M^k} d(f^*\omega') = \int_{\partial M^k} f^*(\omega') = 0$$

since the manifold M^k has no boundary.
b) If M^k is the boundary of N^{k+1} (with allowance made for orientation) and if $F|_{M^k} = f$, then by the Stokes formula

$$\int_{M^k} f^*(\omega) = \int_{N^{k+1}} dF^*(\omega) = \int_{N^{k+1}} F^*(d\omega) = 0$$

and the theorem follows.

We shall now make a statement: *if the periods of a closed form with respect to all cycles are equal to zero, then the form is exact.*

From this fact we can draw several conclusions. We shall give simple examples.

1. If $M^n = S^n$(a sphere), then $H^k(S^n) = 0$ for $k \neq 0, n$.

Proof. If $k > n$, then the statement is obvious immediately from the definition. If $0 < k < n$ and (M^k, f) is an arbitrary cycle, then by Sard's lemma, the image of $f(M^k)$ does not cover at least one point $Q \in S^n$. Therefore, the cycle (M^k, f) lies, in fact, in $\mathbb{R}^n = S^n \backslash Q$. We know already (the Poincaré lemma or Corollary 2) that in $\mathbb{R}^n$ any form is exact. Accordingly, all the periods are zero when $0 < k < n$. Consequently, $H^k(S^n) = 0$ when $0 < k < n$.

2. If $M^n = \mathbb{R}^n \backslash (Q_1 \cup \ldots \cup Q_l)$, then there exists l independent cycles of dimension $n - 1$ (these are spheres surrounding the points $Q_1, \ldots, Q_l$). From Sard's lemma we can readily deduce here that all the periods of the forms of degree $0 < k < n - 1$ are equal to zero by analogy with item 1. Therefore, we have $H^k(\mathbb{R}^n \backslash (Q_1 \cup \ldots \cup Q_l)) = 0$, $0 < k < n - 1$, and next $H^{n-1}(\mathbb{R}^n \backslash (Q_1 \cup \ldots \cup Q_l)) = \mathbb{R}^{l-1}$ (prove this).

3. Suppose M^3 is a region in $\mathbb{R}^3$ from which we discarded a set of points $Q_1, \ldots, Q_l$ and pairwise non-intersecting circles $\Gamma_1, \ldots, \Gamma_s$.

Here we can see cycles of dimension 1 and 2:

a) one-dimensional cycles $A_1, \ldots, A_s$ linked with circles $\Gamma_1, \ldots, \Gamma_s$ (Figure 108);

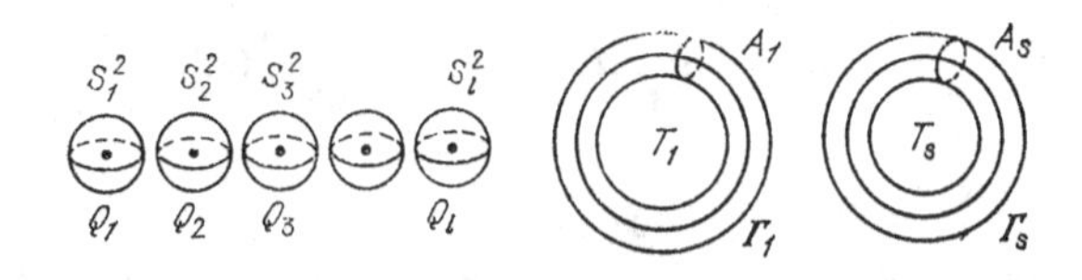

Figure 108.

b) two-dimensional cycles, i.e. spheres $S_1^2, \ldots, S_l^2$ embracing points $Q_1, \ldots, Q_l$ and tori $T_1^2, \ldots, T_s^2$ embracing circles $\Gamma_1 \ldots, \Gamma_s$ (Figure 108). The answer is clear:

$$H^1(M^3) = \mathbb{R}^s \quad (s \text{ independent periods})$$

$$H^2(M^3) = \mathbb{R}^{l+s} \quad (l + s \text{ independent periods}).$$

The answer does not depend on the mutual positions of circles $\Gamma_1, \dots, \Gamma_s$ (this is already not obvious). These facts are not, however, very easily provable. It is immediately seen that $H^1(M^3) = \mathbb{R}^p$, where $p > s$ and $H^2(M^3) = \mathbb{R}^q$, where $q \geq l + s$ (prove this).

REMARK. As distinct from homology groups, the fundamental group depends strongly (due to its non-commutativity) on the position of the discarded circles $\Gamma_1, \dots, \Gamma_s$. For example, let $M_1^3 = \mathbb{R}^3 \backslash \Gamma_1$ (Figure 109) and let $M_2^3 = \mathbb{R}^3 \backslash \Gamma_2$ (Figure 110). Then it turns out that $\pi_1(M_1^3, P)$ is an infinite cyclic group with generatrix a (see Figure 109). Prove it. Next $\pi_1(M_2^3, P)$ has three generatrices a, b and c (see Figure 110) and is given by some relations. Calculate these relations.

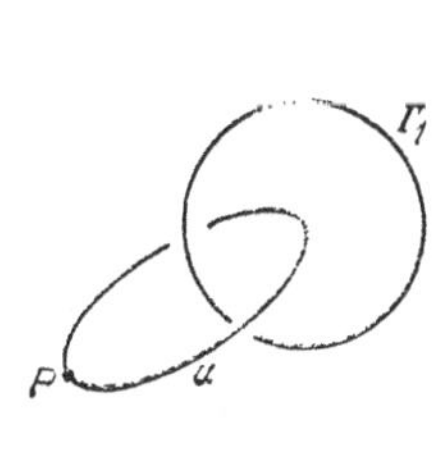

Figure 109.

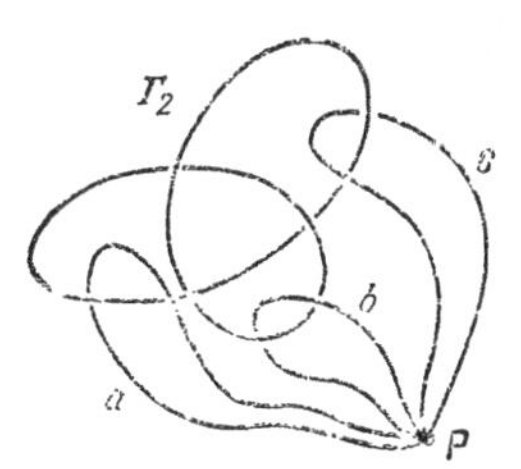

Figure 110.

We shall now present another approach to the definition and investigation of homology groups (simplicial and cell homologies) which allows their simple calculation. It will not however be very easy to establish relationship between this and the preceding approaches (this comes within the scope of the difficult de Rahm theorem which we omit here).

What is a simplex? A zero-dimensional simplex is a point α_0. A one-dimensional simplex is a segment $[\alpha_0, \alpha_1]$, whose boundary is a union of two zero-dimensional simplexes: a simplex $[\alpha_1]$ with the plus sign and a simplex $[\alpha_0]$ with the minus sign. A two-dimensional simplex is a triangle, $[\alpha_0\, \alpha_1\, \alpha_2]$ (Figure 111). A three-dimensional simplex is a tetrahedron, $[\alpha_0\, \alpha_1\, \alpha_2\, \alpha_3]$ (Figure 111). By induction, if an n-dimensional simplex $[\alpha_0\, \alpha_1 \dots \alpha_n]$ is defined and lies in an n-dimensional space $\mathbb{R}^n$, then to construct an $(n+1)$-dimensional simplex, we should take an $(n+1)$-th vertex outside this hyper-plane $\mathbb{R}^n \subset \mathbb{R}^{n+1}$ and consider the set of all points lying on the segments which join this new vertex α_{n+1} with points of the simplex $[\alpha_0 \alpha_1 \dots \alpha_n]$. The body obtained will just be an $(n+1)$-dimensional simplex.

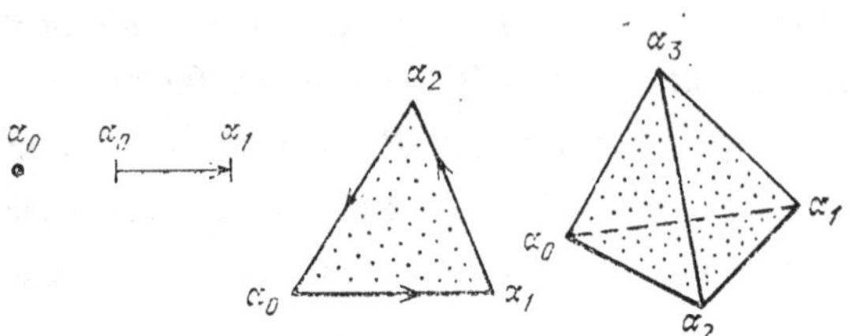

Figure 111.

The faces of the n-dimensional simplex $[\alpha_0 \alpha_1 \dots \alpha_n]$ are spanned by the vertices $[\alpha_0 \alpha_1 \dots \alpha_{n-1}]$, $[\alpha_0 \alpha_1 \dots \alpha_{n-2} \alpha_n]$, ... , $[\alpha_0 \dots \alpha_{n-s} \dots \alpha_n]$, ... , $[\alpha_1, \dots, \alpha_n]$. Thus, the i-th face is obtained by means of the removal of the i-th vertex from the set $[\alpha_0 \dots \alpha_n]$ and is opposite to this vertex: the i-th face $\Gamma_i = [\alpha_0 \alpha_1 \dots \hat{\alpha}_i \dots \alpha_n]$ (the i-th vertex is removed). We shall ascribe to the i-th face Γ_i the sign $(-1)^i$. The oriented boundary of the simplex $[\alpha_0 \dots \alpha_n]$ has the form

$$\partial[\alpha_0 \dots \alpha_n] = \sum_{i=0}^{n} (-1)^i [\alpha_0 \dots \hat{\alpha}_i \dots \alpha_n]$$

We can write the faces of smaller dimensions formally from the simplex $[\alpha_0 \dots \alpha_n]$ by removing a certain number of any vertices.

For simplex boundaries we have

$$\partial[\alpha_0] = 0,$$

$$\partial[\alpha_0 \alpha_1] = [\alpha_1] - [\alpha_0],$$

$$\partial[\alpha_0 \alpha_1 \alpha_2] = [\alpha_1 \alpha_2] - [\alpha_0 \alpha_2] + [\alpha_0 \alpha_1].$$

Compare these formulae with the figure. Indeed, the faces enter with regular signs (we understand here the linear combination formally — in a linear space where the simplexes $[\alpha_{i_0} \dots \alpha_{i_n}]$ themselves are basis elements). A simple but important lemma holds.

LEMMA 2. *For an n-dimensional simplex the formula* $\partial\partial[\alpha_0 \dots \alpha_n] = 0$ *is valid.*

The reader may verify the lemma by a direct calculation. For example, for $n = 2$, we have

$$\partial[\alpha_0\,\alpha_1\,\alpha_2] = [\alpha_1\,\alpha_2] - [\alpha_0\,\alpha_2] + [\alpha_0\,\alpha_1],$$

$$\partial\partial[\alpha_0\,\alpha_1\,\alpha_2] = \{[\alpha_2] - [\alpha_1]\} - \{[\alpha_2] - [\alpha_0]\} + \{[\alpha_1] - [\alpha_0]\} = 0.$$

The situation is similar for all n. The lemma is quite natural since the boundary of something has itself no boundary. A simplicial polyhedron (complex) is, by definition, a set of simplexes of arbitrary dimension, with the following properties:

1) all the faces, of all dimensions, of this polyhedron together with any simplex belong to this set;

2) two simplexes may intersect (have common points) only along a whole face of a certain dimension and only along a single face (or one of them is a face of the other).

We shall number in an arbitrary manner all the vertices of a simplicial polyhedron (a simplicial complex): $\alpha_0, \alpha_1, \dots, \alpha_N$. Then the simplexes are some sub-sets of vertices within a given numeration of the form $[\alpha_{i_0}\,\alpha_{i_1} \dots \alpha_{i_r}]$.

Now suppose that G is any commutative group, where the group law is written as summation (+).

The chains of dimension k in a simplicial complex are formal linear combinations of the form $C_k = \sum g_i\,\sigma_i$, where σ_i are different k-dimensional simplexes written in a given numeration of vertices of the complex, g_i are arbitrary elements of the group G.

The *chain boundary* is a chain of dimension $k - 1$:

$$\partial C_k = \sum_i g_i\,(\partial\sigma_i).$$

The formula $\partial\partial C_k = 0$ is obvious (by the lemma).

Cycles are such chains C_k that $\partial C_k = 0$. *Cycles* form a group.

Cycles homologic to zero (*limit cycles*) are such sycles C_k that $C_k = \partial C_{k+1}$.

DEFINITION 4. The *homology group* $H_k(M, G)$, where M is a simplicial complex, is a quotient group of all cycles of dimension k with respect to the cycles homologic to zero (two cycles are equivalent if and only if $C'_k - C''_k = \partial C_{k+1}$).

Of interest are the cases $G = \mathbb{R}$ (real numbers), $G = \mathbb{C}$ (complex numbers, $G = \mathbb{Z}$ (integers), $G = \mathbb{Z}_2$ (residues modulo 2) and generally $G = \mathbb{Z}_m$ (residues modulo m, especially when m is a simple number and $\mathbb{Z}_m$ is a field).

The conjugate objects are defined to be: the *cochain* C^k is a *linear function on chains* which associates k-dimensional simplexes with the elements of the group G:

$C^k(\sigma_i)$ is the element of the group G;

$$C^k(a\sigma_{i_1} + b\sigma_{i_2}) = aC^k(\sigma_{i_1}) + bC^k(\sigma_{i_2}),$$

where a and b are integers.

The *coboundary* δC^k of any cochain C^k is given by the formula

$$\delta C^k(\sigma_i) = C^k(\partial\sigma_i),$$

where σ_i is any simplex of dimension $k + 1$.

The *cocycle* $\delta C^k = 0$.

The *cocycle equivalent* (*cohomologic*) *to zero* $C^k = \delta C^{k-1}$.

The *cohomology group* $H^k(M; G)$ is a group of cocycles to an accuracy of cocycles equivalent to zero: $\tilde{C}^k = \tilde{\tilde{C}}^k$ if $\tilde{C}^k - \tilde{\tilde{C}}^k = \delta C^{k-1}$.

If any manifold M^n is divided into simplexes and is transformed into a simplicial complex (polyhedron), then for this manifold we can define and calculate the *homology and cohomology groups*.

The *differentiable simplex of dimension k* is a smooth embedding (of maximal rank) of a simplex into a manifold M^n (the map should be smooth up to the boundary of the simplex; it is preferable that the map be defined on a somewhat wider region in $\mathbb{R}^k$).

We shall assume a manifold to be *triangulated* if it is divided into a simplicial complex by means of differentiable simplexes. In this case we can define the homology and cohomology groups of the manifold M^n as those of a simplicial complex.

We have the following (not simple) facts:

a) the homology and cohomology groups do not depend on triangulation of the manifold ("surface") and are homotopically invariant;

b) for $G = \mathbb{R}$ the cohomology groups coincide with those introduced via differential forms.

For a finite simplicial complex (not necessarily a manifold) the Euler characteristic introduced by Poincaré is as follows:

if γ_i is the number of simplexes of dimension i in a complex, then the *Euler characteristic of the complex M* is given by

$$\chi(M) = \sum_{i \geq 0} (-1)^i \gamma_i .$$

We have the following simple theorem:

THEOREM 3. *If* b_i *(the Betti numbers) are dimensions of the homology groups* $H_i(M; \mathbb{R})$, *then the following equality holds*

$$\chi(M) = \sum_{i \geq 0} (-1)^i \gamma_i = \sum_{j \geq 0} (-1)^j b_j .$$

Proof. A group of chains of dimension i is a linear space of dimension γ_i; we shall denote by Z_i the group of cycles of dimension i and by B_i the sub-group of cycles homologic to zero. Obviously, we have

$$\dim B_i = \gamma_{i+1} - \dim Z_{i+1},$$

$$b_i = \dim H^i(M; \mathbb{R}) = \dim Z_i - \dim B_i =$$

$$= \dim Z_i - (\gamma_{i+1} - \dim Z_{i+1}),\ \gamma_0 = \dim Z_0,$$

whence

$$b_0 - b_1 + b_2 - b_3 + \ldots = \gamma_0 - \gamma_1 + \gamma_2 - \gamma_3 \ldots$$

and the theorem follows.

REMARK. The characteristic $\chi(M)$ has already appeared above as the sum of singularities of a vector field (or of a smooth function). This is the same quantity.

A version of the simplicial definition of homology groups are continuous (or singular) homology groups.

Let $\mathcal{H}$ be any topological space (e.g. the manifold and space of all continuous (smooth, piecewise smooth) maps of one manifold into another).

A *singular simplex of dimension* k is a pair (σ, f), where $f\colon \sigma \to \mathcal{H}$ is a continuous (smooth, piecewise smooth) map of a usual simplex $[\alpha_0 \ldots \alpha_k]$ into the space $\mathcal{H}$.

A *singular chain of dimension k* is a formal linear combination $C_k = \sum_i g_i(\sigma_i, f_i)$, where g_i are, as before, elements of the group G and (σ_i, f_i) are singular simplexes of dimension k.

The *boundary of a singular simplex* is a formal expression (chain)

$$\partial C_k = \sum_i g_i \, \partial(\sigma, f_i)$$

where

$$\partial(\sigma, f) = \sum_q (-1)^q \, (\Gamma_q, f|_{\Gamma_q}).$$

Here $\Gamma_q = [\alpha_0 \ldots \hat{\alpha}_q \ldots \alpha_k]$ is the side of the simplex, $f|_{\Gamma_q}$ is the same map f but restricted to the side (this is the side of a singular simplex which itself is a singular simplex).

The *singular cycle* is a chain C_k, such that $\partial C_k = 0$.

The *singular boundary* is a chain C_k, such that $C_k = \partial C_{k+1}$.

Singular homologies are groups $H_k(M; G)$ which consist, as before, of cycles to an accuracy of boundaries.

It turns out that for manifolds (and for all simplicial complexes) singular homology groups $H_k(M; G)$ coincide with those which we introduced earlier (the continual character of this definition is in a sense illusive since the number of homology classes is the same as we had earlier).

This definition is convenient, for example, in the proof of invariance of homology groups under homotopic equivalence and in the operations with functional spaces.

But as far as direct calculations are concerned, all the preceding definitions are inconvenient. We shall give another definition of homology groups (in terms of *cells*).

Let M^n be a manifold (or a more general simplicial complex).

A *cell of dimension k* is a continuous map of a disc $f: D^k \to M^n$, such that it is a smooth regular (of rank k) embedding into the manifold of the interior of the disc D^k and is continuous up to the boundary.

The cell complex: A manifold will be called a *cell complex* if it is divided into a finite number of cells (generally, of different dimensions) $f_i: D^{k_i} \to M^n$ with the following properties:

a) each point of the manifold M^n is an interior point of one (and only one) cell of a certain dimension;

b) for any cell $f\colon D^k \to M^n$ the image of the boundary $f|_{S^{k-1}}\colon S^{k-1} \to M^n$ gets into the union of cells of dimensions smaller than k.

The union of all cells of dimensions smaller than k we shall call the *cell frame* of dimension $k-1$ and denote by $K^{k-1} \subset M^n$. We have $K^0 \subset K^1 \subset K^2 \subset \ldots$ $\ldots \subset K^n = M^n$.

If we take K^q and contract all K^{q-1} to a single point (i.e. simply assume it to be a single point), then we obtain a *bouquet of spheres* of dimension q:

$$K^q/K^{q-1} = \sigma_1^q \cup \sigma_2^q \cup \ldots \cup \sigma_{l(q)}^q$$

where $l(q)$ is the number of cells of dimension q and s^q_α denotes the sphere S^q_α obtained for a q-dimensional cell by contracting its boundary to a point.

The *cell chain* is a formal linear combination which, as before, is given by $c_k = \sum_i q_i \sigma_i^k$, where σ_i^k are cells of dimension k.

What is the *boundary of a cell* σ^q (or of a chain)? By the definition of cell $f\colon D^q \to M^n$, its boundary $f\colon S^{q-1} \to M^n$ gets, in fact, into a $(q-1)$-dimensional cell frame K^{q-1}:

$$f\colon S^{q-1} \to K^{q-1}.$$

Contracting the frame K^{q-2} to a point, we obtain the map

$$S^{q-1} \to K^{q-1}/K^{q-2} = \sigma_1^{q-1} \cup \ldots \cup \sigma_{l(q-1)}^{q-1},$$

where each cell σ_α^{q-1} has become a sphere S_α^{q-1}, and all the K^{q-1}/K^{q-2} is a bouquet of spheres (Figure 112) since the whole frame K^{q-2} is assumed to be one point θ.

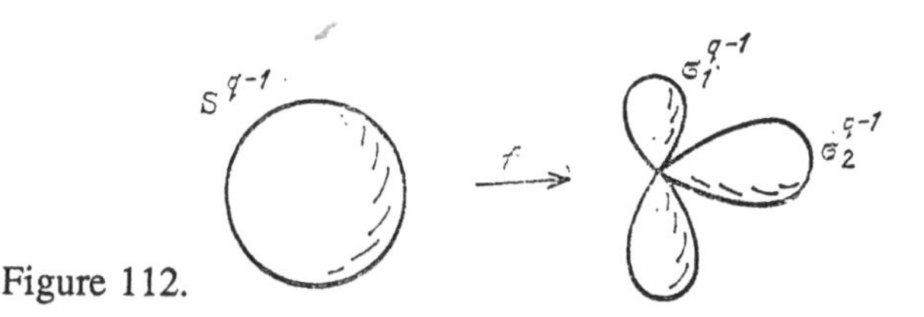

Figure 112.

Let us define the quantity λ_α for a pair of cells σ^q and σ_α^{q-1}: λ_α is the degree of the map f on the sphere S^{q-1} onto the summand σ_α^{q-1} of this bouquet $\sigma_\alpha^{q-1} \cup \ldots$ $\ldots \cup \sigma_{l(q-1)}^{q-1} = K^{q-1}/K^{q-2}$.

The *boundary of the cell* σ^q is defined to be

$$\partial\sigma^q = \sum_\alpha \lambda_\alpha \sigma_\alpha^{q-1},$$

where λ_α is the degree of the map of the boundary S^{q-1} of the cell σ^q onto the sphere S_α^{q-1} obtained from the cell σ_α^{q-1} by contracting the frame K^{q-2} (and, for example, all the other cells σ_β^{q-1} for $\beta \neq \alpha$) to a single point.

Sometimes λ_α may be expressed as

$$\lambda_\alpha = [\sigma^q \colon \sigma_\alpha^{q-1}]$$

and is called the *incidence coefficient of the cell* σ_α^{q-1} in the boundary of the cell σ^q. Now, already for the boundary of any chain we have

$$\partial C_k = \sum_{i,\alpha} q_i \lambda_{\alpha,i} \sigma_\alpha^{q-1},$$

$$C_k = \sum q_i \sigma_i^q, \ \lambda_{\alpha,i} = [\sigma_i^q \colon \sigma_\alpha^{q-1}]; \ \partial\partial C_k = 0.$$

We now introduce the definition of *cycles*: $\partial C_k = 0$, *boundaries*: $C_k = \partial C_{k+1}$ and *homology groups* $H_k(M^n; G)$ referred to as *cell* groups. These homology groups also coincide with those defined earlier.

Examples of cell complexes.

1. *The sphere* S^n. For this sphere, the most economic cell division is as follows: there exists one cell σ^0 of dimension 0, (the "vertex") and one cell σ^n of dimension n, where $\sigma^n = S^n - \sigma^0$. Here we have $\partial\sigma^0 = 0$, $\partial\sigma^n = 0$ (this is an obvious fact for all $n > 1$. Verify it for $n = 1$.)

From this we have

$$H_0(S^n; G) = G,$$

$$H_n(S^n; G) = G,$$

$$H_0(S^n; G) = 0, \quad k \neq 0, n.$$

For $G = \mathbb{R}$ we obtain the same result as for differential forms.

If we have a *bouquet q of spheres* S^{n+1} joined at one point, then there exists one vertex σ^0 and several cells σ_α^{n+1}, where $\partial\sigma_\alpha^{n+1} = 0$. The region obtained from $\mathbb{R}^{n+2}$ contracts to such a bouquet by discarding a set of points. We shall denote this bouquet by K_q^{n+1} (Figure 113). We have

$$H_0(K_q^{n+1}; G) = G,$$

$$H_{n+1}(K_q^{n+1}; G) = G + G + \ldots + G,$$

$$H_l(K_q^{n+1}; G) = 0, \quad l \neq 0, \quad n+1.$$

2. Cell division of the *torus* and of the *Klein bottle*. The case of the torus is depicted in Figure 114.

Figure 113.

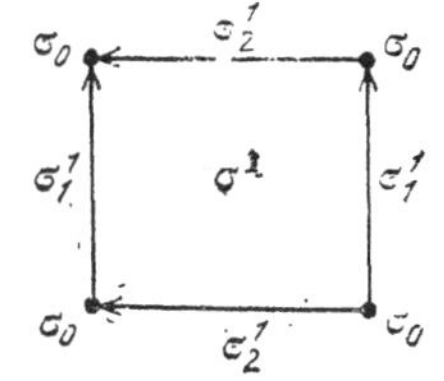

Figure 114.

Here we have the following cells: $\sigma^0, \sigma_1^1, \sigma_2^1, \sigma^2$,

$$\partial\sigma^0 = \partial\,\sigma_1^1 = \partial\,\sigma_2^1 = \partial\,\sigma^2 = 0;$$

for $G = \mathbb{Z}$

$$H_0 = \mathbb{Z}, \quad H_1 = \mathbb{Z} \oplus \mathbb{Z}, \quad H_2 = \mathbb{Z}.$$

Figure 115 shows the Klein bottle K^2. The cells are: $\sigma_1^0, \sigma_1^1, \sigma_2^1, \sigma^2$,

$$\partial\sigma^0 = \partial\sigma_1^1 = \partial\sigma_2^1 = 0, \quad \partial\sigma^2 = 2\sigma_1^1,$$

$$H_0(K^2, \mathbb{Z}) = \mathbb{Z}, \quad H_2(K^2, \mathbb{Z}) = 0,$$

$$H_1(K^2, \mathbb{Z}) = \mathbb{Z} + \mathbb{Z}_2.$$

3. *The projective plane* $\mathbb{R}P^2$ (Figure 116). The cells are: $\sigma^0, \sigma^1, \sigma^2, \partial\sigma^0 = 0$, $\partial\sigma^1 = 0, \partial\sigma^2 = 2\sigma^1$; for $G = \mathbb{Z}$

$$H_0 = \mathbb{Z}, \quad H_1 = \mathbb{Z}_2, \quad H_2 = 0.$$

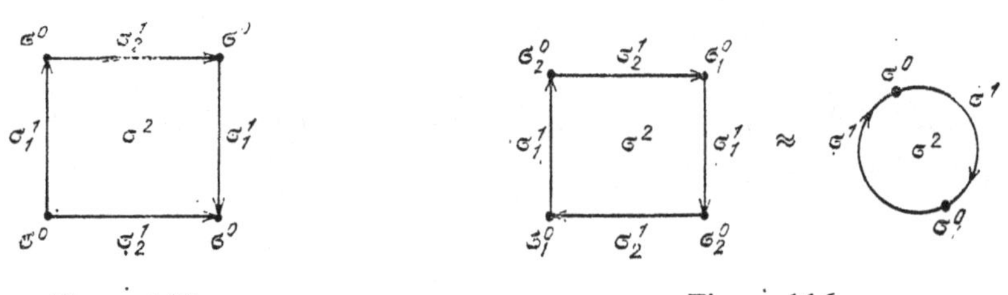

Figure 115. Figure 116.

4. *The orientable surface.* We restrict our consideration to the case of a cracknel ($g = 2$). Take an octagon (Figure 117). The cells here are $\sigma^0, \sigma_1^1, \sigma_2^1, \sigma_3^1$, σ_4^1, σ^2. All the boundaries are equal to zero. For $G = \mathbb{Z}$

$$H_0 = \mathbb{Z}, \quad H_1 = \mathbb{Z} \oplus \mathbb{Z} \oplus \mathbb{Z} \oplus \mathbb{Z}, \quad H_2 = \mathbb{Z}$$

(the latter equality is indicative of surface orientability).

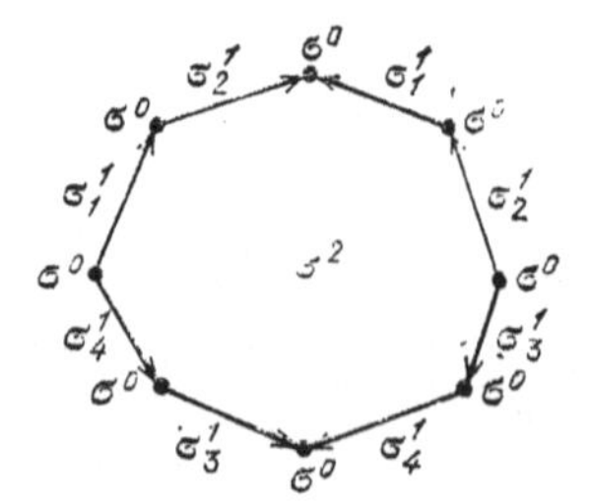

Figure 117.

5. *The projective space* $\mathbb{R}P^n$. The cells are $\sigma^0, \sigma^1, \sigma^2, \ldots, \sigma^n$,

$$\partial\sigma^0 = 0, \ \partial\sigma^1 = 0, \ \partial\sigma^2 = 2\sigma^1, \ \partial\sigma^3 = 0,$$

$$\partial\sigma^4 = 2\sigma^3, \ldots, \partial\sigma^{2k+1} = 0, \ \partial\sigma^{2k+2} = 2\sigma^{2k+1}$$

(construct the cell division). For $G = \mathbb{Z}$

$$H_0 = \mathbb{Z}, H_1 = \mathbb{Z}_2, \ H_2 = 0,$$

$$H_3 = \mathbb{Z}_2, \ldots, H_n = \begin{cases} 0 & \text{if } n \text{ is even } (> 0), \\ \mathbb{Z} & \text{if } n \text{ is odd.} \end{cases}$$

6. *The complex projective space* $\mathbb{C}P^n$. The cells are $\sigma^0, \ \sigma^2, \ldots, \sigma^{2n}$, all the boundaries are zero. (Construct the cell division.) For $G = \mathbb{Z}$

$$H_0 = H_2 = \ldots = H_{2n} = \mathbb{Z}, \ \text{the rest } H_i = 0.$$

For cell complexes with one vertex it is convenient to calculate the fundamental group π_1 (a manifold can always be divided into a cell complex with one vertex). Namely, all the cells σ_α^1 are closed paths q_α since the vertex is single. The set of paths q_α yields the set of generatrices of the fundamental group π_1.

The relation for the generatrices q_α is obtained from the two-dimensional cells: the boundary of each two-dimensional cell is a path homotopic to zero, and this is just the complete set of relations.

EXAMPLES.

1. $\mathbb{R}P^2$: the cell σ^1 is a generatrix a and the boundary of the cell σ^2, that is, $a^2 = 1$.

2. The torus: the cells σ_1^1, σ_2^1 are generatrices a, b and the cell σ^2 yields the relation $aba^{-1}b^{-1} = 1$ (or $ab = ba$).

3. The Klein bottle: the cells σ_1^1, σ_2^1 are generatrices a, b ; the cell σ^2 gives the relation $abab^{-1} = 1$ (or $aba = b$).

4. The cracknel: the cells $\sigma_1^1, \sigma_2^1, \sigma_3^1, \sigma_4^1$ are generatrices a, b, c, d; the cell σ^2 gives the relation $aba^{-1}b^{-1} cdc^{-1} d^{-1} = 1$; we see that this group is noncommutative.

5. The figure-of-eight: here we have no two-dimensional cells; the group π_1 is therefore free (has no relations).

APPENDIX 5

THE THEORY OF GEODESICS, SECOND VARIATION AND VARIATIONAL CALCULUS

We have introduced the Euler-Lagrange equations whose validity is the necessary condition for the functional

$$S(\gamma) = \int_P^Q L(z, \dot{z})\, dt, \quad \gamma = z(t)$$

to have, along a certain curve γ, the minimum among all the curves beginning at a point P and terminating at a point Q.

These equations were of the form

$$\frac{d}{dt}\left(\frac{\partial L}{\partial \dot{z}^i}\right) = \frac{\partial L}{\partial z^i}, \quad i = 1, \dots, n,$$

and for any vector field $\eta^i(t)$ defined at points of the curve γ: $z^i = z^i(t)$ the following identity held

$$\left[\frac{d}{d\varepsilon} S(\gamma + \varepsilon\eta)\right]_{\varepsilon=0} = \left[\frac{d}{d\varepsilon}\int_a^b L(z^i + \varepsilon\eta^i, \dot{z}^i + \varepsilon\dot{\eta}^i)\, dt\right]_{\varepsilon=0},$$

where $a \leq t \leq b$ and $\eta^i(a) = \eta^i(b) = 0$ (respectively at points $P = \gamma(a)$ and $Q = \gamma(b)$).

Furthermore, we derived the equality

$$\left(\frac{d}{d\varepsilon}\int_a^b L(z + \varepsilon\eta, \dot{z} + \varepsilon\dot{\eta})\, dt\right)_{\varepsilon=0} = \int_a^b \left(-\frac{d}{dt}\left(\frac{\partial L}{\partial \dot{z}^i}\right) + \frac{\partial L}{\partial z^i}\right)\eta^i\, dt$$

for all $\eta(t)$, where $\eta(a) = \eta(b)$. How shall we find the condition under which the curve $\{z^i = z^i(t)\}$ actually gives the minimum if it satisfies the Euler-Lagrange equation?

As is known, for the functions of many variables $f(x^1, \dots, x^n)$ the necessary condition of (local) minimum is $\partial f/\partial x^i|_P = 0$, $t = 1, \dots, N$, and the sufficient condition is positiveness of the quadratic form $\frac{\partial^2 f}{\partial x^i \partial x^j} dx^i dx^j$ at the same point P.

Therefore, to find the conditions of minimum for $S(\gamma)$, in the case where γ satisfies the Euler-Lagrange equations, we should necessarily calculate the bilinear form (second variation), analogous to the second differential, namely

$$\left[\frac{\partial^2 S}{\partial\lambda\,\partial\mu}(\gamma+\lambda\bar{\eta}+\mu\bar{\bar{\eta}})\right]_{\substack{\lambda=0\\ \mu=0}} = G_\gamma(\bar{\eta},\bar{\bar{\eta}}),$$

where $\bar{\eta}=\bar{\eta}(t)$, $\bar{\bar{\eta}}=\bar{\bar{\eta}}(t)$ are vector fields on the curve $\gamma(t)$ and vanishing at the end-points $\gamma(a)=P$ and $\gamma(b)=Q$.

PROPOSITION 1. *If $\gamma=\{z^i(t)\}$ satisfies the Euler-Lagrange equations, the following formula holds*

$$\left[\frac{\partial^2 S}{\partial\lambda\,\partial\mu}(\gamma+\lambda\bar{\eta}(t)+\mu\bar{\bar{\eta}}(t))\right]_{\substack{\lambda=0\\ \mu=0}} = \int_a^b (J_{ji}\bar{\eta}^j)\bar{\bar{\eta}}^i\,dt = G_\gamma(\bar{\eta},\bar{\bar{\eta}}),$$

where

$$J_{ji}\bar{\eta}^j = \frac{d}{dt}\left(\frac{\partial^2 L}{\partial\dot{z}^j\,\partial\dot{z}^i}\dot{\bar{\eta}}^j+\frac{\partial^2 L}{\partial\dot{z}^i\,\partial z^j}\bar{\eta}^j\right)-\frac{\partial^2 L}{\partial z^i\,\partial\dot{z}^j}\dot{\bar{\eta}}^j-\frac{\partial^2 L}{\partial z^i\,\partial z^j}\bar{\eta}^j.$$

The proof is carried out by direct calculation proceeding from the formula

$$\left[\frac{\partial^2 S(\gamma+\lambda\bar{\eta}+\mu\bar{\bar{\eta}})}{\partial\lambda\partial\mu}\right]_{\substack{\lambda=0\\ \mu=0}} = \left[\frac{\partial}{\partial\lambda}\left[\frac{\partial S}{\partial\lambda}(\gamma+\lambda\bar{\eta}+\mu\bar{\bar{\eta}})\right]_{\mu=0}\right]_{\lambda=0} =$$

$$= \left[\frac{\partial}{\partial\lambda}\int_a^b\left[-\frac{d}{dt}\left(\frac{\partial L}{\partial\dot{z}^i}\right)+\frac{\partial L}{\partial z^i}\right]\bar{\bar{\eta}}^i\,dt\right]_\lambda$$

where $L=L(z+\lambda\bar{\eta},\ \dot{z}+\lambda\dot{\bar{\eta}})$.

Now we shall turn to the case where the Euler-Lagrange equations coincide with the equation for geodesics. It is convenient here to choose the action

$$S = \int_a^b g_{ij}\,\dot{z}^i\dot{z}^j\,dt,$$

rather than the length

$$l = \int_a^b \left(g_{ij}\dot{z}^i \dot{z}^j\right)^{1/2} dt.$$

The length l and the action S have identical extremals (from the geometric point of view), but the action S is more convenient for analysis.

In the two-dimensional case, in a special system of coordinates (x, y), $x = z^1$, $y = z^2$, near the geodesic $\gamma(t)$, such that the line $x = t$, $y = 0$ is the geodesic $\gamma(t)$ itself, the bilinear form $G_\gamma(\bar{\eta}, \bar{\bar{\eta}})$ is given by

$$G_\gamma(\bar{\eta}, \bar{\bar{\eta}}) = -\int_a^b \left(\frac{d^2}{dt^2}\bar{\eta}^i + K(t)\,\bar{\eta}^i\right)\bar{\bar{\eta}}_i\,dt\,,$$

where K is the Gaussian curvature. Note that $g_{ij}(t) = \delta_{ij}$ for $z^2 = y$, $\dot{z}^1 \equiv 1$, $\dot{z}^2 = 0$, and therefore $\bar{\bar{\eta}}_i = g_{ij}\bar{\bar{\eta}}^j = \bar{\bar{\eta}}^i$.

The minimality condition for the geodesic $\gamma(t)$ suggests that the quadratic form $G_\gamma(\bar{\eta}, \bar{\eta})$ be positive for all the vector fields $\bar{\eta}$ vanishing at the end-points (P and Q) of the geodesic.

This implies the corollary. On a sufficiently small interval, geodesics (among all the smooth curves joining the same points) yield the minimum of the action functional $S(\gamma)$ and therefore also of the length functional.

In a region Ω of a space $\mathbb{R}^n$, or of a manifold M^n, we often have to consider an extremum of the following form.

Suppose we are given a class of smooth or piecewise smooth curves (e.g. the class of curves joining two points P and Q) and a function $L(x, \xi)$, where x is a point of a manifold and ξ is an arbitrary tangent vector at this point. The function $L(x, \xi)$ will be called a "Lagrangian". We shall examine a path $\gamma(t)$, where $a \le t \le b$, such that $\gamma(a) = P$ and $\gamma(b) = Q$, and the integral (the "action") $S(\gamma) = \int_a^b L(\gamma(t), \dot{\gamma}(t))\,dt$.

The question is, on which curve γ of the given class there exists the minimum (the extremum) of the functional $S(\gamma)$. For simplicity we assume that a neighbourhood of the curve is coordinatized by local coordinates $(x^1, \dots, x^n)$ and, in fact, the curve lies in a region of the space $\mathbb{R}^n$. Then the functional of the action $S(\gamma)$ has the form

$$S(\gamma) = \int_a^b L\left(x^1(t), \dots, x^n(t), \dot{x}^1(t), \dots, \dot{x}^n(t)\right) dt.$$

EXAMPLES.

1. $S(\gamma)$ is the length of a curve γ in the Riemannian metric on a manifold, $L(x, \dot{x}) = (g_{ij}\dot{x}^i\dot{x}^j)^{1/2}$.

2. $S(\gamma) = \int_a^b (g_{ij}\dot{x}^i\dot{x}^j)\,dt, \quad L = \langle \dot{x}, \dot{x} \rangle$.

3. A function (a "potential") $U(x)$ is defined and $L = \langle \dot{x}, \dot{x} \rangle + U(x)$.

4. A differential form $A_\alpha\, dx^\alpha$ is defined and $L = \langle \dot{x}, \dot{x} \rangle + A_\alpha \dot{x}^\alpha$. Then the form $\omega = A_\alpha\, dx^\alpha$ is called vector-potential.

If $r^\alpha(t)$ is a vector field, defined at points of a curve $\gamma(t)$, such that $r(a) = r(b) = 0$, then the following formula holds

$$\frac{\partial S(\gamma + \varepsilon r)}{\partial \varepsilon}\Big|_{\varepsilon=0} = -\int_a^b \left\{ \frac{d}{dt}\left(\frac{\partial L}{\partial \dot{x}^\alpha}\right) - \frac{\partial L}{\partial x^\alpha} \right\} r^\alpha(t)\, dt.$$

In particular, if γ is the extremum (or the minimum) for $S(\gamma)$, then $\frac{\partial S(\gamma + \varepsilon r)}{\partial \varepsilon}\Big|_{\varepsilon=0} = 0$ for all of the fields $r(t)$ which are equal at the end-points. We arrive at the Euler-Lagrange equations for extreme curves $\gamma^\alpha(t) = x^\alpha(t)$, $a \le t \le b$ (the necessary condition for the minimum):

$$\frac{d}{dt}\left(\frac{\partial L}{\partial \dot{x}^\alpha}\right) = \frac{\partial L}{\partial x^\alpha}.$$

According to the standard terminology

$\dot{x}^\alpha$ is *velocity*, $\frac{\partial L}{\partial \dot{x}^\alpha} = P_\alpha$ is *momentum* (covector), $\frac{\partial L}{\partial x^\alpha} = f_\alpha$ is *force*, $\dot{x}^\alpha \frac{\partial L}{\partial \dot{x}^\alpha} - L = E$ is *energy*.

The sufficient condition for the extreme curve $\gamma(t)$ to yield the minimum of the functional $S(\gamma)$ among all close to its curves of the same class (i.e. among all curves joining the same points P and Q) is given by the second variation:

$$G(\bar{r}, \bar{\bar{r}}) = \frac{\partial^2 S(\gamma + \varepsilon \bar{r} + \varepsilon\beta \bar{\bar{r}})}{\partial\varepsilon\partial\beta}\Big|_{\varepsilon=0} = \int_a^b (J^{ij}\bar{r}^j)\bar{\bar{r}}^j\, dt,$$

where

$$J^{ij}\bar{r}^j = \frac{d}{dt}\left(\frac{\partial^2 L}{\partial \dot{x}^j \partial \dot{x}^j}\dot{\bar{r}}^j + \frac{\partial^2 L}{\partial \dot{x}^i \partial x^j}\bar{r}^j\right) - \frac{\partial^2 L}{\partial x^i \partial \dot{x}^j}\bar{r}^j - \frac{\partial^2 L}{\partial x^i \partial x^j}\dot{\bar{r}}^j$$

the second variation $G(\bar{r},\bar{\bar{r}})$ being analogous to the second differential $d^2 f$ of ordinary functions. The sufficient condition for the minimum follows the positiveness of the second variation $G(\bar{r},\bar{r}) > 0$ for any fields $\bar{r}(t)$ formed at the 0 at the ends. We shall not go on investigating in detail the theory of the second variation, but examine more thoroughly the structure of the original Euler-Lagrange equations. It is relevant here to make the following remark. We can state various variational problems. We shall present the simplest ones.

EXERCISE 1 (the end-points are fixed). Find the extrema (the minima) of the functional $S(\gamma) = \int_P^Q L\,dt$ among all the piecewise smooth curves joining the points P and Q on a complete manifold (e.g. on a closed on).

EXERCISE 2 (periodic). Find the extrema (the local minima) of the functional $S(\gamma) = \int_\gamma L\,dt$ among all the cyclic paths γ: $S^1 \to M^n$.

The simplest functional is, for instance, the length of a curve γ. Let the manifold M^n be closed.

In Exercise 1 we have: there is the same number of homotopy classes of the paths joining the points P and Q as there are elements of the group $\pi_1(M^n)$. In each homotopy class of the paths there exists at least one (local) minimum of the length functional on a complete (e.g. closed) Riemannian manifold. Besides minima, there may be other extrema of the length functional, which are also geodesics from the point P to the point Q.

In Exercise 2 we have: there are as many homotopy clsses of the cyclic paths as there are classes of conjugate elements in the group $\pi_1(M^n)$. In each homotopy class there exists the minimum of length, i.e. a closed geodesic. Other extrema are also possible, which are also closd geodesics.

It has been proved above that the extremal of the length functional $L = \left(g_{ij}\dot{x}^i\dot{x}^j\right)^{1/2}$ coincides with that of the action functional $L = g_{ij}\dot{x}^i\dot{x}^j$ if we introduce the natural parameter on the curve. These extremals coincide with the geodesics of a single symmetric connection compatible with the metric:

$$\frac{d^2x^\alpha}{dt^2} + \Gamma^\alpha_{ij}\dot{x}^i\dot{x}^j = 0.$$

Let us also pay attention to the functional from Example 4, $L = g_{ij}\dot{x}^i\dot{x}^j + A_\alpha \dot{x}_\alpha$, where $A_\alpha\, dx^\alpha$ is the differential form (the vector-potential)). The action functional of such a form describes the motion of particles (e.g. charged particles) in the special (Minkowski g_{ij}-metric) and general theories of relativity. In this case, the momentum $p_\alpha = g_{i\alpha}\dot{x}^i$ and the Euler-Lagrange equation has the form (verify it!)

$$\dot{p}_\alpha = F_{\alpha\beta}\dot{x}^\beta,$$

where

$$F_{\alpha\beta} = \frac{\partial A_\alpha}{\partial x^\beta} - \frac{\partial A_\beta}{\partial x^\alpha},$$

$$F_{\alpha\beta} = -F_{\alpha\beta},$$

$$\sum F_{\alpha\beta}\, dx^\alpha\, dx^\beta = \Omega = d\omega.$$

Thus, the forces $f = \partial L/\partial x$ are expressed in terms of the 2-form $d\omega = \Omega = (F_{\alpha\beta})$ which is called the *field tensor*.

EXAMPLE 5. If the metric of a three-dimensional space is Euclidean, $g_{ij} = \delta_{ij}$ and if $U(x^1 x^2, x^3) = \dfrac{\text{const}}{|r|}$, then we are dealing with the problem from mechanics on the motion of a point in a field of forces with potential $U(x) = c/|r|$, $r^2 = \sum (x^a)^2$. This results from *Kepler's problem on the planet motion in the gravitational field of the Sun* (where c is always positive) as well as from the charge motion by the Coulomb law (where c may be either positive or negative). Let us pay attention to the following fact: if $L = \alpha|\dot{x}|^2 + \beta/|r|$, then there exists the similarity (homothetic) transformation

$$x \to \lambda x, \quad t \to \mu t.$$

Then $L \to \lambda^{-1} L$ if $\lambda^3 = \mu^2$, where λ and μ are constant numbers. However, the transformation $L \to \text{const.} \cdot L$ does not alter the Euler-Lagrange equations. From this, we can draw a conclusion that the homothetic transformations $x \to \lambda x$, $t \to \lambda^{3/2} t$ carry integral trajectories into trajectories. This is called the "*Kepler law*". (The ratio of cubes of the linear sizes of planet orbits is equal to the ratio of squares of the time sizes since the function x^3/t^2 remains unaltered under homothetic transformations sending an orbit into an orbit.)

What transformations can be made on the Lagrangian without changing the Euler-Lagrange equations? There are two types of such transformations:

$$\text{a) } L(x, \xi) \to \text{const} \cdot L(x, \xi) = L'(x, \xi),$$

$$\text{b) } L(x, \xi) \to L(x, \xi) + \frac{\partial f(x)}{\partial t} = L''(x, \xi).$$

The Lagrangians $L'(x, \xi)$ and $L''(x, \xi)$ are equivalent to the original one.

Suppose that the Lagrangian is invariant under some transformation group of coordinates x (the vector ξ transforms as a tangent vector or a tensor of type $(1, 0)$). The precise meaning of this assumption is that the transformations of this group send the Lagrangian L either exactly into itself or into an equivalent Lagrangian.

For example, suppose this group is one-parameter and preserves the Lagrangian $x(t) = S_t(x)$ or $x^\alpha(t) = S_t^\alpha(x^1, \dots, x^n)$,

$$S_{t+\tau}(x) = S_t(s_\tau(x)) = S_\tau(s_t(x))$$

and

$$S_{-t}(x) = S_t^{-1}(x).$$

Consider a vector field

$$\left(\frac{dx^\alpha(t)}{dt}\right)_{t=0} = A^\alpha(x) = \left(\frac{dS_t^\alpha(x)}{dt}\right)_{t=0}, \quad A = (A^\alpha).$$

The total derivative of the Lagrangian $L(x, \xi)$ along the vector field A^α has the form

$$\left[\frac{dL}{dt}\right]_{t=0} = A^\alpha \frac{\partial L}{\partial x^\alpha} + \frac{\partial A^\alpha}{\partial x^\beta} \xi^\beta \frac{\partial L}{\xi^\alpha}$$

Obviously,

$$\frac{dL}{dt} = \frac{\partial L}{\partial x^\alpha}\frac{dx^\alpha}{dt} + \frac{\partial L}{\partial \xi^\alpha}\frac{d\xi^\alpha}{dt};$$

if we have a small transformation $S_{\Delta t}$, then

$$S_{\Delta t}\colon x^\alpha \to x^\alpha + A^\alpha(\Delta t) = S^\alpha_{\Delta t}(x^1 \ldots x^n),$$

$$\xi^\alpha \to \frac{\partial S^\alpha_{\Delta t}(x)}{\partial x^\beta}\xi^\beta = \left\{\delta_{\alpha\beta} + \frac{\partial A^\alpha}{\partial x^\beta}(\Delta t)\right\}\xi^\beta,$$

which implies the unknown formula for dL/td along the field A^α.

The condition $dL/dt = 0$ means that the Lagrangian is preserved under this group of transformations

$$A^\alpha \frac{dL}{\partial x^\alpha} + \frac{\partial A^\alpha}{\partial x^\beta}\xi^\alpha \frac{\partial L}{\partial \xi^\alpha} = 0. \tag{1}$$

We now return to the Euler-Lagrange equation $\dot{P}_\alpha = f_\alpha$, where

$$p_\alpha = \frac{\partial L}{\partial \dot{x}^\alpha},\quad f_\alpha = \frac{\partial L}{\partial x^\alpha},\quad \xi^\alpha = \dot{x}^\alpha.$$

Obviously, we have

$$\left(A^\alpha, \frac{\partial L}{\partial \dot{x}^\alpha}\right)^{\boldsymbol\cdot} = \frac{\partial A^\alpha}{\partial x^\beta}\dot{x}^\beta \frac{\partial L}{\partial \dot{x}^\alpha} + A^\alpha\left(\frac{\partial L}{\partial \dot{x}^\alpha}\right)^{\boldsymbol\cdot}.$$

Since along the extremals we have $\left(\frac{\partial L}{\partial \dot{x}^\alpha}\right)^{\boldsymbol\cdot} = \dot{p}_\alpha = f_\alpha = \frac{\partial L}{\partial x^\alpha}$, it follows that

$$\left(A^\alpha \frac{\partial L}{\partial \dot{x}^\alpha}\right)^{\boldsymbol\cdot} = (A^\alpha p_\alpha)^{\boldsymbol\cdot} = -A^\alpha \frac{\partial L}{\partial x^\alpha} + A^\alpha\left(\frac{\partial L}{\partial \dot{x}^\alpha}\right)^{\boldsymbol\cdot} = 0$$

if the vector field A^α preserves the Lagrangian (i.e. $\partial L/\partial t$ is equal to zero along the vector field A^α).

We can finally formulate

THEOREM 1. *If a one-parameter transformation group $S_1(x) = x(t)$ preserves the Lagrangian $L(x, \xi)$, i.e. $\frac{dL}{dt} \equiv 0$ and the vector field $A^\alpha(x) = \left[\frac{dx^\alpha(t)}{dt}\right]_{t=0}$ determines this one-parameter group, then the "momentum conservation law" holds*

$$(p, A)^* = (P_\alpha, A^\alpha)^* \equiv 0$$

by virtue of the Euler-Lagrange equations (i.e. the component of momentum along the field A^α is conserved).

In principle, one vector field can always be (locally) regarded as a coordinate on which (under the conditions of the theorem) the Lagrangian does not depend: if $x = x^1, \ldots, x^n$, $\xi = \xi^1, \ldots, \xi^n$, $L = L(x^2, \ldots, x^n, \xi^1, \ldots, \xi^n)$, then we have

$$\dot{p}_1 = \left(\frac{\partial L}{\partial \xi^1}\right)^{\cdot} = \frac{\partial L}{\partial x^1} \equiv 0 \quad (\xi^\alpha = \dot{x}^\alpha), \; p_1 = \text{const.}$$

So long as the Lagrangian $L(x, \xi)$ does not depend explicitly on time, the *energy conservation law holds* (verify it!):

$$\frac{dE}{dt} = \frac{d}{dt}\left(\dot{x}^\alpha \frac{\partial L}{\partial \dot{x}^\alpha} - L\right) = \frac{d}{dt}(\dot{x}^\alpha p_\alpha - L) \equiv 0.$$

What other conservation laws do we know? For example, for the Lagrangian $1/2\,|\dot{x}|^2 + U(r) = L(x, \dot{x})$ in a Euclidean space (say, in a three-dimensional one), where $r^2 = \sum_{\alpha=1}^{3} (x^\alpha)^2$, the transformation under which the Lagrangian is preserved contains all rotations from SO_3 since $U = U(r)$. For geodesics on the sphere S^2, the group under which the Lagrangian $L = \langle \dot{x}, \dot{x} \rangle$ is preserved also contains SO_3, for the Lobachevskian plane the group of motions contains $SO_{2,1} = SL(2, R)/\pm 1$, which is known from the structure of the groups of motion.

Let, for example, the group be SO_3. Here we have three distinct one-parameter groups:

1) rotations around the z-axis by an angle ϕ_1 — the group $S^{(1)}_{\phi_1}$,

2) rotation around the y-axis — the group $S^{(2)}_{\phi_2}$,

3) rotation around the x-axis — the group $S^{(3)}_{\phi_3}$.

Correspondingly, we have three vector fields $X_1 = (X_1^\alpha)$, $X_2 = (X_2^\alpha)$, $X_3 = (X_3^\alpha)$. We can always choose a coordinate system (for, example, cylindrical or spherical), such that the angle of rotation around one axis (say, the z-axis) be the coordinate not entering in the Lagrangian: $\partial L/\partial\phi_1 \equiv 0$. Is it possible to choose two angles ϕ_1, ϕ_2 as coordinates where $\dfrac{\partial L}{\partial\phi_1} \equiv \dfrac{\partial L}{\partial\phi_2} \equiv 0$? From the theorem we know the laws of conservation

$$(X_1^\alpha p_\alpha) = 0, \quad (X_2^\alpha p_\alpha) = 0, \quad (X_3^\alpha p_\alpha) = 0$$

for all the three rotational groups (z-, x-, y-axes). It turns out, however, that due to noncommutativity of the group SO_3 the angles ϕ_1 and ϕ_2 are incompatible in the framework of one coordinate system. What is the reason for that?

For any field A, the "differentiation with respect to direction" is defined to be

$$\nabla_A f = A^\alpha \frac{\partial f}{\partial x^\alpha} = \left[\frac{df(S_t(x))}{dt}\right]_{t=0}$$

where S_t is a one-parameter group generated by the vector field A. The differential equation $\dot{x}^\alpha = A^\alpha(x)$ describes the motion of the points $x^\alpha(t)$. It is conveneint to trace out the motion of the functions $f_t(x) = f(x(t))$: for the functions $f_t(x) = \mathrm{f}(x(t))$ we have the equation

$$\frac{df}{dt} = \frac{\partial f}{\partial x^\alpha}\dot{x}^\alpha = \nabla_A f.$$

We shall denote the operator ∇_A (which acts on the functions) by A. The equation $df/dt = A(f)$, where $f_{t=0} = f_0(x)$, is readily solvable:

$$f(x(t)) = f_t(x) = e^{At}(f_0(x)),$$

$$e^{At} = \left(1 + At + \frac{A^2}{2!}t^2 + \ldots + \frac{A^n}{n!}t^n + \ldots\right)$$

This is the operator on functions, i.e. the operator of translations along trajectories of the vector fields A^α. Indeed, we have

$$\frac{df_t(x)}{dt} = \frac{d}{dt}[e^{At}(\mathrm{f}_0(x))] = A[e^{At}\{f_0(x))\}] = A(f_t(x)).$$

This is not surprising since for the simplest coordinate vector field $e_1 = A$ (say, on a straight line with coordinates $x = x^1$) the operator

$$A = \frac{d}{dx}$$

and

$$e^{At}(f) = \left(1 + At + \frac{A^2}{2!}t^2 + \dots\right)f(x) =$$

$$= f + t\frac{df}{dx} + \frac{t^2}{2}\frac{d^2f}{dx^2} + \dots + \frac{t^n}{n!}\frac{d^nf}{dx^n} + \dots$$

This, obviously, the Taylor series for the quantity

$$f(x + t) = e^{At}(f(x)),$$

at least for all analytic functions.

Thus, the one-parameter group S_t on the functions $f(x)$ acts as e^{At}. Suppose we are given two vector fields (A^α) and (B^α), and the operators $\nabla_A = A$ and $\nabla_B = B$. We have two one-parameter groups e^{At} and e^{Bt} ("translations"). In what case can the fields (A^α), (B^α) be simultaneously included in the coordinate system where $e_1 = (A)$ and $e_2 = (B)$? Obviously, the translation along the axes x^1 and x^2 must commute: $e^{At}e^{Bt} = e^{Bt}e^{At}$. This is, obviously, not the case for rotations around z- and y- axes. It is necessary that the operations ∇_A and ∇_B commute, i.e. that the theorem on mixed derivatives $\frac{\partial^2 f}{\partial x^1 \partial x^2} = \frac{\partial^2 f}{\partial x^2 \partial x^1}$ for any function $f(x)$ holds.

Let us consider the commutator $[\nabla_A, \nabla_B]$. We have (calculate it)

$$(\nabla_A\nabla_B - \nabla_B\nabla_A)f = \nabla_C f; \quad c_\alpha = \sum_\gamma \left(A_\gamma \frac{\partial B^\alpha}{\partial x^\gamma} B^\gamma \frac{\partial A^\alpha}{\partial x^\gamma}\right).$$

The vector field C is called the *commutator* (the *Poisson bracket*) of the fields A and B. The following properties are obvious.

$$[\nabla_A, \nabla_B] = -[\nabla_B, \nabla_A],$$

$$[\nabla_A, [\nabla_B, \nabla_C]] + [\nabla_C, [\nabla_A, \nabla_B]] + [\nabla_B, [\nabla_C, \nabla_A]] = 0.$$

The set of all vector fields is said to form a Lie algebra with respect to the operation $[\dots, \ \dots]$.

EXAMPLE. Rotation around the z-axis: the field $X_1 = (X_1^1, X_1^2, X_1^3)$; $X_1^1 = y$, $X_1^2 = x$, $X_1^3 = 0$ or $\nabla_{\bar{X}_1} = x\frac{\partial}{\partial y} - y\frac{\partial}{\partial x}$. Similarly, making a permutation we obtain the fields X_2 (rotation around the y-axis) and X_3 (rotation around the x-axis). Verify the formulae:

$$[X_\alpha, X_\beta] = \pm X_\gamma, \quad \alpha \neq \beta \neq \gamma.$$

We can see that there arises here a three-dimensional sub-algebra of the Lie algebra of all vector fields (the same as the vector product). In this Lie algebra, for the group SO_3 we cannot choose a pair of commuting fields.

The situation is similar with the group $SO_{2,1}$ (for example, in the Lagrangian for geodesics of a Lobachevskian plane). The reader can calculate the Lie algebra of this group himself.

This situation is more interesting for $L = |\dot{x}|^2 + \frac{c}{|r|}$ in a three-dimensional space. Here in fact for all energy levels $E < 0$ the L-preserving group appears to be larger — not SO_3, but SO_4. But it is not so easy to find this group, and it does not act in the space x_1, x_2, x_3.

For the variational problem with a Lagrangian $L(x, \xi)$ we have introduced the energy $E = \xi^\alpha \frac{\partial L}{\partial \xi^\alpha} - L$ and the momentum $P_\alpha = \frac{\partial L}{\partial \xi^\alpha}$. Now we shall give some definitions.

1. A variational problem (a Lagrangian) is called *positive definite* if the quadratic form $\frac{\partial^2 L}{\partial \xi^\alpha \partial \xi^\beta} \xi^\alpha \xi^\beta$ is positive for all (x, ξ).

2. A Lagrangian is called *non*-degenerate if the equation $p = \frac{\partial L}{\partial \xi}(x, \xi)$ has for any x a unique solution with respect to ξ: $\xi^\alpha = \xi^\alpha(p, x)$.

3. The energy $E = \xi^\alpha \frac{\partial L}{\partial \xi^\alpha} - L$ of a non-degenerate Lagrangian $L(x, \xi)$ is called a *Hamiltonian*, $H(p, x)$, if it is expressed in terms of the variables (p, x).

Obviously, we have $L = p^\alpha \dfrac{\partial H}{\partial p^\alpha} - H$, and the action for the curve $x(t)$, $p(t)$ is given by

$$S(\gamma) = \int_a^b \left(p^\alpha \frac{\partial H}{\partial p^\alpha} - H\right) dt.$$

It can be verified through a direct calculation that the Euler-Lagrange equations acquire in terms of the new variables the form

$$\begin{aligned} \dot{x}^\alpha &= \frac{\partial H}{\partial p^\alpha}, \\ \dot{p}^\alpha &= -\frac{\partial H}{\partial x^\alpha}. \end{aligned} \tag{1}$$

These are called the *Hamilton equations*.

The functional acquires the form

$$S(\gamma) = \int_a^b (p_\alpha \dot{x}^\alpha - H)\, dt = \int_a^b (p_\alpha dx^\alpha - H\, dt).$$

It is readily seen that the Euler-Lagrange equations for this functional in the space (x, p), where x is a point and p is the covector at this point, are of the form (1), p_α and x^α being thought of as independent coordinates.

Let us examine the differential form

$$\Omega = -\sum_{\alpha=1}^{n} dx^\alpha \wedge dp_\alpha = d\left(\sum_{\alpha=1}^{n} p_\alpha\, dx^\alpha\right).$$

This form determines a non-degenerate skew-symmetric scalar product

$$g_{ij} = \begin{pmatrix} 0 & 1 & & & 0 \\ -1 & 0 & & & \\ & & -0 & 1 & \\ & & -1 & 0 & \\ 0 & & & & \ddots \end{pmatrix}$$

in coordinates (x, p) which we shall denote by $y^1, \dots, y^{2n}$, where $y^{2i-1} = x^i$, $y^{2i} = p_i$.

The following lemma holds.

LEMMA 1. *The Hamilton equations* (1) *have the form*

$$\dot{y}^i = g^{iq}\frac{\partial H}{\partial y^q}$$

or, in other words, the vector field ($\dot{y}_i$) *has the form of the gradient (as the vector) of the function* $H(y)$ *in a skew-symmetric metric defined by the form* Ω.

The proof consists in a direct comparison with formulae (1).
The conservation law is given by

$$\dot{H} = \frac{\partial H}{\partial y^q}\dot{y}^q = g^{pq}\frac{\partial H}{\partial y^p}\frac{\partial H}{\partial y^q} = \langle dH, dH\rangle.$$

Since the scalar product is skew-symmetric, it follows that $\langle dH, dH\rangle = 0$.
The derivative of any function $f(y)$ has the form

$$\dot{f} = g^{pq}\frac{\partial f}{\partial y^p}\frac{\partial H}{\partial y^q} = \langle df, dH\rangle.$$

LEMMA 2. *The form* $\Omega = \sum dp_\alpha \wedge dx^\alpha = \sum_{i=1}^{n} dy^{2i}\wedge dy^{2i-1}$ *is preserved by virtue of the differential equations* (1):

$$\dot{\Omega} = 0.$$

Proof. To calculate the derivative of the form along the vector field (1), we shall make use of the following facts

$$(\Omega_1\wedge\Omega_2)^{\cdot} = \Omega_1\wedge\Omega_2 + \Omega_1\wedge\Omega_2,$$

$$(dx^\alpha)^{\cdot} = d\left(\frac{\partial H}{\partial p_\alpha}\right) = \frac{\partial^2 H}{\partial p_\alpha \partial x^\beta}dx^\beta + \frac{\partial^2 H}{\partial p_\alpha \partial p_\beta}dp_\beta,$$

$$(dp_\alpha)^{\cdot} = -d\left(\frac{\partial H}{\partial x^\alpha}\right) = -\frac{\partial^2 H}{\partial x^\alpha \partial x^\beta}dx^\beta - \frac{\partial^2 H}{\partial x^\alpha \partial p_q}dp_q .$$

Therefore, we obtain $\left(\sum dp_\alpha\wedge dx^\alpha\right) = \sum\left[(dp_\alpha)^{\cdot}\wedge dx^\alpha + dp_\alpha\wedge(dx^\alpha)^{\cdot}\right] = 0$ since the outer product is skew-symmetric, as required.

A space (x, p) with form Ω we shall call a *simplectic space* (it has a skew-symmetric non-degenerate scalar product given by a form Ω such that $d\Omega \equiv 0$). For any two functions $f(x, p)$ and $g(x, p)$ their *commutator* is

$$[f, g] = \langle df, dg \rangle = g^{pq} \frac{\partial f}{\partial y^p} \frac{fg}{\partial y^q},$$

where $y^{2i-1} = x^i, y^{2i} = p^i,\ g_{pq} = \begin{pmatrix} 0 & 1 & \\ -1 & 0 & \\ & & \ddots \end{pmatrix}$. Obviously, we have

$$[f, g] = -[g, f].$$

We can easily verify that the Jacobi identity holds

$$[f, [g, h]] + [h, [f, g]] + [g, [h, f]] = 0.$$

This means that the functions $f(x, p)$ form the *Lie algebra* with respect to the commutation operation [,]. On the functions, the Hamilton equations have the form $f = [H, f]$, where

$$\dot{x}^{\alpha} = -[x^{\alpha}, H] = \frac{\partial H}{\partial p^{\alpha}}, \quad \dot{p}^{\alpha} = [H, p_{\alpha}] = -\frac{\partial H}{\partial x^{\alpha}}$$

and $H(x, p)$ is the Hamiltonian. The quantity $f(x, p)$ is therefore said to be the *integral of motion* if it commutes with the energy $H(x, p)$.

The following theorem holds

THEOREM 2. *Given an arbitrary one-parameter group $S_t(y)$ of transformations of a phase space $y = (x, p)$ with the scalar product $\Omega \sum dp_{\alpha} \wedge dx^{\alpha}$, the condition of preservation of the scalar product of this group, $\Omega = 0$, is equal to the condition that (locally) the vector field*

$$\left[\frac{dS_t(y)}{dt}\right]^q_{t=0} = A^q, \quad q = 1, \dots,$$

has the Hamiltonian form (1).

Proof. The fact that equation (1) preserves the scalar product Ω is proved in Lemma 2. We shall now prove the inverse statement. Suppose we are given a vector field A^q or an operator

$$A = A^q \frac{\partial}{\partial y^q} = \Sigma\Big(A^{2\alpha-1} \frac{\partial}{\partial x^\alpha} + A^{2\alpha} \frac{\partial}{\partial p_\alpha}\Big).$$

Suppose also that the equality $\dot{\Omega} = 0$ (along the field A) or $A(\Omega) = 0$ holds.

The vector field gives the equation

$$\dot{x}^\alpha = A^{2\alpha-1}, \quad \dot{p}_\alpha = A^{2\alpha}.$$

By definition we have

$$(dx^\alpha) = dA^{2\alpha-1}, \quad (dp_\alpha)^{\cdot} = dA^{2\alpha},$$

$$0 = \dot{\Omega} = \sum_{\alpha=1}^{n} [(dp_\alpha)^{\cdot} \wedge dx^\alpha + dp_\alpha \wedge (dx^\alpha)^{\cdot}] =$$

$$= \sum_{\alpha=1}^{n} [dA^{2\alpha} \wedge dx^\alpha + dp_\alpha \wedge dA^{2\alpha-1}] =$$

$$= \sum_{\alpha=1}^{n} \Big(\frac{\partial A^{2\alpha}}{\partial x^\gamma} dx^\gamma \wedge dx^\alpha + \frac{\partial A^{2\alpha}}{\partial p^\gamma} dp^\gamma \wedge dx^\alpha +$$

$$+ dp_\alpha \wedge \frac{\partial A^{2\alpha-1}}{\partial x^\gamma} dx^\gamma + dp^\alpha \wedge \frac{\partial A^{2\alpha-1}}{\partial p^\gamma} dp^\gamma\Big)$$

whence we obtain

$$\frac{\partial A^{2\alpha}}{\partial x^\alpha} = \frac{\partial A^{2\alpha-1}}{\partial x^\alpha}; \quad \frac{\partial A^{2\alpha-1}}{\partial p_\gamma} = \frac{A^{2\gamma-1}}{\partial p_\alpha}; \quad \frac{\partial A^{2\alpha-1}}{\partial x^\gamma} = -\frac{\partial A^{2\gamma}}{\partial p_\alpha}.$$

This is equivalent to the fact that the form

$$w = \sum A^{2\alpha-1} dp_\alpha - A^{2\alpha} dx^\alpha$$

was closed: $dw = 0$. We should find (locally) a function $H(x, p)$ such that

$A^{2\alpha-1} = \frac{\partial H}{\partial p_\alpha}$; $A^{2\alpha} = -\frac{\partial H}{\partial x^\alpha}$ or $w = dH$. Obviously, from commutativity of the mixed derivatives it follows that the above conditions are at least the necessary ones. Locally, they are also the sufficient ones: $w = dH(x, p)$, and the result follows.

Let us consider the form $\sum p_\alpha\, dx^\alpha = \omega$ and the form $\dot{\omega} = f_\alpha\, dx^\alpha + g^\alpha\, dp_\alpha = \sum (p_\alpha\, dx^\alpha)^\cdot = \sum [\dot{p}_\alpha\, dx^\alpha + p_\alpha (dx^\alpha)^\cdot]$. Since $d(\dot{\omega} = (d\omega)^\cdot = \dot{\Omega} = 0$, it follows that $\dot{\omega}$ is a closed form. That the form is closed implies that locally $\dot{\omega} = dF$, where F is a function. It can be easily verified that this function has the form

$$F(x, p) = H - p_\alpha \frac{\partial H}{\partial p_\alpha}$$

(this is L expressed in terms of x and p). Upon a time shift by a small quantity Δt we shall have $S^*_{\Delta t}(\omega) = \omega + \Delta t\, \dot{\omega}$, where $\dot{\omega} = dL$. From this it follows that

$$S^*_t(\omega) - \omega = \int \dot{\omega}\, d\tau = d\int_0^t \left(H - p\frac{\partial H}{\partial p}\right) d\tau = d\int_0^t (p_\alpha\, dx^\alpha - H\, d\tau).$$

Thus, Hamiltonian systems determine one-parameter groups S_t in phase space (x, p), which preserve the skew-symmetric "metric" Ω (and inversely) if the Hamiltonian does not depend on time.

Canonical transformations

DEFINITION 1. A smooth transformation of a phase space, which preserves the 2-form Ω is called a *canonical* transformation.

By virtue of what has been said above, autonomous Hamiltonian systems determine one-parameter groups of canonical transformations. Non-autonomous Hamiltonian systems determine one-parameter families of canonical transformations which do not form a group.

We shall consider an arbitrary transformation F: $(p_i, q^i) \to (p'^i, q'^i)$, where

$$p'_i = f_i(p, q), \quad q'^i = g'^i(p, q), \quad i = 1, \dots, n. \tag{2}$$

A function $S(p', q) = S(p'_1, \dots, p'_n, q^i, \dots, q^n)$ is said to be the gnenerating function for the transformation F provided that the following equalities hold

$$p_i = \frac{\partial S}{\partial q^i}, \quad q'^i = -\frac{\partial S}{\partial p'_i}. \tag{3}$$

The following theorem holds

THEOREM 3. *If the system of equations* (3) *is non-degenerate and uniquely* (*locally*) *soluble in a neighbourhood of a certain point and if the transformation F determined by this function possesses a generating function S, then this transformation is* (*locally*) *canonical.*

Proof. We have to prove the equality $\Omega' = dp'_i \wedge dq'^i = \Omega = dp_i \wedge dq^i$. From (3) we have

$$\Omega = d\left(\frac{\partial S}{\partial q^i}\right) \wedge dq^i = \frac{\partial^2 S}{\partial q^i \partial q^j} dq^i \wedge dq^j + \frac{\partial^2 S}{\partial q^i \partial p'} dp'_j \wedge dq^i =$$

$$= \frac{\partial^2 S}{\partial q^i \partial p'_j} dp'_j \wedge dq^i,$$

(the first summand is zero since the summand $\frac{\partial^2 S}{\partial q^i \partial q^j} dq^i \wedge dq^j$ depends on i and j in a skew-symmetric way), and therefore

$$\Omega' = dp'_j \wedge d\left(\frac{\partial S}{\partial p'_j}\right) = \frac{\partial^2 S}{\partial p'_j \partial p'_i} dp'_j \wedge dp'_i + \frac{\partial^2 S}{\partial p'_j \partial q^i} dp'_j \wedge dq^i = \Omega.$$

This completes the proof of the theorem.

EXAMPLE. A class (although not a whole) of a linear canonical transformation is determined by a generating function $S(p', q)$ of the form

$$S = a^i_j p'_i q^j + 1/2\,(b^{ij} p'_i p'_j + c_{ij} q^i q^j).$$

In the case where $\det (a^i_j) \neq 0$, formulae (3) define the transformation $p'_i(p, q)$, $q'^i(p, q)$ well.

REMARK. The generating function S can be taken in a more general form. For example, let the set of indices $(1, 2, \dots, n)$ be divided into two non-intersecting subsets $M \cup N = (1, 2, \dots, n)$, $M \cap N = \varnothing$. Consider a function of the form

$$S(p_M, q_N, q'_M, p'_N)$$

and a map $p'(p, q)$, $q'(p, q)$:

$$p'_i = \frac{\partial S}{\partial q'^i}, \quad q^i = -\frac{\partial S}{\partial p^i}, \quad p_j = \frac{\partial S}{\partial q^j}, \quad q'^j = \frac{\partial S}{\partial p'_j}, \tag{5}$$

where $i \in M, j \in N$.

EXERCISE. Prove that (5) determines a canonical transformation.

It should be noted that the integrals of the Hamiltonian system $\dot{f}(x, p) = [H, f] = 0$ form the corresponding Lie algebra since we have $[f, g]^{\cdot} = [\dot{f}, g] + [f, \dot{g}] = 0$ provided that $\dot{f} = 0$ and $\dot{g} = 0$. Of interest in particular cases are finite-dimensional Lie algebras with respect to integrals. For example, suppose in a three-dimensional space

$$L(x, \xi) = \frac{|\xi|^2}{2} + \frac{a}{r}, \quad \text{where } \left(\sum (x^\alpha)^2\right)^{1/2}.$$

This is a spherically symmetric case (here, obviously, we have the symmetry group SO_3), and therefore there exist three "angular momentum" integrals

$$M_1 = -r^2(\dot{\theta} \sin \phi + \dot{\phi} \sin \theta \cos \theta \cos \phi),$$

$$M_2 = r^2(\dot{\theta} \cos \phi - \dot{\phi} \sin \theta \cos \theta \sin \phi),$$

$$M_3 = r^2(\sin^2 \theta \cdot \dot{\phi}).$$

It turns out that in this problem there exists another integral

$$W_i = [p, M]_i + \frac{ax^i}{r}$$

which, together with M_1, M_2, M_3 gives rise to a corresponding finite-dimensional Lie algebra depending on the energy level E. What is this Lie algebra for the energy levels $E < 0$, $E = 0$ and $E > 0$?

We shall make another remark. The Lagrange submanifold $M^n \subset (x, p)$ in the phase space (x, p) is such that the form $\Omega|_{M^n}$ on tangent vectors to M^n is equal to zero. We shall make the following statement (reader may verify it).

A. If such a manifold is projected regularly, without degeneracies onto the x-space along p, then it has the form of the graph $p_\alpha = \dfrac{\partial S(x)}{\partial x^\alpha}$, $S(x) = \int_{P_0}^{Q} p_\alpha\, dx^\alpha$, where P_0 is a fixed point on M^n, Q is any point on the manifold M^n with the coordinate $x, p(x)$, and the integration path lies also on the manifold M^n; $d(p_\alpha\, dx^\alpha) = 0$.

B. For any Hamiltonian system with Hamiltonian $H(x, p)$ the property $\Omega(M^n(t)) = 0$ is preserved under the motion of the manifold $M^n(t)$. For the function $S(x, t)$ we obtain the Hamilton-Jacobi equation (prove it):

$$-\frac{\partial S}{\partial t} = H\left(x, \frac{\partial S}{\partial x}\right),$$

$$H = H\left(x, \frac{\partial S}{\partial x}\right), \text{ where } p_\alpha = \frac{\partial S}{\partial x^\alpha}(x).$$

In concluding this appendix, we shall consider *Fermat* (or *Maupertuis*) *type principles*. As far back as the XVII century Fermat hypothesized that the path taken by a light ray between any two points is always the shortest, and the time needed for that is correspondingly the least. The speed of light in a medium depends generally on the properties of the medium which change from point to point. Suppose, for example, there is a boundary between two homogeneous media where the speed of light is c_1 and c_2. The Fermat principle implies (this may be verified) the *law of light refraction on the boundary between these two media* (on the interface). Suppose that the medium is *isotropic*. Then the trajectories of light rays are given by a Hamiltonian in a space (x, p) with a Hamiltonian of the form $H = c(x)\,|p|$, where H has the meaning of the light frequency, c is the velocity at a point x and p is the wave vector:

$$\dot{x}^\alpha = \frac{\partial H}{\partial p^\alpha} = c(x)\frac{p_\alpha}{|p|}, \qquad \dot{p}_\alpha = \frac{\partial c}{\partial x^\alpha}\,|p|.$$

Consider the expression $S = \int p_\alpha\, dx^\alpha - H\, dt$; we know that the variation $\delta(S)$ is equal to zero along extremals. We shall consider only those variations under which

a) the energy $E = H$ does not change,
b) the time interval does not change,
c) the beginning and the end of the path do not change.

By virtue of the law of conservation of energy, under such variations we have

$$\delta S = \delta \left[\int_P^Q p_\alpha dx^\alpha - \int_a^b E\, dt \right] = \delta \int_P^Q p_\alpha\, ds^\alpha.$$

Therefore, we can seek the trajectories (for the given energy $H = E$) from the variational principle $\delta S_0 = 0$, where $S_0 = \int_P^Q p_\alpha\, dx^a$. In doing so, we shall be able to find only the trajectories γ of motion, but not the velocity, since this variational principle does not depend on parametrization; to define this variational principle well, i.e. to eliminate dt using the relation $H = E$, we should express the moments p_α in terms of x^α and dx^α.

EXAMPLE 1. $H = c(x)\, |p| = E$ (Fermat). Since $\dfrac{dx^\alpha}{dt} = \dfrac{p^\alpha}{|p|} c(x)$ and $c(x)\, |p| = E$, we have $dt = \dfrac{|dx|}{c}$ or $|\dot{x}| = c(x)$ and

$$\sum p_\alpha\, dx^\alpha = \sum \frac{|p|\, dx^\alpha}{c\, dt}\, dx^\alpha = \frac{E}{c^2} |dx|^2 \frac{1}{dt} = \frac{E}{c} |dx|.$$

Whence

$$S_0 = E \int_P^Q \frac{dx}{c(x)} = E \int_P^Q dt.$$

The minimum condition for S_0 is equivalent, for a constant E, to the minimum condition for the time $T = \int \dfrac{|dx|}{c(x)}$ since $E = \text{const}$.

EXAMPLE 2. $L(x, \dot{x}) = 1/2\, g_{ij} \dot{x}^i \dot{x}^j - U(x)$ or

$$H = 1/2\, g^{ik} p_i p_k + U(x) = E = \text{const}.$$

Let us calculate the expression $\sum_\beta p_\alpha dx^\alpha$. Since $p_\alpha = g_{\alpha\beta} \frac{dx^\beta}{dt}$

$$H = 1/2\, g_{ik} \frac{dx^i dx^k}{dt^2} + U(x) = E$$

we have

$$dt = \left(\frac{g_{ik}\, dx^i dx^k}{2(E - U)}\right)^{1/2}, \quad \sum p_\alpha dx^\alpha = \left(2(E - U)\, g_{ik}\, dx^i dx^k\right)^{1/2}.$$

Therefore, we are finally led to the conclusion that *in the field of force with potential $U(x)$ and a constant energy E, the trajectories of motion of a point are geodesics with respect to the new metric*

$$\tilde{g}_{ij} = 2(E - U)\, g_{ij}$$

since

$$\delta S_0 = \delta \int_P^Q p_\alpha dx^\alpha = \delta \int_P^Q \left(2(E - U)\, g_{ij}\, dx^i dx^j\right)^{1/2}.$$

This is the *Maupertuis principle*. Even if $g_{ij} = \delta H_{ij}$, the new metric is already *not Euclidean*.

APPENDIX 6

BASIC GEOMETRIC PROPERTIES OF THE LOBACHEVSKIAN PLANE

It turns out that some essential properties of the Lobachevskian plane can be illustratively modelled on the surface of a three-dimensional space, called the *Beltrami surface*.

We shall consider on a plane (x, y) a smooth curve Γ characterized by the property that the length of the tangent line segment between the point of tangency to a curve and the point at which this tangent line intersects the x-axis is constant and equals a (Figure 118).

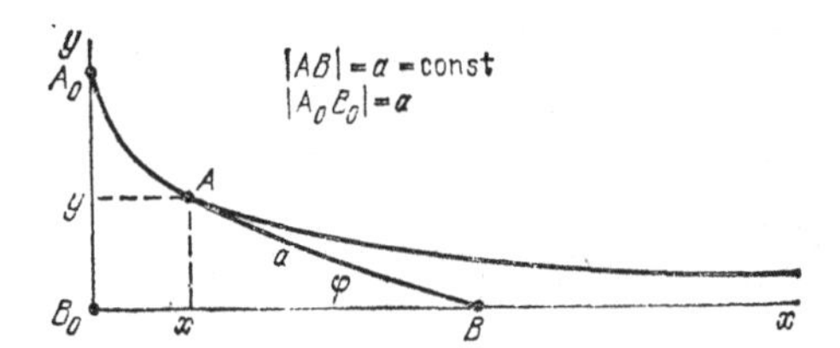

Figure 118.

We assume the curve to be positioned in the first quadrant of the plane. When the point A slides along the curve γ, the point B slides along the x-axis, and the segment AB has a constant length equal to a. The curve γ can be obtained mechanically. To this end we should tie together the points A and B by an inelastic thread of length a, and on placing A and B in the initial positions A_0 and B_0 (Figure 118) begin moving the point B along the x-axis. The point A will then draw a certain curve tangent to the y-axis at the point A_0 and having the x-axis as the asymptote. We shall now find the differential equations for the curve γ. From the triangle ABx (Figure 118) we have $\tan\phi = -y'_x$, where $y = y(x)$ is the graph of the curve γ and $a \sin\phi = y$. From this we obtain

$$\sin\phi = \frac{y'_x}{\left(1 + (y'_x)^2\right)^{1/2}} \quad \text{or} \quad x'_y = -\frac{(a^2 - y^2)^{1/2}}{y}.$$

where $x = x(y)$ is the graph of γ. Therefore,

$$x(y) = -\int_y^a \frac{1}{y}(a^2 - y^2)^{1/2}\,dy = -(a^2 - y^2)^{1/2} + \frac{a}{2}\ln\left(\frac{a + (a^2 - y^2)^{1/2}}{a - (a^2 - y^2)^{1/2}}\right).$$

Thus, we have derived the explicit expression for the curve $x = x(y)$. We shall consider the *surface of revolution* formed by rotation of the curve γ about the horizontal x-axis. We obtain a surface V^2 referred to as the *Beltrami surface* or *pseudo-sphere* (Figure 119). Let us find the Gaussian curvature of the Beltrami surface. To this end we have to calculate the surface of revolution.

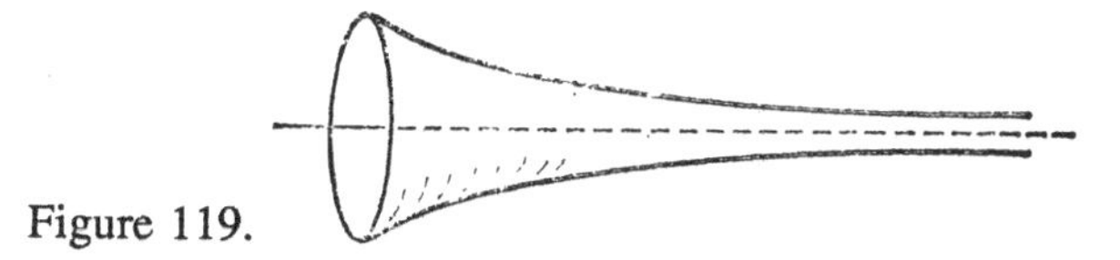

Figure 119.

In a three-dimensional space (x, y, z) we shall consider a surface of revolution M^2 formed by rotation about the x-axis of a certain smooth curve $x = x(y)$ (which we shall call generating) positioned in the (x, y)-plane. On the surface of revolution there arises a natural coordinate net formed by parallels and meridians of the surface. This net has the property that at each point of the surface the coordinate lines intersect at a right angle (Figure 120). Prove the following

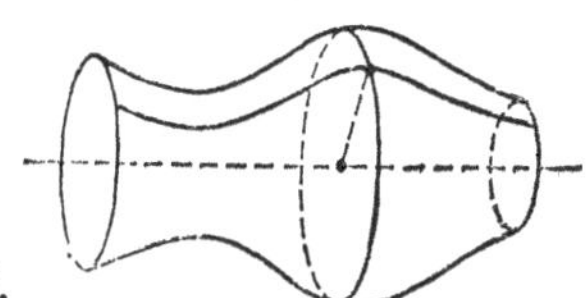

Figure 120.

LEMMA 1. *At each point of a surface of revolution the principal directions, i.e. the directions corresponding to the principal curvatures λ_1 and λ_2, can always be taken as coincident with the directions of the meridian and the parallel passing through this point.*

Using in the lemma the words "can always be assumed" we meant the following. Recall that wnen the principal curvatures are distinct, the principal directions are uniquely defined. In this case they coincide with the directions of the parallel and meridian. If the principal curvatures coincide, then any direction tangent to the surface is principal. In particular, the mutually orthogonal directions of the meridian and parallel are also principal.

Proof of Lemma 1. Recall that principal directions are those and only those mutually orthogonal unit vectors relative to which the matrices of the first and second quadratic forms come as diagonal. The definition of the surface of revolution implies that the first quadratic form is orthogonal in the coordinate system generated by meridians and parallels (as coordinate lines). In the same coordinate system, the second form is automatically diagonal. We shall consider cylindrical coordinates (R, ϕ, x) in a space, where the generatrix of the surface of revolution is given by the equation $R = R(x)$. Therefore, the radius-vector of the surface of revolution is

$$r = r(x, \phi) = (x, R(x)\cos\phi,\ R(x)\sin\phi)$$

(Figure 121). Differentiation yields $r_{x\phi} = (0, -R'_x \sin\phi, R'_x \cos\phi)$.

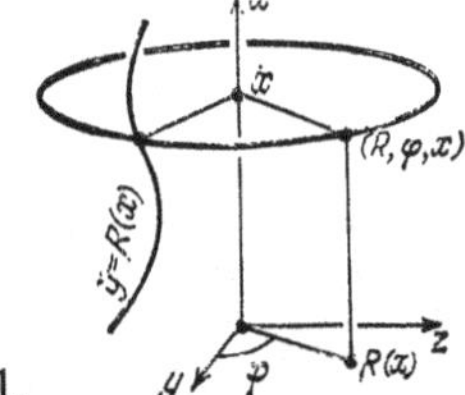

Figure 121.

The normal to the surface of revolution is given by

$$n = \frac{(0, -R'_x \sin\phi, R'_x \cos\phi)}{\left(1 + (R'_x)^2\right)^{1/2}}.$$

This implies that the normal n and the vector $r_{x\phi}$ are orthogonal. This just means that the second quadratic form is diagonal in the basis indicated above, which completes the proof of the lemma.

LEMMA 2. *The Gaussian curvature K of the surface of revolution formed by rotation of a curve $R = R(x)$ about the x-axis has the form:*

$$|K| = \frac{|R''|}{R\left(1 + (R')^2\right)^2}.$$

The proof of the lemma is obtained by a direct calculation carried out on the basis of Lemma 1.

CLAIM 1. *The Beltrami surface is a manifold of a constant negative curvature in a three-dimensional space.*

Proof. Since the Beltrami surface is a surface of revolution, we can use the formula from Lemma 2 to calculate the Gaussian curvatures. Here the function $y = R = R(x)$ is the inverse of the function

$$x = x(y) = -(a^2 - y^2)^{1/2} + \frac{a}{2} \ln \left(\frac{a + (a^2 - y^2)^{1/2}}{a - (a^2 - y^2)^{1/2}}\right).$$

As shown above, $x'_y = -\frac{1}{y}(a^2 - y^2)^{1/2}$, and therefore $x'' = \dfrac{a^2}{R^2(a^2 - R^2)^{1/2}}$.

Substituting this expression into the formula for the Gaussian curvature, we finally come to

$$K = \frac{R''}{R\left(1 + (R')^2\right)^2} = \frac{-x''x'}{R\left(1 + (x')^2\right)^2} = \frac{-1}{a^2} = \text{const.}$$

The minus sign is a result of the fact that the curve $R = R(x)$ is convex down, and therefore the principal curvatures λ_1 and λ_2 have opposite signs with respect to any direction at the point. Hence, $K = -1/a^2$, and the result follows.

Thus in a three-dimensional space there exist three remarkable surfaces of constant curvature.

1. A *manifold of constant zero Gaussian curvature* is a Euclidean plane. More generally, we may consider a cone formed by a family of straight lines coming from a single fixed point (the point may be either in a finite part of the space or at infinity) and sliding along an arbitrary smooth plane curve in space. If the cone vertex is at infinity, the surface is a cylinder.

2. A *manifold of constant positive curvature* is a standard sphere. As distinguished from surfaces of type (a), the sphere is a closed manifold.

3. A *manifold of constant negative curvature* is a Beltrami surface. It has a boundary — a circle with radius a and centre at the origin. It can be shown (but we shall not do this here) that a Beltrami surface cannot be continued outside this circle without violation of the condition that $K = -1/a^2 < 0$. A Beltrami surface is usually completed by adding a surface symmetric to the initial one relative to the (y, z)-plane (Figure 122). The surface obtained has a circle at the points of which the surface is not a smooth submanifold in a three-dimensional space. It turns out that the Beltrami surface is closely connected with the Lobachevskian plane.

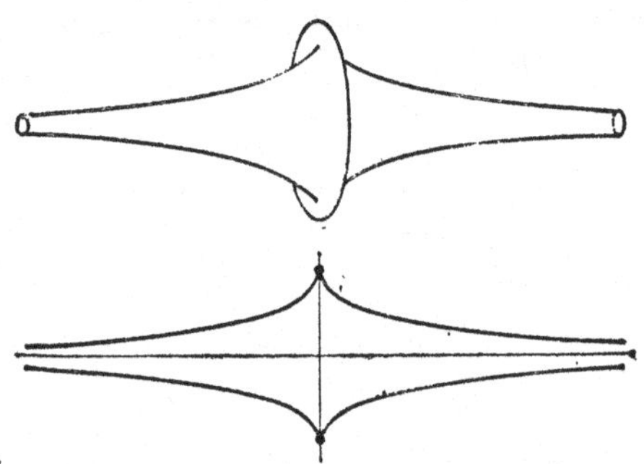

Figure 122.

CLAIM 2. *A Riemannian metric induced on a Beltrami surface by an envelope Euclidean metric is a Lobachevskian metric.*

Proof. We shall introduce in our sphere the cylindrical coordinates (x, R, ϕ), where $x = x$, $y = R\cos\phi$, $z = R\sin\phi$, i.e. the x-axis is the axis of rotation. The metric induced on the surface formed by rotation about the x-axis of the curve $x = x(R)$ has, obviously, the form

$$ds^2 = (dx(R))^2 + dR^2 + R^2\, d\phi^2 = (1 + (x')^2)\, dR^2 + R^2\, d\phi^2.$$

In our case we have $x' = -1/R(a^2 - R^2)^{1/2}$ (see above).

Consequently, $ds^2 = \dfrac{a^2\, dR^2}{R^2} + R^2\, d\phi^2$. We shall consider the following change of variables: $u = \phi/a$, $v = 1/R$. Then the metric transforms like this:

$$ds^2 = \frac{a^2 v^2}{v^4}(dv^2) + \frac{a^2}{v^2}du^2 = \frac{a^2(du^2 + dv^2)}{v^2},$$

which implies our assertion since we have come to the standard notation of the Lobachevskian metric in realization of the upper half-plane.

Thus, the *Beltrami surface is locally isometric to the Lobachevskian plane*. This means that *we have constructed an isometric embedding (i.e. a metric-preserving embedding) of a certain region on a Lobachevskian plane into a three-dimensional Euclidean space*. Which particular part of the Lobachevskian plane admits such an isometric embedding? We shall preliminarily notice that the whole of the Beltrami surface (now we are concerned only with that part of it which is depicted in Figure 119) is not isometric to any piece of the Lobachevskian plane. Indeed, the Beltrami surface is homeomorphic to a disc with a punctured point (i.e. to a ring). If this ring were homeomorphic (with preservation of the metric) to a certain region on a Lobachevskian plane, then an infinitely remote point of a Beltrami funnel should be mapped into a certain finite point of a Lobachevskian plane (Figure 123).

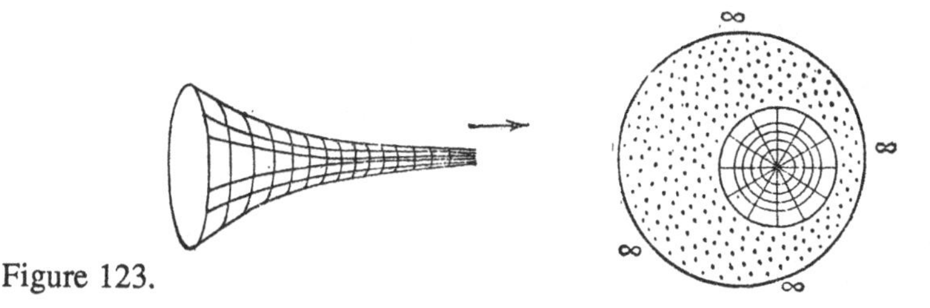

Figure 123.

But this would contradict the fact that an infinitely remote part of a Beltrami funnel is separated by an infinite distance from the funnel neck, i.e. from a singular circle of radius a.

It is convenient to cut a Beltrami funnel along any of its generatrices (Figure 124). As a result, we obtain a surface which admits an isometric embedding into a Lobachevskian plane, in the form of a certain region.

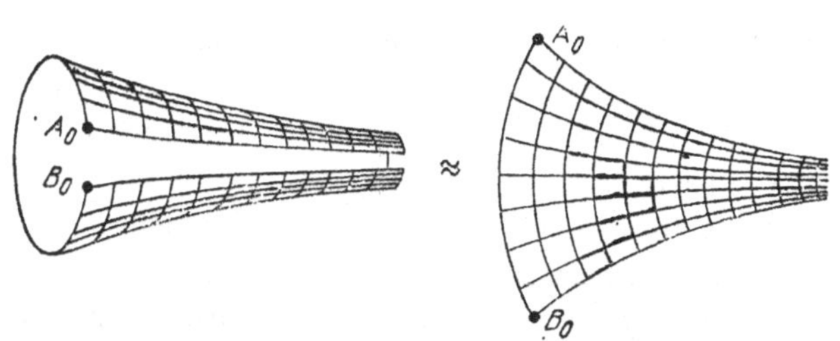

Figure 124.

Indeed such a region in a Lobachevskian plane (we are working, for convenience, with the Poincaré model) is illustrated in Figure 125. This region has the form of a curvilinear rectangle with vertices ∞, A_0, B_0. The sides of the triangle are formed by two parallel straight (in the sense of Lobachevsky geometry) lines coming from one point ∞ at the absolute. The third side of the triangle is the arc $A_0 B_0$ with length equal to $2\pi a$. This arc is a portion of the circumference (in the Euclidean sense) tangent to the point ∞ at the absolute (i.e. on the boundary of the Poincaré model). Consequently, the region (∞, A_0, B_0) is an infinite band between two parallel straight lines on the Lobachevskian plane and limited on one side to the arc A_0, B_0.

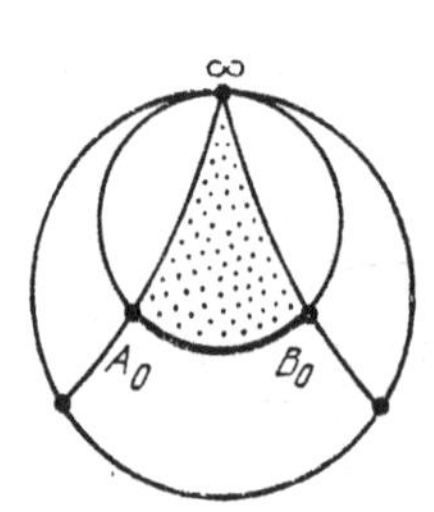

Figure 125.

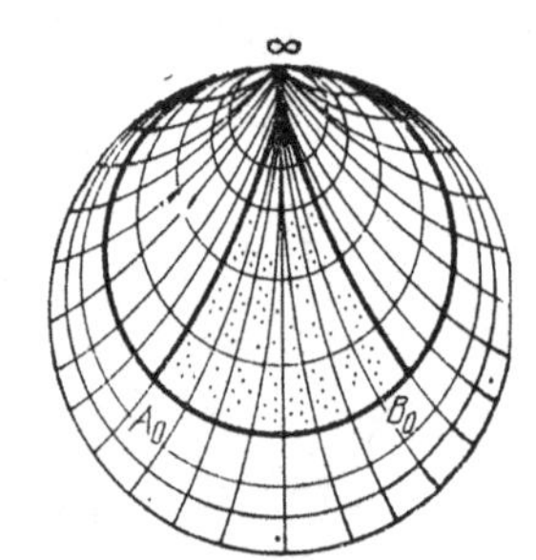

Figure 126.

On the Lobachevskian plane we shall now consider two families of coordinate lines that form an orthogonal net (both in the Euclidean sense, and in the sense of the Lobachevskian metric, since these two metrics differ only by a conformal factor which has no effect upon the orthogonality of intersecting curves). One of these two families of curves is a set of parallel straight lines (in the sense of the Lobachevskian plane) coming from a single point ∞ at the absolute. In the Poincaré model this is a sheaf of circle arcs (in the Euclidean sense) going onto the absolute at right angles. The other family of curves is a set of Euclidean circles which in the Poincaré model touch the absolute, as shown in Figure 126.

We have obtained two families of mutually orthogonal curves. Curves of the first family are straight lines in Lobachevsky geometry. Curves of the second family are not straight lines in Lobachevsky geometry. They possess, however, an important property. These lines are uniquely defined by the condition that all "perpendiculars" going from points of one line are parallel to one another and intersect at one and the same point at the absolute. It can be easily proved that any two lines of the second family are congruent in the sense that they can be mapped into each other through the isometry of the Lobachevskian plane, i.e. through a linear fractional transformation.

We shall now consider an arbitrary line from the second family, i.e. a Euclidean circle tangent to the absolute at the point ∞. On this line we mark a pair of points separated by a distance 2π. For simplicity we assume a to be equal to unity and the radius of the circle (the Poincaré model) to which the Lobachevsky geometry is applied also to be equal to unity. Then the band between two perpendiculars (A_0, ∞) and (B_0, ∞) is isometric to the Beltrami funnel cut along its meridian (i.e. along its generatrix). Under this isometry, the orthogonal net of meridians and parallels on the Beltrami surface transforms into an orthogonal net of curves of the first and second families on the Poincaré model in the band (∞, A_0, B). On a Lobachevskian plane (the same as on an ordinary Euclidean plane) there always exists a reflection (isometry) relative to an arbitrary straight line. In particular, we can reflect the band (∞, A_0, B) relative to the straight line (∞, A_0). As a result we shall obtain a new band isometric to the initial one and, therefore, to the cut Beltrami funnel. Again reflecting this new band (∞, A_1, A_0) relative to the straight line (∞, A_1), we obtain a band (∞, A_2, A_1) with the same properties etc., as shown in Figure 127.

Note that the reflection relative to a straight line on a Lobachevskian plane is an isometry. Consequently, the curve from the second family through a pair of points A_0, B_0 will be sent to itself since any isometry which preserves the point ∞ sends curves to the second family again into curves of the same family. Figure 127 illustrates the result of this infinite sequence of reflections. It is clear that all the segments $A_i A_{i-1}$ (where $0 \leq i < \infty$) have one and the same length 2π. A similar procedure gives rise to bands (∞, B_i, B_{i-1}) with the same properties. Thus, we obtain a disc D^2 (Figure 127) limited to a curve from the second family (i.e. by a circumference) and subdivided into an infinite number of bands convergent at the point ∞ at the absolute. Now we are in a position to construct a locally isometric map of the whole disc D^2 onto a Beltrami funnel (already without a cut). Given this, each band of the type (∞, A_i, A_{i-1}), (∞, A_0, B_0) and (∞, B_i, B_{i-1}) isometrically winds round the Beltrami funnel covering it exactly one time. Consequently, the disc D^2 will wind round the Beltrami funnel infinitely many times, as shown in Figure 128. Thus, *we have constructed an infinite-sheeted covering of a Beltrami surface.*

The arguments above give rise to a natural question of whether or not a whole Lobachevskian plane (and not only a part of it, e.g. the band descibed above) can be isometrically realized in a three-dimensional space in the form of a smooth two-dimensional surface of *constant* negative curvature. The answer appears to be *negative* (D. Hilbert).

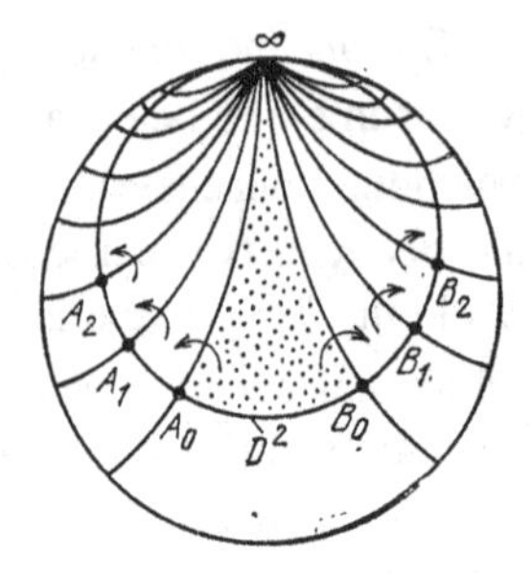

Figure 127.

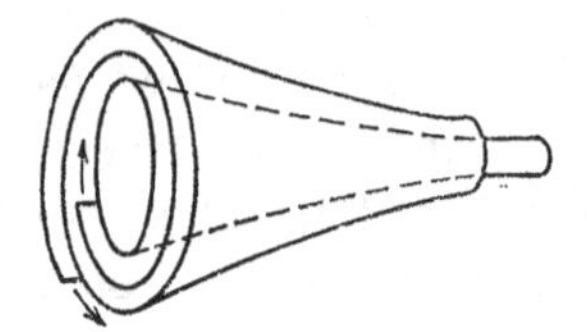

Figure 128.

The generalization of this theorem is due to N.V. Efimov who shows that a two-dimensional plane endowed with an arbitrary complete smooth Riemannian metric with a curvature restricted from above to a negative number also admits no global isometric embedding into a three-dimensional space.

The Lobachevskian plane is closely connected with two-dimensional closed orientable surfaces which, as we know, are homeomorphic to a sphere with g handles (where g is the genus of the surface). The point is that all such surfaces can be represented as a quotient space of a Lobachevskian plane with respect to a certain discrete isometry group.

DEFINITION 1. Let Γ be a certain discrete group of a Lobachevskian plane. A subset D of the Lobachevskian plane is called the *fundamental region for the group* Γ provided there following conditions hold: 1) D is a closed set; 2) the union of sets of the form $\gamma(D)$, where $\gamma \in \Gamma$ coincides with the whole of the Lobachevskian plane; 3) this covering of the Lobachevskian plane by the sets $\gamma(D)$ is such that with a sufficiently small neigbourhood of an arbitrary point there intersect only a finite number of sets of the form $\gamma(D)$; 4) the image of the set of interior points of D does not intersect the set of interior points of D under any other than the identity transformation from the group Γ.

It can be easily proved that as a fundamental region on a Lobachevskian plane for an arbitrary discrete isometry group we can choose a convex polygon with a finitie number of sides.

We shall now give an example of a discrete isometry group (the group of motions) of a Lobachevskian plane whose fundamental region is a $4g$-gon (with angles equal to $\pi/2g$) with centre at the centre of the unit circle (in the Poincaré model) (see Figure 129).

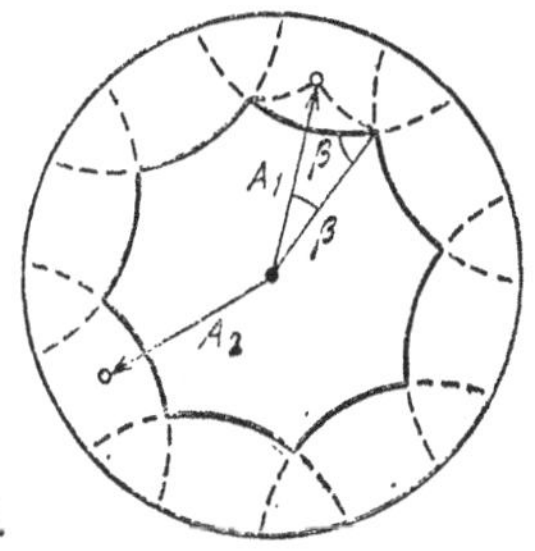

Figure 129.

We shall divide the sides of our $4g$-gon in pairs and shall consider pairs of opposite sides. Suppose $A_1, \dots, A_{2g}$ are "translations" of the Lobachevskian plane under which pairs of opposite sides exchange places (Figure 129). Each subsequent transformation A_{i+1} is obtained from the previous one A_i by a rotation of the "translation" direction by the angle $\pi - \pi/2g$, i.e. by conjugation using the matrix B_g of rotation by the angle $\pi - \pi/2g$. It can be readily verified that the transformations $A_1, \dots, A_{2g}$ are linked by the relation $A_1 \cdot \dots \cdot A_{2g} \cdot A_1^{-1} \cdot \dots \cdot A_{2g}^{-1} = 1$. Proceeding from this we can derive explicit formulae for matrices of transformations $A_1, \dots, A_{2g}$ in realization of a Lobachevskian plane on the upper half plane. In doing so, we shall write the transformations $A_1, \dots, A_{2g}$ by means of some matrices from the group $SL(2, \mathbb{R})$.

We may assume that the motion A_1 (in realization on the upper half plane) sends an imaginary semi-axis into itself. Then it has the form $w \to \lambda w$, $\lambda = e^a$, where a is s doubled leg of a triangle with angles $\pi/2$, $\pi/4g$, $\pi/4g$ (Figure 129). The above-said leg can be easily calculated. For the quantitiy a we obtain

$$a = 2 \ln \frac{\cos \beta + \cos 2\beta}{\sin \beta} \quad \beta = \frac{\pi}{4g}.$$

The matrices $A_2, \dots, A_{2g}$ are obtained from the first matrix A_1 through its conjugations using the matrix B_g, i.e. $A_k = B_g^{-k+1} A_1 B_g^{k-1}$, where B_g is the matrix of rotation through the angle $\pi \frac{2g-1}{2g}$ around the point i (on the upper half plane). This means that

$$B_g = \begin{pmatrix} \cos\pi\frac{2g-1}{4g} & \sin\pi\frac{2g-1}{4g} \\ -\sin\pi\frac{2g-1}{4g} & \cos\pi\frac{2g-1}{4g} \end{pmatrix}.$$

And finally we are led to

$$A_k = \begin{pmatrix} \cos\alpha & \sin\alpha \\ -\sin\alpha & \cos\alpha \end{pmatrix}^{-k+1} \begin{bmatrix} \frac{\cos\beta + (\cos 2\beta)^{1/2}}{\sin\beta} & 0 \\ 0 & \frac{\sin\beta}{\cos\beta + (\cos 2\beta)^{1/2}} \end{bmatrix} \times$$

$$\times \begin{pmatrix} \cos\alpha & \sin\alpha \\ -\sin\alpha & \cos\alpha \end{pmatrix}^{k-1}$$

$$\alpha = \pi\frac{2g-1}{2g}, \quad \beta = \frac{\pi}{4g}, \quad k = 1, 2, \dots, 2g.$$

CLAIM 3. *A group with generatrices $A_1, \dots, A_{2g}$ and with the relation $A_1 A_2 \dots A_{2g} A_1^{-1} \dots A_{2g}^{-1} = 1$ is isomorphic to a group with generatrices $a_1, b_1, \dots, a_g, b_g$ and with the relation $a_1b_1a_1^{-1}b_1^{-1} \dots a_gb_ga_g^{-1}b_g^{-1} = 1$. In particular, this group is isomorphic to the fundamental group of a sphere with g handles (i.e. a two-dimensional closed connected orientable surface of genus g). This surface is obtained from the fundamental region D if on the boundary of this region we identify points corresponding to one another under isometries $A_1, A_2, \dots, A_{2g}$. This implies that on any oriented surface of genus g (where $g > 1$) we can determine a Riemannian metric of constant negative curvatures. This metric is induced on the surface by the metric of a Lobachevskian plane under the factorization described above. A Lobachevskian plane covers a two-dimensional surface of genus g in an infinite-sheeted and locally isometric manner.*

APPENDIX 7

SELECTED EXERCISES ON THE MATERIAL OF THE COURSE (*)

Section 1.1

1. A point M is moving uniformly along a straight line ON which rotates uniformly about a point O. Construct the equation of the trajectory of the point M (the spiral of Archimedes).

2. Verify that the length of a smooth curve can be calculated as the limit of the lengths of broken lines which consist of segments joining successively a finite number of points on the curve, the maximum length of the segments tending to zero.

3. Prove that in a Euclidean space a straight line segment has the minimal length among the lengths of curves joining its two end-points.

4. What is the angle of intersection of curve lines given by the following equations in Cartesian coordinates on a plane

a) $x^2 + y^2 = 8x, \quad y^2 = x^3/(2-x);$
b) $x^2 + y^2 = 8, \quad y^2 = 2x;$
c) $x^2 = 4y, \quad y = 8/(x^2 + 4).$

5. Prove that the length of the segment of a tangent to the astroid $x^{2/3} + y^{2/3} = a^{2/3}$, bounded by the axes of Cartesian coordinates is constant and equal to a.

6. Prove that the segment of a tangent to the tractrix

$$y = \frac{a}{2} \ln \frac{a + (a^2 - x^2)^{1/2}}{a - (a^2 - x^2)^{1/2}} - (a^2 - x^2)^{1/2},$$

bounded by the y-axis and the point of tangency has a constant length equal to a. The tractrix is used in constructing a Beltrami surface which models in a three-dimensional space part of a Lobachevskian plane (to do so, we should rotate the tractrix around its asymptote).

(*) The most difficult exercises are marked with an asterisk.

Section 1.2

1. Prove that the family of functions $u = x + \sin y$; $v = y - 1/2 \sin x$ on a Euclidean plane is a regular coordinate system.

2. Write the Laplacian operator $\Delta u = \frac{\partial^2 u}{\partial x^2} + \frac{\partial^2 u}{\partial y^2}$ in a polar coordinate system on a plane.

3. Let $Q(x) = b_{ij} x^i x^j$, where $b_{ij} b_{ji}$ is a quadratic form, $B(x, y) = b_{ij} x^i y^j$ is the corresponding bilinear form. Prove that the linear transformation A in $\mathbb{R}^n$ preserves the bilinear form, $B(x, y) = B(Ax, Ay)$ if and only if it preserves the quadratic form, $Q(x) = Q(Ax)$ (the vectors x and y are arbitrary).

Section 1.3

1. Consider a stereographic projection of a sphere of radius R in a space $\mathbb{R}^3$ onto a plane passing through the centre of the sphere. The projection is defined as follows. We join a variable point P on the sphere with its north pole and continue the segment till it meets the equatorial plane. Then we associate the intersection point P' with the initial point P. Let the sphere be coordinatized by spherical coordinates θ, ϕ and the plane by polar coordinates r, ϕ. Find the dependence between (θ, ϕ) and (r, ϕ) under a stereographic projection.

2. Prove that a stereographic projection of a sphere onto a plane is a conformal map, that is, preserves the angles between the intersecting curves.

3. How shall we write the metric of a sphere after the change of coordinates $(\theta, \phi) \to (r, \phi)$ induced by a stereographic projection?

4. Prove that under a stereographic projection of a sphere onto a plane each flat-cross-section of the sphere (i.e. the circle resulting from the intersection of the sphere by the plane) is sent either into a circle or into a straight line (on the plane).

Section 1.4

1. Define the "vector product" in pseudo-Euclidean three-dimensional space of index 1, i.e. in $\mathbb{R}^3_1$, assuming

$$\xi \times \eta = (\xi^1 \eta^2 - \xi^2 \eta^1, \ \xi^0 \eta^2 - \xi^2 \eta^0, \ \xi^1 \eta^0 - \xi^0 \eta^1),$$

where $\xi = (\xi^0, \xi^1, \xi^2)$, $\eta = (\eta^0, \eta^1, \eta^2)$.

a) Verify that for basis vectors e_0, e_1, e_2 (where e_0 is time-like) pairwise vector products have the form $e_0 \times e_1 = -e_2$, $e_0 \times e_2 = e_1$, $e_1 \times e_2 = e_0$.

b) Prove that $\times$ is a bilinear anti-symmetric (i.e. skew-symmetric operation and that for it the Jacobi identity holds

$$\xi_1 \times (\xi_2 \times \xi_3) + \xi_3 \times (\xi_1 \times \xi_2) + \xi_2 \times (\xi_3 \times \xi_1) = 0.$$

2. Prove that in a space $\mathbb{R}_1^n$ an orthogonal complement of the time-like vector is a space-like hyperplane. What may an orthogonal complement of a space-like (light-like) vector be?

3. In a pseudo-Euclidean space $\mathbb{R}_1^3$ consider a pseudo-sphere of real radius, i.e. a one-sheeted hyperboloid (instead of the two-sheeted one which we considered in Section 1.4).

a) Write the formulae of a stereoographic projection of a one-sheeted hyperboloid onto the coordinate plane. Describe the set of points of the plane which compose this image. b) Calculate the pseudo-Riemannian metric induced by an envelope pseudo-Euclidean metric on a one-sheeted hyperboloid.

4. Suppose a Lobachevskian plane is realized as the upper half plane $y > 0$ of a Euclidean plane x, y. We shall call "straight lines" in a Lobachevskian plane the Euclidean semi-circles with centres on the x-axis (i.e. "at the absolute of the Lobachevskian plane) and the Euclidean half lines bearing on the x-axis and orthogonal to it. We shall call a triangle in a Lobachevskian plane a figure formed by three points and by the segments of "straight lines" joining these points. Prove that the sum of the angles of a triangle in a Lobachevskian plane is always less than π (if the triangle is non-degenerate).

5. Calculate the circumference on a Lobachevskian plane (as a function of its radius measured in the Lobachevskian metric). For comparison, calculate the circumference on a two-dimensional sphere.

Section 1.5

1. Find the curvature of an ellipse in its vertices if its semi-axes are equal to a and b.

2. Prove that if the curvature of a curve is identically zero, the curve is a straight line.

3*. Let S be the area between a flat curve and a secant at a distance h from a tangent, the secant being parallel to the tangent. Express the quantity $\lim_{h \to 0} \frac{S^2}{h^3}$ in terms of the curvature of the curve.

4. A straight lin OL rotates about the point O with a constant angular velocity ω. A point M moves along the straight line OL with a velocity proportional to the distance $|OM|$. Construct the trajectory described by the point M (a logarithmic spiral).

5. A ring of radius a is rolling rectilinearly without sliding. Compose the equation of the trajectory of a point M braced with the ring and separated from its centre by a distance d. For $d = a$ we obtain the so-called cycloid, for $d > a$ a lengthened cycloid, for $d < a$ a shortened cycloid.

6. A circle of radius r rolls without sliding along a circle of radius R remaining all the time outside the latter. Compose the equation of the path of a point M of the rolling circle (epicycloid). Do the same for a circle of radius r rolling inside a circle of radius R.

7. Find the curve given by the vector equation $r = r(t)$, where $-\infty < t < \infty$ if $r''(t) = a$ is a constant non-zero vector.

8. A flat curve is given by the equation $r = (\phi(t), t\,\phi(t))$. Under what condition will this equation define a straight line?

9. Find the function $r = r(\phi)$ knowing that in polar coordinates on a plane this equation defines a straight line.

10. Calculate the curvature of the following flat curves:

a) $y = \sin x$ in the vertex (sinusoid),

b) $x = a(1 + m) \cos mt - am \cos (1 + m)t$,
$y = a(1 + m) \sin mt - am \sin (1 + m)t$ (an epicycloid),

c) $y = a \operatorname{ch} x/a$ (a chain line, i.e. a curve formed by a heavy sagging chain fixed at the end-points),

d) $x^2y^2 = (a^2 - y^2)(b + y)^2$ (a conchoid),

e) $r^2 = a^2 \cos 2\phi$ (a lemniscate),

f) $r = a(1 + \cos \phi)$ (a cardioid),

g) $r = a\phi$ (a spiral of Archimedes),

h) $r - (a \cos^3 t, a \sin^3 t)$ (an astroid),

i) $y - -\ln \cos x$,

j) $x = 3t^2$, $y = 3t - t^3$ for $t = 1$.

11. Find the curvature of the following curves given in polar coordinates:

a) $r = a\phi^k$,

b) $r = a^\phi$ at the point $\phi = 0$.

12. Find the curvature of a flat curve given by the equation $F(x, y) = 0$.

13. Suppose a family of curves be given by the differential equation $P(x, y)\,dx + Q(x, y)\,dy = 0$. Find their curvature.

14. Natural equations of a flat curve are equations of the form: 1) $k = k(l)$, 2) $F(k, l) = 0$, 3) $k = k(t)$, where $l = l(t)$ is the arc length (counted from a certain fixed

point) and k is the curvature of the curve. Compose natural equations for the following curves:

a) $x = a\cos^3 t,\ \ y = a\sin^3 t$,

b) $y = x^{3/2}$,

c) $y = x^2$,

d) $y = \ln x$,

e) $y = a \operatorname{ch} x/a$,

f) $y = e^x$.

Section 1.6

1. For a helix $r = (a\cos t, a\sin t, bt)$ find the Frenet frame, curvature and torsion.

2. Find the curvature and torsion of the curves:

a) $r = e^t(\sin t, \cos t, 1)$,

b) $r = a(\operatorname{ch} t, \operatorname{sh} t, 1)$.

3. Find the curvature and torsion of the curves:

a) $r = \left(t^2(3/2)^{1/2},\ 2 - t,\ t^3\right)$,

b) $r = (3t - t^3,\ 3t^2,\ 3t + t)$.

4. Prove that if the torsion $\kappa(l)$ of a curve is identically zero, the curve lies in a plane (i.e. the curve is flat). Find the equation of this plane in space.

5. Describe the class of curves with a constant curvature and torsion: $k(l) = \text{const.}$, $\kappa(l) = \text{const.}$

6. Describe the class of curves with a constant torsion: $\kappa(l) = \text{const.}$

7. Prove that the curve $r = r(t)$ is flat if and only if $(\dot{r}, \ddot{r}, \dddot{r}) = 0$, where $(\ ,\ ,\ )$ denotes the mixed product of the three vectors.

8. Prove that for a smooth closed curve the following equality always holds $\int (r\, dk + \kappa b\, dl) = 0$.

9. Prove that the Frenet formulae can be represented in the form $\dot{v} = [\zeta, v]$, $\dot{n} = [\zeta, n]$, $\dot{b} = [\zeta, b]$. Find the vector ζ (the so-called Darboux vector).

10. Solve the vector equation $r' = [\omega, r]$ where ω is a constant vector in space.

11. Prove that the curvature and torsion are proportional (i.e. $k = c\kappa$, where $k \neq 0$ and c is a constant) if and only if there exists a constant vector u such that $\langle u, v\rangle = \text{const.}$

12. Let normal planes to a curve, spanned by vectors n, b pass through a fixed point x_0. Show that the curve lies on a sphere centred at this point.

13*. Prove that a curve lies on a sphere of radius R if and only if the following relation holds

$$R^2 = \frac{1}{k^2}\left(1 + \frac{(k')^2}{(\kappa k)^2}\right),$$

where k is the curvature of the curve.

14. Prove that $\kappa = \dfrac{(\dot r, \ddot r, \dddot r)}{\langle \ddot r, \ddot r\rangle}$.

15. For a smooth curve $r = r(l)$ consider a curve $n(l)$ (where n is the normal vector to the curve at a given point); l^* is the natural parameter on the curve $n(l)$. Prove that $dl^*/dl = (k^2 + \kappa^2)^{1/2}$.

16. Let

$$A = A(l) = \begin{pmatrix} 0 & k(l) & 0 \\ -k(l) & 0 & \kappa(l) \\ 0 & -\kappa(l) & 0 \end{pmatrix} = (a^i_j(l).$$

Let the vectors $t_j = r_j(l)$ be solutions of the system of equations $dr_j/dl = a^i_j\, r_i$, $j = 1, 2, 3$, where $r_1(0)$, $r_2(0)$, $r_3(0)$ is a given orthonormal frame.

a) Prove that the frame $r_1(l)$, $r_2(l)$, $r_3(l)$ is orthonormal for any l.

b) Let $r(l) = r_0 + \int_0^l r_1(l)\, dl$. Prove that $r_1(l) = v(l)$, $r_2(l) = n(l)$, $r_3(l) = b(l)$, where v, n, b are the tangent, the normal and the binormal to the curve $r(l)$, the curvature and torsion of this curve being equal to $k(l)$ and $\kappa(l)$.

17. Let a curve lie on a sphere and have constant curvature. Prove that this curvature is a circumference.

18*. Let $r = r(l)$ be a time-like curve in a pseudo-Euclidean space $\mathbb{R}^3_1$ and $(\dot r(l))^2 = (\dot r^0)^2 - (\dot r^1)^2 - (r^2)^2 \equiv 1$, $\dot r^0$ being greater than zero. We introduce vectors v, n, b assuming $v = \dot r$, $v^{\cdot} = kn$, $b = n \times v$. Prove the pseudo-Euclidean analogue of the Frenet formulae:

$$\dot v = kn,$$

$$\dot n = kv - \kappa b,$$

$$\dot b = \kappa n.$$

19. In a pseudo-Euclidean space $\mathbb{R}_1^3$ solve the equation $\dot{r} = \omega \times r$, where ω is a constant vector.

20. Prove that the trajectories of motion of a material point in a central field of force are flat curves.

21. A curve lying on a sphere and intersecting all the meridians of the sphere at a given angle is called a loxodrome. Compose the equations of a loxodrome. Find the vectors of the Frenet frame of this curve at its arbitrary point. Calculate the curvature and torsion of this curve.

22. Given a curve $r = (v \cos u, v \sin u, kv)$ with $v = v(u)$, prove that it lies on a cone. Define the function $v(u)$ so that the curve intersects the generatrices of the cone at a constant angle θ.

23. For what b value does the torsion of the helix $r = (a \cos t, a \sin t, bt)$ have the maximum value?

24. Prove that if all normal flat lines contain a vector e, the given line is flat.

Section 1.7

1. A two-dimensional torus in a three-dimensional Euclidean space can be given in the form of a surface of revolution of a circle about a straight line lying in the plane of the circle (and not intersecting it). Write the parametric equations of the torus and the induced metric on the torus.

2. Find the metric induced on an ellipsoid of revolution $\frac{x^2}{a^2} + \frac{y^2 + z^2}{b^2} = 1$ by an envelope Euclidean metric, i.e. find the first quadratic form of the ellipsoid.

3. Find the metric induced on an ellipsoid of revolution $r(u, \phi) = (\rho(u) \cos \phi, \rho(u) \sin \phi, z(u))$. Verify that its meridians (given by the equations $\phi = \text{const.}$) and parallels (given by the equations $u = \text{const.}$) form an orthogonal net on the surface. Find the bi-sectrices of the angles between the meridians and parallels.

4. Recall that the lines intersecting the meridians of a sphere at a given angle α are called loxodromes. Find the length of a loxodrome.

5. Let $F(x, y, z)$ be a smooth homogeneous function, i.e. one satisfying the equation $F(cx, cy, cz) = c^n F(x, y, z)$. Prove that on the conic surface $F(x, y, z) = 0$ the metric is Euclidean outside the origin.

6. Construct the parametric equation for a cylinder for which the curve $\rho = \rho(u)$ is a directrix and the generatrices are parallel to the vector e.

7. Construct the parametric equation of a cone with the vertex at the tail of the radius-vector, for which (the cone) the curve $\rho = \rho(u)$ is a directrix.

8. Construct the parametric equation of a surface formed by tangents to a given curve $\rho = \rho(u)$. Such a surface is called an involute surface.

9. A circle $x = a + b \cos v$, $z = b \sin v$, $0 < b < a$, rotates about the z-axis. Construct the equation of the surface of revolution (this is a torus). Prove that the coordinate lines (i.e. the parallels and meridians) form an orthogonal net on the surface.

10. Construct the equation of a surface formed by rotation of a chain line $y = a \cosh x/a$ about the x-axis. Such a surface is called a catenoid. Find its principal curvatures.

11. Construct the equation of a surface formed by rotations of a tractrix $\rho = \left(a \ln \tan\left(\frac{\pi}{4} + \frac{t}{2}\right) - a \sin t, a \cos t\right)$ around its asymptote. This surface is called a Beltrami surface (pseudo-sphere). In a three-dimensional Euclidean space it models a part of a Lobachevskian plane. Calculate its first quadratic form. Prove that the induced metric coincides with the metric of the Lobachevskian plane.

12. A surface is called ruled if it is given by the parametric equation $r = r(u, v) = \rho(u) + va(u)$, where $\rho = \rho(u)$ is the vector function determining the distribution of straight-line generatrices of the ruled surface. Construct the equation of a ruled surface whose generatrices are parallel to the plane $y = z$ and intersect the parabolas $y^2 = 2px$, $z = 0$ and $z^2 = -2px$, $y = 0$.

13. Calculate the first quadratic form of the following surfaces:

a) $r = (a \cos u \cos v, b \sin u \cos v, c \sin v)$ (an ellipsoid),

b) $r = (v \cos u, v \sin u, ku)$ (a helicoid).

14. Suppose that the first quadratic form of a surface is known to be of the form $dl^2 = du^2 + (u^2 + a^2)\, dv^2$. Calculate the angle at which the curves $u + v = 0$ and $u - v = 0$ intersect.

Section 1.8

1. Show that on a standard two-dimensional sphere, the sum of the angles of a triangle composed of arcs of large circles is greater than π.

2. Express the sum of the angles of a triangle on a two-dimensional sphere in terms of the area of the triangle (the triangle is composed of arcs of large circles).

3. Prove that for any Riemannian metric there exists such a local coordinate system with respect to which the matrix of the Riemannian metric is unit at a given point. Note that it is generally impossible to reduce the metric tensor to the unit form simultaneously at all points of a hole neighbourhood of a point. An obstacle to this may appear to be a non-zero Riemannian curvature tensor.

4. Prove that on a pseudo-sphere (i.e. on a Lobachevskian plane) the sum of the angles of a triangle composed of segments of "straight lines" is less than π. Find the relation between the sum of the angles of the triangle (on a Lobachevskian plane) and its area.

5. A surface is given by the equation $r = (u \sin v, u \cos v, v)$. Find
a) the area of a curvilinear triangle $0 \le u \le \sinh v$, $0 \le v \le v_0$;
b) the lengths of the sides of the triangle;
c) the angles of the triangle.

6. Prove that the first quadratic form of a surface of revolution can be reduced, through the appropriate choice of curvilinear coordinates, to the form $dl^2 = du^2 + G(u)\, dv^2$. Perform this operation for a sphere, a torus, a catenoid and a pseudo-sphere.

7. The system of curvilinear coordinates on a surface is called isothermal if the first quadratic form of the surface relative to these coordinates is expressible as $dl^2 = \lambda(u, v)\,(du^2 + dv^2)$. Find the isothermal coordinates on a pseudo-sphere.

8. A spherical lune is a figure formed by two large semi-circles with common end-points (at the extremes of the diameter) on a sphere. Calculate the area of a spherical lune with an angle α at the vertex.

9*. The Liouville surface is a surface whose first quadratic form is representable as $dl^2 = (f(u) + g(v))\,(du^2 + dv^2)$. Prove that a surface locally isometric to a surface of revolution is a Liouville surface.

Section 1.10

1. Calculate the area of a circle on:
a) a Euclidean plane,
b) a sphere,
c) a Lobachevskian plane.

2*. Let a Lobachevskian plane be realized on the upper half plane of a Euclidean plane. As "straight lines" we should take here either Euclidean half lines orthogonal to the real axis, or semi-circles with centres at the real axis. Let ABC be an arbitrary triangle in a Lobachevskian plane, a, b, c— non-Euclidean lengths of the sides BC, AC, AB and let α, β, γ be the magnitudes of its angles at the vertices A, B, C. Prove the equalities

$$\text{a) } \cosh a = \frac{\cos\alpha + \cos\beta\cos\gamma}{\sin\beta\sin\gamma},$$

b) $\cosh b = \dfrac{\cos\beta + \cos\gamma\cos\alpha}{\sin\gamma\sin\alpha}$,

c) $\cosh c = \dfrac{\cos\gamma + \cos\alpha\cos\beta}{\sin\alpha\sin\beta}$.

3. Prove the analogue of the theorem of sines for the Lobachevskian plane:

$$\frac{\sinh a}{\sin\alpha} = \frac{\sinh b}{\sin\beta} = \frac{\sinh c}{\sin\gamma} = \frac{(Q)^{1/2}}{\sin\alpha\sin\beta\sin\gamma},$$

where $Q = \cos^2\alpha + \cos^2\beta + \cos^2\gamma + 2\cos\alpha\cos\beta\cos\gamma - 1$.

Section 1.11

1. Calculate the second quadratic form of a right helicoid $x = u\cos v$, $y = u\sin v$, $z = av$.

2. Given a surface of revolution

$$r(u, v) = (x(u),\ \rho(u)\cos\phi,\ \rho(u)\sin\phi),\ \rho(u) > 0,$$

a) find the second quadratic form,

b) find the Gaussian curvature K at an arbitrary point of the surface. Find out the dependence of the sign of K on the direction of convexity of the meridian.

c) calculate the curvature K in the particular case $\rho(u) = u$,

$$x(u) = \pm\Big(a\ln\frac{a + (a^2 - u^2)^{1/2}}{u} - (a^2 - u^2)^{1/2}\Big),\ a > 0$$

(a pseudo-sphere).

3. Find a surface all normals to which intersect at one point.

4. Calculate the Gaussian and the mean curvatures on a surface given by the equation $z = f(x) + g(y)$.

5. Prove that if the Gaussian and the mean curvature of a surface embedded in a three-dimensional Euclidean space are identically zero, the surface is plane.

6. Prove that on the surface $z = f(x, y)$ the mean curvature is equal to

$$H = \operatorname{div}\Big(\frac{\operatorname{grad} f}{(1 + |\operatorname{grad} f|^2)^{1/2}}\Big).$$

7. Suppose a surface S is formed by tangent straight lines to a given curve with curvature $k(l)$. Prove that if the curve preserves the curvature $k(l)$, then the surface S preserves the metric (i.e. is changed by an isometric one).

8. Prove that if the metric of a surface is given by $dl^2 = A^2\,du^2 + B^2\,dv^2$, $A = A(u, v)$, $B = B(u, v)$, the Gaussian curvature has the form

$$K = -\frac{1}{AB}\left[\left(\frac{A_v}{b}\right)_v + \left(\frac{B_u}{A}\right)_u\right].$$

9. Prove that the only surfaces of revolution having zero mean curvature are the plane and the catenoid. (Recall that the catenoid is obtained by rotation of a curve $y = a \cosh x/a$.) We mean here full surfaces, i.e. such as those on which geodesics are infinitely continued.

10. Prove that any cylindrical surface is locally isometric to a plane.

11. Prove that any conic surface is locally isometric to a plane.

12. Calculate the second quadratic form of the surface

$$x = (u^2 + a^2)^{1/2} \cos v, \quad y = (u^2 + a^2)^{1/2} \sin v,$$
$$z = a \ln\left(u + (u^2 + a^2)^{1/2}\right).$$

13. Prove that two surfaces of equal constant Gaussian curvature are locally isometric. In particular, any surface of constant positive Gaussian curvature is locally isometric to a sphere. Any surface of constant negative Gaussian curvature is locally isometric to a pseudo-sphere (a Lobachevskian plane).

14. Prove that for the metric $dl^2 = \lambda(u, v)\,(du^2 + dv^2)$ the Gaussian curvature can be represented in the form $K = \frac{-1}{2\lambda}\Delta \ln \lambda$, where $\Delta = \frac{\partial^2}{\partial u^2} + \frac{\partial^2}{\partial v^2}$ is the Laplacian operator (see Exercise 8).

Section 1.12

1. Suppose a surface S is formed by tangent straight lines to a curve. Express the Gaussian and the mean curvatures of the surface S in terms of the curvature and torsion of the curve.

2. The direction determined by a vector a tangent to a surface is called asymptotic if the second quadratic form on it is equal to zero, i.e. $Q(a, a) = 0$. A line on the surface is called asymptotic if, at each point of this surface, the tangent has an asymptotic direction. These lines are defined by the differential equation $L\,du^2 + 2M\,du\,dv + N\,dv^2 = 0$, where $Q = \begin{pmatrix} L & M \\ M & N \end{pmatrix}$.

Find the asymptotic lines on the surface

a) $z = a(x/y + y/x)$,

b) $z = xy^2$.

3. Prove that for asymptotic directions making a right angle to exist at a given point on a surface, it is necessary and sufficient that the mean curvature be equal to zero at this point.

Section 1.13

1. Let the metric on a surface have the form $dl^2 = dx^2 + f(x)\, dy^2$, $0 < +(x) < \infty$. Prove that this metric can be reduced to the conformal form $dl^2 = g(u, v)\,(du^2 + dv^2)$.

2*. Prove that a two-dimensional pseudo-Riemannian metric (of the type (1, 1)) with analytic coefficients can be reduced, using the change of coordinates, to the form $dl^2 = \lambda(t, x)\,(dt^2 - dx^2)$.

3. Prove that the group of matrices SU_2 is homeomorphic to a standard three-dimensional sphere.

4. Prove that the groups of complex matrices $GL(n, \mathbb{C})$ and $SL(n, \mathbb{C})$ are connected sets. Prove that $GL(n, \mathbb{R})$ consists of two connected components.

Section 1.14

1. Prove that the space of positions of a rigid segment on a plane is a smooth manifold.

2. Prove that the set of all straight lines on a plane is a smooth manifold homeomorphic to a Möbius strip.

3. Prove that the group of matrices SO_3 is homeomorphic to a three-dimensional projective space.

4. Give an example of a smooth one-to-one map of two smooth manifolds which is not a diffeomorphism.

5. Show that on a sphere (and on a circle) there exists no atlas consisting of one chart.

6. Construct the embedding of a torus $T^n = S^1 \times \ldots \times S^1$ (n times) in $\mathbb{R}^{n+1}$.

7. Construct the embedding of a manifold $S^2 \times S^2$ in $\mathbb{R}^5$.

8. Prove that a sphere S^n given in $\mathbb{R}^{n+1}$ by the equation $(x^1)^2 + \ldots + (x^{n+1})^2 = 1$ is a smooth manifold. Construct on this sphere an atlas of two charts.

9. Prove that a two-dimensional torus T^2 (realized e.g. as a surface of revolution in $\mathbb{R}^3$) is a smooth manifold. Construct on this torus an atlas of four charts.

10. Prove that the union of two coordinates axes on a plane is not a manifold.

11. It is possible to endow the following sets with the structure of a smooth manifold?

a) A one-dimensional triangle on a plane (i.e. a closed broken line with three links).

b) Two one-dimensional triangles on a plane with the vertex as a single common point.

12. Prove that an n-dimensional real projective space $\mathbb{R}P^n$ is a smooth (and real analytic) manifold. Construct on this manifold an atlas of $n + 1$ charts.

13. Prove that an n-dimensional complex projective space $\mathbb{C}P^n$ is a smooth (and complex analytic) manifold.

14. Prove that the graph of a smooth function $x^{n+1} = f(x^1, \dots, x^n)$ is a smooth manifold and a smooth manifold in $\mathbb{R}^{n+1}$.

15. Prove that the group of matrices SO_2 is homeomorphic to a circle and the group O_2 is homeomorphic to a disconnected union of two circles.

16. Prove that the groups of matrices $GL(n, \mathbb{R})$, $GL(n, \mathbb{C})$ are smooth manifolds. Find their dimensions.

17. Prove that the set of all straight lines passing through a point on the plane is homeomorphic to a circle.

18. Prove that in the composition of two smooth maps, the Jacobian matrix is the product of the Jacobian matrices of the cofactors.

19. Prove that the rank of Jacobian matrix does not depend on the choice of the local coordinate system.

20. Calculate the rank of the Jacobian matrix of the map $f\colon \mathbb{R}^2 \to \mathbb{R}^2$, where $f(x, y) = (x, 1)$.

21. Construct the explicit formulae for a smooth diffeomorphism between a plane and a two-dimensional open disc (on the plane).

22. Prove that any smooth manifold has such an atlas that each chart is homeomorphic to a Euclidean space.

23. Show that the stereographic projection of a sphere onto a tangent plane from the pole opposite to the point of tangency is a diffeomorphism everywhere except at the pole of the projection.

24. Identify S^2 and $\mathbb{C}P^1$ (construct a diffeomorphism).

Section 1.15

1. Find geodesics on the following Riemannian manifolds:

a) a Euclidean plane,

b) a standard sphere S^2 in $\mathbb{R}^3$,

c) a Lobachevskian plane given either as the Poincaré model in an open circle with the metric $\frac{dr^2 + r^2 d\phi^2}{(1-r^2)^2}$ or on the upper half plane with the metric $\frac{dx^2 + dy^2}{y^2}$.

2. Let $f(u) > 0$, $g(v) > 0$ be smooth functions and let a be an arbitrary constant. Prove that the level lines of the functions

$$z(u, v) = \int \frac{du}{(f(u) - a)^{1/2}} \pm \int \frac{dv}{(g(v) + a)^{1/2}}$$

are geodesics of the metric $dl^2 = (f(u) + g(v))\,(du^2 + dv^2)$ given on the plane.

COMMENT. Suppose a plane on which Cartesian coordinates (u, v) are introduced is filled with a transparent substance having a variable refractive index $\lambda(u, v)$. If at a certain point (u_0, v_0) there exists a source of light, then the light from this source propagates in the (u, v)-plane not along straight lines, but along lines which are geodesics in the conformal metric $dl^2 = \lambda(u, v)\,(du^2 + dv^2)$.

3. Prove that the meridians of a surface of revolution are geodesic lines.

4. Prove that the parallel of a surface of revolution will be a geodesic if and only if the tangent to the meridian at its points is parallel to the axis of rotation.

5. Show that the geodesic lines of a surface with the first quadratic form $dl^2 = v(du^2 + dv^2)$ are represented on the (u, v)-plane as parabolas.

Section 2.1

1. Prove that the trace (spur) of the operator $A = (a^i_j)$, i.e. $\mathrm{Sp}\, A = a^i_j$ does not change under coordinate changes, i.e. is a scalar.

2. Prove that if g_{ij} is a tensor of type (0, 2), where $\det (g_{ij}) \neq 0$, then the inverse matrix (g^{ij}), where $g^{ij}\, g_{jk} = \delta^i_k$, determines a tensor of type (2, 0).

Section 2.2

1. Prove that in a Euclidean space $\mathbb{R}^n$ there exist no tensors of rank 3 invariant under rotations (i.e. such that their components remain unchanged under rotations). Prove the same for tensors of any odd rank.

2. Let there be given an arbitrary linear operator acting from the space of tensors of type (k, s) to the space of tensors of type (p, q). What type of tensor is this?

Section 2.3

1. Calculate the components of the metric tensor on a plane in a polar coordinate system; in $\mathbb{R}^3$ in a) a cylindrical and b) a spherical coordinate system.

2. Assuming the gradient of a function f to be a composition of two operations — taking partial derivatives and raising the indices — write the gradient of the function:

a) in polar coordinates,
b) in cylindrical coordinates,
c) in spherical coordinates.

Section 2.4

1. Calculate the operator * on skew-symmetric rank one and rank two tensors in two- and three-dimensional Euclidean spaces, where $g_{ij} = \delta_{ij}$.

2. Calculate the operator * on skew-symmetric tensors of any rank in a four-dimensional space endowed with the Minkowskian metric

$$(g_{ij}) = \begin{pmatrix} +1 & & & 0 \\ & -1 & & \\ & & -1 & \\ 0 & & & -1 \end{pmatrix}.$$

Show that the skew-symmetric rank two tensor in a Minkowski space is determined by a pair of quantities — a vector in $\mathbb{R}^3$ and a skew-symmetric tensor in $\mathbb{R}^3$ with respect to linear changes in the spatial part, which do not affect time.

3. Express the vector product of two vectors in $\mathbb{R}^3$ by means of algebraic operations on tensors and by means of the operator *.

4. Classify symmetric and skew-symmetric tensors of rank two with respect to pseudo-rotations in Minkowski space. Compare the result with the case of Euclidean space. (It is useful to solve this problem in the two-dimensional case.)

5. Classify tensors of rank one, two, three, four with respect to rotations in $\mathbb{R}^2$ and $\mathbb{R}^3$, which preserve a unit square (respectively, cube in $\mathbb{R}^3$). Do the same for orthogonal transformations preserving the square (cube).

Section 2.8

1. Prove that a connection is compatible with the metric if and only if for any vector fields η_1, ξ_1, ξ_2 the following equality holds

$$\partial_\eta \langle \xi_1, \xi_2 \rangle = \langle \nabla_\eta \xi_1 \xi_2 \rangle + \langle x_1, \nabla_\eta \xi_2 \rangle.$$

2. Prove that under an infinitesimal parallel transport of the vector ξ^i by δx^k its components change as follows (up to small quantities of high order):

$$\xi^i \to \xi^i - \xi^j \Gamma^i_{jk} \delta x^k + o(|\delta x|).$$

3. Suppose in a region U we are given a connection; P is a fixed point of this region, $T = T_P$ is a tangent space to U at this point. We shall define the map $E: T \to U$. Let ξ be a vector from T. We let a geodesic $\gamma(t)$ with the initial velocity vector ξ from the point P and set $E(\xi) = \gamma_\xi(1)$.

a) Show that the map E is defined in a certain neighbourhood of the origin in T and that it is a local diffeomorphism there.

b) Show that in coordinates determined by the map E all the Christoffel symbols Γ^k_{ij} vanish at the point P.

4. The equation of motion of a point electric charge in a magnetic field has the form $\ddot{r} = a \dfrac{[r, \dot{r}]}{|r|^3}$, $a = \text{const}$. Prove that the charge trajectory is a geodesic line of a circular cone.

5. Prove using geodesics, that the motion that leaves motionless a point and the frame at this point is identical.

6. For a symmetric connection Γ^i_{jk} compatible with the metric g_{ij}, prove the validity of the identities:

$$\text{a)}\ g^{kl}\Gamma^i_{kl} = -\frac{1}{(|g|)^{1/2}} \cdot \frac{\partial}{\partial x^k}\left((|g|)^{1/2} g^{ik}\right), \quad g = \det(g_{ij}),$$

$$\text{b)}\ \Gamma^i_{kl} = \frac{1}{2g} \cdot \frac{\partial g}{\partial x^k}.$$

7*. Prove that two sufficiently close points on a Riemannian manifold can be joined by a geodesic which is locally unique (i.e. unique in a small neighbourhood containing both points).

8. Let the metric have the form $dl^2 = g_{rr}\, dr^2 + r^2 + r^2\, d\phi^2$. Prove that the line $\phi = \phi_0$ coming form the centre is a geodesic.

9. Suppose M is a surface in a Euclidean space $\mathbb{R}^n$, π is a linear operator projecting $\mathbb{R}^n$ orthogonally onto the tangent space to the surface M; X and Y are vector fields in $\mathbb{R}^n$, tangent to the surface M. Prove that the connection compatible with the metric induced on the surface M has the form

$$\nabla_X Y = \pi \left(X^k \frac{\partial Y}{\partial x^k}\right).$$

Section 2.9

1. Let there be given a piecewise smooth curve $x^i(t)$, $i = 1, 2$, restricting a region U. Prove that $\Delta\phi = \int_U \int K(g)^{1/2}\, dx^1 \wedge dx^2$ is the angle of rotation during parallel enclosure along the curve $x^i(t)$, where K is a Gaussian curvature.

2. If this curve (see Exercise 1) consists of three geodesic arcs and if the curvature is constant, the sum of the angles of such a geodesic triangle is equal to $\pi + K\sigma$, where σ is the area of this triangle, (prove it!). Consider the cases of a sphere and a Lobachevskian plane.

3. Let $\xi_1, \ldots, \xi_n$ be vector fields in a Riemannian (or pseudo-Riemannian) n-dimensional space: $g_{ij} = \langle \xi_i, \xi_j \rangle$, $[\xi_i, \xi_j] = c^k{}_{ij}\, \xi_k$. Calculate the components of the symmetric connection Γ^k_{ij} (where $\nabla_{\xi_j} \xi_i = \Gamma^k_{ij}\, \xi_k$) compatible with this metric.

4*. Make parallel (counter-clockwise) transport of the vector $\xi = (\xi^k)$ along the contour of a square with side ε spanned by coordinate axes x^i, x^j. Let $\tilde{\xi}(\varepsilon)$ be the result of this enveloping transport. Prove that

$$\lim_{\varepsilon \to 0} \frac{\tilde{\xi}^k(\varepsilon) - \xi^k}{\varepsilon^2} = -R^k_{ij}\, \xi^l$$

5. Prove the validity of the Bianchi identity for the curvature tensor of a symmetric connection compatible with the metric

$$\nabla_m R^n_{jkl} + \nabla_l R^n_{imk} + \nabla_k R^n_{ilm} = 0.$$

6. Derive from the previous formula the following identity for the divergence of the Ricci tensor: $\nabla_l R^l_m = 1/2 \dfrac{\partial R}{\partial x^m}$.

Section 2.10

1*. Let $X_1, \dots, X_n$ be orthonormal vector fields in an n-dimensional Riemannian space and let $\omega_1, \dots, \omega_n$ be the dual basis of the 1-forms $\omega_i(X_j) = \delta_{ij}$ (all indices may be regarded as lower). Define the 1-forms ω_{ij} and the 2-forms Ω_{ij} setting $\omega_{ij} = \Gamma^i_{jk}\,\omega_k$; $\Omega_{ij} = 1/2\, R_{ijkl}\,\omega_k \wedge \omega_l$. Here $\nabla_{X_k} X_j = \Gamma^i_{jk} X_i$, $\langle R(X_k, X_l)\, X_j, X_i \rangle = R_{ijkl}$; summation is over twice repeated indices.

a) Prove that $\omega_{ij} = -\,\omega_{ji}$.

b) Derive the following relations (i.e. structural Cartesian equations):

$$
\begin{aligned}
d\omega_i &= -\,\omega_j \wedge \omega_{ij},\\
d\omega_{ij} &= -\,\omega_{il} \wedge \omega_{lj} - \Omega_{ij},\\
d\Omega_{ij} &= -\,\Omega_{jl} \wedge \omega_{lj} + \omega_{il}.
\end{aligned}
$$

Section 3.1

1. Let $\omega^j = a^j_i\, dx^i$. Derive the formula $\omega^{i_1} \wedge \dots \wedge \omega^{i_k} = J^{i_1 \dots i_k}_{j_1 \dots j_k}\, dx^{j_1} \wedge \dots \wedge dx^{j_k}$, where $J^{i_1 \dots i_k}_{j_1 \dots j_k}$ is the minor of the matrix (a^j_i) positioned at the intersection of rows with numbers $i_1, \dots, i_k$ and columns with numbers $j_1, \dots, j_k$. In particular, $\omega^1 \wedge \dots \wedge \omega^n = \det(a^j_i)\, dx^1 \wedge \dots \wedge dx^n$.

2. Find the dimension of the space of k-forms (at a given point).

3. Let $X_1, \dots, X_n$ be linearly independent vector fields in an n-dimensional region, and let $[X_i, X_j] = 0$. Prove that there exists (locally) a system of coordinates $(x^1, \dots, x^n)$ such that the field X_i is tangent $\partial X_i(x^k) = \delta^k_i$ to the i-th coordinate axis.

Section 3.2

1. Let $f\colon S^n \to \mathbb{R}P^n$ be a map associating a point $x \in S^n$ with the straight line through the point x and through the origin in $\mathbb{R}^{n+1}$. Prove that all the values of the map f are regular.

2. Let $f\colon SO_n \to S^{n-1}$ assign to each orthogonal matrix its first column. Prove that all the values of the map f are regular. Find the pre-image $f^{-1}(y)$.

3. Let $f\colon U_n \to S^{2n-1}$ assign to each unitary matrix its first column. Prove that all the values of the map f are regular. Find the pre-image $f^{-1}(y)$.

4. Prove orientability of the following manifolds: a) a torus T^2, b) a sphere S^n, c) projective spaces $\mathbb{R}P^{2n+1}$ and $\mathbb{C}P^n$. Prove non-orientability of the following manifolds: a) $\mathbb{R}P^{2n}$, b) a Möbius strip.

5. Cut a Möbius strip along its middle line (i.e. along the circle). Is the manifold obtained orientable? Repeat this process several times.

6. Prove that a Euclidean space contracts continuously along itself into a point. Prove that on a sphere S^n, $n > 1$, any two paths with coincident end-points are homotopic (the end-points are the same, and the homotopy is motionless at the end-points).

Section 3.6

1. Prove that by glueing together a Möbius strip and a disc along identical diffeomorphism of their boundaries, we obtain a projective plane.

2. Prove that by glueing a Möbius strip into a torus, we obtain the so-called "Klein bottle", i.e. a two-dimensional non-orientable manifold which is also obtainable by glueing together two Möbius strips (by way of identification of their boundary circles).

Appendix 1

1. Prove that any motion of a Euclidean space is given by an affine transformation of the form $y = Ax + b$, where A is an orthogonal matrix (i.e. by a composition of rotation and translation by a constant vector b).

2. Prove that the matrix group of affine transformations in an n-dimensional Euclidean space is isomorphic to the group of matrices of the order $n + 1$ of the form $\begin{pmatrix} A & b \\ 0 & 1 \end{pmatrix}$, where A is a non-degenerate $n \times n$ matrix and b is an arbitrary n-dimensional vector-column.

Appendix 3

1. Find all symmetry groups of all regular polygons on the plane.

2*. Find all symmetry groups (groups of motions) of all regular convex polyhedrons in three-dimensional space. Among these groups, point out those that are non-commutative.

ADDITIONAL MATERIAL

1. Addition to Appendix 2

The Einstein equation for a gravitational field in the absence of matter and for all other physical fields has the form $R_{ij} = 0$, where R_{ij} is the Ricci tensor. In the presence of matter the Einstein equation changes. It has the universal form $R_{ij} - 1/2\, Rg_{ij} = \lambda T_{ij}$, where $\lambda = 8\pi Gc^{-4}$, G is the gravitational Newton's constant, $G = 6.67 \cdot 10^{-11}$ $\mathrm{N \cdot m^2/kg^2}$, c is the speed of light in a vacuum, $c = 2.9979 \cdot 10^8$ m/s. The tensor T_{ij} is called the energy-momentum tensor. If the "matter" is either a fluid or an electro-magnetic field, then the following formula holds (see [1], [29])

$$T_{ij} = \frac{1}{4\pi c}\left(-F_{il}F_j^l + \frac{1}{4} g_{ij} F_{kl} F^{kl}\right) \quad \text{(the field)},$$

$$T_{ij} = (p + \varepsilon)\, u_i\, u_j - p g_{ij} \quad \text{(the fluid)}.$$

The usual matter ("fluid") is characterized by the 4-velocity u, pressure p and energy density ε, where $\varepsilon = \varepsilon(p)$. The Maxwell equations for the electro-magnetic field remain the same in their geometrical meaning as in the Minkowski space where the gravitational field is trivial (see Section 2.11). If $F = F_{ab}\, dx^a \wedge dx^b$ is the differential 2-form corresponding to the field strength tensor, then we have (the metric and the covariant derivatives are determined by the gravitational field) $dF = 0$ (the first pair of Maxwell equations), $F_{a;b}^{b} = \frac{4\pi}{c} j_a$ (the second pair of equations; here j_a is the 4-vector of current) or $*d*F = \frac{4\pi}{c} j$, where $j = j_a\, dx^a$. So we can say that a gravitational field is described by a pseudo-Riemannian metric on M^4, and an electro-magnetic field is described by the simplest one-component gauge field (see the end of Secion 2.7), where $N = 1$, and the field curvature tensor R coincides with the field strength F.

Taken together, they form the geometry of a five-dimensional space, as indicated in Example 3 at the end of Section 2.10. It is remarkable (and it was an important discovery in the sixties and seventies) that nuclear forces and strong interactions effective at distances of the order of 10^{-13} cm and smaller are also described using geometric objects, namely, multi-component non-Abelian gauge fields (connections) with groups $SU_3, SU_4, \ldots$ and some other fields affected by these connections.

The unified theory of strong and weak interactions uses the same methods; if not quantized, individual geometric fields do not have direct physical meaning here: their set yields a base underlying the difficult and not at all purely geometric quantization procedure in which there are more questions than answers. So there appeared internal degrees of freedom corresponding to so-called quarks which are essential only at small distances. Now we observe processes already at distances of the order of 10^{-17} to 10^{-18} cm, i.e. deep in the nucleus. It should be noted that in all these processes the gravitational field is not subjected to any quantization procedures (and generally does not play a role in them). The reason is very simple — the gravitational forces are much weaker than all other forces on these scales. They are essential as classical, non-quantum forces at large distances: the other forces (including electro-magnetic, due to the attraction of opposite charges only) are effectively more short-range than gravitational ones. Gravity becomes essentially "quantum" on a very small "Planck" scale — the characteristic unit length which can be composed of the product of powers of three fundamental constants: The Newton constant G (the symbol of gravity), the speed of light c (the symbol of relativity) and the Planck constant $h = 6.6262 \cdot 10^{-34}$ J/Hz (the symbol of quantum theory), $l_{pl} = G^{1/2}\, h^{1/2}\, c^{-3/2} \approx 10^{-33}$cm.

It should be noted here that the Planck time $t_{pl} \approx 10^{-43}$ s shows (by order of magnitude) during what time the global evolution of the Universe was determined by purely quantum laws. The results of direct observations of experimental physics now differ from Planck's scales by a great many orders of magnitue. The search for indirect observational consequences for quantum-gravitational phenomena has not yet yielded any definite results. There exists, in modern literature, a considerable number of (sometimes mathematically very elegant) papers showing attempts to formulate a theory interpolating nuclear, weak and quantum-gravitational processes. Such papers are based entirely on mathematical intuition and should therefore be regarded as purely mathematical. Who knows what physical phenomena will come out "on the way" from nuclear to Planck's scales?

Some people ("conservatives") think that the existing theories (string theory and others) are too daring. There exists a serious objection to such type of theories because the latter suggest that nothing will happen "on the way" from nuclear to Planck's scales. In any case, there is no need for hurry until some observational data appears.

On the contrary, quite recently, there has appeared an idea that as the scales decrease, the number of degrees of freedom necessary for a convenient systematization of physical "events" (i.e. dimension) may change, and even the concepts of locally Euclidean topology of space is not necessary. Although the idea of discreteness of space in the naive sense as a lattice already does not satisfy the intellect of theoreticians, more complicated spatial and analytic models based, for instance, on discrete normalizable fields of p-adic type may, perhaps, appear to be of use. But at the present time there are no serious models of this type. These ideas have appeared within recent decades for purely mathematical reasons, namely, due to the evolution of topological methods of algebraic geometry towards arithmetical discrete structures. Very many people, however, share the idea that the ensemble of mathematical concepts, notions and methods used in contemporary physics will surely be insufficient for the physics of the XXIst century. According to modern considerations, the structure of the Universe on a large scale is determined only by gravity— the other factors are now unknown. Suppose that on scales of galactic clusters ($\sim 10^{22}$ cm) the Universe is approximately spatially homogeneous.

Mathematically, this is expressed by the fact that in the approximation the space-time manifold M^4 ("the cosmological model") admits the group of motions G with three-dimensional space-like orbits ("spatial cross-sections" $t = \text{const.}$). If, besides, the angular distribution of matter in the Universe is approximately isotropic at all its points, then the number of parameters of the group G should be equal to 6. The metric should satisfy the Einstein equation with the energy-momentum tensor of the usual matter (fluid), where either $p = 0$ ("dust") or $p = \varepsilon/3$ (relativistic fluid, radiation). The solution of the equations (see [1], [29]) shows that the Universe is non-stationary (A.A. Friedman, the early 20s).

Astronomical observations of the 30s led to the conclusion that the Universe is actually expanding, the galaxies are receding , and the farther the faster. Comparing the observational data with the solutions of the Einstein equations, we come to the conclusion that it took the processes proceeding to the Universe up to now not more than 10 to 20 billion years. This is a remarkable conclusion of the 30s; later it became clear that no other observations (e.g. the age of objects of the solar system) contradict this one. The consideration of anisotropic cosmological models in the framework if GTR also left these conclusions unaffected and revealed the possibility of interesting phenomena at early stages of the evolution.

The discovery of background radiation in the 70s confirmed the idea that the Universe has been monotonically expanding for a very long time, and has changed scales by many orders of magnitude. So, the opinion that the Universe expanded infinitely in all the four directions is the only obvious possibility compatible with physical laws, has been shared for no more than about three hundred years.

2. Addition to Appendix 3. On quasi-crystals

Quite recently, a new type of crystals — "quasi-crystals" — has been discovered experimentally by physicists. The atoms of these quasi-crystals are positioned in $\mathbb{R}^2$ or $\mathbb{R}^3$ in a translation-invariant manner. Their lattice $R \subset \mathbb{R}^n$ ($n = 2, 3$) is such that there exists no finite number of atoms of such that the rest could be obtained by integer combinations of n basis translations. Such lattices are not already determined by crystallographic groups — discrete sub-groups of the groups of motions of $\mathbb{R}^n$. These lattices are described in one of the following two ways.

WAY 1. We are given a multi-dimensional crystallographic lattice $\bar{R}$ in $\mathbb{R}^N$, where $N > n$ and the physical space is positioned as a sub-space $\mathbb{R}^n \subset \mathbb{R}^N$ perhaps in an irrational manner. We set the radius $d > 0$ and associate with the atoms all the points of the lattice $\bar{R}$ separated from the sub-space $\mathbb{R}^n \subset \mathbb{R}^N$ by a distance smaller than d. Their position in $\mathbb{R}^3$ is determined by the orthogonal projection.

DEFINITION 1. A *quasi-lattice* R is a set of points in $\mathbb{R}^n$ obtained by an orthogonal projection onto $\mathbb{R}^n$ from points of the lattice $\bar{R}$ lying close to $\mathbb{R}^n$ (i.e. at a distance smaller than d).

The choice of the number N, of the multi-dimensional lattice, of the sub-space $\mathbb{R}^n$ and of the number d should be discussed in each particular case. In the most interesting cases there exist finite symmetry groups of the lattice in $\mathbb{R}^N$ which leave invariant the sub-space $\mathbb{R}_n$ (for example, fifth-order symmetry which is not realized in ordinary crystallographic groups in $\mathbb{R}^2$ and $\mathbb{R}^3$ is realized for $n = 3, N = 6$).

WAY 2. We determine in $\mathbb{R}^2$ or in $\mathbb{R}^3$ a finite number of convex polyherons $K_1, \dots, K_m$.

DEFINITION 2. The *Penrose lattice* is a partition of $\mathbb{R}^n$ ($n = 2, 3$) into polyhedrons congruent to $K_1, \dots, K_m$, where two polyhedrons either have a common side, or a common vertex, or do not intersect. The lattice itself is a set of veritces.

Not any Penrose lattice is a quasi-crystal. On a polyhedron K_j we determine a function $f_j(x)$ constant on the boundary. The partition of $\mathbb{R}^n$ into polyhedrons congruent to $K_1, \dots, K_m$ naturally give rise to a unique function $F(x)$ on $\mathbb{R}^n$, equal to the translation of $f_j(x)$ on each polyhedron.

DEFINITION 3. A Penrose lattice is said to be a *quasi-crystal* if for any $f_j(x)$ the function $F(x)$ is quasi-periodic, i.e. is expansible into a Fourier series with some finite set of basis frequencies $(\omega_1, \dots, \omega_N)$, l_j being integers:

$$F(x) = \sum A_{l_1, \dots l_N} \exp \left[2\pi i \sum_{j=1}^{N} l_j \omega_j \right].$$

The mathematical theory of quasi-crystals is still at the initial stage of its development. It is very likely that very interesting geometrical problems exist here.

3. Addition to Appendix 5. Multi-valued functionals and the Dirac monopole

An interesting topological situation is observed for the motion of a charged particle in a topologically non-trivial magnetic field. Recall that the magnetic field is defined as a skew-symmetric 2-tensor (or a differential 2-form) in a three-dimensional space $\mathbb{R}^3$ or in its region $V \subset \mathbb{R}^3$. We denote this form $H = H_{ab}\, dx^a \wedge dx^b$ $(a, b = 1, 2, 3)$. This is the spatial part of the electro-magnetic field tensor $F_{\alpha\beta}$ $(a, \beta = 0, 1, 2, 3)$, $F_{ab} = H_{ab}$. The form H is always closed: $dH = 0$. This allows us to introduce locally the vector potential $A = A_a\, dx^a$, where $H = dA$, and to define the Lagrangian and the action (see the corresponding example from Appendix 5):

$$L = g_{ab}\dot{x}^a\dot{x}^b + eA_a\dot{x}^a, \quad S(\gamma) = \int L\, dt.$$

Here e is the particle charge, the metric g_{ab} in the Euclidean case has the form $2g_{ab} = m\delta_{ab}$, m is the particle mass.

What is to be done if the field H is topologically non-trivial, i.e. if there exists a two-dimensional cycle (a closed surface) $Q \subset V$ such that the flux is not equal to zero: $\iint_Q H \neq 0$?

EXAMPLE. Suppose $V = \mathbb{R}^3 \backslash 0$, the cycle Q coincides with the sphere S^2 and $\Sigma(x^a)^2 = 1$. For a spherically symmetric "monopole" we have $H^a = \text{const} \cdot x^a/r^2$, $H^1 = H_{23}, H^2, = -H_{13}, H^3 = H_{12}$. Since the vector-potential does not exist in the entire region V, the action functional is not defined as functional on all (e.g. closed) trajectories γ. We shall denote by F the set of all smooth closed parametrized curves in the region (manifold) V. Here V may be not only a region in $\mathbb{R}^3$, it may be any n-dimensional manifold on which a closed 2-form H_i is defined: $dH = 0$. We shall cover V with a set of regions V_x, i.e. $V = \cup_x V_x$ with the properties that 1) for any smooth curve γ there exists a "number" x such that γ lies wholly in the region V_x; 2) on each region V_x there globally exists a vector-potential $H = dA_x$.

We shall denote the set of all curves $\gamma \subset V_x$ by F_x. Obviously, we have $F = \cup_x F_x$. For example, for $V = \mathbb{R}^3 \backslash 0$ the index x can be associated with the ray l_x

from the origin to infinity. The region V_x has the form $V_x = \mathbb{R}^3 \setminus l_x$. The region V_x is contractible. Any smooth curve γ in $\mathbb{R}^3\setminus 0$ does not meet a single ray l_x (even a continuum). The vector-potentials A_x are defined. For curves $\gamma \subset V_x$ we have the "action"

$$S_x(\gamma) = \oint_\gamma (g_{ab}\dot{x}^a\dot{x}^b + eA_a^{(x)}\dot{x}^a)\,dt,$$

defined in the region F_x of the functional space F.

LEMMA. *For curves $\gamma \in F_{x_1} \cap F_{x_2}$ (i.e. the curves γ lie in the intersection of the regions $\gamma \subset V_{x_1} \cap V_{x_2}$) the difference of the actions is locally constant.*

Proof. Let the closed curve $\gamma(t)$ depend on the parameter τ, i.e. $\gamma_\tau = \gamma(\tau, t)$. For $\gamma_\tau \subset V_{x_1} \cap V_{x_2}$ we have

$$[S_{x_1}(\gamma_\tau) - S_{x_2}(\gamma_\tau)] = e\oint_{\gamma_t} (A_a^{(x_1)} - A_a^{(x_2)})\,\dot{x}^a\,dt.$$

Since $dA^{(x_1)} = dA^{(x_2)}$, the difference does not depend on τ, which implies the lemma.

Thus, on all regions F_x of the topological space F the functionals S_x are given whose difference is locally constant. In this case, the set (S_x) determines the one-dimensional class of cohomologies $[S] \in H^1(F; \mathbb{R})$ and we speak of "multi-valued functionals". In other words, the variation δS is a closed but, possibly, not exact 1-form on the infinite-dimensional manifold F. The requirement of "quantization" suggests that a single-valued functional — the Feynman "amplitude" $\exp[2\pi i S(\gamma)]$ be defined on F. This implies that the class $[S]$ should be integer: $[S] \in H^1(F; \mathbb{Z}) \subset H^1(F; \mathbb{R})$. In other words, the "contour integrals" of the 1-form δS on F over the contours in F should be integer.

In the above example of Dirac monopole the contour in F is a surface in $\mathbb{R}^3$. We arrive at the condition of "quantization of the magneic field flux" through a sphere $S^2 \subset \mathbb{R}^3$. The topological analysis of this type of situation and extension to multi-dimensional problems of field theory appeared only in the early 80s (S.P. Novikov).

Section 4. Addition to Section 1.12. Minimum surfaces and boundaries between physical media

In Part I of the book we acquainted ourselves with an important concept of mean curvature $H = \lambda_1 + \lambda_2$, where λ_1, λ_2 are the principal curvatures of a two-dimensional surface M^2 in $\mathbb{R}^3$. The mean curvature occurs naturally in many physical problems. As an example we shall present the Poisson theorem. Suppose a smooth surface M^2 in $\mathbb{R}^3$ is a boundary (interface) between two media (e.g. two liquids, two gases or between a liquid and a gas, etc.) which are in equilibrium. Let p_1 and p_2 be pressures in the media. *Then the mean curvature H of the surface M^2 is constant (does not depend on the point) and equal to $h(p_1 - p_2)$, where the constant $\lambda = 1/h$ is called the surface tension coefficient and $p_1 - p_2$ is the pressure difference.*

We shall apply this result, for example, to the well-known physical object — soap bubbles. They occur on wire contours when the latter are taken out of a soap solution. We shall discuss two cases:

a) a closed soap film — a bubble, i.e. a film without boundary;

b) a film bounded by a wire contour.

In Case a) the film separates two media with distinct pressures (inside and outside the bubble). As a model we can take a soap bubble blown out of a tube. Consequently, here $H = h(p_1 - p_2) = \text{const} > 0$. In Case b) the film separates two media with equal pressures, therefore, $H = \text{const} \equiv 0$. Here the gas on both sides of the film is, in fact, one and the same medium. If we neglect gravity, then in Case a) the condition of constancy of mean curvature implies the statement that a soap bubble homeomorphic to a sphere is a standard sphere (of a constant radius). This is a non-trivial theorem. Freely falling soap bubbles acquire, therefore, the shape of a sphere. We shall concentrate our attention on Case b) of zero mean curvature. It turns out that surfaces of zero mean curvature are locally minimum in the following sense. Let us consider all possible small enough perturbations of the surface M^2. We shall call a perturbation small if it is is small in amplitude and concentrated inside a small region (i.e. outside a certain small ball the surface remains unchanged). We shall call a surface locally minimum if no small perturbation of this surface decreases its area.

THEOREM. *A surface M^2 in $\mathbb{R}^3$ is locally minimum if and only if its mean curvature is identically zero.*

Soap bubbles pulled on contours do not decrease their area under small perturbations. We shall write analytically the local minimum condition.

From the definition of H it follows that

$$H = \frac{GL - 2MF + EN}{EG - F^2},$$

where $E\,du^2 + 2\,F du\,dv + G\,dv^2$ is the first quadratic form and $L\,du^2 + 2M\,du\,d\,v +$ $+ N\,dv^2$ is the second quadratic form of the surface. It is mentioned in Section 1.13 that on a smooth surface we can always choose (locally) conformal coordinates u, v such that with respect to which $E = G$, $F = 0$. Consequently, in these coordinates the condition $H = 0$ is equivalent to the identity $L + N = 0$. From this it follows that $\frac{\partial^2 r}{\partial u^2} + \frac{\partial^2 r}{\partial v^2} = 0$, where $r(u, v)$ is the radius-vector of the surface. Thus, in conformal coordinates $\Delta r = 0$, where Δ is the Laplace operator, i.e. the radius-vector of the minimum surface is a harmonic vector-function with respect to our conformal coordinates. Hence, the equality $H = 0$ can be regarded as the differential equation of the minimum surface. We shall write it with respect to coordinates. Following Section 1.12, we shall choose on M^2 local coordinates in a neighbourhood of a regular value, such that x and y change in the tangent plane to the surface and z be directed along the normal to the surface. We shall determine the surface locally in the form of a graph $z = f(x, y)$. Then the equation of the minimum surface will become $(1 + f_x^2)\,f_{yy} - 2f_x f_y f_{xy} + (1 + f_y^2)\,f_{xx} = 0$. We shall now give several examples.

EXAMPLE 1. We consider a surface formed by rotation about the x-axis of a curve given by the equation $y = a \cosh x/a$, where a is a constant. This curve determines the form of a heavy sagging chain fixed at two points. The surface obtained is locally minimum and is called a *catenoid*. In Euclidean coordinates x, y, z, a catenoid can also be given in $\mathbb{R}^3$ by the equation $a^2(x^2 + y^2) = \cosh^2(az)$, $a = \text{const}$.

EXAMPLE 2. A *helicoid* is given by the graph of the function $z = \arctan x/y$. Geometrically, this surface is obtained when a straight line A which intersects orthogonally a vertical straight line B moves uniformly up this straight line B (with a constant velocity) and at the same time uniformly rotates about the B.

EXAMPLE 3. The *Sherk surface* is given by the equation

$$z = \frac{1}{a} \ln \frac{\cos ay}{\cos ax},$$

where $a = \text{const}$, or (for $a = 1$) $e^z = \frac{\cos y}{\cos x}$. Minimum surfaces in a multi-dimensional space can be defined as surfaces which do not decrease in area (volume) under any sufficiently small perturbation with small support.

EXAMPLE 4. Consider $\mathbb{R}^{2m} \approx \mathbb{C}^m$ and let $z = x + iy$. Let $g_1(z), \dots, g_m(z)$ be complex analytic functions. Then the surface given by the radius-vector

$$r(z) = (f_1(z), \dots, f_{2m}(z)),$$

where $f_{2p+1}(z) = \mathrm{Re}\, g_p(z), f_{2p}(z) = \mathrm{Im}\, g_p(z)$, $1 \le p \le m$, is minimum. Thus, a *complex analytic curve in* $\mathbb{C}^m$ *regarded as a two-dimensional surface in* $\mathbb{R}^{2m} \approx \mathbb{C}^m$ *is a minimum surface.*

Let $M^2 \subset \mathbb{R}^3$ and let E, F, G be coefficients of the first quadratic form of the surface. If M^2 is given by the radius-vector $r(u, v) = (x^1(u, v), x^2(u, v), x^3(u, v))$, then we may consider the complex functions $\phi_k(z) = \dfrac{\partial x^k}{\partial u} - i\dfrac{\partial x^k}{\partial v}$, where $z = u + iv$.

We can easily make sure that the following two equalities hold:

$$\sum_{k=1}^{3} \phi_k^2(z) = E - G - 2iF \quad \text{and} \quad \sum_{k=1}^{3} |\phi_k(z)|^2 = E + G.$$

From this it follows that 1) the functions $\phi_k(z)$ are complex analytic if and only if $x^k(u, v)$ are harmonic functions of u and v; 2) the coordinates u and v are conformal on M^2 if and only if $\sum_{k=1}^{3} \phi_k^2(z) \equiv 0$ (the condition (1)); 3) if u and v are conformal coordinates on M^2, then the surface M^2 is regular if and only if $\sum_{k=1}^{3} |\phi_k(z)|^2 \neq 0$ (the condition (2)). Formulae (1) and (2) were pointed out by Weierstrass.

THEOREM. *Suppose the radius-vector $r(u, v)$ defines locally the minimum surface M^2 in $\mathbb{R}^3$, u and v being conformal coordinates. Then the functions $\phi_k(z)$ are complex analytic and satisfy the conditions* (1) *and* (2). *Inversely, let ϕ_1, ϕ_2, ϕ_3 be complex analytic functions satisfying the conditions* (1) *and* (2) *in a simply-connected domain D on a plane $\mathbb{R}^2(u, v)$. Then there exists a regular minimum surface given by the radius-vector $r = (x^1(u, v), x^2(u, v), x^3(u, v))$ defined on the domain D and*

$$\phi_k(z) = \frac{\partial x^k}{\partial u} - i\frac{\partial x^k}{\partial v}.$$

It turns out that equation (1) can be solved explicitly.

THEOREM. *Let D be a simply-connected domain in a complex z-plane and let $g(z)$ be an arbitrary function meromorphic in the domain D. Let the function $f(z)$ be* *analytic in D. Suppose at those points where* $g(z)$ *has a pole of order m the function* $f(z)$ *has zero of order not less than 2m. Then the functions* $\phi_1 = 1/2\, f(1 - g^2)$, $\phi_2 = i/2\, f(1 + g^2)$ *and* $\phi_3 = fg$ (3) *are analytic in D and satisfy equation* (1), *that is,* $\phi_1^2 + \phi_2^2 + \phi_3^2 = 0$. *Inversely, any triple of functions analytic in D and satisfying the equation* $\phi_1^2 + \phi_2^2 + \phi_3^2 = 0$ *is representable in the form* (3), *except in the case where* $\phi_1 \equiv i\,\phi_2$, $\phi_2 \equiv 0$.

This implies the following representation for a minimum surface:

$$
\begin{aligned}
2x^1 &= \operatorname{Re} \int f(1 - g^2)\, dz + c_1, \\
2x^2 &= \operatorname{Re} \int if(1 + g^2)\, dz + c_2, \\
2x^3 &= \operatorname{Re} \int fg\, dz + c_3.
\end{aligned}
$$

As an independent complex variable we can take the function g. Then for minimum simply-connected surfaces we obtain the *classical Weierstrass-Annepert representation*

$$
\begin{aligned}
x^1 &= \operatorname{Re} \int F(g)\,(1 - g^2)\, dg + a_1, \\
x^2 &= \operatorname{Re} \int iF(g)\,(1 + g^2)\, dg + a_2, \\
x^3 &= \operatorname{Re} \int 2F(g)\, g\, dg + a_3,
\end{aligned}
$$

where $F(g) = 1/2\, f\, dz/dg$ is called the Weierstrass function. If, for example, we set $F(g) \equiv 1$, we obtain the known minimum *Annepert surface*. For $F = -1/2\, g^2$ we obtain a *catenoid*, for $F = 1/(1 - 14g^4 + g^8)^{1/2}$ the *Schwarz surface*, and for $F = 1 - 1/g^4$ the non-orientable *Henneberg surface*.

It is clear that when the boundary contour is deformed, the minimum surface (modelled by a soap film) is deformed too. What is the character of this dependence? What is the solution of the equation for minimum surfaces on the boundary conditions? We shall consider for simplicity some two-parameter family of boundary contours (i.e. a two-parameter family of deformations of a given contour). As an example we shall take the following contour. On a standard torus $T^2 = S^1(\phi) \times S^1(\psi)$ we take ordinary angular coordinates ϕ and ψ, where $|\phi| \le \pi$ and $|\psi| \le \pi$. Then on this torus we take a union of two strips (bands) given by the inequalities $|\phi| \le u$ and $|\psi| \le v$. In other words, the region Q_{uv} is obtained by a cross-wise glueing of two flat rings of width $2u$ and $2v$, respectively.

As the contour Γ_{uv} we take the boundary of the region Q_{uv} on the torus. Changing u and v, we change the position of the contour Γ_{uv} in space. The contour Γ_{uv} is homeomorphic to a circle. We fix arbitrary values of the parameters u and v and consider all locally minimum two-dimesional surfaces with boundary Γ_{uv}. From among these surfaces we select only those homeomorphic to a disc. They are, generally speaking, few. By calculating their areas we obtain a set of numbers depending on u and v. For some values u and v these numbers may coincide.

We have obtained a certain multi-valued function which may be naturally called the area function of minimum surfaces. This function is defined on the domain where the parameters u and v vary. This domain may be regarded as a square on a plane. We can therefore, construct the graph of this multi-valued function in $\mathbb{R}^3$. It illustrates the character of the dependence of the areas of minimum surfaces on the boundary contour. Separate branches (leaves) of the multi-valued function may flow together, may have branching points on the graph, etc. In other words, this graph characterizes the topology of the space of solutions of the minimum surface equation (under variation of boundary conditions).

THEOREM. *In the above example the graph of the multi-valued function of the areas of minimum films is represented as a surface referred to as a "dovetail".*

This surface and the corresponding singularity are well known in the modern theory of singularities. It is also known that the "dovetail" can be represented as a surface in a three-dimensional space of polynomials of the form $x^4 + ax^2 + bx + c$, which consists of points (a, b, c) corresponding to polynomials with multiple roots.

The appearance of this surface (and analogous ones) in the theory of minimum surfaces is a reflection of deep topological properties of minimum surfaces discovered of late. A systematic study of the topology of minimum surfaces (including minimum surfaces of arbitrary dimension) has started rather recently. For the review of the progress in this field see, for example, the books by A.T. Fomenko, [25], [35] and by Dao Chong Thi and A.T. Fomenko [36]. The same books elucidate the role of minimum surfaces in physics, chemistry, biology, animate nature, etc.

REFERENCES

1. Aleksandrov, A.D. *Intrinsic Geometry of Convex Surfaces*. Gostekhizdat: Moscow, Leningrad (1948) (in Russian).
2. Arnol'd, V.I. *Mathematical Methods of Classical Mechanics*. Graduate Texts in Math., Springer Verlag: New York, Heidelberg, Berlin (1978).
3. Bishop, R.L. and Crittenden, R.J. *Geometry of Manifolds*. Pure and Applied Math., Vol. **25**. Academic Press: New York, London (1964).
4. Bogoljubov, N.N. and Shirokov, D.B. *Introduction to the Theory of Quantum Fields*. Interscience: New York (1959).
5. Dao Chong Thi and Fomenko, A.T. *Minimum Surfaces and the Plateau Problem*. American Mathem. Society (in press).
6. Delone, B.N., Aleksandrov, A.D. and Padurov, N.N. *Mathematical Foundations of the Lattice Analysis of Crystals*. ONTI: Leningrad, Moscow (1934) (in Russian).
7. Dubrovin, B.A., Fomenko, A.T. and Novikov, S.P. *Modern Geometry – Methods and Applications*. Springer Verlag: New York, Berlin, Heidelberg, Tokyo (1984).
8. Efimov, N.V. *Higher Geometry*. Nauka: Moscow (1971) (in Russian).
9. Finikov, S.P. *A Course in Differential Geometry*. Gostekhizdat: Moscow (1952) (in Russian).
10. Fomenko, A.T. *Differential Geometry and Topology*. Plenum Publishing Corporation: New York (1987).
11. Fomenko, A.T. *Variational Problems in Topology*. Gordon & Breach (in press).
12. Fomenko, A.T. *Variational Methods in Topology*. Kluwer Acad. Publ. (in press).
13. Gromoll, D., Klingenburg, W. and Meyer, W. *Riemannische Geometrie im Grossen*. Lecture Notes in Mathematics, No. 55. Springer Verlag: Berlin, New York (1968).
14. Helgason, S. *Differential Geometry and Symmetric Spaces*. Academic Press: New York (1962).
15. Hilbert, D. and Cohn-Vossen, S. *Geometry and the Imagination*. Chelsea Publishing Co.: New York (1952).
16. Landau, L.D. and Lifschitz, E.M. *Mechanics: Course of Theoretical Physics*. Pergamon Press: Oxford, London, New York, Paris, and Addison-Wesley, Reading, Mass. (1960).
17. Landau, L.D. and Lifschitz, E.M. *The Classical Theory of Fields*. Pergamon Press, London (1971).

18. Landau, L.D. and Lifschitz, E.M. *Electrodynamics of Continuous Media.* Pergamon Press: Oxford, London, New York, Paris (1960).
19. Milnor, J.W. *Morse Theory.* Ann. of Math. Studies, No. 51. Princeton Univ. Press: Princeton, N.J. (1963).
20. Nomidzu, K. *Lie Groups and Differential Geometry.* Math. Soc. Japan (1956).
21. Norden, A.P. *The Theory of Surfaces.* Gostekhizdat: Moscow (1956) (in Russian).
22. Novikov, S.P., Mishchenko, A.S., Solov'ev, Yu. P. and Fomenko, A.T. *Problems in Geometry.* Moscow State Univ. Press: Moscow (1978) (in Russian).
23. Pogorelov, A.V. *Differential Geometry.* P. Noordhoff: Groningen (1967).
24. Pogorelov, A.V. *Extrinsic Geometry of Convex Surfaces.* Translations of Math. Monographs, vol. **35**. A.M.S.: Providence, R.I. (1973).
25. Pontryagin, L.S. *Smooth Manifolds and Their Application to Homotopy Theory.* Am. Math. Soc. Translations, Series 2, Vol. **11**, pp. 1-114. A.M.S.: providence, R.I. (1959).
26. Pontryagin, L.S. *Topological Groups.* Gordon & Breach: New York, London, Paris (1966).
27. Rashevskii, P.K. *Riemannian Geometry and Tensor Analysis.* Nauka: Moscow (1967) (in Russian).
28. Rashevskii, P.K. *A Course in Differential Geometry.* Nauka: Moscow (1956) (in Russian).
29. Rohlin, V.A. and Fuks, D.B. *A Beginning Course in Topology.* Chapters in Geometry. Nauka: Moscow (1977) (in Russian).
30. Rozendorn, E.R. *Problems in Differential Geometry.* Nauka: Moscow (1971) (in Russian).
31. Sedov, L.I. *Mechanics of a Continuous Medium.* Nauka: Moscow (1976) (in Russian).
32. Seifert, H. and Threlfall, W. *A Textbook of Topology.* Academic Press: New York (1980).
33. Seifert, H. and Threlfall, W. *Variationsrechnung im Grossen.* Hamburger Math. Einzelschr., No. 24. Teubner: Leipzig (1932).
34. Slavov, A.A. and Faddeev, L.D. *Introduction to the Quantum Theory of Gauge Fields.* Nauka: Moscow (1978) (in Russian).
35. Springer, G. *Introduction to Riemann Surfaces.* Addison-Wesley: Reading, Mass. (1957)
36. Steenrod, N.E. *The Topology of Fibre Bundles.* Princeton Math. Series, Vol. **14**, Princeton Univ. Press: Princeton, N.J. (1951).

INDEX